MODERN PLASMA PHYSICS
VOLUME 1: PHYSICAL KINETICS OF TURBULENT PLASMAS

This three-volume series presents the ideas, models and approaches essential to understanding plasma dynamics and self-organization for researchers and graduate students in plasma physics, controlled fusion and related fields such as plasma astrophysics.

Volume 1 develops the physical kinetics of plasma turbulence through a focus on quasi-particle models and dynamics. It discusses the essential physics concepts and theoretical methods for describing weak and strong fluid and phase space turbulence in plasma systems far from equilibrium. The book connects the traditionally "plasma" topic of weak or wave turbulence theory to more familiar fluid turbulence theory, and extends both to the realm of collisionless phase space turbulence. This gives readers a deeper understanding of these related fields, and builds a foundation for future applications to multi-scale processes of self-organization in tokamaks and other confined plasmas. This book emphasizes the conceptual foundations and physical intuition underpinnings of plasma turbulence theory.

PATRICK H. DIAMOND is a Professor of Physics and Distinguished Professor at the Center for Astrophysics and Space Sciences and the Department of Physics at the University of California at San Diego, USA.

SANAE-I. ITOH is a Distinguished Professor at the Research Institute for Applied Mechanics at Kyushu University, Japan.

KIMITAKA ITOH is a Fellow and Professor at the National Institute for Fusion Science, Japan.

All three authors have extensive experience in turbulence theory and plasma physics.

MODERN PLASMA PHYSICS

Volume 1: Physical Kinetics of Turbulent Plasmas

PATRICK H. DIAMOND
University of California at San Diego, USA

SANAE-I. ITOH
Kyushu University, Japan

KIMITAKA ITOH
National Institute for Fusion Science, Japan

CAMBRIDGE
UNIVERSITY PRESS

CAMBRIDGE
UNIVERSITY PRESS

University Printing House, Cambridge CB2 8BS, United Kingdom

Published in the United States of America by Cambridge University Press, New York

Cambridge University Press is part of the University of Cambridge.

It furthers the University's mission by disseminating knowledge in the pursuit of education, learning and research at the highest international levels of excellence.

www.cambridge.org
Information on this title: www.cambridge.org/9781107424562

First published 2010
First paperback edition 2014

A catalogue record for this publication is available from the British Library

Library of Congress Cataloguing in Publication data

Diamond, Patrick H.
Modern plasma physics / Patrick H. Diamond, Sanae-I. Itoh, Kimitaka Itoh.
p. cm.
ISBN 978-0-521-86920-1 (Hardback)
1. Plasma turbulence. 2. Kinetic theory of matter. I. Itoh, S. I. (Sanae I), 1952– II. Itoh, K. (Kimitaka) III. Title.
QC718.5.T8D53 2010
530.4′4–dc22

2009044418

ISBN 978-0-521-86920-1 Hardback
ISBN 978-1-107-42456-2 Paperback

Contents

Preface

The universe abounds with plasma turbulence. Most of the matter that we can observe directly is in the plasma state. Research on plasmas is an active scientific area, motivated by energy research, astrophysics and technology. In nuclear fusion research, studies of confinement of turbulent plasmas have lead to a new era, namely that of the international thermonuclear (fusion) experimental reactor, ITER. In space physics and in astrophysics, numerous data from measurements have been heavily analyzed. In addition, plasmas play important roles in the development of new materials with special industrial applications.

The plasmas that we encounter in research are often far from thermodynamic equilibrium: hence various dynamical behaviours and structures are generated because of that deviation. The deviation is often sufficient for observable mesoscale structures to be generated. Turbulence plays a key role in producing and defining observable structures. An important area of modern science has been recognized in this research field, namely, research on structure formation in turbulent plasmas associated with electromagnetic field evolution and its associated selection rules. Surrounded by increasing and detailed information on plasmas, some unified and distilled understanding of plasma dynamics is indeed necessary – "*Knowledge must be developed into understanding*". The understanding of turbulent plasma is a goal for scientific research in plasma physics in the twenty-first century.

The objective of this series on modern plasma physics is to provide the viewpoint and methods which are essential to understanding the phenomena that researchers on plasmas have encountered (and may encounter), i.e., the mutually regulating interaction of strong turbulence and structure formation mechanisms in various strongly non-equilibrium circumstances. Recent explosive growth in the knowledge of plasmas (in nature as well as in the laboratory) requires a systematic explanation of the methods for studying turbulence and structure formation.

The rapid growth of experimental and simulation data has far exceeded the evolution of published monographs and textbooks. In this series of books, we aim to provide systematic descriptions (1) for the theoretical methods for describing turbulence and turbulent structure formation, (2) for the construction of useful physics models of far-from-equilibrium plasmas and (3) for the experimental methods with which to study turbulence and structure formation in plasmas. This series will fulfil needs that are widely recognized and stimulated by discoveries of new astrophysical plasmas and through advancement of laboratory plasma experiments related to fusion research. For this purpose, the series constitutes three volumes: Volume 1: Physical kinetics of turbulent plasmas, Volume 2: Turbulence theory for structure formation in plasmas and Volume 3: Experimental methods for the study of turbulent plasmas. This series is designed as follows.

Volume 1: *Physical Kinetics of Turbulent Plasmas* The objective of this volume is to provide a systematic presentation of the theoretical methods for describing turbulence and turbulent transport in strongly non-equilibrium plasmas. We emphasize the explanation of the progress of theory for strong turbulence. A viewpoint, i.e., that of the "quasi-particle plasma" is chosen for this book. Thus we describe 'plasmas of excitons, dressed by collective interaction', which enable us to understand the evolution and balance of plasma turbulence.

We stress (a) test field response (particles and waves, respectively), taking into account screening and dressing, as well as noise, (b) disparate scale interaction and (c) mean field evolution of the screened element gas. These three are essential building blocks with which to construct a physics picture of plasma turbulence in a strongly non-equilibrium state. In the past several decades, distinct progress has been made in this field, and verification and validation of nonlinear simulations are becoming more important and more intensively pursued. This is a good time to set forth a systematic explanation of the progress in methodology.

Volume 2: *Turbulence Theory for Structure Formation in Plasmas* This volume presents the description of the physics pictures and methods to understand the formation of structures in plasmas. The main theme has two aspects. The first is to present ways of viewing the system of turbulent plasmas (such as toroidal laboratory plasmas, etc.), in which the dynamics for both self-sustaining structure and turbulence coexist. The other is to illustrate key organizing principles and to explain appropriate methods for their utilization. The competition (e.g., global inhomogeneity, turbulent transport, quenching of turbulence, etc.) and self-sustaining mechanisms are described.

One particular emphasis is on a self-consistent description of the mechanisms of structure formation. The historical recognition of the proverb *"All things flow"* means that structures, which disappear within finite lifetimes, can also be, and are usually, continuously generated. Through the systematic description of plasma turbulence and structure formation mechanisms, this book illuminates principles that govern evolution of laboratory and astrophysical plasmas.

Volume 3: *Experimental Methods for the Study of Turbulent Plasmas* The main objective is this volume is to explain methods for the experimental study of turbulent plasmas. Basic methods to identify elementary processes in turbulent plasmas are explained. In addition, the design of experiments for the investigation of plasma turbulence is also discussed with the aim of future extension of experimental studies. This volume has a special feature. While many books and reviews have been published on plasma diagnostics, i.e., how to obtain experimental signals in high temperature plasmas, little has been published on how one analyzes the data in order to identify and extract the physics of nonlinear processes and nonlinear mechanisms. In addition, the experimental study of nonlinear phenomena requires a large amount of data processing. This volume explains the methods for performing quantitative studies of experiments on plasma turbulence.

Structure formation in turbulent media has been studied for a long time, and the proper methodology to model (and to formulate) has been elusive. This series of books will offer a perspective on how to understand plasma turbulence and structure formation processes, using advanced methods.

Regarding readership, this book series is aimed at the more advanced graduate student in plasma physics, fluid dynamics, astrophysics and astrophysical fluids, nonlinear dynamics, applied mathematics and statistical mechanics. Only minimal familiarity with elementary plasma physics at the level of a standard introductory text is presumed. Indeed, a significant part of this book is an outgrowth of advanced lectures given by the authors at the University of California, San Diego, at Kyushu University and at other institutions. We hope the book may be of interest and accessible to postdoctoral researchers, to experimentalists and to scientists in related fields who wish to learn more about this fascinating subject of plasma turbulence.

In preparing this manuscript, we owe much to our colleagues for our scientific understanding. For this, we express our sincere gratitude in the Acknowledgements. There, we also acknowledge the funding agencies that have supported our research. We wish to show our thanks to young researchers and students who have helped in preparing this book, by typing and formatting the manuscript while providing invaluable feedback: in particular, Dr. N. Kasuya of NIFS and Mr. S. Sugita of Kyushu University for their devotion, Dr. F. Otsuka, Dr. S. Nishimura, Mr. A. Froese, Dr. K. Kamataki and Mr. S. Tokunaga of Kyushu University also deserve mention. A significant part of the material for this book was developed in the Nonlinear Plasma Theory (Physics 235) course at UCSD in 2005. We thank the students in this class, O. Gurcan, S. Keating, C. McDevitt, H. Xu and A. Walczak for their penetrating questions and insights. We would like to express our gratitude to all of these young scientists for their help and stimulating interactions during the preparation of this book. It is our great pleasure to thank Kyushu University, the University of California, San Diego, and National Institute for Fusion Science for their hospitality while the manuscript of the book was prepared. Last but not least, we thank Dr. S. Capelin and his staff for their patience during the process of writing this book.

Acknowledgements

The authors acknowledge their mentors, for guiding their evolution as plasma physicists: Thomas H. Dupree, Marshall N. Rosenbluth, Tihiro Ohkawa, Fritz Wagner and Akira Yoshizawa: the training and challenges they gave us form the basis of this volume.

The authors are also grateful to their teachers and colleagues (in alphabetical order), R. Balescu, K. H. Burrell, B. A. Carreras, B. Coppi, R. Dashen, A. Fujisawa, A. Fukuyama, X. Garbet, T. S. Hahm, A. Hasegawa, D. W. Hughes, K. Ida, B. B. Kadomtsev, H. Mori, K. Nishikawa, S. Tobias, G. R. Tynan, M. Yagi, M. Wakatani and S. Yoshikawa. Their instruction, collaboration and many discussions have been essential and highly beneficial to the authors.

We also wish to express our sincere gratitude to those who have given us material for the preparation of the book. In alphabetical order, J. Candy, Y. Gotoh, O. Gurcan, K. Hallatschek, F. L. Hinton, C. W. Horton, S. Inagaki, F. Jenko, N. Kasuya, S. Keating, Z. Lin, C. McDevitt, Y. Nagashima, H. Sugama, P. W. Terry, S. Toda, A. Walczak, R. Waltz, T.-H. Watanabe, H. Xu, T. Yamada and N. Yokoi.

We wish to thank funding agencies that have given us support during the course of writing this book. We were partially supported by Grant-in-Aid for Specially-Promoted Research (16002005) of MEXT, Japan [Itoh project], by Department of Energy Grant Nos. DE-FG02-04ER54738, DEFC02-08ER54959 and DE-FC02-08ER54983, by Grant-in-Aid for Scientific Research (19360418, 21224014) of the Japan Society for the Promotion of Science, by the Asada Eiichi Research Foundation and by the collaboration programmes of the Research Institute for Applied Mechanics of Kyushu University, and of the National Institute for Fusion Science.

1
Introduction

The beginning is the most important part of the work.

(Plato)

In this introduction, we set out directly to answer the many questions the reader may have in mind about this book, such as:

(1) *Why* is this book being written? Why study theory in the age of high performance computing and experimental observations in unparalleled detail? In what way does it usefully augment the existing literature? *Who* is the target readership?
(2) *What* does it cover? What is the logic behind our particular choice of topics? *Where* will a reader stand and *how* will he or she benefit after completing this book?
(3) *What* is *not* included and *why* was it omitted? What alternative sources are recommended to the reader?

We now proceed to answer these questions.

1.1 Why?

Surely the need for study of plasma turbulence requires no explanation.

Turbulence pervades the dynamics of both laboratory and astrophysical plasmas. Turbulent transport and its associated confinement degradation are *the* main obstacles to achieving ignition in magnetically confined plasma (i.e., for magnetic confinement fusion (MCF) research), while transport bifurcations and self-generated shear flows are the principal means for controlling such drift wave turbulence. Indeed, predictions of degradation of confinement by turbulence have been used to (unjustifiably) challenge plans for ITER (International Thermonuclear Experimental Reactor). In the case of inertial confinement fusion (ICF) research, turbulent mixing driven by Rayleigh–Taylor growth processes limit implosion performance for indirect drive systems, while the nonlinear evolution of laser–plasma instabilities (such as filamentation – note, these are examples of turbulence in disparate

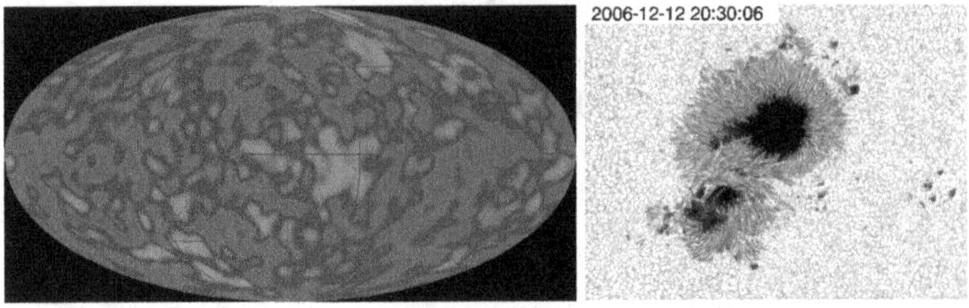

Fig. 1.1. The cosmic microwave background fluctuations (left). Fluctuations pervade the universe. The cosmic microwave background radiation is a remnant of the Big Bang and the fluctuations are the imprint of the density contrast in the early universe. [http://aether.lbl.gov/www/projects/cobe/COBE_Home/DMR_Images.html] Turbulent dynamics are observed in solar plasmas near the sunspot (right). [Observation by Hinode, courtesy NAOJ/JAXA.]

scale interaction) must be controlled in order to achieve fast ignition. In space and astrophysical plasma dynamics, turbulence is everywhere, i.e. it drives inter-stellar medium (ISM) scintillations, stirs the galactic and stellar dynamos, scatters particles to facilitate shock acceleration of cosmic rays, appears in strongly driven 3D magnetic reconnection, drives angular momentum transport to allow accretion in disks around protostars and active galactic nuclei (AGNs), helps form the solar tachocline, etc. – the list is indeed endless. (Some examples are illustrated in Figure 1.1.) Moreover, this large menu of MCF, ICF and astrophysical applications offers an immensely diverse assortment of turbulence from which to choose, i.e. strong turbulence, wave turbulence, collisional and very collisionless turbulence, strongly magnetized systems, weakly magnetized systems, multi-component systems with energetic particles, systems with sheared flow, etc., all are offered. Indeed, virtually *any possible* type of plasma turbulence finds some practical application in the realm of plasma physics.

Thus, while even the most hardened sceptic must surely grant the merits of plasma turbulence and its study, one might more plausibly ask, "Why study plasma turbulence *theory*, in the age of computation and detailed experimental observations? Can't we learn all we need from direct numerical simulation?" This question is best dealt with by considering the insights in the following set of quotations from notable individuals. Their collective wisdom speaks for itself.

"Theory gives meaning to our understanding of the empirical facts."

(John Lumley)

"Without simple models, you can't get anything out of numerical simulation."

(Mitchell J. Feigenbaum)

"When still photography was invented, it soon became so popular that it was expected to mark the end of drawing and painting. Instead, photography made artists honest, requiring more of them than mere representation."

(Peter B. Rhines)

In short, theory provides a necessary intellectual framework – a structure and a system from within which to derive meaning and/or a message from experiment, be it physical or digital. Theory defines the simple models used to *understand* simulations and experiments, and to extract more general lessons from them.

This process of extraction and distillation is a prerequisite for development of predictive capacity. Theory also forms the basis for both *verification* and *validation* of simulation codes. It defines exactly solvable mathematical models needed for verification and also provides the intellectual framework for a programme of validation. After all, any meaningful comparison of simulation and experiment requires specification of physically relevant questions or comparisons which must be addressed. It is unlikely this can be achieved in the absence of guidance from theory. It is surely the case that the rise of the computer has indeed made the task of the theorists more of a challenge. As suggested by Rhines, the advent of large-scale computation has forced theory to define ideas or to teach a conceptual lesson, rather than merely to crunch out numbers. Theory must constitute the knowledge necessary to make use of the raw information obtained from simulation and experiment. *Theory must then lead the scientist from knowledge to understanding*. It must identify, define and teach us a simple, compact lesson. As Rhines states, it must do more than merely represent. Indeed, the danger here is that in this data-rich age, without distillation of a message, a simulation or representation or experimental data-acquisition will grow as large and complex as the object being represented, as imagined in the following short fiction by the incomparable Jorge Luis Borges.

. . . In that Empire, the Art of Cartography attained such Perfection that the map of a single Province occupied the entirety of a City, and the map of the Empire, the entirety of a Province. In time, those Unconscionable Maps no longer satisfied, and the Cartographers Guilds struck a Map of the Empire whose size was that of the Empire, and which coincided point for point with it. The following Generations, who were not so fond of the Study of Cartography as their Forebears had been, saw that that vast Map was Useless, and not without some Pitilessness was it, that they delivered it up to the Inclemencies of Sun and Winters. In the Deserts of the West, still today, there are Tattered Ruins of that Map, inhabited by Animals and Beggars; in all the Land there is no other Relic of the Disciplines of Geography.

(Suarez Miranda, Viajes de varones pudentes,
Libro IV, Cap. XLV, Lerida, 1658.
Jorge Luis Borges)

Without theory, we are indeed doomed to a life amidst a useless pile of data and information.

1.2 The purpose of this book

With generalities now behind us, we proceed to state that this book has two principal motivations, which are:

(1) to serve as an up-to-date and advanced, yet accessible, monograph on the basic physics of plasma turbulence, from the perspective of the physical kinetics of quasi-particles,
(2) to stand as the first book in a three-volume series on the emerging science of structure formation and self-organization in turbulent plasma.

Our ultimate aim is not only to present developments in the *theory* but also to describe how these elements are applied to the understanding of structure formation phenomena, in real plasma, such as tokamaks, other confinement devices and in the universe, Thus, this series forces theory to confront reality! These dual motivations are best served by an approach in the spirit of Lifshitz and Pitaevski's *Physical Kinetics*, namely with an emphasis on quasi-particle descriptions and their associated kinetics. We feel this is the optimal philosophy within which to organize the concepts and theoretical methods needed for understanding *ongoing* research in structure formation in plasma, since it naturally unites resonant and non-resonant particle dynamics.

This long-term goal motivates much of the choice of topical content of the book, in particular:

(1) the discussion of dynamics in both real space and wave-number space; i.e. explanation of Prandtl's theory of turbulent boundary layers in parallel with Kolmogorov's cascade theory (K41 theory), in Chapter 2, where the basic notions of turbulence are surveyed. Prandtl mixing length theory is an important paradigm for profile stiffness, etc. and other commonplace ideas in MFE (magnetic fusion energy) research;
(2) the contrast between the zero spectral flux in "near equilibrium" theory (i.e. the dressed test particle model) and the large spectral flux inertial range theory (as by Kolmogorov), discussed in Chapter 2. These two cases bound the dynamically relevant limit of weak or moderate turbulence, which we usually encounter in the real world of confined plasmas;
(3) the treatment of quasi-linear theory in Chapter 3, which focuses on the energetics of the interaction of resonant particles with quasi-particles. This is, without a doubt, the most useful approach to mean field theory for collisionless relaxation;
(4) the renormalized or dressed resonant particles response, discussed at length in Chapter 4. In plasma, both the particle and collective responses are nonlinear and require detailed, individual treatment. The renormalized particle propagator defines a key, novel time-scale;

(5) the extensive discussion of disparate scale interaction, in Chapters 5–7:
   (a) from the viewpoint of nonlocal wave–wave interactions, such as induced diffusion, in Chapter 5,
   (b) from the perspective of Mori–Zwanzig theory in Chapter 6,
   (c) in the context of adiabatic theory for Langmuir turbulence (both mean field theory for random phase wave kinetics *and* the coherent Zakharov equations) in Chapter 7.

   We remark here that disparate scale interaction is fundamental to the dynamics of "negative viscosity phenomena" and so is extremely important to structure formation. Thus, it merits the very detailed description accorded to it here;

(6) the detailed and extensive discussion of phase space density granulation and its role in the description of mean field relaxation, which we present in Chapter 8. In this chapter, the notion of the "quasi-particle in turbulence" is expanded to encompass the screened "clump" or phase space vortex. An important consequence of this conceptual extension is the manifestation of *dynamical friction* in the mean field theory for the Vlasov plasma. Note that dynamical friction is *not* accounted for in standard quasi-linear theory, which is the traditional backbone of mean field methodology for plasma turbulence;

(7) the discussion of quenching of diffusion in 2D MHD (magnetohydrodynamics), presented in Chapter 9. Here, we encounter the principle of a quasi-particle with a dressing that confers a *memory* to the dynamics. This memory which follows from the familiar MHD freezing-in law, quenches the diffusion of fluid relative to the fluid, and so severely constrains relaxation.

All of (1)–(7) represent new approaches not discussed in existing texts on plasma turbulence.

Throughout this book, we have placed special emphasis on identifying and explaining the physics of key time and space scales. These are usually summarized in an offset table which is an essential and prominent part of the chapter in which they are developed and defined. Essential time-scale orderings are also clarified and tabulated. We also construct several tables which compare and contrast the contents of different problems. We deem these useful in demonstrating the relevance of lessons learned from simple problems to more complicated applications. More generally, understanding of the various nonlinear time scales and their interplay is essential to the process of construction of tractable simple models, such as spectral tranfer scalings, from more complicated frameworks such as wave kinetic formalisms. Thus, we place great emphasis on the physics of basic time scales.

The poet T. S. Eliot once wrote, "Dante and Shakespeare divide the world between them. There is no third." So it is with introductory books on plasma turbulence theory – the two classics of the late 1960s, namely R. Z. Sagdeev and A. A. Galeev's *Nonlinear Plasma Theory*[1] and B. B. Kadomtsev's *Plasma Turbulence* are the twin giants of this field and are still quite viable guides to the subject. Any new monograph must meet their standards. This is a challenge for all monographs prepared on the subject of plasma turbulence.

Nevertheless, we presumptuously argue that now there *is* indeed room for a 'third'. In particular (the following list does not intend to enumerate the shortcomings of these two classics but rather to observe the significant advancement of plasma physics in the past two decades) we aim to elucidate the following important issues:

(1) the smooth passage from one limit (weak turbulence theory – Sagdeev and Galeev) to the other (strong turbulence based on ideas from hydrodynamics – Kadomtsev), and the duality of these two approaches in collisionless regimes;
(2) the important relation between self-similarity in *space* (i.e. turbulent mixing), and self-similarity in *scale* (turbulent cascade), including the inverse cascades or MHD turbulence dynamics;
(3) the resonance broadening theory or Vlasov response renormalization, both of which extend the concept of eddy viscosity into phase space in an important way;
(4) the important subject of the theory of disparate scale interaction or "negative viscosity phenomena", which is crucial for describing self-organization and structure formation in turbulent plasma. This class of phenomena is *the* central focus of our series;
(5) the theory of phase space density granulation. This important topic is required for understanding and describing stationary phase space turbulence with resonant heating, etc., where dynamical friction is a *must*;
(6) the important problem of the structure of two-point correlation in turbulent plasma;
(7) applications to MHD turbulence or transport, or to the quasi-geostrophic/drift wave turbulence problem.

Thus, even before we come to more advanced subjects such as zonal flow formation, phase-space density holes, solitons and collisionless shocks and transport barriers and bifurcations – all subjects for out next volume – it seems clear that a fresh look at the basics of plasma turbulence is indeed warranted. This book is our attempt to realize this vision.

1.3 Readership and background literature

We have consciously written this book so as to be accessible to more advanced graduate students in plasma physics, fluid dynamics, astrophysics and astrophysical fluids, nonlinear dynamics, applied mathematics and statistical mechanics. Only minimal familiarity with elementary plasma physics – at the level of a

[1] The longer version published in *Reviews of Plasma Physics*, Vol. VII is more complete and in many ways superior to the short monograph.

standard introductory text such as Kulsrud (2005), Sturrock (1994) or Miyamoto (1976) – is presumed.

This series of volumes is designed to provide a focused explanation of the physics of plasma turbulence. Introductions to many elementary processes in plasmas, such as the dynamics of particle motion, varieties of linear plasma eigenmodes, instabilities, the MHD dynamics of confined plasmas and the systems for plasma confinement, etc. are in the literature and are already widely available to readers. For instance, the basic properties of plasmas are explained in Krall and Trivelpiece (1973), Ichimaru (1973), Miyamoto (1976) and Goldston and Rutherford (1995); waves are thoroughly explained in Stix (1992); MHD equations, equilibrium and stability are explained in Freidberg (1989) and Hazeltine and Meiss (1992); an introduction to tokamaks is given in White (1989), Kadomtsev (1992), Wesson (1997) and Miyamoto (2007); drift wave instabilities are reviewed in Mikailovski (1992), Horton (1999) and Weiland (2000); issues in astrophysical plasmas are discussed in Sturrock (1994), Tajima and Shibata (2002) and Kulsrud (2005) and subjects of chaos are explained in Lichtenberg and Lieberman (1983) and Ott (1993). Further explanation is available by reference to the literature. The reader may also find it helpful to refer to books on plasma turbulence which precede this volume, e.g. Kadomtsev (1965), Galeev and Sagdeev (1965), Sagdeev and Galeev (1969), R. C. Davidson (1972), Itoh *et al.* (1999), Moiseev *et al.* (2000), Yoshizawa *et al.* (2003), Elskens and Escande (2003) and Balescu (2005). Advanced material on neutral fluid turbulence dynamics, which is related to the contents of this volume, can be found in Lighthill (1978), McComb (1990), Frisch (1995), Moiseev *et al.* (2000), Pope (2000), Yoshizawa *et al.* (2003) and P. A. Davidson (2004).

1.4 Contents and structure of this book

Having completed our discussion of motivation, we now turn to presenting the actual contents of this book.

Chapter 2 deals with foundations. Since most realizations of plasma turbulence are limits of "weak turbulence" or intermediate regime cases where the mode self-correlation time τ_c is longer than (weak) or comparable to (intermediate) the mode frequency $\omega(\tau_c\omega > 1)$, we address foundations by discussing the opposite extremes of:

(1) states of *zero* spectral flux – i.e. fluctuations at equilibrium, as described by the test particle model. In this limit, linear emission and absorption balance locally, at each **k**, to define the thermal equilibrium fluctuation spectrum. Moreover, the theory of dressed test particle dynamics is a simple, instructive example of the impact of collective screening effects on fluctuations. The related Lenard–Balescu theory, which we also discuss, defines *the* prototypical formal structure for a mean field theory of

transport and relaxation. These basic paradigms are fundamental to the subsequent discussions of quasi-linear theory in Chapter 3, nonlinear wave–particle interaction in Chapter 4, and the theory of phase space density granulations in Chapter 8;

(2) states dominated by a large spectral flux, where nonlinear transfer exceeds all other elements of the dynamics. Such states correspond to turbulent cascades, in which nonlinear interactions couple sources and sinks at very different scales by a sequence of local transfer events. Indeed, the classic Kolmogorov cascade is defined in the limit where the dissipation rate ϵ is the *sole* relevant rate in the inertial range. Since confined plasmas are usually strongly magnetized, so that the parallel degrees of freedom are severely constrained, we discuss both the 3D forward cascade and the 2D inverse cascade in equal depth. We also discuss pertinent related topics such as Richardson's calculation of two-particle dispersion.

Taken together, (1) and (2) in a sense "bound" most plasma turbulence applications of practical relevance. However, given our motivations rooted in magnetic confinement fusion physics, we also devote substantial attention to *spatial transport* as well as spectral transfer. To this end, then, the Introduction also presents the Prandtl theory of pipe flow profiles in space on an equal footing with the Kolmogorov spectral cascade in scale. Indeed the Prandtl boundary layer theory is the prototype of familiar MFE concepts such as profile "stiffness", mixing length concepts, and dimensionless similarity. It is the natural example of self-similarity in space, with which to complement self-similarity in scale. For these and other reasons, it merits inclusion in the lead-in chapter on fundamentals, and is summarized in Table 2.4, at the conclusion of this chapter.

Chapter 3 presents quasi-linear theory, which is *the* practical, workhorse tool for mean field calculations of relaxation and transport for plasma turbulence. Despite quasi-linear theory's celebrated status and the fact that it appears in nearly every basic textbook on plasma physics, we were at a loss to find a satisfactory treatment of its foundations, and, in particular, one which does justice to their depth and subtlety. Quasi-linear theory is simple but *not* trivial. Thus, we have sought to rectify this situation in Chapter 3. In particular, we have devoted considerable effort to:

(1) a basic discussion of the origin of irreversibility – which underpins the coarse-graining intrinsic to quasi-linear theory – in particle stochasticity due to phase space island overlap;

(2) a careful introductory presentation of the many time scales in play in quasi-linear theory, and the orderings they must satisfy. The identification and ordering of pertinent time scales is one of the themes of this book. Special attention is devoted to the distinction between the wave–particle correlation time and the spectral auto-correlation time. This distinction is especially important for the case of the quasi-linear theory of 3D drift wave turbulence, which we discuss in detail. We also "locate" quasi-linear theory in the realm of possible Kubo number orderings;

(3) presentation of the multiple forms of conservation laws (i.e. resonant particles versus waves or particles versus fields) in quasi-linear theory, along with their physical meaning. These form the foundation for subsequent quasi-particle formulations of transport, stresses, etc. The concept of the plasma as coupled populations of resonant particles and quasi-particles (waves) is one of the most intriguing features of quasi-linear theory;
(4) an introduction to up-gradient transport (i.e. the idea of a thermodynamic inward flux or "pinch"), as it appears in the quasi-linear theory of transport. As part of this discussion, we address the entropy production constraint on the magnitude of up-gradient fluxes;
(5) an introduction to nonlinear Landau damping as 'higher-order quasi-linear theory', in which $\langle f \rangle$ relaxes via beat–wave resonances.

The aim of Chapter 3 is to give the reader a working introduction to mean field methods in plasma turbulence. The methodology of quasi-linear theory, developed in this chapter, is used throughout the rest of this book, especially in Chapter 7, 8 and 9. Specific applications of quasi-linear theory to advanced problems in tokamak confinement are deferred to Volume 2.

Chapter 4 continues the thematic exploration of resonant particle dynamics by an introduction to *nonlinear* wave–particle interaction. Here we focus on selected topics, which are:

(1) resonance broadening theory, i.e. how finite fluctuation levels broaden the wave–particle resonance and define a nonlinear decorrelation time for the response δf. The characteristic scale of the broadened resonance width is identified and discussed. We present applications to 1D Vlasov dynamics, drift wave turbulence in a sheared magnetic field, and enhanced decorrelation of fluid elements in a sheared flow;
(2) perturbative or iterative renormalization of the 1D Vlasov response function. Together with (1), this discussion presents propagator renormalization or – in the language of field theory – "mass renormalization" in the context of Vlasov plasma dynamics. We discuss the role of background distribution counter-terms (absent in resonance broadening theory) and the physical significance of the non-Markovian character of the renormalization. The aim here is to connect the more intuitive approach of resonance broadening theory to the more formal and systematic approach of perturbative renormalization.
(3) The application of renormalization of the drift wave problem, at the level of drift kinetics. The analysis here aims to illustrate the role of energy conservation in constraining the structure of the renormalized response. This instructive example illustrates the hazards of naive application of resonance-broadening theory.

Further study of nonlinear wave–particle interaction is deferred until Chapter 8.

Chapter 5 introduces the important topic of nonlinear wave–wave interaction. Both the integrable dynamics of coherent interaction in discrete mode triads, as well as the stochastic, random phase interactions as occur for a broad spectrum of dispersive waves are discussed. This chapter is fundamental to all that follow. Specific attention is devoted to:

(1) the coherent, resonant interaction of three drift waves. Due to the dual constraints of conservation of energy and enstrophy (mean squared vorticity), this problem is demonstrated to be isomorphic to that of the motion of the free asymmetric top, and so can be integrated by the Poinsot construction. We also show that a variant of the Poinsot construction can be used to describe the coherent coupled motion of three modes which conserve energy and obey the Manley–Rowe relations. Characteristic time scales for parametric interaction are identified;
(2) the derivation of the random-phase spectral evolution equation (i.e. the *wave kinetic equation*) which is presented in detail. The stochastic nature of the wave population evolution is identified and traced to overlap of triad resonances. We explain the modification of the characteristic energy transfer time scales by stochastic scattering;
(3) basic concepts of wave cascades. Here, we discuss the cascade of energy in gravity wave interaction. In Chapter 9, we discuss the related application of the Alfvén wave cascade. The goal here is to demonstrate how a tractable scaling argument is constructed using the structure of the wave kinetic equation;
(4) non-local (in **k**) wave coupling processes. Given our over-arching interest in the dynamics of structure formation, we naturally place a great deal of emphasis on non-local interactions in **k**, especially the direct interactions of small scales with large, since these drive stresses, transport etc. (which are quadratic in fluctuation amplitude), which directly impact macro-structure. Indeed, significant parts of Chapters 5 and 6, and *all* of Chapter 7 deal with non-local, disparate scale interaction. In this chapter, we identify three types of non-local (in **k**) interaction processes (induced diffusion, parametric subharmonic interaction and elastic scattering), which arise naturally in wave interaction theory. Of these, induced diffusion is especially important and is discussed at some length.

Chapter 6 presents renormalized turbulence closure theory for wave–wave interactions. Key concepts such as the nonlinear scrambling or self-coherence time, the interplay of nonlinear noise emission with nonlinear damping and the non-Markovian structure of the closure theory are discussed in detail. Non-standard aspects of this chapter include:

(1) a discussion of Kraichnan's random coupling model, which is the paradigm for understanding the essential physics content of the closure models, since it defines a physical realization of the closure theory equations;
(2) the development of the Mori–Zwanzig theory of problem reduction in parallel with the more familiar direct interaction approximation (DIA). The merits of this approach are two-fold. First, the Mori–Zwanzig memory function constitutes a well-defined limit of the DIA response function and so defines a critical benchmark for that closure method. Second the Mori–Zwanzig theory is a rigorous but technically challenging solution to the problem of disparate scale interaction. It goes further than the induced

diffusion model (as discussed in Chapter 5), since it systematically separates the resolved degrees of freedom from the unresolved (on the basis of relaxation time disparity) by projecting the latter into a "noise" field and then grafting the entire problem onto a fluctuation–dissipation theorem structure. By this, the unresolved modes are tacitly assumed to thermalize and so produce a noise bath and a memory decay time. Thus, the Mori–Zwanzig approach is a natural and useful tool for the study of disparate scale interaction;

(3) explicit calculation of both positive and negative turbulent eddy viscosity examples.

Chapter 7 presents the theory of disparate scale interaction, in the context of Langmuir turbulence. Both the coherent, envelope ('Zakharov equation') approach and stochastic, wave kinetic theory are presented. This simple problem is a fundamental paradigm for structure formation by the simplification of local symmetry-breaking perturbations by wave radiation stresses. The mechanism is often referred to as one of 'modulational instability', since in its course, local modulations in the wave population field are amplified and induce structure formation. This chapter complements and extends the more formal procedures and analyses of Chapters 5 and 6. We will later build extensively on this chapter in our discussion of zonal flow generation in Volume 2. Chapter 7 also presents the theory of Langmuir collapse, which predicts the formation in finite time of a density cavity (i.e. 'caviton') singularity from the evolution of modulational instability in 3D. This is surely the simplest and most accessible example of a theory of finite time singularity in a nonlinear continuum system. We remark here that a rigorous answer to the crucial question of finite time singularity in the Navier–Stokes equations in 3D remains elusive.

Chapter 8 sets forth the theory of phase space density granulation. Phase space granulations are eddies or vortices formed in the Vlasov phase space fluid as a result of nonlinear mode–mode coupling. They are distinguished from usual fluid vortices by their incidence in *phase space*, at wave–particle resonance. Thus, granulations formation may be thought of as the turbulent, multi-wave analogue of trapping in a single, large amplitude wave. To this end, it is useful to note that the collisionless Vlasov fluid satisfies a variant of the Kelvin circulation theorem, thus suggesting the notion of a phase space eddy. Granulations impact relaxation and transport by introducing *dynamical friction*, by radiation and Cerenkov emission into damped collective modes. This mechanism, which obviously is analogous to dynamical friction induced by particle discreteness near thermal equilibrium, originates from the finite spatial and velocity scales characteristic of the phase space eddy. As a consequence, novel routes to transport and relaxation open via scattering off localized structures. These are often complementary to more familiar linear instability mechanisms, and so may be thought of as routes to subcritical, nonlinear instability.

To the best of our knowledge, Chapter 8 is the first pedagogical discussion of phase space density granulations available. Some particularly novel aspects of this chapter are:

(1) motivation of the concept of a phase space eddy via a Kelvin's theorem for a Vlasov fluid, and consideration of the effect of collisions on Vlasov turbulence;
(2) the parallel development of phase space granulations and quasi-geostropic eddy dynamics. Both are governed by evolution equations for a locally conserved quantity, which is decomposed into a mean and fluctuating part;
(3) the discussion of nonlinear growth dynamics for a localized structure, both in terms of coherent phase space vortexes and a Lenard–Balescu-like formulation which describes statistical phase space eddies;
(4) the discussion of possible turbulent states predicted by the two-point correlation equation, including nonlinear noise enhanced waves, "clump" instability, etc.

Connections to other sections of the book, especially Chapters 2, 3 and 4, are discussed throughout this chapter.

Chapter 9, the final chapter, deals with MHD turbulence, and is organized into three sections, dealing with:

(1) MHD cascades;
(2) derivative nonlinear Schroedinger equation (DNLS) wave packets for compressible MHD;
(3) turbulent diffusion of magnetic fields in 2D.

The choice of MHD is motivated by its status as a fairly simple, yet relevant model, and one which also forces both author and reader to synthesize various concepts encountered along the way, in this book. The choice of particular topic, (1), (2) or (3), is explained in the course of their description below.

The first part – MHD cascades – deals with extension of the turbulent cascade to MHD, where both eddies and Alfvén waves co-exist as fluctuation constituents. An analogy with counter-propagating wave beat resonance with particles – as in nonlinear Landau damping in Chapter 3 – is used to develop a unified treatment of both the strongly magnetized (i.e. Goldreich–Sridhar) and weakly magnetized (i.e. Kraichnan–Iroshnikov) cascades, which are characteristic of both 2D and 3D MHD turbulence. Here, counter-propagating Alfvén excitations are the "waves", and zero mean frequency eddies are the "particles". Both relevant limits are recovered, depending on the degree of anisotropy.

The second part – DNLS Alfvénic solitons – deals with the complementary limit of uni-directional wave group propagation while admitting weak parallel compressibility. As a result, modulational instability of wave trains becomes possible, resulting in the formation of strong dipolar parallel flows along with the formation of steepened wave packet phase fronts. Thus this mechanism enables coupling to small scale by direct, nonlinear steepening – rather than by sequential eddy mitoses – and so is complementary to the theory of cascades discussed in (1). These two processes are also complementary in that (1) requires bi-directional wave streams while (2) applies to uni-directional streams.

The third part deals with turbulent diffusion in 2D MHD. This topic is of interest because:

(a) it is perhaps the simplest possible illustration of the constraint of the freezing-in law on transport and relaxation;
(b) it also illustrates the impact of a "topological" conservation law – namely that of mean square magnetic potential – on macroscopic transport processes;
(c) it demonstrates the importance of dynamical regulation of the transport cross-phase.

This section forms an important part of the foundation for our discussion of dynamo theory.

Note that in each of Chapters 2–9, we encounter the quasi-particle concept in different forms – from dressed test particle, to eddy, to phase space eddy, to wave packet, to caviton, etc. Indeed, the concept of the quasi-particle runs throughout this entire book. Table 1.1 presents a condensed summary of the different quasi-particle concepts encountered in each chapter. It lists the chapter topic, the relevant quasi-particle concept, and the physics ideas which motivate the theoretical development. Thus, we recommend Table 1.1 to the reader as a concentrated outline of the key contents of this book.

A reader who works through this book, thinks the ideas over, and does some practice calculations, will have a good, basic introduction to fluid and plasma turbulence theory. He or she will be well prepared for further study in drift wave turbulence, tokamak transport theory and secondary structure formation, which together form the nucleus of Volume 2 of this series. The reader will also be prepared for study of advanced topics in dynamo theory, MHD relaxation and phase space structures in space plasmas. Whatever the reader's future direction, we think that this experience with the fundamentals of the subject will continue to be of value.

After this discussion of what we do cover, we should briefly comment on the principle topics we *do not* address in detail in this book.

Even if we focus on only the basic description of the physics of plasma turbulence, the dual constraints of manageable length and the broad coverage required by the nature of this subject necessitate many painful omissions. Alternative sources are noted here to fill in the many gaps we have left. These issues include, but are not limited to: (i) intermittency models and multi-fractal scaling, (ii) details of weak turbulence nonlinear wave–particle interaction, (iii) advanced treatment of the Kolmogorov spectra of wave turbulence, (iv) mathematical theory of nonlinear

Table 1.1. *Summary of the issues explained in Chapters 2–9*

Chapter – Topic	Quasi-particle concept	Physics issue
2 – Foundations	(a) Dressed test particle	(a) Near-equilibrium fluctuations, transport
	(b) Eddy	(b) Turbulence cascade
	(c) Slug or blob	(c) Turbulent mixing
3 – Mean field, quasi-linear theory	Resonant particle and wave/quasi-particle populations	(a) Energy–momentum conservation in quasi-linear theory
		(b) Dynamics as that of interpenetrating fluids
4 – Nonlinear wave–particle interaction	(a) Phase space fluid element, characteristic scales	(a) Scattering and resonance broadening
	(b) Dressed particle propagator	(b) Response to test wave in turbulence
5 – Wave–wave interaction	(a) Quasi-particle population density	(a) Kinetics of local and non-local wave interaction
	(b) Test wave	(b) Wave energy cascade
	(c) Modal amplitude	(c) Reduced, integrable system model
6 – Wave turbulence	(a) $\mathbf{k} \leftrightarrow \omega_k$, $\Delta\omega_{\mathbf{k}}$	(a) Wave–eddy unification, wave auto-coherence time
	(b) Test wave in scrambling background	(b) Strong wave–wave turbulence
	(c) Evolving resolved degrees of freedom in noisy background of coarse-grained fast modes (Mori–Zwanzig Theory)	(c) Disparate scale interaction with fast modes eliminated and thermalized
7 – Langmuir turbulence	(a) Plasmon gas with phonon (wave kinetics)	(a) Disparate scale interaction with fast modes adiabatically varying (wave kinetics)
	(b) Collapsing caviton with acoustic response (Zakharov equation)	(b) Disparate scale interaction with fast modes supporting envelope
8 – Phase space density granulations	(a) Phase space eddy	(a) Circulation for Vlasov fluid
	(b) Clump	(b) Granulation formed by mode–mode coupling via resonant particles. Dynamics of screened test particle
	(c) Phase space density hole	(c) Jeans equilibrium and self-bound phase space structure

Table 1.1. (*cont.*)

Chapter – Topic	Quasi-particle concept	Physics issue
9 – MHD turbulence	(a) Alfvén wave	(a) Basic wave of incompressible MHD
	(b) Eddy	(b) Fluid excitation – virtual mode
	(c) Shocklet	(c) Compressible MHD soliton evolving uni-directional from wave-packet

Schroedinger equations and modulations, (v) details of turbulence closure theory, (vi) general aspects of wave turbulence in stratified media, etc. For an advanced treatment of some of these issues, we refer the reader to books on plasma turbulence which precede this volume, e.g., Kadomtsev (1965), Galeev and Sagdeev (1965), Sagdeev and Galeev (1969), R. C. Davidson (1972), Ichimaru (1973), Itoh *et al.* (1999), Moiseev *et al.* (2000), Yoshizawa *et al.* (2003), Elskens and Escande (2003) and Balescu (2005), and to those on nonlinear Schroedinger equations and modulations, Newell (1985), Trullinger *et al.* (1986) and Sulem and Sulem (1999), and to those on neutral fluids, Lighthill (1978), Craik (1985), McComb (1990), Zakharov *et al.* (1992), Frisch (1995), Lesieur (1997), Pope (2000) and P. A. Davidson (2004).

1.5 On using this book

This book probably contains more material than any given reader needs or wants to assimilate, especially on the first pass. Thus, anticipating the needs of different readers, it seems appropriate to outline different possible approaches to the use of this book.

(1) Core programme – for serious readers

Chapters 2–6 form the essential core of this book. Most readers will want to at least survey these chapters, which can also serve as the nucleus of a one-semester advanced graduate course on Nonlinear Plasma Theory or Plasma Turbulence Theory. This core material includes fluctuation theory, self-similar cascades and transport, mean field quasi-linear theory, resonance broadening and nonlinear wave–particle interaction, wave–wave interaction and wave turbulence, and strong turbulence theory and renormalization.

More ambitious advanced courses could supplement this core with any or all of Chapters 7–9, with topics from Volume 2, or with material from the research literature.

(2) Shorter pedagogical introduction

A briefer, less detailed programme which at the same time introduces the reader to the essential physical concepts is also helpful. For such purposes, a "turbulence theory light" course could include Sections 2.1, 2.2.1, 2.2.2, 2.3, 3.1–3.3, 4.1–4.3, 5.1–5.4, 5.5.1, 6.1, 7.1, 7.2 and 7.3. This "introductory tour" could be supplemented by other material in this book and in Volume 2.

(3) For experimentalists

We anticipate that experimentalists may desire simple explanations of essential physical concepts. For such readers, we recommend the "turbulence theory light" course, described above.

In particular, MFE experimentalists will no doubt be interested in issues pertinent to drift wave turbulence. These are discussed in Sections 3.5 (quasi-linear theory), 4.2.2 (resonance broadening theory), 4.4 (nonlinear wave–particle interaction), 5.2.4 (three-wave interaction and isomorphism to the asymmetric top), 5.3.4 (fluid wave interaction) and 6.4 (strong turbulence theory). Chapter 7 is also a necessary prerequisite for the extensive discussion of zonal flows planned for Volume 2.

(4) For readers from related fields

We hope and anticipate that this book will interest readers from outside of plasma physics. Possible subgroups include readers with a special interest in phase space dynamics, readers interested in astrophysical fluid dynamics, readers from space and astrophysics and readers from geophysical fluid dynamics.

(a) For phase space dynamics
Readers familiar with fluid turbulence theory who desire to learn about phase space dynamics should focus on Sections 2.1–2.2, Chapters 3, 4 and 8. These discuss fluctuation and relaxation theory, quasi-linear theory, nonlinear wave–particle interaction and phase space granulations and cascades, respectively.

(b) For astrophysical fluid dynamics
Readers primarily interested in astrophysical fluid dynamics and MHD should examine Section 2.3, Chapters 3, 5, 6 and 7 and most especially Chapter 9. These present the basics of turbulence theory (Section 2.3), foundations of mean field theory – with special emphasis on the fundamental origins of irreversibility (Chapter 3), wave turbulence (Chapter 5), strong turbulence (Chapter 6), disparate-scale interaction (Chapter 7) – which is very closely related to dynamo theory – and MHD turbulence and transport (Chapter 9).

(c) For space and astrophysical plasma physicists
Readers primarily interested in space physics should visit Chapters 2, 3, 4, 5 and 7. This would acquaint them with basic concepts (Chapter 2), quasi-linear theory (Chapter 3), nonlinear wave–particle (Chapter 4) and wave–wave (Chapter 5) interaction and disparate scale interaction (Chapter 7).

(d) For readers from geophysical fluid dynamics
Readers from geophysical fluid dynamics will no doubt be interested in the topics listed under "(a) For phase space dynamics", above. They may also be interested in the analogy between drift waves in confined plasmas and wave dynamics in GFD. This is discussed in Sections 5.2.5, 5.3.4, 5.5.3, 8.1 and Appendix 1. Volume 2 will develop this analogy further.

2

Conceptual foundations

學而不思則罔，思而不學則殆

If one studies but does not think, one will be bewildered. If one thinks but does not study, one will be in peril.

(Confucius)

2.1 Introduction

This chapter presents the conceptual foundations of plasma turbulence theory from the perspective of physical kinetics of quasi-particles. It is divided into two sections:

(1) Dressed test particle model of fluctuations in a plasma near equilibrium;
(2) K41 beyond dimensional analysis – revisiting the theory of hydrodynamic turbulence.

The reason for this admittedly schizophrenic beginning is the rather unusual and atypical niche that plasma turbulence occupies in the pantheon of turbulent and chaotic systems. In many ways, most (though not all) cases of plasma turbulence may be thought of as weak turbulence, spatiotemporal chaos or wave turbulence, as opposed to fully developed turbulence in neutral fluids. Dynamic range is large, but nonlinearity is usually *not* overwhelmingly strong. Frequently, several aspects of the linear dynamics persist in the turbulent state, though wave breaking is possible, too. While a scale-to-scale transfer is significant, local emission and absorption, at a particular scale, are not negligible. Scale invariance is usually only approximate, even in the absence of dissipation. Indeed, it is fair to say that plasma turbulence lacks the elements of simplicity, clarity and universality which have attracted many researchers to the study of high Reynolds number fluid turbulence. In contrast to that famous example, plasma turbulence is a problem in the dynamics of a *multi-scale and complex system*. It challenges the researcher to isolate, define and solve interesting and relevant thematic or idealized problems which illuminate the more complex and intractable whole. To this end, then, it is useful to begin by discussing two rather different 'limiting case paradigms', which in some sense 'bound' the position of most plasma turbulence problems in the intellectual realm. These limiting cases are:

- The test particle model (TPM) of a near-equilibrium plasma, for which the relevant quasi-particle is a dressed test particle;
- The Kolmogorov (K41) model of a high Reynolds number fluid, very far from equilibrium, for which the relevant quasi-particle is the fluid eddy.

The TPM illustrates important plasma concepts such as local emission and absorption, screening response and the interaction of waves and sources (Balescu, 1963; Ichimaru, 1973). The K41 model illustrates important turbulence theory concepts such as scale similarity, cascades, strong energy transfer between scales and turbulent dispersion (Kolmogorov, 1941). We also briefly discuss turbulence in two dimensions – very relevant to strongly magnetized plasmas – and turbulence in pipe flows. The example of turbulent pipe flow, usually neglected by physicists in deference to homogeneous turbulence in a periodic box, is especially relevant to plasma confinement, as it constitutes *the* prototypical example of eddy viscosity and mixing length theory, and of profile formation by turbulent transport. The prominent place that engineering texts accord to this deceptively simple example is no accident – engineers, after all, need answers to real world problems. More fundamentally, just as the Kolmogorov theory is a basic example of self-similarity in scale, the Prandtl mixing length theory nicely illustrates self-similarity in space (Prandtl, 1932). The choice of these two particular paradigmatic examples is motivated by the huge disparity in the roles of spectral transfer and energy flux in their respective dynamics. In the TPM, spectral transport is ignorable, so the excitation at each scale k is determined by the local balance of excitation and damping at that scale. In the inertial range of turbulence, local excitation and damping are negligible, and all scales are driven by spectral energy flux – i.e. the cascade – set by the dissipation rate. (See Figure 2.1 for illustration.) These two extremes correspond, respectively, to a state with no flux and to a flux-driven state, in some sense 'bracket' most realizations of (laboratory) plasma turbulence, where excitation, damping and transfer are all roughly comparable. For this reason, they stand

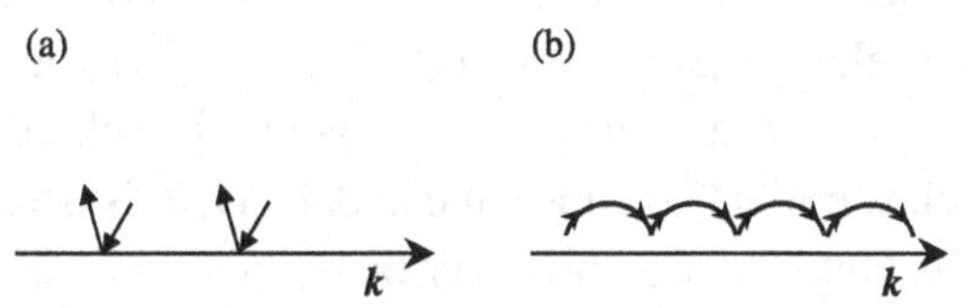

Fig. 2.1. (a) Local in k emission and absorption near equilibrium. (b) Spectral transport from emission at k_1, to absorption at k_2 via nonlinear coupling in a non-equilibrium plasma.

out as conceptual foundations, and so we begin out study of plasma turbulence with them.

2.2 Dressed test particle model of fluctuations in a plasma near equilibrium

2.2.1 Basic ideas

Virtually *all* theories of plasma kinetics and plasma turbulence are concerned, in varying degrees, with calculating the fluctuation spectrum and relaxation rate for plasmas under diverse circumstances. The simplest, most successful and best known theory of plasma kinetics is the *dressed test particle model* of fluctuations and relaxation in a plasma near equilibrium. This model, as presented here, is a synthesis of the pioneering contributions and insights of Rostoker and Rosenbluth (1960), Balescu (1963), Lenard (1960), Klimontovich (1967), Dupree (1961), and others. The unique and attractive feature of the test particle model is that it offers us a physically motivated and appealing picture of dynamics near equilibrium which is *entirely consistent* with Kubo's linear response theory and the fluctuation–dissipation theorem (Kubo, 1957; Callen and Welton, 1951), but does *not* rely upon the abstract symmetry arguments and operator properties that are employed in the more formal presentations of generalized fluctuation theory, as discussed in texts such as Landau and Lifshitz's *Statisical Physics* (1980). Thus, *the test particle model is consistent with formal fluctuation theory, but affords the user far greater physical insight.* Though its applicability is limited to the rather simple and seemingly dull case of a stable plasma 'near' thermal equilibrium, the test particle model nevertheless constitutes a vital piece of the conceptual foundation upon which all the more exotic kinetic theories are built. For this reason we accord it a prominent place in our study, and begin our journey by discussing it in some depth.

Two questions of definition appear immediately at the outset. These are as follows:

(a) What is a plasma?
(b) What does 'near equilibrium' mean?

For our purposes, a plasma is a quasi-neutral gas of charged particles with thermal energy far in excess of electrostatic energy (i.e. $k_B T \gg q^2/\bar{r}$), and with many particles within a Debye sphere (i.e. $1/n\lambda_D^3 \ll 1$), where q is a charge, $\bar{r}$ is a mean distance between particles, $\bar{r} \sim n^{-1/3}$, n is a mean density, T is a temperature, and k_B is the Boltzmann constant. The first property distinguishes a gaseous plasma from a liquid or crystal, while the second allows its description by a Boltzmann equation. Specifically, the condition $1/n\lambda_D^3 \ll 1$ means that discrete particle effects are, in some sense, 'small' and so allows truncation of the BBGKY (Bogoliubov,

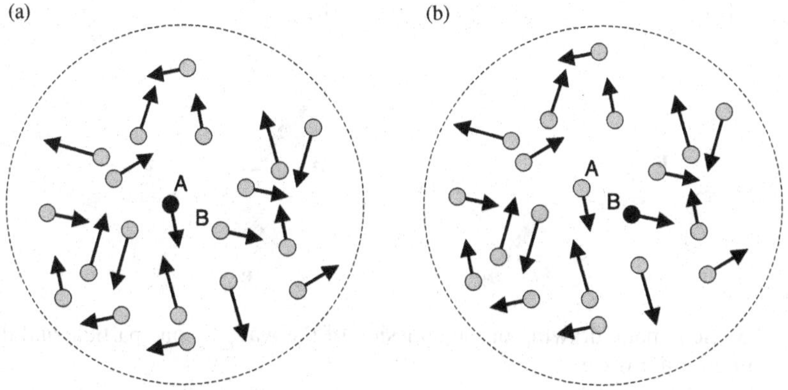

Fig. 2.2. A large number of particles exist within a Debye sphere of particle A (shown in black) in (a). Other particles provide a screening on the particle A. When the particle B is chosen as a test particle, others (including A) produce screening on B, (b). Each particle acts the role of test particle and the role of screening for the other test particle.

Born, Green, Kirkwood, Yvon) hierarchy at the level of a Boltzmann equation. This is equivalent to stating that if the two body correlation $f(1, 2)$ is written in a cluster expansion as $f(1)f(2) + g(1, 2)$, then $g(1, 2)$ is of $O(1/n\lambda_D^3)$ with respect to $f(1)f(2)$, and that higher order correlations are negligible. Figure 2.2 illustrates a test particle surrounded by many particle in a Debye sphere. The screening on the particle A is induced by other particles. When the particle B is chosen as a test particle, others (including A) produce screening of B. Each particle acts in the dual roles of a test particle and as part of the screening for other test particles.

The definition of 'near-equilibrium' is more subtle. A near-equilibrium plasma is one characterized by:

(1) a balance of emission and absorption by particles at a rate related to the temperature, T;
(2) the viability of linear response theory and the use of linearized particle trajectories.

Condition (1) necessarily implies the absence of linear instability of collective modes, but *does not* preclude collectively enhanced relaxation to states of higher entropy. Thus, a near-equilibrium state need not be one of maximum entropy. Condition (2) *does* preclude zero frequency convective cells driven by thermal fluctuations via mode–mode coupling, such as those that occur in the case of transport in 2D hydrodynamics. Such low frequency cells are usually associated with long time tails and require a renormalized theory of the nonlinear response for their description, as is discussed in later chapters.

The essential element of the test particle model is the compelling physical picture it affords us of the balance of emission and absorption which is intrinsic to

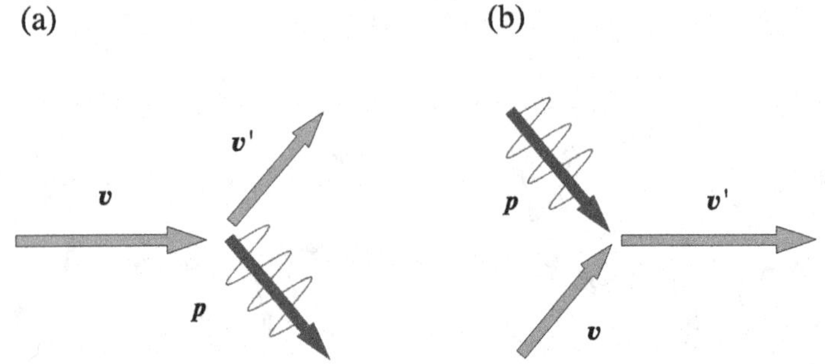

Fig. 2.3. Schematic drawing of the emission of the wave by one particle and the absorption of the wave.

thermal equilibrium. In the test particle model (TPM), *emission* occurs as a discrete particle (i.e. electron or ion) moves through the plasma, Cerenkov emitting electrostatic waves in the process. This emission process creates *fluctuations* in the plasma and converts particle kinetic energy (i.e. thermal energy) to collective mode energy. Wave radiation also induces a *drag* or dynamical friction on the emitter, just as the emission of waves in the wake of a boat induces a wave drag force on the boat. Proximity to equilibrium implies that emission is, in turn, balanced by *absorption*. Absorption occurs via Landau damping of the emitted plasma waves, and constitutes a wave energy dissipation process which heats the resonant particles in the plasma. Note that this absorption process ultimately returns the energy which was radiated by the particles to the thermal bath. The physics of wave emission and absorption which defines the thermal equilibrium balance intrinsic to the TPM is shown in Figure 2.3.

A distinctive feature of the TPM is that in it, each plasma particle has a 'dual identity', both as an 'emitter' and an 'absorber'. As an emitter, each particle radiates plasma waves, which are moving along some specified, linear, unperturbed orbit. Note that each emitter is identifiable (i.e. as a discrete particle) while moving through the Vlasov fluid, which is composed of other particles. As an absorber, each particle helps to define an element of the Vlasov fluid responding to, and (Landau) damping the emission from, *other* discrete particles. In this role, particle discreteness is smoothed over. Thus, the basic picture of an equilibrium plasma is one of a soup or gas of *dressed test particles*. In this picture, each particle:

(i) stimulates a collective response from the other particles by its discreteness;
(ii) responds to or 'dresses' other discrete particles by forming part of the background Vlasov fluid.

Thus, if one views the plasma as a pea soup, then the TPM is built on the idea that 'each pea in the soup acts like soup for all the other peas'. The *dressed test particle* is the *fundamental quasi-particle* in the description of near-equilibrium

(a) (b)

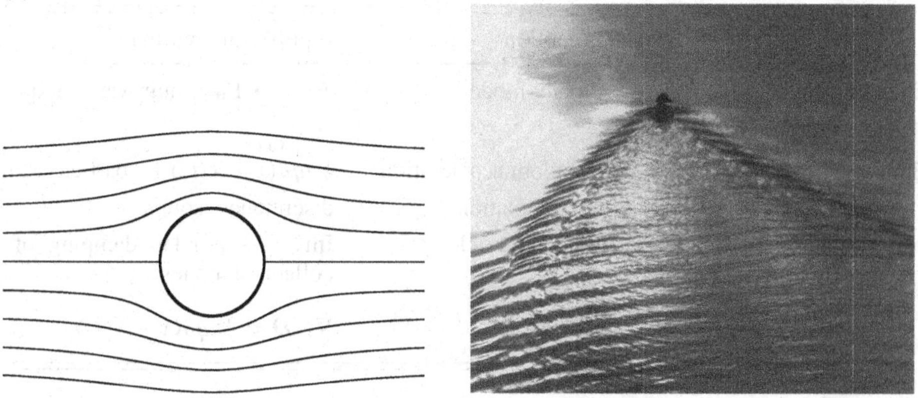

Fig. 2.4. Dressing of moving objects. Examples like a sphere in a fluid (a) and a supersonic object (b) are illustrated. In the case of a sphere, the surrounding fluid moves with it, so that the effective mass of the sphere (measured by the ratio between the acceleration to the external force) increases. The supersonic object radiates the wake of waves.

plasmas. Examples of dressing by surrounding media are illustrated in Figure 2.4. In the case of a sphere in a fluid, the surrounding fluid moves with it, so that the effective mass of the sphere (defined by the ratio between the external force to the acceleration) increases by an amount $(2\pi/3)\,\rho a^3$, where a is the radius of the sphere and ρ is the mass density of the surrounding fluid. The supersonic object radiates the wake of waves (b), thus its motion deviates from one in a vacuum.

At this point, it is instructive to compare the test particle model of thermal equilibrium to the well-known elementary model of Brownian fluctuations of a particle in a thermally fluctuating fluid. This comparison is given in Table 2.1, below.

Predictably, while there are many similarities between Brownian particles and thermal plasma fluctuations, a key *difference* is that in the case of Brownian motion, the roles of emission and absorption are clearly distinct and played, respectively, by random forces driven by thermal fluctuations in the fluid and by Stokes drag of the fluid on the finite size particle. In contrast, for the plasma the roles of both the emitter and absorber are played by the *plasma particles themselves* in the differing guises of discreteness and as chunks of the Vlasov fluid. In the cases of both the Brownian particle and the plasma, the well-known fluctuation–dissipation theorem of statistical dynamics near equilibrium applies, and relates the fluctuation spectrum to the temperature and the dissipation via the collective mode dissipation, i.e. $\operatorname{Im}\epsilon(k, \omega)$, the imaginary part of the collective response function.

Table 2.1. *Comparison of Brownian particle motion and plasma fluctuations*

	Brownian motion	Equilibrium plasma
Excitation	$v_\omega \to$ velocity mode	$E_{k,\omega} \to$ Langmuir wave mode
Fluctuation spectrum	$\langle \tilde{v}^2 \rangle_\omega$	$\langle E^2 \rangle_{\boldsymbol{k},\omega}$
Emission noise	$\langle \tilde{a}^2 \rangle_\omega \to$ random acceleration by thermal fluctuations	$4\pi q \delta(x - x(t)) \to$ particle discreteness source
Absorption	Stokes drag on particle	$\mathrm{Im}\,\epsilon \to$ Landau damping of collective modes
Governing equations	$\dfrac{\mathrm{d}\tilde{v}}{\mathrm{d}t} + \beta \tilde{v} = \tilde{a}$	$\nabla \cdot \boldsymbol{D} = 4\pi q \delta(x - x(t))$

2.2.2 Fluctuation spectrum

Having discussed the essential physics of the TPM and having identified the dressed test particle as the quasi-particle of interest for the dynamics of near-equilibrium plasma, we now proceed to calculate the plasma fluctuation spectrum near thermal equilibrium. We also show that this spectrum is that required to satisfy the fluctuation–dissipation theorem (F–DT). Subsequently, we use the spectrum to calculate plasma relaxation.

2.2.2.1 Coherent response and particle discreteness noise

As discussed above, the central tenets of the TPM are that each particle is both a discrete emitter as well as a participant in the screening or dressing cloud of other particles and that the fluctuations are *weak*, so that linear response theory applies. Thus, the total phase space fluctuation δf is written as

$$\delta f = f^{\mathrm{c}} + \tilde{f}, \tag{2.1}$$

where f^{c} is the coherent Vlasov response to an electric field fluctuation, i.e.

$$f^{\mathrm{c}}_{k,\omega} = R_{k,\omega} E_{k,\omega},$$

where $R_{k,\omega}$ is a linear response function and $\tilde{f}$ is the particle discreteness noise source, i.e.

$$\tilde{f}(x, v, t) = \frac{1}{n} \sum_{i=1}^{N} \delta(x - x_i(t)) \delta(v - v_i(t)) \tag{2.2}$$

(see, Fig. 2.5). For simplicity, we consider high frequency fluctuations in an electron–proton plasma, and assume the protons are simply a static background.

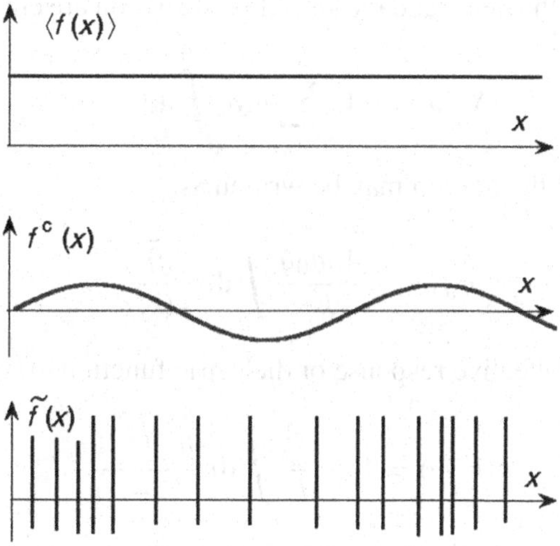

Fig. 2.5. Schematic drawing of the distribution of plasma particles. The distribution function, $f(x, v)$, is divided into the mean $\langle f \rangle$, the coherent response f^c, and the fluctuation part owing to the particle discreteness $\tilde{f}$.

Consistent with linear response theory, we use unperturbed orbits to approximate $x_i(t)$, $v_i(t)$ as:

$$v_i(t) = v_i(0), \tag{2.3a}$$

$$x_i(t) = x_i(0) + v_i t. \tag{2.3b}$$

Since $k_B T \gg q^2/\bar{r}$, the fundamental ensemble here is one of uncorrelated, discrete test particles. Thus, we can define the ensemble average of a quantity A to be

$$\langle A \rangle = n \int \mathrm{d}\boldsymbol{x}_i \int \mathrm{d}\boldsymbol{v}_i \, f_0(\boldsymbol{v}_i, \boldsymbol{x}_i) \, A, \tag{2.4}$$

where $\boldsymbol{x}_i$ and $\boldsymbol{v}_i$ are the phase space coordinates of the particles and f_0 is same near-equilibrium distribution, such as a local Maxwellian. For a Vlasov plasma, which obeys

$$\frac{\partial f}{\partial t} + v \frac{\partial f}{\partial x} + \frac{q}{m} E \frac{\partial f}{\partial v} = 0, \tag{2.5a}$$

the linear response function $R_{k,\omega}$ is

$$R_{k,\omega} = -i \frac{q}{m} \frac{\partial \langle f \rangle / \partial v}{\omega - kv}. \tag{2.5b}$$

Self-consistency of the fields and the particle distribution are enforced by Poisson's equation

$$\nabla^2 \phi = -4\pi \sum_s n_s q_s \int \mathrm{d}v\, \delta f_s, \tag{2.6a}$$

so that the potential fluctuation may be written as:

$$\phi_{k,\omega} = -\frac{4\pi n_0 q}{k^2} \int \mathrm{d}v\, \frac{\tilde{f}_{k,\omega}}{\epsilon(k,\omega)}, \tag{2.6b}$$

where the plasma collective response or dielectric function $\epsilon(k, \omega)$ is given by:

$$\epsilon(k,\omega) = 1 + \frac{\omega_p^2}{k} \int \mathrm{d}v\, \frac{\partial \langle f \rangle / \partial v}{\omega - kv}. \tag{2.6c}$$

Note that Eqs.(2.6) relate the fluctuation level to the discreteness noise emission and to $\epsilon(k, \omega)$, the linear collective response function.

2.2.2.2 *Fluctuations driven by particle discreteness noise*

A heuristic explanation is given here that the 'discreteness' of particles induces fluctuations. Consider a case that charged particles (charge q) are moving as shown in Figure 2.6(a). The distance between particles is given by d, and particles are moving at the velocity u. (The train of particles in Fig. 2.6(a) is a part of the distribution of particles. Of course, the net field is calculated by accumulating contributions from all particles.) Charged particles generate the electric field. The time-varying electric field (measured at the position A) is shown in Figure 2.6(b). When we make one particle smaller, but keeping the average density constant, the oscillating field at A becomes smaller. For instance, if the charge of one particle becomes half $q/2$ while the distance between particles is halved, we see that the amplitude of the varying electric field becomes smaller while the frequency becomes higher. This situation is shown in Figure 2.6(b) by a dashed line. In the limit of continuity, i.e.,

Charge per particle $\to 0$
Distance between particle $\to 0$,

while the average density is kept constant, the amplitude of the fluctuating field goes to zero. This example illustrates why 'discreteness' induces fluctuations.

Before proceeding with calculating the spectrum, we briefly discuss an important assumption we have made concerning collective resonances. For a discrete test particle moving on unperturbed orbits,

$$\tilde{f} \sim q\delta(x - x_{0i}(t))\delta(v - v_{0i})$$

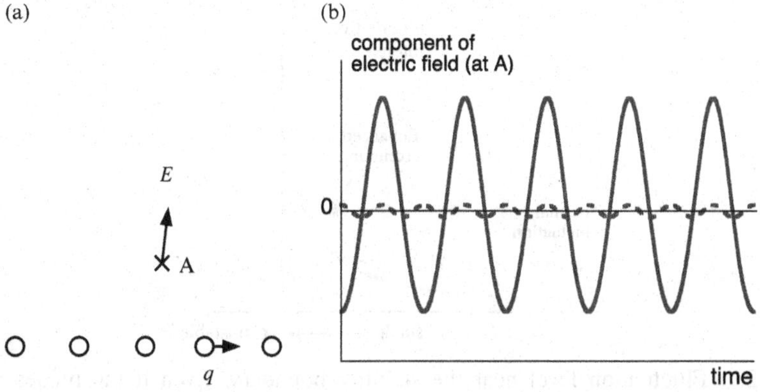

Fig. 2.6. Schematic illustration that the discreteness of particles is the origin of radiations. A train of charged particles (charge q, distance d) is moving near by the observation point A (a). The vertical component of the electric field observed at point A (b). When each particle is divided into two particles, (i.e. charge per particle is $q/2$ and distance between particles is $d/2$), the amplitude of the observed field becomes smaller (dashed curve in (b)).

so

$$\int \mathrm{d}v\, \tilde{f}_k \sim q\mathrm{e}^{-ikvt}$$

and

$$\epsilon(k,t)\phi_k(t) = \frac{4\pi}{k^2} q\mathrm{e}^{-ikvt}.$$

Here, the dielectric is written as an operator ϵ, to emphasize the fact that the response is non-local in time, on account of *memory* in the dynamics. Then strictly speaking, we have

$$\phi_k(t) = \epsilon^{-1} \cdot \left[\frac{4\pi q}{k^2}\mathrm{e}^{-ikvt}\right] + \phi_{k,\omega_k}\mathrm{e}^{-i\omega_k t}. \tag{2.7}$$

In Eq.(2.7), the first term on the right-hand side is the inhomogeneous solution driven by discreteness noise, while the second term is the homogeneous solution (i.e. solution of $\epsilon\phi = 0$), which corresponds to an eigenmode of the system (i.e. a fluctuation at k, ω which satisfies $\epsilon(k,\omega) \simeq 0$, so $\omega = \omega_k$). However, the condition that the plasma be 'near equilibrium' requires that all collective modes be damped (i.e. $\mathrm{Im}\,\omega_k < 0$), so the homogeneous solutions decay in time. Thus, in a near-equilibrium plasma,

$$\phi_k(t) \xrightarrow[t\to\infty]{} \epsilon^{-1} \cdot \left[\frac{4\pi q}{k^2}\mathrm{e}^{-ikvt}\right],$$

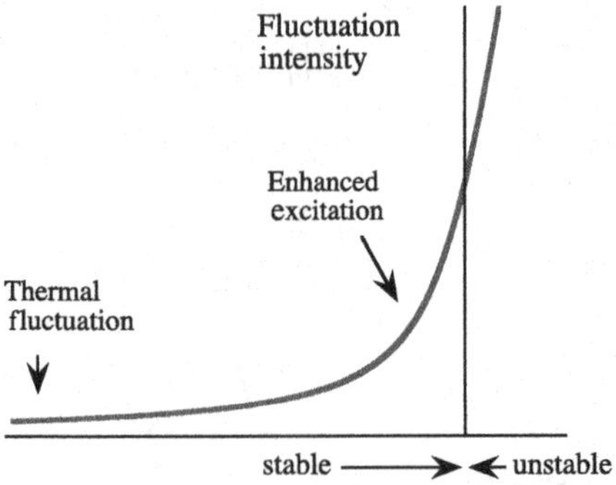

Fig. 2.7. Fluctuation level near the stability boundary. Even if the modes are stable, enhanced excitation of eigenmodes is possible when the controlling parameter approaches the boundary of stability. Linear theory could be violated even if the eigenmodes are stable. Nonlinear noise is no longer negligible.

so only the inhomogeneous solution survives. Two important caveats are necessary here. First, for weakly damped modes with $\mathrm{Im}\,\omega_k \lessapprox 0$, one may need to wait quite a long time indeed to actually arrive at asymptopia. Thus the homogeneous solutions may be important, in practice. Second, for weakly damped ('soft') modes, the inhomogeneous solution $\hat{\phi} \sim |\mathrm{Im}\,\epsilon|^{-1}$ can become large and produce significant orbit scattering and deflection. The relaxation times of such 'soft modes', thus, increase significantly. This regime of approach to marginality from below is analogous to the approach to criticality in a phase transition, where relaxation times and correlation lengths diverge, and fluctuation levels increase. As in the case of critical phenomena, renormalization is required for an accurate theoretical treatment of this regime. The moral of the story related in this small digression is that the TPM's validity actually fails *prior* to the onset of linear instability, rather than *at* the instability threshold, as is frequently stated. The behaviour of fluctuation levels near the stability boundary is schematically illustrated in Figure 2.7. Even if the modes are stable, enhanced excitation of eigenmodes is possible when the controlling parameter is sufficiently close to the boundary of stability, approaching from below. Linear response theory could be violated even if the eigenmodes are stable.

2.2.2.3 *Potential fluctuations*

Proceeding with the calculation of the spectrum, we first define the spectral density of the potential fluctuation as the transform of the potential fluctuation correlation function, i.e.

$$\left\langle \phi^2 \right\rangle_{\boldsymbol{k},\omega} = \int_{-\infty}^{\infty} \mathrm{d}\boldsymbol{x} \int_0^{\infty} \mathrm{d}t\, \mathrm{e}^{i(\omega t - \boldsymbol{k}\cdot\boldsymbol{x})} \left\langle \phi(0,0)\phi(\boldsymbol{x},t) \right\rangle \tag{2.8a}$$

and

$$\langle \phi(0,0)\phi(\boldsymbol{x},t)\rangle = \int_{-\infty}^{\infty} \frac{\mathrm{d}\boldsymbol{k}}{(2\pi)^3} \int_{-\infty}^{\infty} \frac{\mathrm{d}\omega}{2\pi} \mathrm{e}^{i(\boldsymbol{k}\cdot\boldsymbol{x}-\omega t)} \left\langle \phi^2 \right\rangle_{\boldsymbol{k},\omega}. \tag{2.8b}$$

Note that the transformation is a Fourier transform in space but a Laplace transform in time. The "one-sided" Laplace transform is intrinsic to fluctuation and TPM theory, as both are built upon the idea of causality, along with assumptions of stationarity and linear response. As linear response theory applies here, the fluctuation modes are uncorrelated, so

$$\left\langle \phi_{\boldsymbol{k}\omega}\phi_{\boldsymbol{k}'\omega'} \right\rangle = (2\pi)^4 \left\langle \phi^2 \right\rangle_{\boldsymbol{k},\omega} \delta(\boldsymbol{k}+\boldsymbol{k}')\delta(\omega+\omega'). \tag{2.8c}$$

Therefore, from Eqs.(2.8a)–(2.8c) and Eq.(2.6b), we can immediately pass to the expression for the potential fluctuation spectrum,

$$\left\langle \phi^2 \right\rangle_{\boldsymbol{k},\omega} = \left(\frac{4\pi n_0}{k^2} q\right)^2 \int \mathrm{d}v_1 \int \mathrm{d}v_2 \frac{\left\langle \tilde{f}(1)\tilde{f}(2)\right\rangle_{\boldsymbol{k},\omega}}{|\epsilon(k,\omega)|^2}. \tag{2.9}$$

Here (1) and (2) refer to points in phase space. Observe that the fluctuation spectrum is entirely determined by the discreteness correlation function $\left\langle \tilde{f}(1)\tilde{f}(2)\right\rangle$ and the dielectric function $\epsilon(\boldsymbol{k},\omega)$. Moreover, we know *ab initio* that since the plasma is in equilibrium at temperature T, the fluctuation–dissipation theorem applies, so that the TPM spectrum calculation *must* recover the general FDT result, which is,

$$\frac{\left\langle D^2 \right\rangle_{k,\omega}}{4\pi} = \frac{2T}{\omega} \operatorname{Im}\epsilon(k,\omega). \tag{2.10}$$

Here $\boldsymbol{D}_{k,\omega} = \epsilon(k,\omega)\boldsymbol{E}_{k,\omega}$ is the electric displacement vector. Note that the FDT quite severely constrains the form of the particle discreteness noise.

2.2.2.4 *Correlation of particles and fluctuation spectrum*

To calculate $\left\langle \tilde{f}(1)\tilde{f}(2)\right\rangle_{\boldsymbol{k},\omega}$, we must first determine $\left\langle \tilde{f}(1)\tilde{f}(2)\right\rangle$. Since $\tilde{f}$ is the distribution of discrete uncorrelated test particles, we have:

$$\tilde{f}(x,v,t) = \frac{1}{n}\sum_{i=1}^{N} \delta(x - x_i(t))\delta(v - v_i(t)). \tag{2.11}$$

From Eqs.(2.3), (2.4) and (2.11), we obtain

$$\left\langle \tilde{f}(1)\tilde{f}(2)\right\rangle = \int \mathrm{d}\boldsymbol{x}_i \int \mathrm{d}\boldsymbol{v}_i \frac{\langle f\rangle}{n} \sum_{i,j=1}^{N} [\delta(\boldsymbol{x}_1 - \boldsymbol{x}_i(t))$$
$$\times\, \delta(\boldsymbol{x}_2 - \boldsymbol{x}_j(t))\delta(\boldsymbol{v}_1 - \boldsymbol{v}_i(t))\delta(\boldsymbol{v}_2 - \boldsymbol{v}_j(t))]. \qquad (2.12)$$

Since the product of δs is non-zero only if the arguments are interchangeable, we obtain immediately the discreteness correlation function

$$\left\langle \tilde{f}(1)\tilde{f}(2)\right\rangle = \frac{\langle f\rangle}{n}\delta(\boldsymbol{x}_1 - \boldsymbol{x}_2)\delta(\boldsymbol{v}_1 - \boldsymbol{v}_2). \qquad (2.13)$$

Equation (2.13) gives the *discreteness correlation function* in *phase space*. Since the physical model is one of an ensemble of discrete, uncorrelated test particles, it is no surprise that $\left\langle \tilde{f}(1)\tilde{f}(2)\right\rangle$ is *singular*, and *vanishes unless the two points in phase space are coincident*. Calculation of the fluctuation spectrum requires the velocity integrated discreteness correlation function $C(\boldsymbol{k}, \omega)$ which, using spatial homogeneity, is given by:

$$C(\boldsymbol{k}, \omega) = \int \mathrm{d}v_1 \int \mathrm{d}v_2 \left\langle \tilde{f}(1)\tilde{f}(2)\right\rangle_{\boldsymbol{k},\omega}$$
$$= \int \mathrm{d}v_1 \int \mathrm{d}v_2 \left[\int_0^\infty \mathrm{d}\tau\, \mathrm{e}^{i\omega\tau} \int \mathrm{d}x\, \mathrm{e}^{-i\boldsymbol{k}\cdot\boldsymbol{x}} \left\langle \tilde{f}(0)\tilde{f}(\boldsymbol{x},\tau)\right\rangle\right.$$
$$\left. + \int_{-\infty}^0 \mathrm{d}\tau\, \mathrm{e}^{-i\omega\tau} \int \mathrm{d}x\, \mathrm{e}^{-i\boldsymbol{k}\cdot\boldsymbol{x}} \left\langle \tilde{f}(\boldsymbol{x},-\tau)\tilde{f}(0)\right\rangle\right]. \qquad (2.14)$$

Note that the time history which determines the frequency dependence of $C(\boldsymbol{k}, \omega)$ is extracted by propagating particle (2) *forward* in time and particle (1) *backward* in time, resulting in the *two* Laplace transforms in the expression for $C(\boldsymbol{k}, \omega)$. The expression for $C(\boldsymbol{k}, \omega)$ can be further simplified by replacing v_1, v_2 with $(v_1 \pm v_2)/2$, performing the trivial v_--integration, and using unperturbed orbits to obtain

$$C(\boldsymbol{k}, \omega) = 2\int \mathrm{d}v \int_0^\infty \mathrm{d}\tau\, \mathrm{e}^{i\omega\tau} \int \mathrm{d}x\, \mathrm{e}^{-i\boldsymbol{k}\cdot\boldsymbol{x}} \frac{\langle f(v)\rangle}{n}\delta(\boldsymbol{x} - \boldsymbol{v}\tau)$$
$$= 2\int \mathrm{d}v \frac{\langle f(v)\rangle}{n} \int_0^\infty \mathrm{d}\tau\, \mathrm{e}^{i(\omega - \boldsymbol{k}\cdot\boldsymbol{v})\tau}$$
$$= \int \mathrm{d}v\, 2\pi \frac{\langle f(v)\rangle}{n}\delta(\omega - \boldsymbol{k}\cdot\boldsymbol{v}). \qquad (2.15)$$

Equation (2.15) gives the well-known result for the density fluctuation correlation function in $\boldsymbol{k}$, ω of an ensemble of discrete, uncorrelated test particles. $C(\boldsymbol{k}, \omega)$ is also the particle discreteness noise spectrum. Note that $C(\boldsymbol{k}, \omega)$ is composed of the unperturbed orbit propagator $\delta(\omega - \boldsymbol{k} \cdot \boldsymbol{v})$, a weighting function $\langle f \rangle$ giving the distribution of test particle velocities, and the factor of $1/n$, which is a generic measure of discreteness effects. Substitution of $C(\boldsymbol{k}, \omega)$ into Eq.(2.9) then finally yields the explicit general result for the TPM potential fluctuation spectrum:

$$\left\langle \phi^2 \right\rangle_{\boldsymbol{k},\omega} = \left(\frac{4\pi n_0}{k^2} q \right)^2 \frac{\int \mathrm{d}\boldsymbol{v} \, \frac{2\pi}{n} \langle f \rangle \, \delta(\omega - \boldsymbol{k} \cdot \boldsymbol{v})}{|\epsilon(\boldsymbol{k}, \omega)|^2}. \tag{2.16}$$

2.2.2.5 One-dimensional plasma

In order to elucidate the physics content of the fluctuation spectrum, it is convenient to specialize the discussion to the case of a 1D plasma, for which:

$$C(k, \omega) = \frac{2\pi}{n|k|v_{\mathrm{T}}} F(\omega/k) \tag{2.17a}$$

$$\left\langle \phi^2 \right\rangle_{k,\omega} = n_0 \left(\frac{4\pi q}{k^2} \right)^2 \frac{2\pi}{|k|v_{\mathrm{T}}} \frac{F(\omega/k)}{|\epsilon(k, \omega)|^2}. \tag{2.17b}$$

Here, F refers to the average distribution function, with the normalization factor of v_{T} extracted, $\langle f(v) \rangle = (n/v_{\mathrm{T}}) \, F(v)$, and v_{T} is a thermal velocity. It is interesting to observe that $\left\langle \phi^2 \right\rangle_{k,\omega} \sim$ (density) $\times$ (Coulomb spectrum) $\times$ (propagator) $\times$ (particle emission spectrum)$\times$(screening). Thus, spectral line structure in the TPM is determined by the distribution of Cerenkov emission from the ensemble of discrete particles (set by $\langle f \rangle$) and the collective resonances (where $\epsilon(k, \omega)$ becomes small).

In particular, for the case of an electron plasma with stationary ions, the natural collective mode is the electron plasma wave, with frequency $\omega \simeq \omega_{\mathrm{p}}(1 + \gamma k^2 \lambda_{\mathrm{D}}^2)^{1/2}$ (γ: specific heat ratio of electrons) (Ichimaru, 1973; Krall and Trivelpiece, 1973). So for $\omega \gg \omega_{\mathrm{p}}, kv_{\mathrm{T}}, \epsilon \to 1$, we have;

$$\left\langle \phi^2 \right\rangle_{k,\omega} \simeq n_0 \left(\frac{4\pi q}{k^2} \right)^2 \frac{2\pi}{|k|v_{\mathrm{T}}} F(\omega/k), \tag{2.18a}$$

where $F \sim \exp[-\omega^2/k^2 v_{\mathrm{T}}^2]$ for a Maxwellian distribution, while in the opposite limit of $\omega \ll \omega_{\mathrm{p}}, kv_{\mathrm{T}}$ where $\epsilon \to 1 + k^{-2}\lambda_{\mathrm{D}}^{-2}$, the spectrum becomes

$$\left\langle \phi^2 \right\rangle_{k,\omega} \simeq n_0 (4\pi q)^2 \frac{2\pi}{|k|v_{\mathrm{T}}} \left(1 + 1/k^2 \lambda_{\mathrm{D}}^2 \right)^{-2}. \tag{2.18b}$$

In the first, super-celeric limit, the discrete test particle effectively decouples from the collective response, while in the second, quasi-static limit, the spectrum is that of an ensemble of Debye-screened test particles. This region also corresponds to the $k\lambda_{\mathrm{De}} \gg 1$ range, where the scales are too small to support collective modes. In the case where collective modes are weakly damped, one can simplify the structure of the screening response via the pole approximation, which is obtained by expanding $\epsilon(k,\omega)$ about ω_k, i.e.

$$\begin{aligned}\epsilon(k,\omega) &= \epsilon_{\mathrm{r}}(k,\omega) + i\mathrm{Im}\,\epsilon(k,\omega)\\ &\simeq \epsilon_{\mathrm{r}}(k,\omega_k) + (\omega-\omega_k)\left.\frac{\partial\epsilon}{\partial\omega}\right|_{\omega_k} + i\mathrm{Im}\,\epsilon(k,\omega_k).\end{aligned} \tag{2.19}$$

So since $\epsilon_{\mathrm{r}}(k,\omega_k) \approx 0$,

$$\begin{aligned}1/|\epsilon|^2 &\simeq \frac{1}{|\mathrm{Im}\,\epsilon|}\left\{\frac{|\mathrm{Im}\,\epsilon|}{(\omega-\omega_k)^2|\frac{\partial\epsilon}{\partial\omega}|^2 + |\mathrm{Im}\,\epsilon|^2}\right\}\\ &\simeq \frac{1}{|\mathrm{Im}\,\epsilon|\,|\partial\epsilon_{\mathrm{r}}/\partial\omega|}\cdot\delta(\omega-\omega_k).\end{aligned} \tag{2.20}$$

Here it is understood that $\mathrm{Im}\,\epsilon$ and $\partial\epsilon_{\mathrm{r}}/\partial\omega$ are evaluated at ω_k. Notice that in the pole approximation, eigenmode spectral lines are weighted by the dissipation $\mathrm{Im}\,\epsilon$.

The fluctuation spectrum of plasma oscillations in thermal equailibrium is shown in Figure 2.8. The real frequency and the damping rate of the plasma oscillation are shown as a function of the wavenumber in (a). In the regime of $k\lambda_{\mathrm{De}} \ll 1$, the real frequency is close to the plasma frequency, $\omega \sim \omega_{\mathrm{p}}$, and the damping rate is exponentially small. The power spectrum of fluctuations as a function of the frequency is illustrated in (b) for various values of the wave number. In the regime of $k\lambda_{\mathrm{De}} \ll 1$, a sharp peak appears near the eigenfrequency $\omega \sim \omega_k$. Owing to the very weak damping, the line width is narrow. As the mode number becomes large (in the regime of $k\lambda_{\mathrm{De}} \sim 1$), the bandwidth becomes broader, showing the fact that fluctuations are generated and absorbed very rapidly.

2.2.2.6 *Fluctuation–dissipation theorem and energy partition*

By now, the reader is surely anxious to see if the results obtained using the test particle model are in fact consistent with the requirements and expectations of generalized fluctuation theory, as advertised. First, we check that the fluctuation–dissipation theorem is satisfied. This is most easily accomplished for the case of a Maxwellian plasma. There,

$$\mathrm{Im}\,\epsilon = -\frac{\omega_{\mathrm{p}}^2\pi}{k|k|}\left.\frac{\partial\langle f\rangle}{\partial v}\right|_{\omega/k} = \frac{2\pi\omega}{k^2v_{\mathrm{T}}^2}\frac{\omega_{\mathrm{p}}^2}{|k|v_{\mathrm{T}}}F_{\mathrm{M}}\left(\frac{\omega}{k}\right), \tag{2.21a}$$

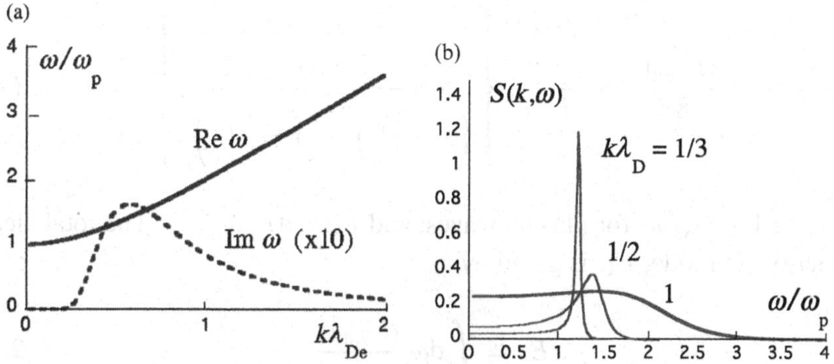

Fig. 2.8. Illustration of the fluctuation spectrum of plasma oscillations. The real frequency and the damping rate of the plasma oscillation are shown as a function of the wavenumber in (a). In the regime of $k\lambda_{De} \ll 1$, $\omega \sim \omega_p$ holds and that the damping rate is exponentially small. The power spectrum of fluctuation as a function of the frequency is illustrated in (b) as a function of the wave number. In the regime of $k\lambda_{De} \ll 1$, a high and sharp peak appears near the eigenfrequency $\omega \sim \omega_k$. Owing to the very weak damping, the line width is narrow. As the mode number becomes large an in the regime of $k\lambda_{De} \sim 1$, the bandwidth becomes broader, and the fluctuation intensity becomes high.

so using Eq.(2.21a) to relate $\mathrm{Im}\,\epsilon(k,\omega)$ to $F(\omega/kv_T)$ in Eq.(2.17b) gives

$$\left\langle \phi^2 \right\rangle_{k,\omega} = \frac{8\pi T}{k^2\omega} \frac{\mathrm{Im}\,\epsilon}{|\epsilon|^2}, \tag{2.21b}$$

so we finally obtain

$$\frac{\left\langle D^2 \right\rangle_{k,\omega}}{4\pi} = \frac{2T}{\omega} \mathrm{Im}\,\epsilon \tag{2.21c}$$

which is in precise agreement with the statement of the FDT for a classical, plasma at temperature T. It is important to reiterate here that applicability of the FDT rests upon to the applicability of linear response theory for the emission and absorption of each mode. Both fail as the instability marginal point is approached (from below).

Second, we also examine the k-spectrum of energy, with the aim of comparing the TPM prediction to standard expectations for thermal equilibrium, i.e. to see whether energy is distributed according to the conventional wisdom of "$T/2$ per degree-of-freedom". To this end, it is useful to write (using Eq.(2.21)) the electric field energy as:

$$\frac{|E_{k,\omega}|^2}{8\pi} = \frac{4\pi n q^2}{k|k|}\left\{\frac{F(\omega/k)}{\left(1-\frac{\omega_{\mathrm{p}}^2}{\omega^2}\right)^2+\left(\frac{\pi\omega_{\mathrm{p}}^2}{k|k|}F'\right)^2}\right\}, \tag{2.22}$$

where $\epsilon_{\mathrm{r}} \simeq 1-\omega_{\mathrm{p}}^2/\omega^2$ for plasma waves, and $F' = \mathrm{d}F/\mathrm{d}u|_{\omega/k}$. The total electric field energy per mode, E_k, is given by

$$E_k = \int \mathrm{d}\omega \frac{|E_{k,\omega}|^2}{8\pi}, \tag{2.23a}$$

so that use of the pole approximation to the collective resonance and a short calculation then gives,

$$E_k = \frac{n_e\omega_{\mathrm{p}}}{2|k|}\frac{F}{|F'|} = \frac{T}{2}. \tag{2.23b}$$

So, yes, the electric field energy for plasma waves is indeed at equipartition. Since for plasma waves the particle kinetic energy density E_{kin} equals the electric field energy density E_k (i.e. $E_{\mathrm{kin}} = E_k$), the total wave energy density per mode W_k is constant at T. Note that Eq.(2.23b) does not imply the divergence of total energy density. Of course, some fluctuation energy is present at very small scales ($k\lambda_{\mathrm{De}} \gtrsim 1$) which cannot support collective modes. However, on such scales, the pole expansion is not valid and simple static screening is a better approximation. A short calculation gives, for $k^2\lambda_{\mathrm{De}}^2 > 1$, $E_k \cong (T/2)/k^2\lambda_{\mathrm{De}}^2$, so that the *total* electric energy density is

$$\begin{aligned}\left\langle\frac{E^2}{8\pi}\right\rangle &= \int \mathrm{d}k\, E_k \\ &= \int_{-\infty}^{\infty}\frac{\mathrm{d}k}{2\pi}\frac{T/2}{\left(1+k^2\lambda_{\mathrm{De}}^2\right)} \sim \left(\frac{nT}{2}\right)\left(\frac{1}{n\lambda_{\mathrm{De}}}\right).\end{aligned} \tag{2.24}$$

As Eq.(2.24) is for 1D, there n has the dimensions of particles-per-distance. In 3D, the analogue of this result is

$$\left\langle\frac{E^2}{8\pi}\right\rangle \sim \left(\frac{nT}{2}\right)\left(\frac{1}{n\lambda_{\mathrm{De}}^3}\right), \tag{2.25}$$

so that the total electric field energy equals the total thermal energy times the discreteness factor $1/n\lambda_{\mathrm{De}}^3 \sim 1/N$, where N is the number of particles in a Debye sphere. Hence $\langle E^2/8\pi\rangle \ll nT/2$, as is required for a plasma with weak correlations.

2.2.3 Relaxation near equilibrium and the Balescu–Lenard equation

Having determined the equilibrium fluctuation spectrum using the TPM, we now turn to the question of how to *use it to calculate relaxation near equilibrium.* By "relaxation" we mean the long time evolution of the mean (i.e. ensemble averaged) distribution function $\langle f \rangle$. Here long time means long or slow evolution in comparison to fluctuation time scales. Generally, we expect the mean field equation for the prototypical example of a 1D electrostatic plasma to have the form of a continuity equation in velocity space, i.e.

$$\frac{\partial \langle f \rangle}{\partial t} = -\frac{\partial}{\partial v} J(v). \tag{2.26}$$

Here, $J(v)$ is a flux or current and $\langle f \rangle$ is the corresponding coarse-grained phase space density; $J \xrightarrow[v \to \pm\infty]{} 0$ assures conservation of total $\langle f \rangle$. *The essence of the problem* at hand is *how to actually calculate* $J(v)$! Of course it is clear from the Vlasov equation that $J(v)$ is simply the average acceleration $\langle (q/m) E \delta f \rangle$ due to the phase space density fluctuation δf. Not surprisingly, then, $J(v)$ is most directly calculated using a mean field approach. Specifically, simply substitute the *total* δf into $\langle (q/m) E \delta f \rangle$ to calculate the current $J(v)$. Since $\delta f = f^{\mathrm{c}} + \tilde{f}$, $J(v)$ will necessarily consist of *two* pieces. The *first piece*, $\langle (q/m) E f^{\mathrm{c}} \rangle$, accounts for the *diffusion in velocity* driven by the TPM potential fluctuation spectrum. This contribution can be obtained from a Fokker–Planck calculation using the TPM spectrum as the noise. The *second piece*, $\left\langle (q/m) E \tilde{f} \right\rangle$, accounts for relaxation driven by the *dynamic friction* between the ensemble of discrete test particles and the Vlasov fluid. It accounts for the evolution of $\langle f \rangle$ which must accompany the slowing down of a test particle by wave drag. The second piece has the structure of a drag term. As is shown in the derivation of Eq.(2.16), $\langle E f^{\mathrm{c}} \rangle$ ultimately arises from the discreteness of particles, $\tilde{f}_1$.

The kinetic equation for $\langle f \rangle$ which results from this mean-field calculation was first derived by R. Balescu and A. Lenard, and so is named in their honour (Lenard, 1960; Balescu, 1963). The diffusion term of the Balescu–Lenard (B–L) equation is very similar to the quasi-linear diffusion operator, discussed in Chapter 3, although the electric field fluctuation spectrum is prescribed by the TPM and the frequency spectrum is not restricted to eigenmode frequency lines, as in the quasi-linear theory. The total phase space current $J(v)$ is similar in structure to that produced by the glancing, small angle Coulomb interactions which the Landau collision integral calculates. (See Rosenbluth *et al.* (1957), Rostoker and Rosenbluth (1960) and Ichimaru and Rosenbluth (1970) for the Fokker–Planck approach to Coulomb collisions.) However, in contrast to the Landau theory, the B–L equation incorporates *both* static and dynamic screening, and so treats the interaction of collective

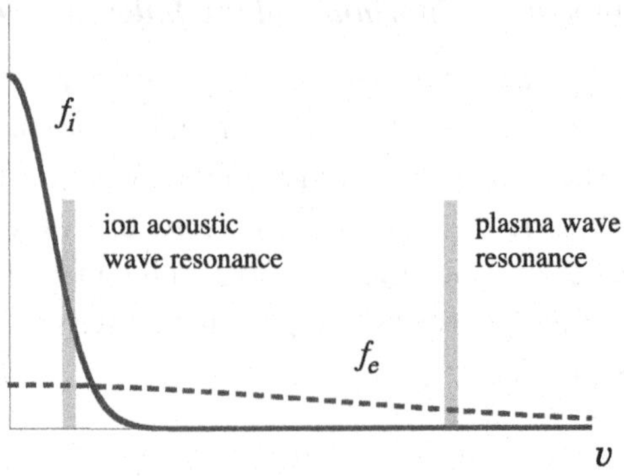

Fig. 2.9. Structure of $\langle f_i \rangle$, $\langle f_e \rangle$ for stable plasma. Velocity space configuration, showing electron plasma wave resonance on the tail of f_e and ion acoustic wave resonance in the bulk of f_e, f_i. In the case with $T_e \gg T_i$, waves can resonate with the bulk of the electron distribution while avoiding ion Landau damping. The slope of f_e is small at resonance, so ion acoustic waves are only weakly damped. In the case with $T_e \sim T_i$, ion acoustic waves resonant in the bulk of f_e cannot avoid ion Landau damping, so the collective modes are heavily damped.

processes with binary encounters. Screening also eliminates the divergences in the Landau collision integral (i.e. the Coulomb logarithm) which are due to long range electrostatic interactions. Like the Landau integral, the B–L equation is ultimately *nonlinear* in $\langle f \rangle$.

At this point, the sceptical reader will no doubt be entertaining question like, "What kind of relaxation is possible here?", "how does it differ from the usual collisional relaxation process?" and "just what, precisely, does 'near equilibrium' mean?". One point relevant to all these questions is that it is easy to define states that have finite free energy, but which are *stable* to collective modes. One example is the current-driven ion acoustic (CDIA) system shown in Figure 2.9. Here the non-zero current, which shifts the electron Maxwellian, constitutes free energy. However, since the shift does not produce a slope in $\langle f_e \rangle$ sufficient to overcome ion Landau damping, the free energy is *not* accessible to linear CDIA instabilities (Ichimaru, 1973). Nevertheless, electron $\rightarrow$ ion momentum transfer *is* possible, and *can* result in electron relaxation, since the initial state is *not* one of maximum entropy. Here, relaxation occurs via binary interactions of *dressed* test particles. Note however, that in this case relaxation rates may be significantly faster than for 'bare' particle collisions, on account of fluctuation enhancement by weakly damped collective modes. Thus, the B–L theory offers both something old and something new relative to its collisional antecedent, the Landau theory.

In order to best elucidate the physics of relaxation processes, we keep the calculations as simple as possible and divide our discussion into three parts. The basic theory is developed for an electrostatic plasma in one dimension, and then applied to single species and two species relaxation processes. Single species relaxation in 3D is then considered followed by a discussion of collective enhancement of momentum exchange.

2.2.3.1 Kinetic equation for mean distribution function

The Balescu–Lenard equation may be derived by a mean-field calculation of the fluctuation-induced acceleration $(q/m)\langle E\delta f\rangle$. Specifically,

$$\begin{aligned}\frac{\partial\langle f\rangle}{\partial t} &= -\frac{\partial}{\partial v}\frac{q}{m}\langle E\delta f\rangle \\ &= -\frac{\partial}{\partial v}J(v),\end{aligned} \tag{2.27}$$

where $J(v)$ must be calculated using the *total* δf, which includes both the linear response f^{c} and the discreteness fluctuation $\tilde{f}$. Thus, substitution of

$$\delta f = f^{\mathrm{c}} + \tilde{f},$$

yields

$$\begin{aligned}J(v) &= -\left(\frac{q}{m}\left\langle E f^{\mathrm{c}}\right\rangle + \frac{q}{m}\left\langle E\tilde{f}\right\rangle\right) \\ &= -D(v)\frac{\partial\langle f\rangle}{\partial v} + F(v),\end{aligned} \tag{2.28}$$

where $D(v)$ is the fluctuation-induced diffusion, while $F(v)$ is the dynamical friction term. Consistent with linear response theory, we can then write:

$$f^{\mathrm{c}}_{k,\omega} = -i\frac{(q/m)E_{k,\omega}}{\omega - kv}\frac{\partial\langle f\rangle}{\partial v}. \tag{2.29a}$$

so for stationary fluctuations, a short calculation gives:

$$D(v) = \sum_{k,\omega}\frac{q^2}{m^2}k^2\left\langle\phi^2\right\rangle_{k,\omega}\pi\delta(\omega - kv). \tag{2.29b}$$

The spectrum $\left\langle\phi^2\right\rangle_{k,\omega}$ is understood to be the test particle model spectrum, i.e. that of Eq.(2.17b). Similarly, the dynamic friction term $F(v)$ is given by

$$F(v) = -\frac{q}{m}\sum_{k,\omega} ik\left\langle\phi\tilde{f}\right\rangle_{k,\omega}, \tag{2.30a}$$

where, via Eq.(2.6b), we have:

$$\left\langle \phi\tilde{f}\right\rangle_{k,\omega} = \frac{4\pi n_0 q}{k^2}\int \mathrm{d}v\,\frac{\left\langle \tilde{f}\tilde{f}\right\rangle_{k,\omega}}{\epsilon(k,\omega)^*}. \tag{2.30b}$$

This result explains that the discreteness of particles is the source of correlations in the excited mode. Since

$$\left\langle \phi^2\right\rangle_{k,\omega} = \left(\frac{4\pi n_0 q}{k^2}\right)^2 \frac{C(k,\omega)}{|\epsilon(k,\omega)|^2},$$

and (from Eq.(2.14))

$$C(k,\omega) = \left\langle \tilde{n}^2\right\rangle_{k,\omega} = \int \mathrm{d}v\, 2\pi\delta(\omega - kv)\,\langle f(v)\rangle,$$

we have

$$\left\langle \phi\tilde{f}\right\rangle_{k,\omega} = \left(\frac{4\pi n_0 q}{k^2}\right)\frac{2\pi\delta(\omega - kv)\,\langle f\rangle}{\epsilon(k,\omega)^*}.$$

Thus, the current $J(v)$ is given by:

$$\begin{aligned} J(v) &= -D(v)\frac{\partial\langle f\rangle}{\partial v} + F_r(v) \\ &= -\sum_{k,\omega}\left(\frac{4\pi n_0 q}{k^2}\right)\frac{q}{m}\left(\frac{2\pi\delta(\omega - kv)}{n_0|\epsilon(k,\omega)|^2}\right)k \\ &\quad\times\left\{\left(\frac{4\pi n_0 q}{k^2}\right)\frac{\pi k}{|k|v_{\mathrm{T}}}\left(\frac{q}{m}\right)F\left(\frac{\omega}{k}\right)\frac{\partial\langle f\rangle}{\partial v} + \mathrm{Im}\,\epsilon(k,\omega)\,\langle f\rangle\right\}. \end{aligned} \tag{2.31}$$

Note that the contributions from the diffusion $D(v)$ and dynamic friction $F(v)$ have been grouped together within the brackets. Poisson's equation relates $\epsilon(k,\omega)$ to the electron and ion susceptibilities $\chi(k,\omega)$ by

$$\epsilon(k,\omega) = 1 + \left(\frac{4\pi n_0 q}{k^2}\right)\left[\chi_{\mathrm{i}}(k,\omega) - \chi_{\mathrm{e}}(k,\omega)\right],$$

where χ_{i}, χ_{e} are the ion and electron susceptibilities defined by

$$n_{k,\omega} = \chi_{\mathrm{i}}(k,\omega)\phi_{k,\omega}.$$

It is straightforward to show that

$$\mathrm{Im}\,\epsilon(k,\omega) = -\frac{\pi\omega_{\mathrm{p}}^2}{k^2}\frac{k}{|k|v_{\mathrm{T}}}F'(\omega/k) + \mathrm{Im}\,\epsilon_{\mathrm{i}}(k,\omega), \tag{2.32a}$$

where

$$\operatorname{Im} \epsilon_{\mathrm{i}}(k, \omega) = \frac{4\pi n_0 q}{k^2} \operatorname{Im} \chi_{\mathrm{i}}(k, \omega). \tag{2.32b}$$

Here $\epsilon_{\mathrm{i}}(k, \omega)$ is the *ion* contribution to the dielectric function. Thus, we finally obtain a simplified expression for $J(v)$:

$$J(v) = -\sum_{k,\omega} \left(\frac{\omega_{\mathrm{p}}^2}{k^2}\right)^2 \left(\frac{2\pi^2}{n_0 k v_{\mathrm{T}}}\right) \frac{\delta(\omega - kv)}{|\epsilon(k,\omega)|^2} \times \left\{ F\left(\frac{\omega}{k}\right) \frac{\partial \langle f \rangle}{\partial v} - \langle f(v) \rangle F'\left(\frac{\omega}{k}\right) + \operatorname{Im} \epsilon_{\mathrm{i}}(k,\omega) \langle f \rangle \right\}. \tag{2.33}$$

Equation (2.33) gives the *general* form of the velocity space electron current in the B–L equation for electron relaxation, as described within the framework of the TPM.

In order to elucidate the physics content of $J(v)$, it is instructive to re-write Eq.(2.33) in alternative forms. One way is to define the fluctuation phase velocity by $u = \omega/k$, so that

$$J(v) = -\sum_{k,\omega} \left(\frac{\omega_{\mathrm{p}}^2}{k^2}\right)^2 \left(\frac{2\pi^2}{n_0 k v_{\mathrm{T}}}\right) \frac{\delta(u - v)}{|k||\epsilon(k,\omega)|^2} \times \left\{ F(u) \langle f(v) \rangle' - F'(u) \langle f(v) \rangle + \operatorname{Im} \epsilon_{\mathrm{i}}(k,\omega) \langle f(v) \rangle \right\}. \tag{2.34}$$

Alternatively, one could just perform the summation over frequency to obtain

$$J(v) = -\sum_{k} \left(\frac{\omega_{\mathrm{p}}^2}{k^2}\right)^2 \left(\frac{\pi}{n_0 k v_{\mathrm{T}}}\right) \frac{1}{|\epsilon(k,kv)|^2} \times \left\{ F(v) \frac{\partial \langle f(v) \rangle}{\partial v} - \langle f(v) \rangle F'(v) + \operatorname{Im} \epsilon_{\mathrm{i}}(k,kv) \langle f(v) \rangle \right\}. \tag{2.35}$$

Finally, it is also useful to remind the reader of the counterpart of Eq.(2.33) in the unscreened Landau collision theory, which we write for 3D as:

$$J_\alpha(\boldsymbol{p}) = -\sum_{\mathrm{e,i}} \int_{q_\alpha} \mathrm{d}^3 q \int \mathrm{d}^3 p'\, W(\boldsymbol{p}, \boldsymbol{p}', \boldsymbol{q}) q_\alpha q_\beta \left\{ f(\boldsymbol{p}') \frac{\partial f(\boldsymbol{p})}{\partial p_\beta} - \frac{\partial f(\boldsymbol{p}')}{\partial p'_\beta} f(\boldsymbol{p}) \right\}. \tag{2.36}$$

In Eq.(2.36), $W(\boldsymbol{p}, \boldsymbol{p}', \boldsymbol{q})$ is the transition probability for a collision (with momentum transfer $\boldsymbol{q}$) between a 'test particle' of momentum $\boldsymbol{p}$ and a 'field particle' of momentum $\boldsymbol{p}'$. Here the condition $|\boldsymbol{q}| \ll |\boldsymbol{p}|, |\boldsymbol{p}|'$ applies, since long range Coulomb collisions are 'glancing'.

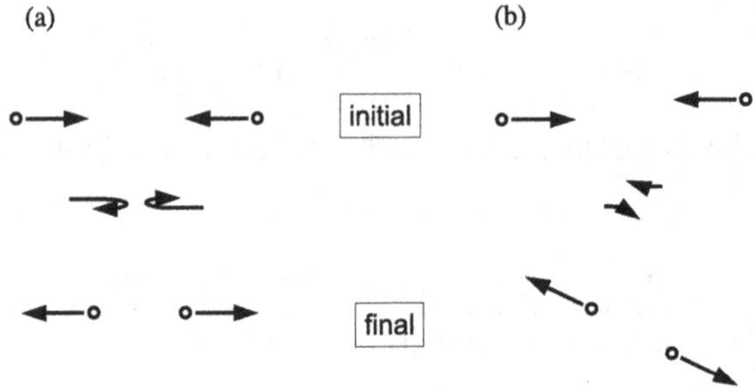

Fig. 2.10. Like-particle collisions in (a) 1D and (b) 3D.

2.2.3.2 *Offset on Landau–Rosenbluth theory*

Several features of $J(v)$ are readily apparent. First, just as in the case of the Landau theory, the current $J(v)$ can be written as a sum of electron–electron and electron–ion scattering terms, i.e.

$$J(v) = -\left[D_{\mathrm{e,e}}(v)\frac{\partial \langle f \rangle}{\partial v} + F_{\mathrm{e,e}}(v) + F_{\mathrm{e,i}}(v)\right]. \tag{2.37}$$

Here $D_{\mathrm{e,e}}(v)$ refers to the diffusion (in velocity) of electrons by fluctuations excited by electron discreteness emission, $F_{\mathrm{e,e}}(v)$ is the dynamical friction on electrons due to fluctuations generated by discreteness, while $F_{\mathrm{e,i}}$ is the electron–ion friction produced by the coupling of emission (again due to electron discreteness) to dissipative ion absorption. Interestingly, in 1D,

$$-D_{\mathrm{e,e}}(v)\frac{\partial \langle f \rangle}{\partial v} + F_{\mathrm{e,e}}(v) \sim \delta(u-v)\{-F(u)\langle f \rangle' + F'\langle f \rangle\} = 0,$$

since $F = \langle f \rangle$ for single species interaction. Thus, we see that *electron–electron friction exactly cancels electron diffusion in 1D*. In this case,

$$J(v) \sim \delta(u-v)\mathrm{Im}\,\epsilon_{\mathrm{i}}(k,\omega)\langle f_e(v)\rangle,$$

so that electron relaxation is determined *solely* by electron–ion friction. This result is easily understood from consideration of the analogy between same-species interaction in a stable, 1D plasma and like-particle collisions in 1D (Figure 2.10). On account of conservation of energy and momentum, it is trivial to show that such collisions leave final state = initial state, so no entropy production is possible and no relaxation can occur. This fact is manifested in the B–L theory by the cancellation between electron–electron terms – since the only way to produce finite momentum transfer in 1D is via inter-species collisions, the only term that survives in $J(v)$ is $F_{\mathrm{e,i}}(v)$. Note that this result is not a purely academic consideration, since a strong magnetic field $\boldsymbol{B}_0$ often imposes a 1D structure on the wave–particle resonance interaction in more complicated problems.

Table 2.2. *Comparison of Landau and Balescu–Lenard relaxation theory*

	Laudau theory	B–L theory
Physical scenario	'Test' particle scattered by distribution of 'field' particles	Test particle scattered by distribution of fluctuations with $v_{\mathrm{ph}} = \omega/k$, produced via discreteness
Scatterer distribution	$f(p')$ Field particles distribution	$F(u),\ u = \omega/k$ Fluctuation phase velocity distribution
Correlation	Uncorrelated particles as assumed molecular chaos $\langle f(1,2)\rangle = \langle f(1)\rangle \langle f(2)\rangle$	Discrete uncorrelated test particles $\left\langle \tilde{f}\tilde{f}\right\rangle = (\langle f\rangle/n)\delta(x\_)\delta(v\_)$
Screening	None Coulomb $\ln\Lambda$ factor put in 'by intuition'	$1/\lvert\epsilon(k,\omega)\rvert^2$
Scattering strength	$\lvert\boldsymbol{q}\rvert \ll \lvert\boldsymbol{p}\rvert$ Weak deflection	Linear response and unperturbed orbits
Interaction selection rule	$W(\boldsymbol{p},\boldsymbol{p}',\boldsymbol{q}) = \delta(p-p')$ in 1D, 1 species	$\delta(u-v)$ in 1D $\delta(\boldsymbol{k}\cdot(\boldsymbol{v}-\boldsymbol{v}'))$ in 3D

A detailed comparison and contrast between the Landau theory of collisions and the B–L theory of near-equilibrium relaxation is presented in Table 2.2.

2.2.3.3 *Resistivity (relaxation in one-dimensional system)*

Having derived the expression for $J(v)$, it can then be used to calculate transport coefficients and to macroscopically characterize relaxation. As an example, we consider the effective resistivity associated with the current driven system of Figure 2.9. To construct an effective Ohm's law for this system, we simply write

$$\frac{\partial\langle f\rangle}{\partial t} + \frac{q}{m}E_0\frac{\partial\langle f\rangle}{\partial v} = -\frac{\partial J(v)}{\partial v}, \tag{2.38a}$$

and then multiply by $n_0 q v$ and integrate to obtain, in the stationary limit,

$$\begin{aligned} E_0 &= -\frac{4\pi n_0 q}{\omega_{\mathrm{p}}^2}\int \mathrm{d}v\, J(v) \\ &= 4\pi n_0 |q| \sum_{k,\omega} \frac{\omega_{\mathrm{p}}^2}{(k^2)^2}\frac{(2\pi/|k|)}{n_0 k v_{\mathrm{T}}}\left(\frac{\mathrm{Im}\,\epsilon_i(k,\omega)}{|\epsilon(k,\omega)|^2}\right)\left\langle f_e\left(\frac{\omega}{k}\right)\right\rangle \\ &\equiv \eta_{\mathrm{eff}} J_0. \end{aligned} \tag{2.38b}$$

Not surprisingly, the response of $\langle f_e \rangle$ to E_0 cannot unambiguously be written as a simple, constant effective resistivity, since the resonance factor $\delta(\omega - kv)$ and the k, ω dependence of the TPM fluctuation spectrum conflate the field particle distribution function with the spectral structure. However, the necessary dependence of the effective resistivity on electron–ion interaction *is* readily apparent from the factor of $\operatorname{Im} \epsilon_i(k, \omega)$. In practice, a non-trivial effect here requires a finite, but not excessively strong, overlap of electron and ion distributions. Note also that collective enhancement of relaxation below the linear instability threshold *is* possible, should $\operatorname{Im} \epsilon(k, \omega)$ become small.

2.2.3.4 Relaxation in three-dimensional system

Having discussed the 1D case at some length, we now turn to relaxation in 3D (Lifshitz and Pitaevskii, 1981). The principal effect of three dimensionality is to relax the tight link between particle velocity $\boldsymbol{v}$ and fluctuation phase velocity $(\omega/|k|)\hat{k}$. Alternatively put, conservation constraints on like-particle collisions in 1D force the final state = initial state, but in 3D, glancing collisions which conserve energy and the magnitude of momentum $|\boldsymbol{p}|$, but change the particle's directions, are possible. The contrast between 1D and 3D is illustrated in Figure 2.10. In 3D, the discreteness correlation function is

$$C(k, \omega) = \langle \tilde{n}\tilde{n} \rangle_{k,\omega} = \int \mathrm{d}^3 v \frac{2\pi}{n_0} \delta(\omega - \boldsymbol{k} \cdot \boldsymbol{v}) \langle f \rangle . \tag{2.39}$$

So the B–L current $\boldsymbol{J}(\boldsymbol{v})$ for like-particle interactions becomes

$$\begin{aligned} \boldsymbol{J}(\boldsymbol{v}) = -\sum_{\mathbf{k},\omega} \left(\frac{\omega_{\mathrm{p}}^2}{k^2} \right)^2 \frac{2\pi^2 \delta(\omega - \boldsymbol{k} \cdot \boldsymbol{v})}{v_{\mathrm{T}} n_0 |\epsilon(k, \omega)|^2} \\ \times \boldsymbol{k} \left\{ \int \mathrm{d}\boldsymbol{v}' \, \delta(\omega - \boldsymbol{k} \cdot \boldsymbol{v}') \langle f(\boldsymbol{v}') \rangle \boldsymbol{k} \cdot \frac{\partial \langle f \rangle}{\partial \boldsymbol{v}} \right. \\ \left. - \int \mathrm{d}\boldsymbol{v}' \, \delta(\omega - \boldsymbol{k} \cdot \boldsymbol{v}') \boldsymbol{k} \cdot \frac{\partial \langle f \rangle}{\partial \boldsymbol{v}'} \langle f(\boldsymbol{v}) \rangle \right\} . \end{aligned} \tag{2.40a}$$

Note that the product of delta functions can be re-written as

$$\delta(\omega - \boldsymbol{k} \cdot \boldsymbol{v}) \delta(\omega - \boldsymbol{k} \cdot \boldsymbol{v}') = \delta(\omega - \boldsymbol{k} \cdot \boldsymbol{v}) \delta(\boldsymbol{k} \cdot \boldsymbol{v} - \boldsymbol{k} \cdot \boldsymbol{v}').$$

We thus obtain an alternate form for $\boldsymbol{J}(\boldsymbol{v})$, which is

$$J(v) = -\sum_{\mathbf{k},\omega}\left(\frac{\omega_p^2}{k^2}\right)^2 \frac{2\pi^2\delta(\omega - \boldsymbol{k}\cdot\boldsymbol{v})}{v_T n_0|\epsilon(k,\omega)|^2}$$
$$\times\left\{\boldsymbol{k}\int d\boldsymbol{v}'\,\delta(\boldsymbol{k}\cdot\boldsymbol{v} - \boldsymbol{k}\cdot\boldsymbol{v}')\left[\langle f(\boldsymbol{v}')\rangle \boldsymbol{k}\cdot\frac{\partial\langle f\rangle}{\partial\boldsymbol{v}} - \boldsymbol{k}\cdot\frac{\partial\langle f\rangle}{\partial\boldsymbol{v}'}\langle f(\boldsymbol{v})\rangle\right]\right\}. \tag{2.40b}$$

This form illustrates an essential aspect of 3D, which is that *only the parallel (to* $\boldsymbol{k}$*) components of test and field particle velocities* $\boldsymbol{v}$ and $\boldsymbol{v}'$ *need be equal for interaction to occur.* This is in distinct contrast to the case of 1D, where *identity*, i.e. $v = v' = u$, is required for interaction. Thus, relaxation by like-particle interaction is possible, and calculations of transport coefficients are possible, following the usual procedures of the Landau theory.

2.2.3.5 *Dynamic screening*

We now come to our final topic in B–L theory, which is *dynamic screening*. It is instructive and enlightening to develop this topic from a direction slightly different from that taken by our discussion up till now. In particular, we will proceed from the Landau theory, but will calculate momentum transfer *including* screening effects and thereby arrive at a B–L equation.

2.2.3.6 *Relaxation in Landau model*

Starting from Eq.(2.36), the Landau theory expression for the collision-induced current (in velocity) may be written as:

$$J_\alpha(\boldsymbol{p}) = \sum_{\text{species}}\int d^3p'\left[f(\boldsymbol{p})\frac{\partial f(\boldsymbol{p}')}{\partial p'_\beta} - f(\boldsymbol{p}')\frac{\partial f(\boldsymbol{p})}{\partial p_\beta}\right]B_{\alpha,\beta}, \tag{2.41a}$$

where

$$B_{\alpha,\beta} = \frac{1}{2}\int d\sigma\; q_\alpha q_\beta|\boldsymbol{v} - \boldsymbol{v}'|. \tag{2.41b}$$

The notation here is standard: $d\sigma$ is the differential cross-section and $\boldsymbol{q}$ is the momentum transfer in the collision. We will calculate $B_{\alpha,\beta}$ directly, using the same physics assumptions as in the TPM. A background or 'field' particle with velocity $\boldsymbol{v}'$, and charge e' produces a potential field

$$\phi_{k,\omega} = \frac{4\pi e'}{k^2\epsilon(k,\omega)}2\pi\delta(\omega - \boldsymbol{k}\cdot\boldsymbol{v}'), \tag{2.42a}$$

so converting the time transform gives

$$\phi_{\boldsymbol{k}}(t) = \frac{4\pi e'}{k^2\epsilon(\boldsymbol{k},\boldsymbol{k}\cdot\boldsymbol{v}')}e^{-i\boldsymbol{k}\cdot\boldsymbol{v}'t}. \tag{2.42b}$$

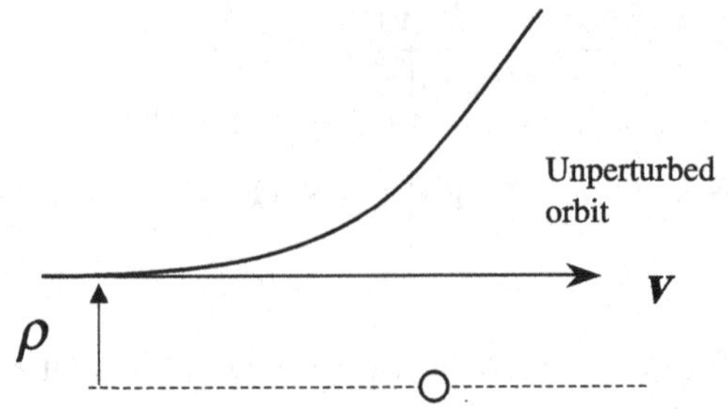

Fig. 2.11. Deflection orbit and unperturbed orbit.

From this, it is straightforward to calculate the net deflection or momentum transfer $\boldsymbol{q}$ by calculating the impulse delivered to a test particle with charge e moving along an unperturbed trajectory of velocity $\boldsymbol{v}$. This impulse is:

$$\boldsymbol{q} = -\int_{\boldsymbol{r}=\boldsymbol{\rho}+\boldsymbol{v}t} \frac{\partial V}{\partial \boldsymbol{r}} \mathrm{d}t, \tag{2.43a}$$

where the potential energy V is just

$$\begin{aligned} V &= e\phi \\ &= 4\pi e e' \int \mathrm{d}^3 k \, \frac{\mathrm{e}^{i\boldsymbol{k}\cdot\boldsymbol{r}} \mathrm{e}^{-i\boldsymbol{k}\cdot\boldsymbol{v}'t}}{k^2 \epsilon(\boldsymbol{k}, \boldsymbol{k}\cdot\boldsymbol{v}')}. \end{aligned} \tag{2.43b}$$

Here $\boldsymbol{\rho}$ is the impact parameter for the collision, a representation of which is sketched in Figure 2.11. A short calculation then gives the net momentum transfer $\boldsymbol{q}$,

$$\begin{aligned} \boldsymbol{q} &= 4\pi e e' \int \frac{\mathrm{d}^3 k}{(2\pi^3)} \frac{-i\boldsymbol{k}\mathrm{e}^{i\boldsymbol{k}\cdot\boldsymbol{\rho}} \, 2\pi\delta(\boldsymbol{k}\cdot(\boldsymbol{v}-\boldsymbol{v}'))}{k^2 \epsilon(\boldsymbol{k}, \boldsymbol{k}\cdot\boldsymbol{v}')} \\ &= 4\pi e e' \int \frac{\mathrm{d}^2 k_\perp}{(2\pi^3)} \frac{-i\boldsymbol{k}_\perp \mathrm{e}^{i\boldsymbol{k}_\perp\cdot\boldsymbol{\rho}}}{k^2 \epsilon(\boldsymbol{k}, \boldsymbol{k}\cdot\boldsymbol{v}')|\boldsymbol{v}-\boldsymbol{v}'|}. \end{aligned} \tag{2.44}$$

To obtain Eq.(2.44), we use

$$\delta(\boldsymbol{k}\cdot(\boldsymbol{v}-\boldsymbol{v}')) = \frac{\delta(k_\parallel)}{|\boldsymbol{v}-\boldsymbol{v}'|},$$

and the directions $\parallel$ and $\perp$ are defined relative to the direction of $\boldsymbol{v}-\boldsymbol{v}'$. Since $J \sim \rho^2$, we may write $B_{\alpha,\beta}$ as,

$$B_{\alpha,\beta} = \int \mathrm{d}^2\rho \, q_\alpha q_\beta |\boldsymbol{v}-\boldsymbol{v}'|. \tag{2.45}$$

Noting that the $\mathrm{d}^2\rho$ integration just produces a factor of $(2\pi)^2\delta\left(\boldsymbol{k}_\perp + \boldsymbol{k}'_\perp\right)$, we can then immediately perform one of the $\int \mathrm{d}^2\boldsymbol{k}_\perp$ integrals in $B_{\alpha,\beta}$ to get,

$$B_{\alpha,\beta} = 2e^2e'^2 \int \mathrm{d}^2\boldsymbol{k}_\perp \frac{k_{\perp\alpha}k_{\perp\beta}}{|k_\perp^2\epsilon(\boldsymbol{k}, \boldsymbol{k}\cdot\boldsymbol{v}')|^2|\boldsymbol{v}-\boldsymbol{v}'|}. \tag{2.46}$$

It is easy to see that Eq.(2.46) for $B_{\alpha,\beta}$ (along with Eq.(2.41a)) is entirely equivalent to the B–L theory for $\boldsymbol{J}(\boldsymbol{v})$. In particular, note the presence of the dynamic screening factor $\epsilon(\boldsymbol{k}, \boldsymbol{k}\cdot\boldsymbol{v}')$. If screening is neglected, $\epsilon \to 1$ and

$$B_{\alpha,\beta} \sim \int \mathrm{d}^2k_\perp \frac{k_\perp^2}{|\epsilon|^2k_\perp^4} \sim \int \mathrm{d}k_\perp/k_\perp \sim \ln(k_{\perp\max}/k_{\perp\min}),$$

which is the familiar Coulomb logarithm from the Landau theory. Note that if $\mathbf{k}, \omega \to 0$,

$$k_\perp^2\epsilon \sim k_\perp^2 + 1/\lambda_\mathrm{D}^2,$$

so that Debye screening eliminates the long-range divergence (associated with $k_{\perp\min}$) without the need for an ad hoc factor. To make the final step toward recovering the explicit B–L result, one can 'undo' the $\mathrm{d}k_\parallel$ integration and the frequency integration to find,

$$B_{\alpha,\beta} = 2(ee')^2 \int_{-\infty}^{\infty} \mathrm{d}\omega \int_{k<k_\mathrm{max}} \mathrm{d}^3k\, \delta(\omega - \boldsymbol{k}\cdot\boldsymbol{v})\delta(\omega - \boldsymbol{k}\cdot\boldsymbol{v}') \frac{k_\alpha k_\beta}{k^4|\epsilon(\boldsymbol{k}, \omega)|^2}. \tag{2.47}$$

Here k_max is set by the distance of closest approach, i.e.

$$k_\mathrm{max} \sim \frac{\mu v_\mathrm{rel}^2}{2ee'}.$$

Substituting Eq.(2.47) for $B_{\alpha,\beta}$ into Eq.(2.41a) recovers Eq.(2.40a). This short digression convincingly demonstrates the equivalence of the B–L theory to the Landau theory with dynamic screening.

2.2.3.7 *Collective mode*

We now explore the enhancement of relaxation by weakly damped collective modes. Consider a stable, two species plasma with electron and ion distribution functions, as shown in Figure 2.9. This plasma has no free energy (i.e. no net current), but is not necessarily a maximum entropy state, if $T_\mathrm{e} \neq T_\mathrm{i}$. Moreover, the plasma supports two types of collective modes, namely

(i) electron plasma waves, with $v_\mathrm{Te} < \omega/k$.
(ii) ion acoustic waves, with $v_\mathrm{Ti} < \omega/k < v_\mathrm{Te}$,

where v_{Te} and v_{Ti} are the electron and ion thermal speeds, respectively. Electron plasma waves are resonant on the tail of $\langle f_e \rangle$, where there are few particles. Hence plasma waves are unlikely to influence relaxation in a significant way. On the other hand, ion acoustic waves are resonant in the *bulk* of the electron distribution. Moreover, if $T_{\text{e}} \gg T_{\text{i}}$, it is easy to identify a band of electron velocities with significant population levels f but for which ion Landau damping is negligible. Waves resonant there will be weakly damped, and so may substantially enhance relaxation of $\langle f_e(v) \rangle$. It is this phenomenon of collectively enhanced relaxation that we seek to explore quantitatively.

To explore collective enhancement of relaxation, we process from Eq.(2.47), make a pole expansion around the ion acoustic wave resonance and note for ion acoustic wave, $\omega < \boldsymbol{k} \cdot \boldsymbol{v}$ (for electrons), so

$$B_{\alpha,\beta} = 2\pi q^4 \int_{-\infty}^{\infty} \mathrm{d}\omega \int \mathrm{d}^3 k \, \delta(\boldsymbol{k} \cdot \boldsymbol{v}) \delta(\boldsymbol{k} \cdot \boldsymbol{v}') \frac{\delta \epsilon_r(\boldsymbol{k}, \omega_{\boldsymbol{k}})}{|\mathrm{Im}\, \epsilon(\boldsymbol{k}, \omega)|}. \tag{2.48}$$

Here $\epsilon_{\text{r}}(\boldsymbol{k}, \omega_{\boldsymbol{k}})$=0 for wave resonance, and e=e'=q, as scattered and field particles are all electrons. The term $\mathrm{Im}\, \epsilon(\mathbf{k}, \omega)$ refers to the collective mode dissipation rate. Now, changing variables according to

$$\begin{aligned} R &= \boldsymbol{k} \cdot \hat{n}, \\ k_1 &= \boldsymbol{k} \cdot \boldsymbol{v}, \quad k_2 = \boldsymbol{k} \cdot \boldsymbol{v}' \\ \hat{n} &= \boldsymbol{v} \times \boldsymbol{v}' / |\boldsymbol{v} \times \boldsymbol{v}'|. \end{aligned}$$

We have

$$\mathrm{d}^3 k = \mathrm{d}R \mathrm{d}k_1 \mathrm{d}k_2 / |\boldsymbol{v} \times \boldsymbol{v}'|,$$

so the k_1, k_2 integrals in Eq.(2.48) may be immediately performed, leaving

$$B_{\alpha,\beta} = \frac{2\pi q^4 n_\alpha n_\beta}{|\boldsymbol{v} \times \boldsymbol{v}'|} 2 \int_{R>0} \mathrm{d}k \int_{-\infty}^{\infty} \mathrm{d}\omega \frac{\delta(\epsilon_{\text{r}}(\boldsymbol{k}, \omega))}{R^2 |\mathrm{Im}\, \epsilon|}. \tag{2.49}$$

We remind the reader that this is the piece of $B_{\alpha,\beta}$ associated with field particle speeds or fluctuation phase speeds $v' \sim \omega/k \ll v_{\text{Te}}$ for which the collective enhancement is negligible and *total* $J(\boldsymbol{v})$ is, of course, the *sum* of both these contributions. Now, the dielectric function for ion acoustic waves is,

$$\mathrm{Re}\, \epsilon(k, \omega) = 1 - \frac{\omega_{\text{pi}}^2}{\omega^2} + \frac{1}{k^2 \lambda_{\text{D}}^2}. \tag{2.50a}$$

$$\mathrm{Im}\, \epsilon(k, \omega) = \sqrt{\frac{\pi}{2}} \frac{\omega}{k^3} \left(\frac{1}{\lambda_{\text{De}}^2 v_{\text{Te}}} + \frac{1}{\lambda_{\text{Di}}^2 v_{\text{Ti}}} \mathrm{e}^{-\omega^2 / 2k^2 v_{\text{Ti}}^2} \right). \tag{2.50b}$$

Here λ_{De} and λ_{Di} are the electron and ion Debye lengths, respectively. Anticipating the result that

$$\omega^2 = \frac{k^2 c_{\mathrm{s}}^2}{1 + k^2 \lambda_{\mathrm{De}}^2}$$

for ion acoustic waves, Eqs.(2.50a, 2.50b) together suggest that the strongest collective enhancement will come from short wavelength (i.e. $k^2\lambda_{\mathrm{De}}^2 \gtrapprox 1$), because $\mathrm{Im}\,\epsilon(k,\omega)$ is smaller for these scales, since

$$\mathrm{Im}\,\epsilon(k,\omega) \sim \frac{1}{k^2\lambda_{\mathrm{De}}^2}\frac{\omega}{k v_{\mathrm{Te}}}.$$

For such short wavelengths, then

$$\begin{aligned}\delta(\epsilon_{\mathrm{r}}) &\cong \delta\left(1 - \omega_{\mathrm{pi}}^2/\omega^2\right)\\ &= \frac{1}{2}\omega_{\mathrm{pi}}\left[\delta(\omega - \omega_{\mathrm{pi}}) + \delta(\omega + \omega_{\mathrm{pi}})\right].\end{aligned}$$

Evaluating $B_{\alpha,\beta}$, as given by Eqs.(2.48) and (2.49), in this limit then finally gives,

$$\begin{aligned}B_{\alpha,\beta} &= \left(\frac{4\pi q^2 \omega_{\mathrm{pi}} n_\alpha n_\beta}{|\boldsymbol{v}\times\boldsymbol{v}'|}\right)\int \frac{\mathrm{d}k}{k^2|\mathrm{Im}\,\epsilon(k,\omega_{\mathrm{pi}})|}\\ &= \frac{2\sqrt{2\pi}q^4 v_{\mathrm{T_e}}\lambda_{\mathrm{De}}^2}{|\boldsymbol{v}\times\boldsymbol{v}'|\lambda_{\mathrm{Di}}^2} n_\alpha n_\beta \int \mathrm{d}\xi\left[1 + \exp\left(-\frac{1}{2\xi} + \frac{L}{2}\right)\right]^{-1},\end{aligned} \tag{2.51a}$$

where:

$$L = \ln\left[\left(\frac{T_{\mathrm{e}}}{T_{\mathrm{i}}}\right)^2 \frac{m_{\mathrm{i}}}{m_{\mathrm{e}}}\right] \tag{2.51b}$$

and $\xi = k^2\lambda_{\mathrm{De}}^2$. Equation (2.51a) quite clearly illustrates that maximal relaxation occurs for *minimal* $\mathrm{Im}\,\epsilon(\xi, L)$, that is when $\exp[-1/2\xi + L/2] \ll 1$. That is, the collective enhancement of discreteness-induced scattering is determined by $\mathrm{Im}\,\epsilon$ for the least damped mode. This occurs when $\xi \lessapprox 1/L$, so that the dominant contribution to $B_{\alpha,\beta}$ comes from scales for which

$$k^2 = (\boldsymbol{k}\cdot\hat{\boldsymbol{n}})^2 < 1/\left(\lambda_{\mathrm{De}}^2 L\right).$$

Note that depending on the values of L and the Coulomb logarithm ($\ln\Lambda$, which appears in the standard Coulombic scattering contribution to $B_{\alpha,\beta}$ from $v' \sim v_{Te}$), the collectively enhanced $B_{\alpha,\beta}$ due to low velocity field particles ($v' \ll v_{Te}$) may even exceed its familiar counterpart Coulomb scattering. Clearly, this is possible only when $T_e/T_i \gg 1$, yet not so large as to violate the tenets of the TPM and B–L theories.

2.2.4 Test particle model: looking back and looking ahead

In this section of the introductory chapter, we have presented the test particle model for fluctuations and transport near thermal equilibrium. As we mentioned at the beginning of the chapter, the TPM is the most basic and most successful fluctuation theory for weakly collisional plasmas. So, despite its limitation to stable, quiescent plasmas, the TPM has served as a basic paradigm for treatments of the far more difficult problems of *non-equilibrium* plasma kinetics, such as plasma turbulence, turbulent transport, self-organization etc. Given this state of affairs, we deem it instructive to review the essential elements of the TPM and place the subsequent chapters of this book in the context of the TPM and its elements. In this way, we hope to provide the reader with a framework from which to approach the complex and sometimes bewildering subject of the physical kinetics of non-equilibrium plasmas. The discussion that follows is summarized in Table 2.3. We discuss and compare the test particle model to its non-equilibrium descendents in terms of both physics concepts and theoretical constructs.

Regarding *physics concepts*, the TPM is fundamentally a "near-equilibrium" theory, which presumes a balance of emission and absorption at each k. In a turbulent plasma, non-linear interaction produces spectral transfer and a spectral cascade, which can de-localize the location of absorption from the region of emission in $\boldsymbol{k}, \omega$ space. A spectral cascade transfers turbulence energy from one region in $\boldsymbol{k}$ (i.e. emission) to another (i.e. damping). There two cases are contrasted in Figure 2.1.

A second key concept in the TPM is that emission occurs *only* via Cerenkov radiation from discrete test particles. Thus, since the only source for collective modes is discreteness, we always have

$$\nabla \cdot \epsilon E = 4\pi q \delta(\boldsymbol{x} - \boldsymbol{x}(t)),$$

so

$$\left\langle \phi^2 \right\rangle_{\mathbf{k},\omega} = \frac{\left\langle \tilde{n}^2 \right\rangle}{|\epsilon(\mathbf{k}, \omega)|^2}.$$

In contrast, for non-equilibrium plasmas, nonlinear coupling produces incoherent emission so the energy in mode $\boldsymbol{k}$ evolves according to,

Table 2.3. *Test particle model and its non-equilibrium descendents: physical concepts and theoretical constructs*

Test particle model	Non-equilibrium descendent
Physics concepts	
Emission versus absorption balance per $\boldsymbol{k}$	Spectral cascade, transfer, inertial range (Chapter 5, 6)
Discreteness noise	Incoherent mode-coupling (Chapter 5, 6), granulation emission (Chapter 8)
Relaxation by screened collisions	Collective instability-driven relaxation, quasi-linear theory, granulation interaction (Chapter 3, 8)
Theoretical constructs	
Linear response unperturbed orbit	Turbulence response, turbulent diffusion, resonance broadening (Chapter 4, 6)
Damped mode response	Nonlinear dielectric, wave–wave interaction, wave kinetics (Chapter 5, 6)
Mean-field theory	Mean-field theory without and with granulations (Chapter 3, 8)
Discreteness-driven stationary spectrum	Wave kinetics, renormalized mode coupling, disparate scale interaction (Chapter 5 – 7)
Balescu–Lenard, screened Landau equations	Quasi-linear theory, Granulation relaxation theory (Chapter 3, 8)

$$\frac{\partial}{\partial t}\left\langle E^2\right\rangle_{\boldsymbol{k}} + \left(\sum_{\boldsymbol{k}'} C(\boldsymbol{k},\boldsymbol{k}')\left\langle E^2\right\rangle_{\boldsymbol{k}'} \tau_{c_{\boldsymbol{k},\boldsymbol{k}'}}\right)\left\langle E^2\right\rangle_{\boldsymbol{k}} + \gamma_{\mathrm{d}\boldsymbol{k}}\left\langle E^2\right\rangle_{\boldsymbol{k}}$$
$$= \sum_{\substack{\boldsymbol{p},\boldsymbol{q}\\ \boldsymbol{p}+\boldsymbol{q}=\boldsymbol{k}}} C(\boldsymbol{p},\boldsymbol{q})\tau_{c\,\boldsymbol{p},\boldsymbol{q}}\left\langle E^2\right\rangle_{\boldsymbol{p}}\left\langle E^2\right\rangle_{\boldsymbol{q}} + S_{\mathrm{D}\boldsymbol{k}}\left\langle E^2\right\rangle_{\boldsymbol{k}}, \tag{2.52}$$

where $S_{\mathrm{D}\boldsymbol{k}}$ is the discreteness source and $\gamma_{\mathrm{d}\boldsymbol{k}}$ is the linear damping for the mode $\boldsymbol{k}$ (Kadomtsev, 1965). For sufficient fluctuation levels, the nonlinear noise term (i.e. the first on the right-hand side) will clearly dominate over discreteness. A similar comment can be made in the context of the left-hand side of Eq.(2.52), written above. Nonlinear damping will similarly eclipse linear response damping for sufficiently large fluctuation levels.

A third physics concept is concerned with the mechanism of relaxation and transport. In the TPM, these occur only via screened collisions. Collective effects associated with weakly damped modes may enhance relaxation but do not fundamentally change this picture. In a non-equilibrium plasma, collective modes can drive relaxation of the unstable $\langle f \rangle$, and nonlinear transfer can couple the relaxation process, thus enhancing its rate.

In the realm of *theoretical constructs* and methods, *both* the test particle model and its non-equilibrium counterparts are fundamentally mean-field type approaches. However, they differ dramatically with respect to particle and model responses, nature of the wave spectrum and in how relaxation is calculated. The TPM assumes linear response theory is valid, so particle response functions exhibit only 'bare' Landau resonances. In contrast, scattering by strong electric field fluctuations will *broaden* the Landau resonance and *renormalize* the Landau propagator, so that,

$$\begin{aligned} R_{k,\omega} &\sim \mathrm{e}^{ikx} \int_0^\infty \mathrm{e}^{i\omega\tau} \mathrm{e}^{-ikx(-\tau)} \mathrm{d}\tau \\ &\sim \int_0^\infty \mathrm{e}^{i(\omega - kv)\tau} \mathrm{d}\tau = i/(\omega - kv) \end{aligned} \tag{2.53a}$$

becomes,

$$\begin{aligned} R_{k,\omega} &\sim \int_0^\infty \mathrm{e}^{i\omega\tau}\, \mathrm{e}^{-ikx_0(-\tau)} \left\langle \mathrm{e}^{-ik\delta x(-\tau)} \right\rangle \mathrm{d}\tau \\ &\sim \int_0^\infty \mathrm{e}^{i(\omega - kv)\tau - \frac{k^2 D}{3}\tau^3} \mathrm{d}\tau \sim i/(\omega - kv + i/\tau_c), \end{aligned} \tag{2.53b}$$

where $1/\tau_c = (k^2 D/3)^{1/3}$. This is equivalent to the renormalization

$$\left[-i(\omega - kv) \right]^{-1} \rightarrow \left[-i(\omega - kv) - \frac{\partial}{\partial v} D \frac{\partial}{\partial v} \right]^{-1}. \tag{2.53c}$$

Here $D = D\left[\langle E^2 \rangle\right]$ is a functional of the turbulence spectrum. In a similar way to that sketched in Eq.(2.53), collective responses are renormalized and broadened by nonlinear wave interaction. Moreover, in the non-equilibrium case, a separate wave kinetic equation for $N(k, \boldsymbol{x}, t)$, the wave population density, is required to evolve the wave population in the presence of sources, non-linear interaction and refraction, etc. by disparate scales. This wave kinetic equation is usually written in the form,

$$\frac{\partial N}{\partial t} + (\boldsymbol{v}_{\mathrm{g}} + \boldsymbol{v}) \cdot \nabla N - \frac{\partial}{\partial \boldsymbol{x}}(\omega + \boldsymbol{k} \cdot \boldsymbol{v}) \cdot \frac{\partial N}{\partial \boldsymbol{k}} = S_{\boldsymbol{k}} N + C_{\boldsymbol{k}}(N). \tag{2.54}$$

Since in practical problems, the mean field or coarse grained wave population density $\langle N \rangle$ is of primary interest, a similar arsenal of quasi-linear type closure techniques has been developed for the purpose of extracting $\langle N \rangle$ from the wave kinetic equation. We conclude by noting that this discussion, which began with the TPM, comes full circle when one considers the effect of nonlinear mode coupling on processes of relaxation and transport. In particular, mode localized coupling produces phase space density vortexes or eddies in the phase space fluid. These phase space eddies are called *granulations*, and resemble a macroparticle (Lynden-Bell, 1967; Kadomtsev and Pogutse, 1970; Dupree, 1970; Dupree, 1972; Diamond *et al.*, 1982). Such granulations are associated with peaks in the phase space density correlation function. Since these granulations resemble macroparticles, it should not be too surprising that they drive relaxation via a mechanism similar to that of dressed test particles. Hence, the mean field equation for $\langle f \rangle$ in the presence of granulations has the *structure* of a Balescu–Lenard equation, although of course its components differ from those discussed in this chapter.

2.3 Turbulence: dimensional analysis and beyond – revisiting the theory of hydrodynamic turbulence

So, naturalists observe, a flea
Hath smaller fleas that on him prey,
And those have smaller yet to bite 'em,
And so proceed ad infinitum:
Thus every poet in his kind,
Is bit by him that comes behind.

(Jonathan Swift, from "On poetry: a Rhapsody")

We now turn to our second paradigm, namely Navier–Stokes turbulence, and the famous Kolmogorov cascade through the inertial range. This is *the* classic example of a system with dynamics controlled by a self-similar spectral flux. It constitutes the ideal complement to the TPM, in that it features the role of transfer, rather than emission and absorption. We also discuss related issues in particle dispersion, two-dimensional turbulence and turbulent pipe flows.

2.3.1 Key elements in Kolmogorov theory of cascade

2.3.1.1 Kolmogorov theory

Surely everyone has encountered the basic ideas of Kolmogorov's theory of high Reynolds number turbulence! (McComb, 1990; Frisch, 1995; Falkovich *et al.*, 2001; Yoshizawa *et al.*, 2003). Loosely put, it consists of empirically motivated assumptions of:

(1) spatial homogeneity – i.e. the turbulence is uniformly distributed in space;
(2) isotropy – i.e. the turbulence exhibits no preferred spatial orientation;

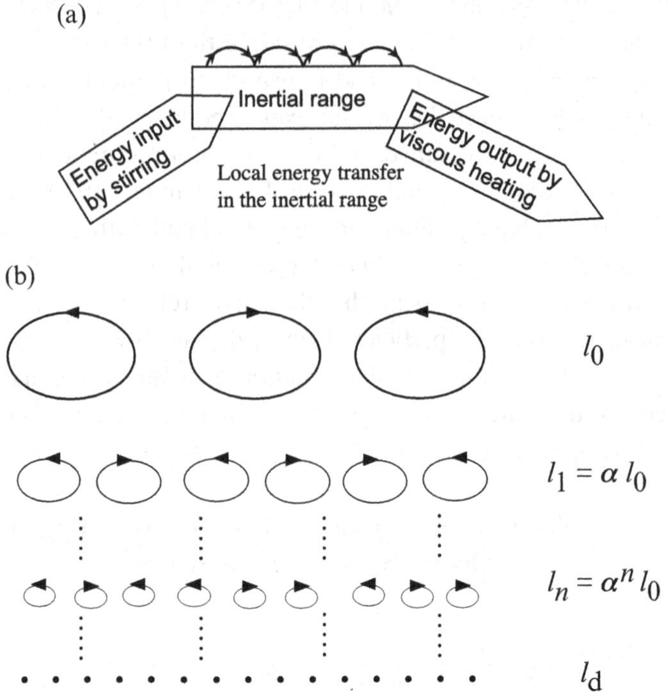

Fig. 2.12. Basic cartoon explanation of the Richardson–Kolmogorov cascade. Energy transfer in Fourier–space (a), and real space (b).

(3) self-similarity – i.e. all inertial range scales exhibit the same physics and are equivalent. Here "inertial range" refers to the range of scales ℓ smaller than the stirring scale ℓ_0 but larger than the dissipation scale ($\ell_d < \ell < \ell_0$);

(4) locality of interaction – i.e. the (dominant) nonlinear interactions in the inertial range are local in scale; that is, while large scales advect small scales, they cannot distort or destroy small scales, only sweep them around. Inertial range transfer occurs via like-scale straining, *only*.

Assumptions (1)–(4) and the basic idea of an inertial range cascade are summarized in Figure 2.12. Using assumptions (1)–(4), we can state that energy throughput must be constant for all inertial range scales, so,

$$\epsilon \sim v_0^3/\ell_0 \sim v(\ell)^3/\ell, \tag{2.55a}$$

and

$$v(\ell) \sim (\epsilon \ell)^{1/3}, \tag{2.55b}$$

$$E(k) \sim \epsilon^{2/3} k^{-5/3}, \tag{2.55c}$$

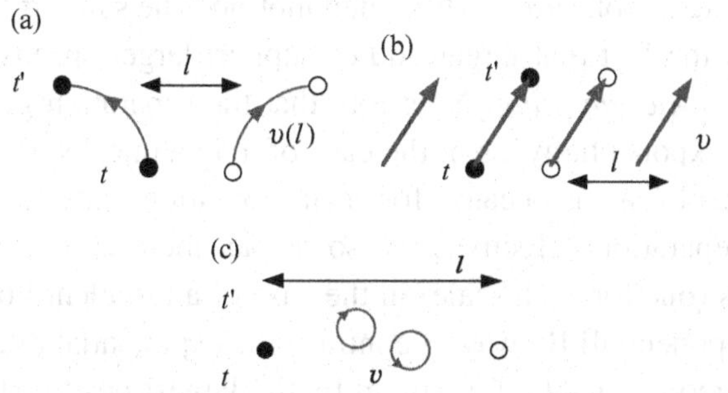

Fig. 2.13. Basic idea of the Richardson dispersion problem. The evolution of the separation of the two points (black and white dots) l follows the relation $\mathrm{d}l/\mathrm{d}t = v$ (a). If the advection field scale exceeds l, the particle pair swept together, so l is unchanged (b). If the advection field scale is less than l, there is no effect (except diffusion) on particle dispersion (c).

which are the familiar K41 results. The dissipation scale ℓ_{d} is obtained by balancing the eddy straining rate $\epsilon^{1/3}/\ell^{2/3}$ with the viscous dissipation rate ν/ℓ^2 to find the Kolmogorov microscale,

$$\ell_{\mathrm{d}} \sim \nu^{3/4}/\epsilon^{1/4}. \tag{2.56}$$

2.3.1.2 Richardson theory of particle separation

A related and important phenomenon, which may also be illuminated by scaling arguments, is how the distance between two test particles grows in time in a turbulent flow. This problem was first considered by Louis Fry Richardson, who was stimulated by observations of the rate at which pairs of weather balloons drifted apart from one another in the (turbulent) atmosphere (Richardson, 1926). Consistent with the assumption of locality of interaction in scale, Richardson postulated that the distance between two points in a turbulent flow increases at the speed set by the eddy velocity on scales corresponding (and comparable) to the distance of separation (Fig. 2.13). Thus, for distance ℓ,

$$\frac{\mathrm{d}\ell}{\mathrm{d}t} = v(\ell), \tag{2.57a}$$

so using the K41 results (2.55b) gives,

$$\ell(t) \sim \epsilon^{1/2} t^{3/2}; \tag{2.57b}$$

a result that Richardson found to be in good agreement with observations. Notice that the distance of separation grows *super-diffusively*, i.e. $\ell(t) \sim t^{3/2}$, and not

$\sim t^{1/2}$, as for the textbook case of Brownian motion. The super-diffusive character of $\ell(t)$ is due to the fact that larger eddies support larger speeds, so the separation process is *self-accelerating*. Note too, that the separation grows as a power of time, and not exponentially, as in the case of a dynamical system with positive Lyapunov exponent. This is because for each separation scale ℓ, there is a *unique* corresponding separation velocity $v(\ell)$, so in fact there is a *continuum* of Lyapunov exponents (one for each scale) in the case of a turbulent flow. Thus, $\ell(t)$ is algebraic, not exponential! By way of contrast, the exponential rate of particle pair separation in a smooth chaotic flow is set by the largest positive Lyapunov exponent. We also remark here that while intermittency corrections to the K41 theory, based upon the notion of a dissipative attractor with a fractal dimension less than three, have been extensively discussed in the literature, the effects of intermittency in the corresponding Richardson problem have received relatively little attention. This is unfortunate, since, though it may seem heretical to say so, the Richardson problem is, in many ways, more fundamental than the Kolmogorov problem, since unphysical effects due to sweeping by large scales are eliminated by definition in the Richardson problem. Moreover, the Richardson problem is of interest to calculating the rate of turbulent dispersion and the lifetime of particles or quasi-particles of turbulent fluid. An exception to the lack of advanced discussion of the Richardson problem is the excellent review article by Falkovich, Gawedski and Vergassola (2001).

2.3.1.3 Stretching and generation of enstrophy

Of course, 'truth in advertising' compels us to emphasize that the scaling arguments presented here contain no more physics than what was inserted ab initio. To understand the *physical mechanism* underpinning the Kolmogorov energy cascade, one must consider the dynamics of structures in the flow. As is well known, the key mechanism in 3D Navier–Stokes turbulence is *vortex tube stretching*, shown schematically in Figure 2.14. There, we see that alignment of strain $\nabla \boldsymbol{v}$ with vorticity $\boldsymbol{\omega}$ (i.e. $\boldsymbol{\omega} \cdot \nabla \boldsymbol{v} \neq 0$) *generates* small-scale vorticity, as dictated by angular momentum conservation in incompressible flows. The enstrophy (mean squared vorticity) thus diverges as,

$$\langle \omega^2 \rangle \sim \epsilon / \nu, \tag{2.58}$$

for $\nu \to 0$. This indicates that *enstrophy is produced in 3D turbulence*, and suggests that there may be a *finite time singularity* in the system, an issue to which we shall return later. By finite time singularity of enstrophy, we mean that the enstrophy diverges within a finite time (i.e. with a growth rate which is faster than exponential). In a related vein, we note that finiteness of ϵ as $\nu \to 0$ constitutes what is called an *anomaly* in quantum field theory. An anomaly occurs

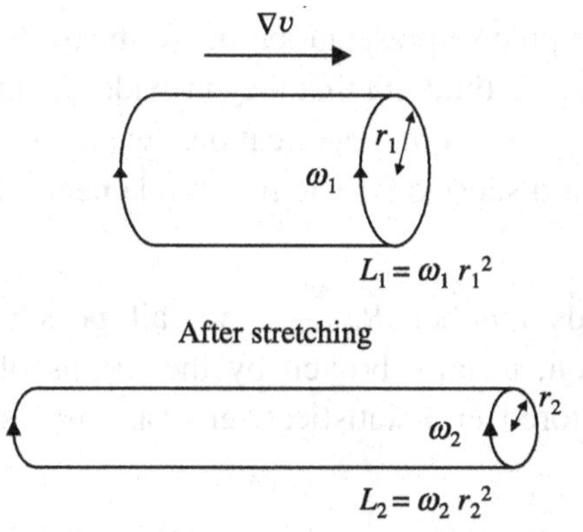

Fig. 2.14. The mechanism of enstrophy generation by vortex tube stretching. The vortex tube stretching vigorously produces small scale vorticity.

when symmetry breaking (in this case, breaking of time reversal symmetry by viscous dissipation) persists as the symmetry breaking term in the field equation asymptotes to zero. The scaling $\langle \omega^2 \rangle \sim 1/\nu$ is suggestive of an anomaly. So is the familiar simple argument using the Euler vorticity equation (for $\nu \to 0$),

$$\frac{\mathrm{d}\boldsymbol{\omega}}{\mathrm{d}t} = \boldsymbol{\omega} \cdot \nabla \boldsymbol{v}, \tag{2.59a}$$

$$\frac{\mathrm{d}}{\mathrm{d}t}\omega^2 \sim \omega^3. \tag{2.59b}$$

Of course, this "simple argument" is grossly over-simplified, and incorrect.[1] In two dimensions $\boldsymbol{\omega} \cdot \nabla \boldsymbol{v} = 0$, so enstrophy is conserved. As first shown by Kraichnan, this necessitates a *dual cascade*, in which enstrophy *forward cascades* to small scales, while energy *inverse cascades* to large scales. The mechanism by which the dual conservation of energy and enstrophy force a dual cascade in 2D turbulence is discussed further later in this chapter.

2.3.1.4 Fundamental hypothesis for K41 theory

As elegantly and concisely discussed by U. Frisch in his superb monograph "Turbulence – The Legacy of A.N. Kolmogorov" (Frisch, 1995), the K41 theory can be systematically developed from a few fundamental hypotheses or postulates. Upon proceeding, the cynical reader will no doubt conclude that the hypotheses (H1)–(H4) stated below are simply restatements of assumptions (1)–(4). While it is difficult to refute such a statement, we remark here that (H1)–(H4), *are* indeed

[1] In fact, a mathematical proof of finite time singularity of enstrophy remains an elusive goal, with an as-yet-unclaimed Clay prize of $1,000,000. (2007)

of value, both for their precise presentation of Kolmogorov's deep understanding and for the insights into his thinking that they provide. As these postulates involve concepts of great relevance to other applications, we revisit them here in preparation for our subsequent discussions. The first fundamental hypothesis of the K41 theory is:

(H1) As the Reynolds number $R_e \rightarrow \infty$, all possible symmetries of the Navier–Stokes equation, usually broken by the means of turbulence initiation or production, are restored in a statistical sense at small scales, and away from boundaries.

The reader should note that (H1) is a deceptively simple, and fundamentally quite profound hypothesis! The onset or production of turbulence nearly always involves symmetry breaking. Some examples are:

(i) shear flow turbulence: the initial Kelvin–Helmholtz instability results from breaking of translation and rotation symmetry.
(ii) turbulence in a pipe with a rough boundary: the driving pressure drop, the wall and roughenings break symmetry.
(iii) turbulence in a flushing toilet: the multiphase flow has finite chirality and is non-stationary.

Naively, one might expect the turbulent state to have some memory of this broken symmetry. Indeed, the essence of β-model and multi-fractal theories of intermittency is the persistence of some memory of the large, stirring scales into the smallest inertial range scales. Yet, the universal character of K41 turbulence follows directly from, and implies a restoration of, initially broken symmetry at small scales. Assumptions (i) and (ii) really follow from hypothesis (H1).

The second K41 hypothesis is:

(H2) Under the assumptions of (H1), the flow is self-similar at small scales and has a unique scaling exponent h, such that,

$$\boldsymbol{v}(\boldsymbol{r}, \lambda\ell) = \lambda^h \boldsymbol{v}(\boldsymbol{r}, \ell).$$

Here, $\boldsymbol{v}(\boldsymbol{r}, \ell)$ refers to the velocity wavelet field at position $\boldsymbol{r}$ and scale ℓ. Clearly, (H2) implies assumptions (3) and (4), concerning self-similarity and locality of interaction.

Hypotheses (H1) and (H2) pertain to flow structure and scaling properties. Two additional postulates pertain to dynamics. These are:

(H3) Given the assumptions of (H1) and (H2), turbulent flow has a finite, non-vanishing mean rate of dissipation per unit mass ϵ, as $\nu \rightarrow 0$,

and

(H4) In the limit of high but finite R_e, all small-scale statistical properties are uniquely and universally determined by ϵ and ℓ.

Hypothesis (H3) is tacitly equivalent to stating that an anomaly exists in K41 turbulence. Note that ϵ is independent of ν. However, notice also that ϵ, the "mean rate of dissipation per unit mass" is not related to physical, calculable quantities, and is left as a more-than-slightly ambiguous concept. Introduction of fluctuations (which relax the statement 'uniquely' in (H4) in the local dissipation rate (which in reality are usually associated with localized dissipative structures such as strong vortex tubes) and of a statistical distribution of dissipation, leads down the path to intermittency modelling, a topic which is beyond the scope of this book. The reader is referred to Frisch (1995), for an overview, and to seminal references such as Frisch *et al.* (1978), She and Leveque (1994), Falkovich *et al.* (2001), and others for an in depth discussion of intermittency modifications to the K41 theory. Finally, hypothesis (H4) relates all statistics to ϵ and ℓ, the only two possible relevant parameters, given (H1), (H4).

2.3.2 Two-dimensional fluid turbulence

In this subsection, we briefly summarize certain key features of the theory of two-dimensional (2D) fluid turbulence. Our attention will focus upon the dual cascades of energy and enstrophy in 2D turbulence, the dispersion of particle pairs (i.e., the Richardson problem), and on the emergence of long-lived coherent structures in turbulent 2D flow. Two-dimensional fluid dynamics has many features in common with those of magnetized plasmas, and so is of great interest to us (Hasegawa, 1985). The 2D fluid turbulence is a critically important paradigm for plasma turbulence. The literature of 2D turbulence theory and experiment is vast, so here we survey only the most basic and fundamental elements of this interesting story.

2.3.2.1 Forward and inverse cascade

As we have already discussed, the defining feature of 2D fluid dynamics is the absence of vortex tube stretching (i.e. $\boldsymbol{\omega} \cdot \nabla \boldsymbol{v} = 0$). Thus, vorticity is conserved locally, up to viscous dissipation, i.e.

$$\frac{\partial}{\partial t}\omega + \boldsymbol{v} \cdot \nabla \omega - \nu \nabla^2 \omega = 0, \tag{2.60a}$$

or, representing $\boldsymbol{v}$ using a stream function, $\boldsymbol{v} = \nabla\phi \times \hat{z}$ (where $\hat{z}$ is the coordinate in the direction of uniformity) and,

$$\frac{\partial}{\partial t}\nabla^2\phi + \nabla\phi \times \hat{z} \cdot \nabla\nabla^2\phi - \nu\nabla^4\phi = 0. \tag{2.60b}$$

The local, inviscid conservation of the vorticity *underlies many of the similarities* between 2D fluid dynamics and Vlasov plasma dynamics. In particular, we note that the equation for an inviscid 2D fluid is just,

$$\frac{\mathrm{d}\rho}{\mathrm{d}t} = 0,$$

(for $\rho = \nabla^2\phi$) which is similar in structure to the Vlasov equation,

$$\frac{\mathrm{d}f}{\mathrm{d}t} = 0.$$

Both state that phase space density is conserved along particle orbits. Hence, from Eq.(2.60b), it follows that in two dimensions both energy,

$$E = \iint \mathrm{d}^2x \frac{v^2}{2} = \iint \mathrm{d}^2x \frac{1}{2}|\nabla\phi|^2,$$

and enstrophy,

$$\Omega = \iint \mathrm{d}^2x \frac{\omega^2}{2} = \iint \mathrm{d}^2x \frac{1}{2}\left|\nabla^2\phi\right|^2$$

together are quadratic inviscid invariants. The existence of *two* conserved quantities complicates the construction of the theory of turbulent cascade for 2D turbulence. As we shall show, the resolution of this quandary is a dual cascade (Kraichnan, 1967): that is, for forcing at some intermediate scale with wave number k_f such that $k_{\mathrm{min}} < k_f < k_{\mathrm{max}}$, there is:

(i) a self-similar, local enstrophy flux from k_f toward viscous damping at high k. This is called the forward *enstrophy cascade*;
(ii) a self-similar, local energy flux from k_f toward *low* k and *large scale*. This is called the *inverse energy cascade*.

Obviously, the forward and inverse cascades must have distinct spectral power law scalings. Also, we remark that energy and enstrophy are each transferred in *both* directions, toward high and low k. What distinguishes the two cascade ranges is that the directions for *self-similar* transfer differ.

The need for a dual cascade picture can easily be understood from the following simple argument (Vallis, 2006). Consider some initial spectral energy $E\,(k, t = 0)$

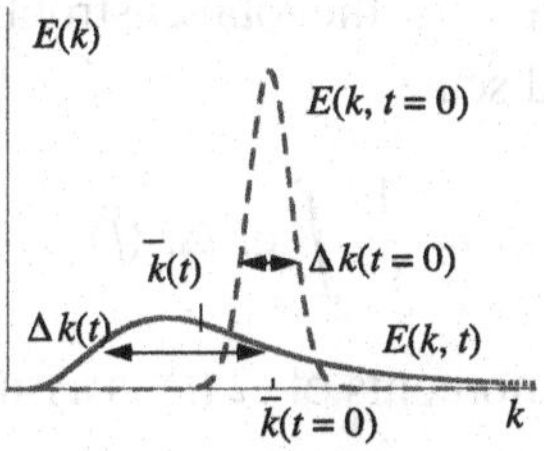

Fig. 2.15. Energy spectral density $E(k)$ shifts toward lower k (schematic illustration). As Δk^2 increases, the centroid $\bar{k}$ decreases.

distributed over a range as shown by a dotted line in Figure 2.15. This initial distribution has variance Δk^2 and centroid wave number $\bar{k}$,

$$\Delta k^2 = \frac{1}{E} \int \mathrm{d}k \left(k - \bar{k}\right)^2 E\,(k)\,, \tag{2.61a}$$

and

$$\bar{k} = \frac{1}{E} \int \mathrm{d}k\; kE\,(k)\,. \tag{2.61b}$$

Now, it is eminently plausible that the turbulence will act to broaden Δk^2 as the spectrum evolves in time. Thus, we expect that, as time increases, Δk^2 will grow,

$$\frac{\partial}{\partial t} \Delta k^2 > 0. \tag{2.62}$$

However, we know that the relation $\int \mathrm{d}k \left(k - \bar{k}\right)^2 E\,(k) = \Omega - \left(\bar{k}\right)^2 E$ holds and that Ω and E are (inviscidly) conserved, i.e.

$$\Delta k^2 = \frac{\Omega\,(t = 0)}{E\,(t = 0)} - \left(\bar{k}\right)^2. \tag{2.63a}$$

Since $\Omega\left(t = 0\right)/E\left(t = 0\right)$ is constant, we see that the growth of Δk^2 (Eq.(2.62)) requires,

$$\frac{\partial}{\partial t} \bar{k} < 0, \tag{2.63b}$$

so that the centroid of the spectrum must shift toward *lower* wave numbers. This is shown in Figure 2.15. This trend is quite suggestive of the *inverse* energy cascade.

We now repeat this type of exercise for the case of enstrophy. Here, it is convenient to work with *scale*, not wave number. Thus, for $l = 1/k$, we can define the variance,

$$\Delta l^2 = \frac{1}{\Omega} \int \mathrm{d}l \left(l - \bar{l}\right)^2 \Omega\,(l)\,, \tag{2.64a}$$

where $\Omega(l)$ is the enstrophy density, the total enstrophy is given by $\Omega = \int \mathrm{d}l\Omega(l)$, and $\bar{l}$ is the enstrophy centroid scale,

$$\bar{l} = \frac{1}{\Omega}\int \mathrm{d}l\; l\Omega(l). \tag{2.64b}$$

Note that convergence of the moments of $\Omega(l)$ is assumed a priori, but *not* proved. Then the change of the variance is given as,

$$\frac{\partial}{\partial t}\Delta l^2 = \frac{\partial}{\partial t}\left\{\frac{1}{\Omega}\int \mathrm{d}l\left(l-\bar{l}\right)^2\Omega(l)\right\} = \frac{\partial}{\partial t}\left\{\frac{1}{\Omega}\int \mathrm{d}l\; l^2\Omega(l) - \left(\bar{l}\right)^2\right\}. \tag{2.65a}$$

However, the integral $\int \mathrm{d}l l^2\Omega(l)$ is just the total energy, which is conserved along with the total enstrophy. Hence,

$$\frac{\partial}{\partial t}\Delta l^2 = -\frac{\partial}{\partial t}\left(\bar{l}\right)^2. \tag{2.65b}$$

For the range of scales to broaden in time (i.e. $\partial\Delta l^2/\partial t > 0$),

$$\frac{\partial}{\partial t}\bar{l} < 0 \tag{2.66}$$

is required, so the centroid of the distribution of enstrophy density (by scale) must move toward *smaller* scale. This is suggestive of a direct cascade of enstrophy to smaller scale. Thus, we see that the simultaneous conditions of spectral broadening and inviscid conservation of energy and enstrophy force the dual cascade model. In this dual cascade scenario, enstrophy is self-similarly transferred to smaller scales while energy is self-similarly transferred to large scales.

2.3.2.2 Self-similar spectral distribution

Simple scaling arguments for the cascade spectra are then easy to construct. To describe the cascade spectra, it is convenient to work with the energy density spectrum $E(k)$, so with a factor of k for density of states, $kE(k)$ has the dimension of $\langle v^2\rangle$. Hence $k^3E(k)$ corresponds to enstrophy density. Spectral self-similarity leads us to the hypothesis that enstrophy cascades locally, with a rate set by the eddy-turn-over time τ_{et} for each k, i.e.,

$$\frac{1}{\tau_{\mathrm{cascade}}} = \frac{1}{\tau_{\mathrm{et}}} = \frac{v(l)}{l} = k(kE(k))^{1/2}. \tag{2.67}$$

Then, a scale-independent enstrophy dissipation rate $\eta = k^3E(k)/\tau_{\mathrm{cascade}}$ requires that

$$\left(k^3E(k)\right)^{3/2} = \eta, \tag{2.68a}$$

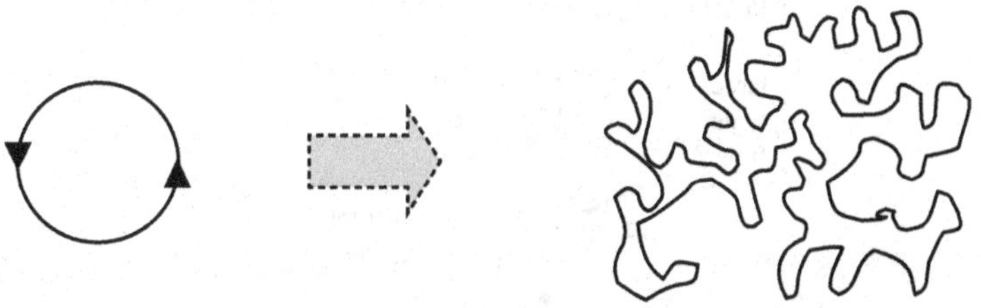

Fig. 2.16. Mean squared vorticity increases as vorticity isocontours stretch in a turbulent flow.

which immediately gives the energy spectrum for the (forward) enstrophy cascade as,

$$E(k) = \eta^{2/3}k^{-3}. \tag{2.68b}$$

Note that the eddy-turn-over rate in the enstrophy cascade range is constant in k from Eqs.(2.67) and (2.68b). The enstrophy spectrum is given by $\Omega(k) = k^2 E(k)$, so that equi-partition holds for $k\Omega(k)$, according to Eq.(6.68b). The physics of the enstrophy cascade is successfully described by the sketch in Figure 2.16. This shows that stretching of iso-contours of vorticity by a turbulent flow necessarily generates smaller scale structure in these contours, thus producing a net increase in mean square vorticity gradient $\left\langle \left(\nabla\nabla^2\phi\right)^2 \right\rangle$. The increase is what underlies the forward enstrophy cascade process. The cascade is ultimately terminated by viscous mixing. The forward cascade of enstrophy in k space is closely related to the homogenization (i.e. mixing and dissipation) of vorticity in configuration space, to be discussed later.

The self-similar inverse cascade of energy is correspondingly described, by balancing the energy dissipation rate ϵ with the flow rate of energy to larger scale, set locally by the eddy-turn-over rate, i.e. $kE(k)/\tau_{\text{cascade}} = \epsilon$. This gives the relation, with the help of Eq.(2.67):

$$k^{5/2}E(k)^{3/2} = \epsilon, \tag{2.69a}$$

so

$$E(k) = \epsilon^{2/3}k^{-5/3}. \tag{2.69b}$$

Of course, the energy cascade spectrum is the same as the K41 spectrum, though the cascade is toward large scale. The dual cascade is represented by the schematic drawing in Figure 2.17. Note that the inverse cascade builds up a large-scale flow from intermediate forcing. The process of large-scale build-up is nicely illustrated by Figure 2.18, which shows the evolution of the spectrum during a simulation of 2D turbulence forced at intermediate scale. Ultimately, this flow occupies the

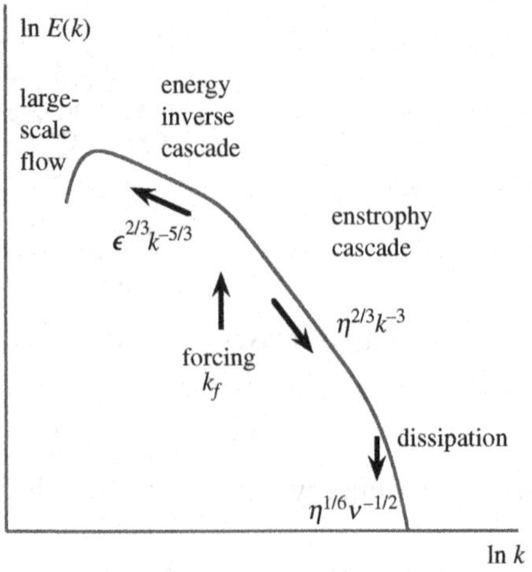

Fig. 2.17. Schematic of energy spectrum for dual cascade.

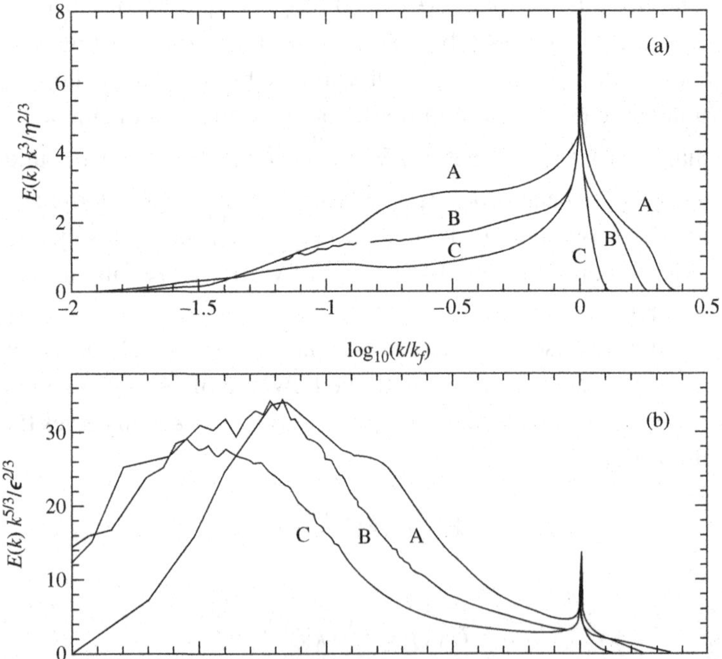

Fig. 2.18. Build-up of a large-scale flow in dual cascade (Borue, 1994). Energy spectra normalized as (a) $E(k)\,k^3\eta^{-2/3}$ and (b) $E(k)\,k^{5/3}\epsilon^{-2/3}$ as the function of $\log_{10}\left(k/k_f\right)$ for three different parameters in simulations. (B and C have finer resolution than A. In C, forcing occurs at finer scale than in B. (See Borue (1994) for details of parameters.)

largest scale of the system, thus generating a macroscopic shear flow on that scale. Such large-scale shears can then directly strain the smaller scales, thus breaking self-similarity and producing strong intermittency in the turbulent flow.

2.3.2.3 Dispersion of particle pairs

The dispersion of particle pairs (i.e. Richardson's problem) in a turbulent 2D flow is strongly tied to the dynamics of the dual cascades. In all cases, the dispersion of particles separated by distance l is determined by the eddies of that size (Eq.(2.57a)), so

$$\frac{\mathrm{d}}{\mathrm{d}t}l = v\left(l\right).$$

For the inverse cascade range, i.e. $l > k_f^{-1}$, Eq.(2.69b) gives $v\left(l\right) = \epsilon^{1/3} l^{1/3}$, so,

$$l^2 \sim \epsilon t^3, \tag{2.70}$$

as in K41. Particle pair separation grows super diffusively. For the forward enstrophy cascade range, $l < k_f^{-1}$, we note that the velocity $v\left(l\right) = \left(kE\left(k\right)\right)^{1/2}$ is given by $\eta^{1/3} l$, because $E\left(k\right) = \eta^{2/3} k^{-3}$ holds as Eq.(2.68b). We immediately have,

$$\frac{\mathrm{d}}{\mathrm{d}t}l = \eta^{1/3} l. \tag{2.71}$$

Thus, particle separation $l\left(t\right)$ grows exponentially in time for separation scales smaller than the forcing scale, but super diffusive growth occurs for scales larger than the forcing scale. The exponential divergence of particles in the enstrophy cascade range resembles the exponential divergence of trajectories in a stochastic system, such as for the case of overlapping resonances between plasma particles and a spectrum of waves.

2.3.2.4 Long-lived vortices

It is interesting to note that long-lived coherent vortices have been observed to emerge from decaying turbulent flows, and even in certain forced turbulent flows. This important phenomenon has long been recognized, but was dramatically emphasized by the seminal work of J. McWilliams and its offshoots. These studies revealed a two-stage evolution for decaying turbulence, namely:

(i) a fast stage of rapid decay and cascading, as shown in Figure 2.19(a);
(ii) a second, slower stage of evolution by binary vortex interaction. In this stage, vortices advect and strain each other, merge and sometimes form persisting pairs. An example of this evolution is shown in Figure 2.19(b).

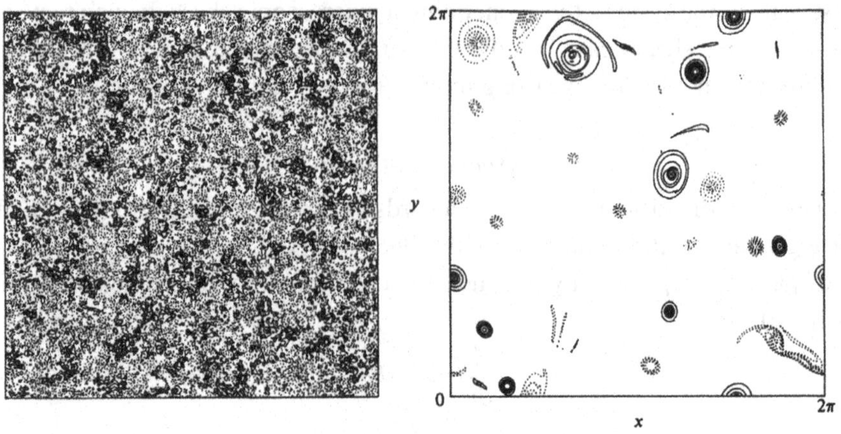

Fig. 2.19. Vorticity contours in the initial condition (a) and long-time evolution at a normalized time of $t = 16.5$ (b), where the eddy-turn-over time increases from 0.5 to 2.0 in the decay process. (McWilliams, 1984)

One of the most interesting aspects of this work is that it confirms the intuitively appealing Okubo–Weiss criterion (Okubo, 1970; Weiss, 1991), which constitutes a plausible answer (for 2D fluids) to the often-asked question, "What makes a coherent structure coherent?"

The Okubo–Weiss criterion emerges from an asymptotic expression for the time evolution of the local vorticity gradient $\nabla\rho$ (where $\rho = \nabla^2\phi$ is the local vorticity), which predicts that

$$\frac{\partial}{\partial t}\nabla\rho = \sqrt{S^2 - \rho^2}. \tag{2.72}$$

Here, $S = \partial^2\phi/\partial x\partial y$ is the local flow shear. The Okubo–Weiss (O–W) criterion thus states that the evolution of the local vorticity gradient is set by the Gaussian curvature of the stream function. In physical terms, the O–W criterion states that when the magnitude of the local shear exceeds the magnitude of local vorticity, the vorticity gradient is steeper and small scales will develop, as they do in the enstrophy cascade. If the local enstrophy density exceeds $|S|$, however, the vorticity gradient will not steepen, and a coherent vortex will simply rotate, without distortion. Locally, the flow will be stable to the cascade process. The O–W criterion is quite plausible, as it is consistent with the expected natural competition between shearing and vortical circulation. Comparisons with simulations of decaying turbulence indicate that the O–W criterion successfully predicts the location of long-lived, coherent vortices, which are, in some sense, stable to cascading in a turbulent flow. Indeed, when applied to a fully turbulent flow, the O–W criterion successfully predicts the subsequent emergence and locations of coherent vortices

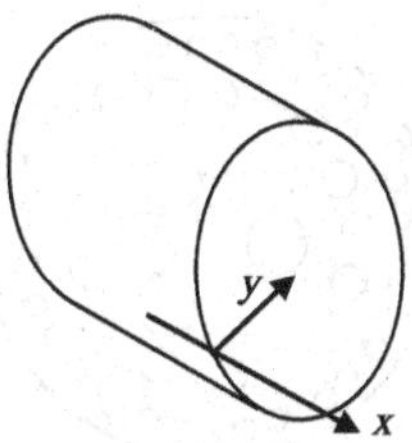

Fig. 2.20. Geometry of pipe flow. The y-axis is measured from the wall (perpendicular to the wall) according to the convention.

after the early phase of rapid decay. Thus, the O–W criterion constitutes one physically plausible approach to predicting intermittency in 2D turbulence.

Here, intermittency refers to breakdown of self-similar transfer by the formation of stable structures. We should caution the reader that many types of intermittency are plausible. (For instance, another origin of intermittency, which is induced by the statistical variance of dissipation rate ϵ from its mean $\langle\epsilon\rangle$, is explained in Arimitsu and Arimitsu (2001) and Yoshizawa *et al.* (2003).) A full discussion of this challenging, forefront problem requires a book in itself.

2.3.3 Turbulence in pipe and channel flows

2.3.3.1 Illustration of problem

We now turn to the interesting and relevant problem of turbulence in pipe and channel flows, which we hereafter refer to simply as 'turbulent pipe flow'. The essence of the pipe flow problem is the calculation of the mean flow profile $V_{(y)}$ for flow of a fluid with viscosity ν through a long pipe with fixed pressure drop per length $\Delta p/L$, assuming no-slip boundary conditions. The geometry and coordinates (after convention) are illustrated in Figure 2.20.

As we shall see, there are many parallels between the K41 paradigm of homogeneous turbulence in a periodic box and the problem of turbulent flow in a pipe. The study of turbulent pipe flow was pioneered by Ludwig Prandtl in seminal works published in 1932 (Prandtl, 1932), hereafter referred to as P32. The parallel between the K41 and P32 problems is summarized in Subsection 2.3.4.

Like K41 turbulence, pipe flow turbulence manifests an element of universality in its phenomenology. In simple terms, pipe flow turbulence is driven by turbulent mixing of the cross-stream shear of the mean flow $\mathrm{d}V_x(y)/\mathrm{d}y$ by turbulent Reynolds stress $\left\langle \tilde{V}_y \tilde{V}_x \right\rangle$, so that turbulent energy production P is given by:

$$P = -\left\langle \tilde{V}_y \tilde{V}_x \right\rangle \frac{\mathrm{d}}{\mathrm{d}y} V_x(y). \qquad (2.73)$$

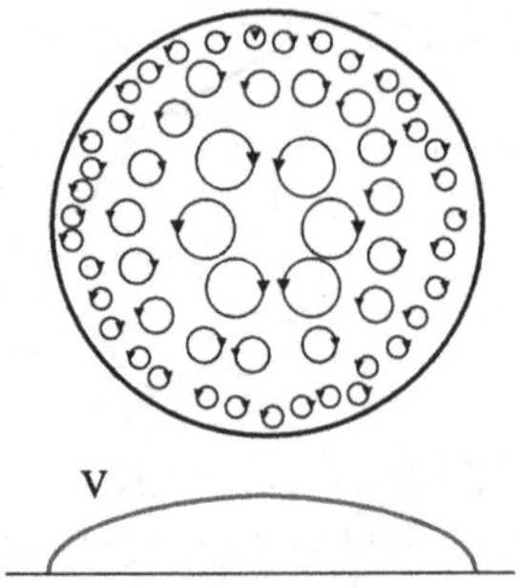

Fig. 2.21. Schematic drawings of turbulent eddies in a cross-section of pipe flow and the mean velocity profile across the mid-plane.

We see therefore that the turbulence is driven by the cross-stream flux of along stream momentum. Pipe flow is perhaps the simplest example of flux-driven turbulence, a ubiquitous paradigm with many applications to tokamaks, solar convection, etc.

The effective drag on the flow, which opposes the driving $\Delta p/L$, results from turbulent transport to the pipe wall, where the no-slip boundary condition forces the stream-wise flow to vanish. Thus, turbulent transport transfers or connects momentum input or drive by pressure drop to dissipation in the viscosity dominated region close to the no-slip boundary. A diagram of this spatial transport process and its implications for the flow profile is given in Figure 2.21.

2.3.3.2 *Viscous sublayer*

The Reynolds stress $\left\langle \tilde{V}_y \tilde{V}_x \right\rangle$ is an effective measure of momentum transport to the wall, or equivalently, the stress exerted on the wall, which we call

$$T_{\mathrm{W}} = \rho \left\langle \tilde{V}_y \tilde{V}_x \right\rangle.$$

Here ρ indicates the mass density and T_{W} is the stress. Clearly, T_{W} is proportional to $\Delta p/L$. Since there is no sink of momentum other than viscous drag at the wall, the force balance on the fluid requires,

$$T_{\mathrm{W}} = \frac{a}{2L}\Delta p.$$

For the turbulent stress near the wall, T_{W} is constant across the flow, and so we can define a constant friction velocity,

$$V_* = \sqrt{T_{\mathrm{W}}/\rho},$$

where the mass density ρ is taken as constant here for the transparency of the argument. V_* is a characteristic turbulent velocity for a pipe flow. (Note that the relation $V_* \propto \sqrt{\Delta p/\rho}$ holds.)

Having defined the characteristic velocity (which is called friction velocity) V_*, we can immediately identify two characteristic scales and Reynolds numbers for pipe flow turbulence. One is the viscous sublayer width y_d,

$$y_d = \nu V_*^{-1}, \tag{2.74}$$

which is a measure of the thickness of the viscosity-dominated range near the wall. In the viscous sublayer, $y < y_d$, the Reynolds number $R_e = V_* y$ satisfies the relation $R_e < 1$. In order to balance the constant wall stress and satisfy the no-slip boundary condition at the wall, the flow profile must be linear, i.e. $V(y) \sim V_* y / y_d$, in the viscous sublayer. Of course, the flow further away from the wall is strongly turbulent, and the Reynolds number computed with the pipe cross-section length a, $R_e = V_* a/\nu$, is much larger than unity. Indeed, in practical applications, R_e is so large that all vestiges of the (subcritical) instability process, which initially triggered the turbulence, are obliterated in the fully evolved turbulent state.

2.3.3.3 *Log law of the wall*

As with the K41 problem, empirical observation plays a key role in defining the problem. In the pipe flow problem, numerous experimental studies over a broad range of turbulent flows indicate that the flow profile has a universal, self-similar structure consisting of three layers, namely:

(a) the core; i.e. $y \sim a$,
(b) an inertial sublayer; i.e. $y_d < y \ll a$,
(c) the viscous sublayer; i.e. $0 < y < y_d$;

and that in the inertial sublayer, the flow gradient is scale independent, with a universal structure of the form,

$$\frac{\mathrm{d}}{\mathrm{d}y} V(y) \simeq \frac{V_*}{y}, \tag{2.75a}$$

so

$$V(y) = \kappa V_* \ln y. \tag{2.75b}$$

This logarithmic profile for the inertial sublayer flow is often referred to as the (Prandtl) Law of the Wall, and is, to reasonable accuracy, a universal feature of high R_e pipe flow. The flow profile and the three regimes are sketched in Figure 2.22. The empirically determined constant, $\kappa = 0.4$, is named the von Karman constant.

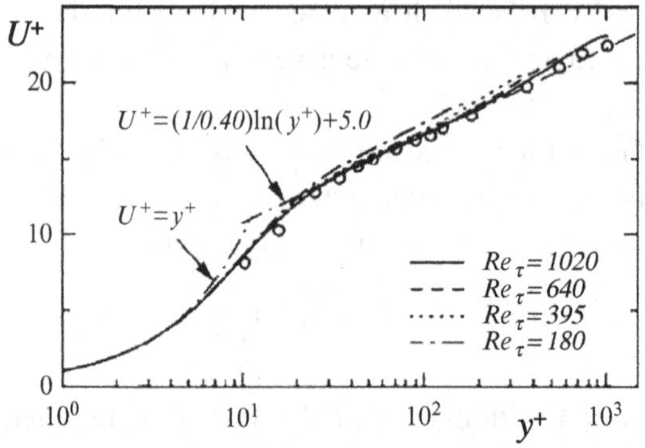

Fig. 2.22. Mean velocity of turbulent channel flows normalized by the friction velocity, $U^+ = V(y)/V_*$ as a function of the normalized distance $y^+ = y/y_d$. Quoted from Yoshizawa (2005), which compiled the lines, DNS (Abe, 2004) and circles, observation (Wei and Willmarth, 1989) at $R_{e\tau} = 1016$. Here, $R_{e\tau}$ is the Reynolds number defined by use of the friction velocity V_*. Viscous flow near the wall, log law and core profile are observed.

We should mention here that although the logarithmic law of the wall profile is the best known feature of turbulent pipe flow, it is perhaps *more instructive to focus on the universality of the flow profile gradient* $\mathrm{d}V(y)/\mathrm{d}y$. Note that the local gradient is determined entirely by the distance from the wall y (a purely local parameter!) and the friction velocity V_*. In some sense, it is more appropriate to focus on the flow gradient instead of flow, since the former is determined purely locally, while the flow at y is affected by physical effects originating at distant points.

A simple, physically appealing model can be constructed to explain the empirical law of the wall. The basic ideas of this model are:

(i) turbulence intensity in the inertial sublayer is determined by a local balance between mean profile relaxation induced by turbulent viscosity ν_T and turbulent dissipation of fluctuation energy;

(ii) turbulence is characterized locally by a simple velocity, namely the friction velocity V_*, and a single length scale l.

Now, turbulence energy E evolves according to a competition between production P and dissipation ϵ, so

$$\frac{\partial}{\partial t}E = P - \epsilon, \tag{2.76a}$$

where

$$P = \nu_T \left(\frac{\partial V_{(y)}}{\partial y}\right)^2 = V_* l \left(\frac{\partial V_{(y)}}{\partial y}\right)^2, \tag{2.76b}$$

and

$$\epsilon = \frac{V_*^3}{l}. \tag{2.76c}$$

Here l is the characteristic length scale of the turbulence. Now, empirically we have $\partial V_{(y)}/\partial y = V_*/y$, it follows that,

$$\frac{\partial}{\partial t}E = V_* l \frac{V_*^2}{y^2} - \frac{V_*^3}{l}. \tag{2.76d}$$

Thus, we see that the most direct way to ensure stationarity in the inertial sublayer is to simply take the characteristic length scale l to be y, the distance from the wall,

$$l \sim y, \quad \text{so,} \quad \nu_{\mathrm{T}} = V_* y.$$

Note this ansatz ensures scale invariance in the inertial sublayer! The length $l \sim y$ is often referred to as the *mixing length*, since by analogy with gas kinetics where viscosity is given by thermal velocity and mean free path, here eddy viscosity $\nu = v_{\mathrm{T}} l_{mfp}$, so $l \sim y$ may be thought of as an effective mean free path, over which fluid momentum is mixed by a random walk with root-mean-square velocity V_*. In other words, the log law of the wall is based on the picture that the length of turbulent mixing l is given by the distance from the wall y (the upper bound by the vortex size in the region between the location y and the wall).

This mixing length model of pipe flow turbulence was first proposed by Prandtl, and thus goes by the name of the Prandtl Mixing Length Theory. Note that mixing length theory also immediately recovers the logarithmic profile, since by making the assumption of diffusive transport,

$$\frac{T_{\mathrm{w}}}{\rho} = \left\langle \tilde{V}_y \tilde{V}_x \right\rangle = \nu_{\mathrm{T}} \frac{\partial}{\partial y} V(y), \tag{2.77a}$$

(note the minus sign is absorbed since y is measured from the wall) and if $\nu_{\mathrm{T}} = V_* l = V_* y$, we have,

$$\frac{\partial}{\partial y} V(y) = \frac{V_*}{y}. \tag{2.77b}$$

2.3.3.4 *Approach to self-similarity*

It is enlightening to briefly review another even simpler approach to the problem of the inertial sublayer profile, assuming similarity methods. To this end, one can formulate the problem by noting that since it is the *mean velocity gradient* which is locally determined self-similar and seemingly 'universal', we know that the dimensionless function $y V_*^{-1} \partial V(y) / \partial y$ is determined exclusively by the dimensionless

Table 2.4. *Parallel studies in self-similarity*

Inertial range spectrum (K41)	Pipe flow profile (P32)
Basic ideas	
self-similarity in *scale*	self-similarity in *space*
inertial range spectrum $V(l)$	inertial sublayer profile $V(y)$
eddy/wavelet	mixing 'slug' or eddy
K41 spectrum	law of the wall
Range	
stirring	core
inertial	inertial sub-layer
dissipation	viscous sub-layer
Element	
$l \rightarrow$ eddy scale	$l_{\mathrm{M}} = y \rightarrow$ mixing length
Throughput	
$\epsilon \rightarrow$ dissipation rate	$V_*^2 = T_{\mathrm{W}}/\rho$ $\rightarrow$ wall stress, friction velocity
Rate	
$1/\tau(l) \sim V(l)/l$ (eddy turn-over)	$\nu_{\mathrm{T}} y^{-2} \sim V_*/y$ (ν_{T} : eddy viscosity)
Balance	
$\epsilon = V(l)^2/\tau(l)$ $V(l) \sim \epsilon^{1/3} l^{1/3}$	$V_*^2 \simeq \nu_{\mathrm{T}} \partial V(y)/\partial y$ $\partial V(y)/\partial y \sim V_*/y \rightarrow$ log profile
Dissipation scale length	
$l_d = \nu^{3/4}\epsilon^{-1/4}$	$y_d = \nu V_*^{-1}$
Fit constant	
Kolmogorov constant	von Karman constant
Theorem	
4/5 law	?

parameters in the problem. Now, since there are two characteristic length scales in pipe flow turbulence, namely the viscous sublayer scale $y_d = \nu V_*^{-1}$ and pipe cross-section a, the relevant dimensionless function can be written as,

$$\frac{y}{V_*}\frac{\partial V(y)}{\partial y} = F\left(\frac{y_d}{y}, \frac{y}{a}\right). \tag{2.78a}$$

For the inertial sublayer of a high Reynolds number pipe flow, $y/y_d \gg 1$ and $a/y \gg 1$. Thus, assuming complete Reynolds number similarity amounts to taking $y_d/y \rightarrow 0$ and $y/a \rightarrow 0$. In this limit,

$$\frac{y}{V_*}\frac{\partial V(y)}{\partial y} = F(0, 0) \rightarrow \text{const}, \tag{2.78b}$$

so once again we arrive at the logarithmic 'law of the wall' profile,

$$V(y) = \kappa V_* \ln y. \tag{2.78c}$$

Thus, we see that Prandtl's law of the wall emerges from extremely simple arguments of complete Reynolds number similarity and scaling methods. The reader should note that study of corrections to the law of the wall induced by incomplete similarity is ongoing and remains an active topic of research.

2.3.4 Parallels between K41 and Prandtl's theory

The parallel between the K41 and P32 problems has been referred to many times during the above discussion. At this point, the reader may wish to visit the summary in Table 2.4, to review the many parallels between the twin studies in self-similarity which constitute Kolmogorov's theory of the inertial range spectrum and Prandtl's theory of turbulent pipe flow. This table is largely self-explanatory. It is interesting, however, to comment on one place where a parallel does *not* exist, namely, in the last entry, which deals with 'rigorous results'. For K41 theory, the '4/5 Law' (Frisch, 1995) is a rigorous asymptotic theorem which links the dissipation rate ϵ, the length scale l, and the triple moment $\langle \delta V^3(l) \rangle$ by the relation,

$$\left\langle \delta V^3(l) \right\rangle = -\frac{4}{5}\epsilon l.$$

The 4/5 Law, derived from the Karman–Howarth relation, is perhaps the one true theorem which is actually *proved* in turbulence theory. Since P32 theory tacitly assumes

$$\left\langle \delta V^3(l) \right\rangle \simeq V_*^3 \simeq \epsilon y,$$

it is naturally desirable to know a theorem for turbulent pipe flow, which corresponds to the 4/5 law. Unfortunately, no such result is available at this time.

3

Quasi-linear theory

A mean field theory of plasma transport

Nothing can be more fatal to progress than too confident reliance on mathematical symbols; for the student is only too apt to take the easier course, and consider the formula and not the fact as the physical reality.

(Lord Kelvin)

3.1 The why and what of quasi-linear theory

In the first part of the previous chapter, we discussed fluctuations and relaxation in a *stable* plasma, *close to equilibrium*. Now we embark on the principal discussion of this book, which deals with the far more difficult, but also more interesting, problem of understanding the dynamics of a *turbulent plasma, far from equilibrium*. The first topic in plasma turbulence we address is *quasi-linear theory*.

Plasma turbulence is usually thought to result from the nonlinear evolution of a spectrum of unstable collective modes. A collective instability is an excitation and a process whereby some available potential energy stored in the initial distribution function (either in its velocity space structure or in the gradients of the parameters which define the local Maxwellian, such as, $n(x)$, $T(x)$, etc) is converted to fluctuating collective electromagnetic fields and kinetic energy. A simple example of this process, familiar to all, is Rayleigh–Benard (R–B) convection, the mechanism whereby hot air rises on time scales faster than that determined by molecular diffusion (Pope, 2000). The starting point is unstably stratified air, which contains gravitational potential energy. Rayleigh–Benard convection taps this available 'free energy', converting some of it to convection rolls. The convection rolls in turn relax the vertical temperature gradient $\mathrm{d}T/\mathrm{d}z$ which drives the instability (i.e. in R–B convection, $\mathrm{d}T/\mathrm{d}z < (\mathrm{d}T/\mathrm{d}z)_{\mathrm{crit}}$). Thus, they exhaust the available free energy and so eliminate the drive of the R–B instability. A diagram schematic of this process is given in Figure 3.1. Examples of paradigmatic velocity space instabilities are the bump-on-tail (BOT) instability and the current-driven

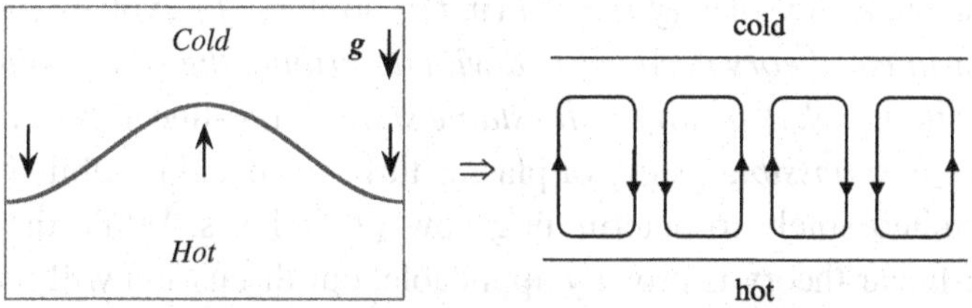

Fig. 3.1. Diagram showing the evolution of a super-critical gradient to convective instability and convection rolls to turbulence and turbulent mixing of the temperature gradient.

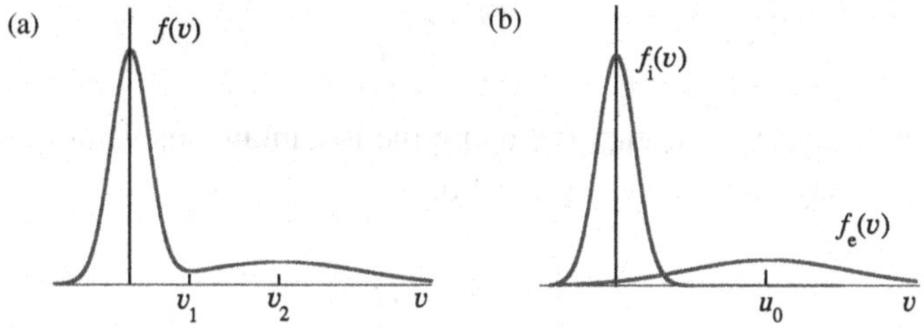

Fig. 3.2. (a) Sketch of the distribution function for the bump-on-tail instability. Phase velocities ω/k such that $v_1 < \omega/k < v_2$ are resonant where $\partial\langle f\rangle/\partial v > 0$, so instability occurs. (b) Sketch of the distribution function for the current-driven ion acoustic instability. Here the electron distribution function has centroid $u_0 \neq 0$, and so carries a net current. Phase velocities $v_{T_i} < \omega/k < u_0$ may be unstable, if electron growth exceeds ion Landau damping.

ion acoustic (CDIA) instability (Krall and Trivelpiece, 1973). In the BOT, the free energy is the kinetic energy of the 'bump' or weak beam population situated on the tail of the Maxwellian. The presence of the bump implies an interval of velocity for which $\partial\langle f\rangle/\partial v > 0$, so that waves resonant in that interval are unstable. The unstable spectrum will grow at the expense of the free energy in the bump, thus decelerating it and 'filling in' the distribution, so that $\partial\langle f\rangle/\partial v \leq 0$, everywhere. To conserve total momentum, heating of the bulk distribution must occur. A sketch of this evolutionary process is given in Figure 3.2(a). In the case of the CDIA, shown in Figure 3.2(b), the current carried by the electrons can produce a region of positive $\partial\langle f\rangle/\partial v$ sufficient to overcome the effects of ion Landau damping, thus triggering instability. The turbulent electric fields will act to reduce $\partial\langle f\rangle/\partial v$ by reducing the shift of, or 'slowing down', the electron distribution function. Again, conservation of momentum requires some bulk heating and some momentum transfer to the ions. *In all cases, the instability-driven turbulence acts to expend the available free energy, thus driving the system back toward a marginally stable state, and extinguishing the instability.* Since this evolution occurs on a time scale which is necessarily longer than the characteristic times of the waves, we may say that $\langle f \rangle = \langle f(v,t) \rangle$, so that $\langle f \rangle$ evolves *on slow* time scales. *Quasi-linear theory is concerned with describing the slow evolution of* $\langle f \rangle$ *and its relaxation back to a marginally stable state*. Quasi-linear theory is, in some sense, the *simplest possible* theory of plasma turbulence and instability saturation, since it is limited solely to determining how $\langle f \rangle$ relaxes. While the methodology of quasi-linear theory is broadly applicable, our discussion will focus first on its application to problems in Vlasov plasma turbulence, and later consider more complicated applications.

In quasi-linear theory, the mean field evolution of $\langle f \rangle$ is taken to be slow, so that,

$$\frac{1}{\langle f \rangle} \frac{\partial \langle f \rangle}{\partial t} \ll \gamma_k.$$

Thus, the growth rate γ_k is computed using the instantaneous value of $\langle f \rangle$, which evolves more slowly than the waves do. So,

$$\gamma_k = \gamma_k[\langle f(v,t) \rangle]$$

is determined by plugging $\langle f \rangle$ at the time of interest into the linear dielectric function,

$$\epsilon(k,\omega) = 1 + \sum_j \frac{\omega_{pj}^2}{k} \int \mathrm{d}v \, \frac{\partial \langle f_j \rangle / \partial v}{\omega - kv}, \tag{3.1}$$

and then computing ω_k, γ_k via $\epsilon(k,\omega) = 0$. The equation for $\langle f \rangle$ is obtained by averaging the Vlasov equation,

$$\frac{\partial f}{\partial t} + v \frac{\partial f}{\partial x} + \frac{q}{m} E \frac{\partial f}{\partial v} = 0, \tag{3.2a}$$

and using the separation $f = \langle f \rangle + \delta f$, so

$$\frac{\partial \langle f \rangle}{\partial t} = -\frac{\partial}{\partial v} \left\langle \frac{q}{m} E \delta f \right\rangle. \tag{3.2b}$$

Note that Eq.(3.2b) constitutes the first of the Vlasov hierarchy, which couples the evolution of the first moment to the second moment, the evolution of the second moment to the third moment, etc. Quasi-linear theory *truncates* this hierarchy by simply approximating the fluctuating distribution function f by the *linear coherent response* f_k^c to E_k, i.e.,

$$\delta f_k = f_k^c = -i \frac{q}{m} \frac{E_k \partial \langle f \rangle / \partial v}{\omega - kv}. \tag{3.2c}$$

Plugging f_k^c into Eq.(3.2b) gives the quasi-linear equation for $\langle f \rangle$ evolution,

$$\frac{\partial \langle f \rangle}{\partial t} = \frac{\partial}{\partial v} D(v) \frac{\partial \langle f \rangle}{\partial v} \tag{3.3a}$$

$$D(v) = \mathrm{Re} \sum_k \frac{q^2}{m^2} |E_k|^2 \frac{i}{\omega_k - kv + i\,|\gamma_k|}. \tag{3.3b}$$

Thus, quasi-linear theory is a straightforward application of mean field theory methodology to the problem of $\langle f \rangle$ evolution. Note that all noise and mode–mode coupling effects are neglected, so *all fluctuations are assumed to be eigenmodes which satisfy* $\omega = \omega(k)$. Other parts of f, i.e. the incoherent part $\tilde{f}$ in Eq.(2.1), have impact on the relaxation. This effect is discussed in Chapter 8. The other issue is a truncation of δf at the linear response. The roles of nonlinear terms mode coupling, etc. will be explained in subsequent chapters. In this chapter, the ω-subscript is superfluous and hereafter dropped. In the language of critical phenomena, quasi-linear theory is concerned with the evolution of the order parameter in a phase of broken symmetry, not with noise-driven fluctuations while criticality is approached from below.

For completeness, then, we now write the full set of equations used in the quasi-linear description of Vlasov turbulence. These are the linear dispersion relation,

$$\epsilon(k, \omega) = 0, \tag{3.4a}$$

the equation for the evolution of the electric field energy, which is just

$$\frac{\partial}{\partial t} |E_k|^2 = 2\gamma_k |E_k|^2, \tag{3.4b}$$

and the equations for $\langle f \rangle$ and $D(v)$, i.e.

$$\frac{\partial \langle f \rangle}{\partial t} = \frac{\partial}{\partial v} D(v) \frac{\partial \langle f \rangle}{\partial v}, \tag{3.4c}$$

$$D(v) = \sum_k \frac{q^2}{m^2} |E_k|^2 \frac{|\gamma_k|}{(\omega - kv)^2 + \gamma_k^2}. \tag{3.4d}$$

Note that the absolute value (i.e. $|\gamma_k|$) is required by causality. Since $D \sim |\gamma|$, negative diffusion is precluded, even if the modes are linearly damped. This is physically plausible, since damped waves of finite amplitude are quite capable of scattering particles and driving diffusion and relaxation. Equations (3.4a–3.4d) constitute the famous "quasi-linear equations", first derived by Vedenov, Velikov and Sagdeev (Vedenov *et al.*, 1961; Vedenov *et al.*, 1962) and by Drummond and Pines in the early 1960s (Drummond and Pines, 1962; Stix, 1992). The quasi-linear

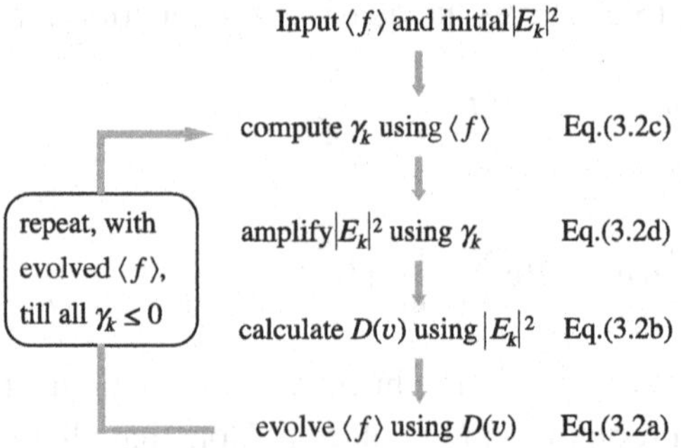

Fig. 3.3. Schematic for implementation of quasi-linear theory.

theory is implemented by solving equations (3.4a–3.4d) to describe the relaxation of $\langle f \rangle$ to a state where all $\gamma_k \leqq 0$. The concomitant evolution and saturation level of $|E_k|^2$ can also be calculated. Figure 3.3 gives a flow chart description of how Eq.(3.4a–3.4d) might actually be solved iteratively, to obtain an $\langle f \rangle$ which is everywhere marginal or submarginal.

At first glance, the quasi-linear theory seems easy, even trivial, and so bound to fail. Yet, quasi-linear theory is often amazingly successful! The key question of *why* this is so is still a subject of research after over 40 years. Indeed, the depth and subtlety of the quasi-linear theory begin to reveal themselves after a few minutes of contemplating Eq.(3.4a–3.4d). Some observations and questions one might raise include, but are not limited to:

(i) The quasi-linear equation for $\langle f \rangle$ Eq.(3.4c) has the form of a diffusion equation. So, what is the origin of irreversibility, inherent to any concept of diffusion, in quasi-linear theory? Can Eq.(3.4c) be derived using Fokker–Planck theory?

(ii) $D(v)$, as given by Eq.(3.4d), varies rapidly with v, as for resonant particles with $\omega \sim kv$,

$$D(v) = \sum_k \frac{q^2}{m^2} |E_k|^2 \pi \delta(\omega - kv),$$

while for non-resonant particles with $\omega \gg kv$,

$$D(v) = \sum_k \frac{q^2}{m^2} |E_k|^2 \frac{|\gamma_k|}{\omega^2}.$$

What is the physics of this distinction between resonant and non-resonant diffusion? What does non-resonant diffusion *mean*, in physical terms?

(iii) When and under what conditions does quasi-linear theory apply or break down? What criteria must be satisfied?

(iv) How does quasi-linear theory balance the energy and momentum budgets for fields and particles?

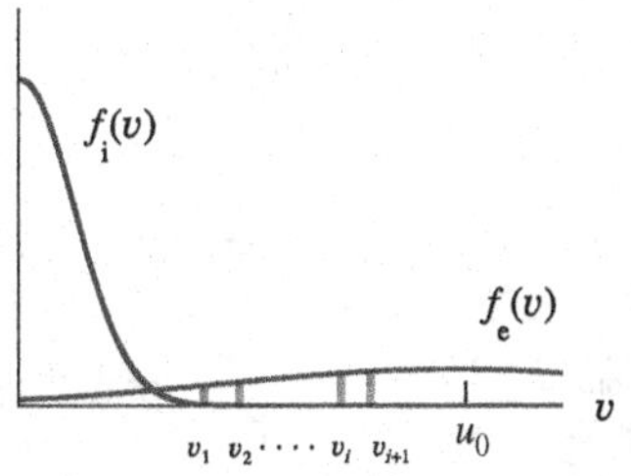

Fig. 3.4. Possible excitations of unstable CDIA modes, resonating to electrons.

(v) How does a spectrum of unstable waves drive $\langle f \rangle$ to evolve toward a marginal state, with $\gamma_k = 0$ for all k?

These questions are addressed in the remainder of this chapter. Applications to some simple examples, such as the BOT and CDIA instabilities, are discussed as well.

3.2 Foundations, applicability and limitations of quasi-linear theory

3.2.1 Irreversibility

We first address the issue of irreversibility. Generally, quasi-linear theory is applied in the context of a broad spectrum of unstable waves. Of course, one important question is, "How broad is 'broad'?". In the case of the CDIA system, the unstable spectrum is sketched in Figure 3.4. Note that, as for any realistic system, k is quantized, so the phase velocities $v_{\mathrm{ph},i} = \omega(k_i)/k_i$ are quantized, as well. Particle motion in such a wave field is entirely deterministic, according to Newton's laws, so that,

$$m\frac{\mathrm{d}^2 x}{\mathrm{d}t^2} = \sum_m qE_m \cos(kx_m - \omega_m t), \tag{3.5a}$$

and if $v \sim \omega_i/k_i$, one resonance dominates:

$$m\frac{\mathrm{d}^2 x}{\mathrm{d}t^2} \simeq qE_i \cos(k_i x + (k_i v - \omega_i)t). \tag{3.5b}$$

Hence, each resonant velocity defines a phase space island, shown in Figure 3.5. The phase space island is defined by a separatrix of width $\Delta v \sim (q\phi_m/m)^{1/2}$, which divides the trajectories into two classes, namely trapped and circulating. In the case with multiple resonances *where the separatrices of neighbouring phase space islands overlap*, the separatrices are destroyed, so that the particle motion becomes *stochastic*, and the particle can wander or 'hop' in velocity, from resonance to resonance. In this case, the *motion is non-integrable* and, in fact, *chaotic* (Chirikov, 1960; Zaslavsky and Filonenko, 1968; Smith and Kaufman, 1975; Fukuyama *et al.*, 1977; Chirikov, 1979; Lichtenberg and Lieberman, 1983;

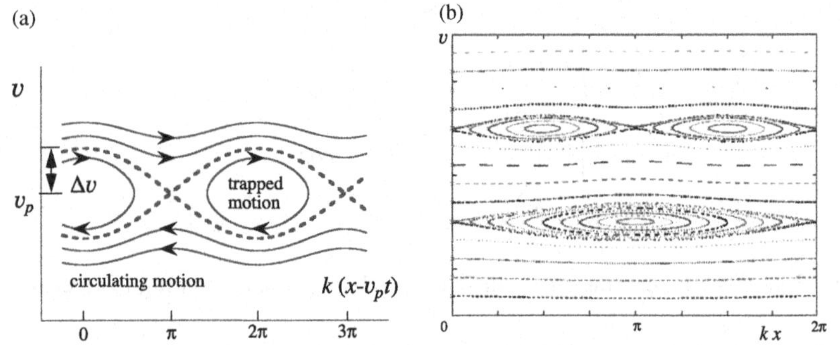

Fig. 3.5. (a) Structure of wave–particle resonance in phase space. The separatrix width is $\Delta v \sim (q\phi_i/m)^{1/2}$. Particles inside the separatrix (region of libration) undergo periodic motion on iso-energy contours and so are said to be trapped. Particles outside the separatrix circulate. (b) For several waves with distinct phase velocities, multiple resonance islands can co-exist and interact. [Courtesy of Prof. A. Fukuyama].

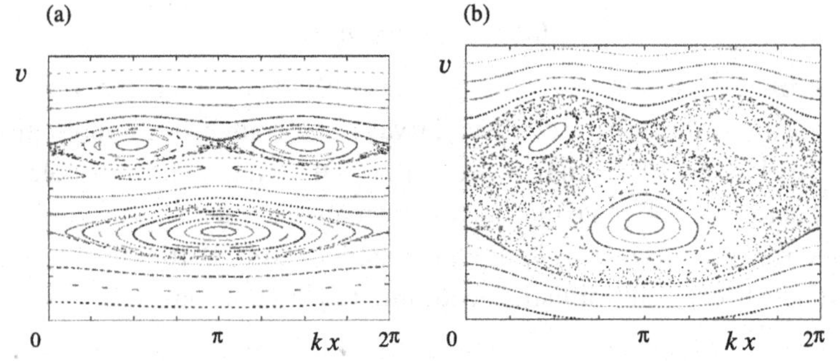

Fig. 3.6. Multiple separated resonances. Two waves (common in amplitude) with different frequencies (ω and $\omega + \Delta\omega$) co-exist. When the amplitude is below the threshold, particles may be trapped in the vicinity of an individual resonance, but cannot interact with multiple resonances (a) $\omega_b^2/\Delta\omega^2 = 0.025$. When the amplitude is above the threshold, particles can stochastically wander or hop from resonance to resonance. This produces diffusion in velocity (b) $\omega_b^2/\Delta\omega^2 = 0.1$. [Courtesy of Prof. A. Fukuyama].

Ott, 1993). A simple criterion for the onset of chaos and stochasticity is the Chirikov island overlap criterion,

$$\frac{1}{2}(\Delta v_i + \Delta v_{i\pm1}) > |v_{\mathrm{ph},i} - v_{\mathrm{ph},i\pm1}|. \tag{3.6}$$

Here Δv is the separatrix width, so that the left-hand side (LHS) of Eq.(3.6) is a measure of the excursion in v due to libration, while the right-hand side (RHS) is the distance in velocity between adjacent resonances. If as shown in Figure 3.6(a),

LHS $\ll$ RHS, separatrix integrity is preserved and the motion is integrable. If, on the other hand, LHS $\gg$ RHS, as shown in Figure 3.6(b), individual separatrices are destroyed and particle orbit stochasticity results.

It is well known that stochastic Hamiltonian motion in velocity may be described by a Fokker–Planck equation, which (in 1D) can often be further simplified to a diffusion equation by using a stochastic variant of Liouville's theorem, because the phase space flow is incompressible on account of the underlying Hamiltonian equations of motion. The resulting equation is identical to the resonant diffusion equation obtained in quasi-linear theory. Thus, we see that *the fundamental origin of the irreversibility presumed by the quasi-linear theory is the stochasticity of resonant particle trajectories.* While research on the question of the precise wave amplitude necessary for stochasticity is still ongoing, *the Chirikov overlap criterion (Eq.(3.6)) is a good 'working rule', and so constitutes a necessary condition for the applicability of the quasi-linear theory of resonant diffusion.* Note that, in contrast to the presentations given in older texts, no assumption of "random wave phases", or "random phase approximation" is necessary, a priori. Particle orbit stochasticity is the ultimate underpinning of the quasi-linear diffusion equation. Further discussion of the relation between the quasi-linear theory and the Fokker–Planck theory can be found in (Escande and Sattin, 2007).

3.2.2 Linear response

At this point, the alert reader may be wondering about the use of linearized trajectories (i.e. unperturbed orbits) in proceeding from Eq.(3.5a) to Eq.(3.5b). Of course, linearization of δf occurs in the derivation of the quasi-linear theory, as well. This question brings us to a second important issue, namely that of the spectral auto-correlation time. The configuration of the electric field $E(x, t)$ which a particle actually "sees" at any particular x, t is a pattern formed by the superposition of the various modes in the spectrum, as depicted in Figure 3.7(a). For an evolving spectrum of (usually) dispersive waves, this pattern will persist for some lifetime τ_L. The pattern lifetime τ_L should be compared to the 'bounce time' of a particle in the pattern. Here the bounce time is simply the time required for a particle to reverse direction and return to the close proximity of its starting point. Two outcomes of the comparison are possible. These are:

i) $\tau_L \ll \tau_b \rightarrow$ field pattern changes prior to particle bouncing,
(Fig. 3.7(b)) so that trajectory linearization is *valid*;

ii) $\tau_b \ll \tau_L \rightarrow$ the particle bounces prior to a change in the field
(Fig. 3.7(c)) pattern. In this case, *trapping* can occur,
so linearized theory *fails*.

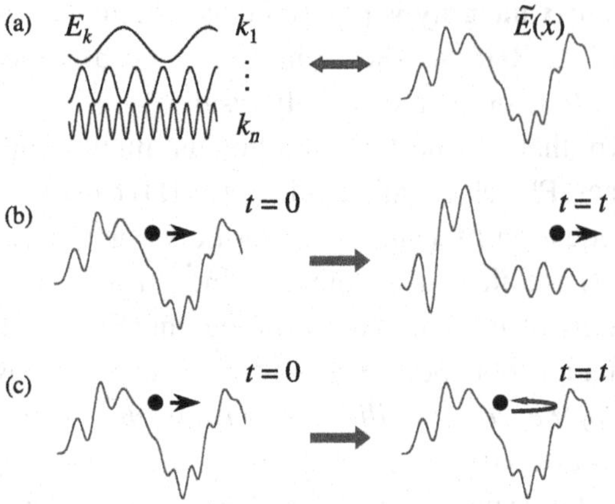

Fig. 3.7. (a) Representation of the instantaneous pattern of an electric field that a particle actually sees. The pattern has an effective duration time of τ_{ac}. (b) Showing that for $\tau_{ac} < \tau_b$, the E-field pattern a particle sees will change before the particle bounces, thus validating the use of unperturbed orbits. (c) Showing that for $\tau_b < \tau_{ac}$, the particle will bounce within a field pattern before the pattern changes. In this case, trapping occurs and the use of unperturbed orbits is not valid.

Not surprisingly, quasi-linear theory is valid when $\tau_L \ll \tau_b$, so that unperturbed orbits are a good approximation. The question that remains is how to relate our conceptual notions of τ_L, τ_b to actual physical quantities which characterize the wave spectrum.

3.2.3 *Characteristic time-scales in resonance processes*

The *key point* for determining the value of τ_L is the *realization that wave dispersion is what limits the pattern lifetime*, τ_L. Note the total electric field may be written (as before) as,

$$E(x,t) = \sum_k E_k e^{i(kx-\omega t)},$$

or as,

$$= \sum_k E_k \exp[i(k[x - v_{ph}(k)t])],$$

where $v_{ph}(k) = \omega(k)/k$. The pattern or packet dispersal speed is $\Delta(\omega_k/k)$, the net spread in the phase velocities in the packet. The net dispersal rate, i.e. the inverse time for a wave packet to disperse one wavelength, is then just,

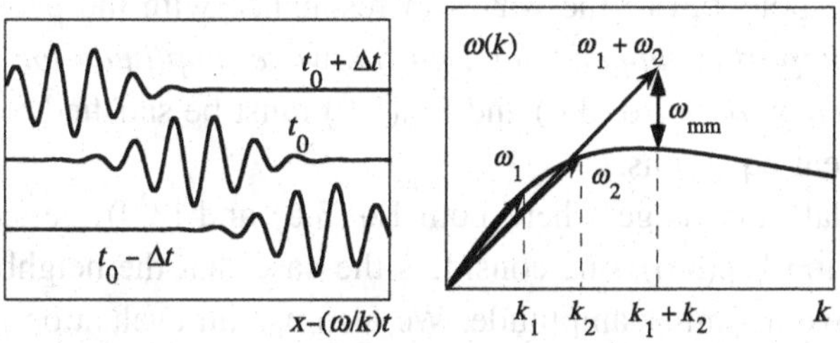

Fig. 3.8. Illustration of finite interaction times. Left: wave packet in the frame of the resonant particles which are moving at the phase velocity ω/k. When the group velocity $\partial\omega/\partial k$ is different from the phase velocity (here, the case of $\partial\omega/\partial k < \omega/k$ is shown), a wave packet passes by the resonant particle. Therefore the interaction time is limited. Right: mis-match of the frequency in the case where modes with k_1, k_2 and $k_1 + k_2$ are nonlinearly coupling.

$$\begin{aligned} 1/\tau_{\mathrm{L}} &= k|\Delta(\omega_k/k)| \\ &= k\left|\left(\frac{\mathrm{d}\omega_k}{\mathrm{d}k}\frac{\Delta k}{k} - \frac{\omega_k}{k^2}\Delta k\right)\right| \\ &= |\left(v_{\mathrm{g}}(k) - v_{\mathrm{ph}}(k)\right)\Delta k|. \end{aligned} \tag{3.7}$$

Equation (3.7) relates the pattern lifetime to Δk, the spectral width in k, and the net dispersion in velocity, which is just the difference between the phase (v_{ph}) and group (v_{g}) velocities. That is, the resonant particle, which has the velocity v_p, feels the difference of phase and group speeds, owing to the change of phase by wave dispersion (see Fig. 3.8.) Note that regardless of Δk, $\tau_{\mathrm{L}} \to \infty$ for non-dispersive waves. In this case, the pattern coherence time must necessarily be set by wave steepening and breaking, or some other strongly nonlinear effect that is outside the scope of quasi-linear theory. Thus, we conclude that the applicability of quasi-linear theory is limited to $\langle f \rangle$ evolution in the presence of a sufficiently broad spectrum of dispersive waves. Interestingly, despite the large volume of research on the validity of quasi-linear theory, this seemingly obvious point has received very little attention. Of course, the quantitative validity of quasi-linear theory requires that $1/\tau_{\mathrm{b}} < 1/\tau_{\mathrm{L}}$, so using,

$$\frac{1}{\tau_{\mathrm{b}}} \simeq k\sqrt{\frac{q\phi_{\mathrm{res}}}{m}} \tag{3.8}$$

gives an upper bound on the bounce frequency $\sim 1/\tau_{\mathrm{b}}$ which is,

$$\sqrt{\frac{q\phi_{\mathrm{res}}}{m}} < |v_{\mathrm{g}} - v_{\mathrm{ph}}|. \tag{3.9}$$

Here ϕ_{res} is the potential of the waves in resonance with the particle. *Equation* (3.9) *gives an important upper bound on the wave amplitude for the validity of quasi-linear theory. Both* Eq.(3.6) and Eq.(3.9) must be satisfied for applicability of the quasi-linear equations.

One can isolate the range where both Eq.(3.6) and (3.9) are satisfied. In the argument deriving Eq.(3.6), one considers the case that the neighbouring modes k_j and k_{j+1} have a similar amplitude. We also use an evaluation $\omega_{j+1} = \omega_j + (k_{j+1} - k_j)\partial\omega/\partial k$, where ω_j is the wave frequency for k_j. The phase velocity for the k_{j+1} mode, $v_{p,j+1}$, is given as,

$$v_{\text{p}.j+1} \simeq v_{\text{p},j} + \left(v_{\text{p},j} - v_{\text{g}.j}\right)\left(k_{j+1} - k_j\right) k_j^{-1}.$$

Thus, Eq.(3.6) is rewritten as,

$$\sqrt{\frac{e\phi}{m}} \geq \left|v_{\text{p},j} - v_{\text{g},j}\right| \left(k_{j+1} k_j^{-1} - 1\right). \tag{3.10}$$

Combining Eqs.(3.9) and (3.10), the range of validity, for the quasi-linear theory, is given as,

$$\frac{|v_{\text{p}} - v_{\text{g}}|}{kL} \leq \sqrt{\frac{e\phi}{m}} \leq \left|v_{\text{p}} - v_{\text{g}}\right|, \tag{3.11}$$

where the difference $k_{j+1} - k_j$ is given by L^{-1} (L : the system size). Therefore, the validity of the quasi-linear theory also requires that the wavelength must be much shorter than the system size.

3.2.4 Two-point and two-time correlations

In order to place the discussion given here on a more solid foundation, we now consider the two-point, two-time correlation $\langle E(x_1, t_1) E(x_2, t_2)\rangle$ along the particle orbit. Here the brackets refer to a space–time average. The goal here is to demonstrate rigorously the equivalence between the heuristic packet dispersal rate given in Eq.(3.7) and the actual spectral auto-correlation rate, *as seen by a resonant particle*. Now for homogeneous, stationary turbulence, the field correlation function simplifies to,

$$\langle E(x_1, t_1) E(x_2, t_2)\rangle = C(x_-, t_-), \tag{3.12}$$

where,

$$x_1 = x_+ + x_-$$

$$x_2 = x_+ - x_- \tag{3.13a}$$

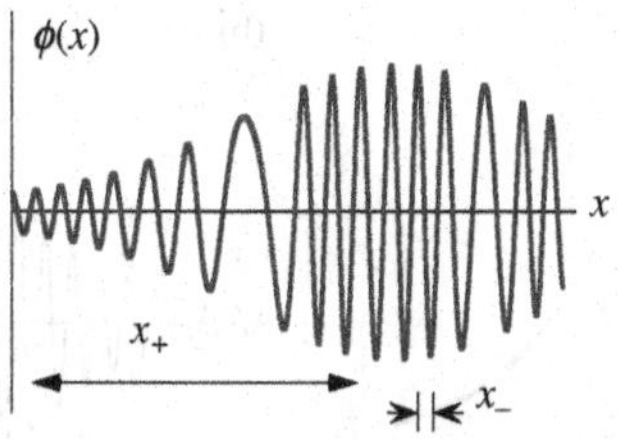

Fig. 3.9. Small-scale variable and large-scale variable for fluctuations.

and

$$t_1 = t_+ + t_-$$
$$t_2 = t_+ - t_-. \tag{3.13b}$$

The variables (x_-, t_-) denote the wave phase and (x_+, t_+) describe the slow variation of the envelope, as illustrated schematically in Figure 3.9. Upon taking the average over x_+, t_+, a short calculation then gives,

$$C(x_-, t_-) = \sum_k |E_k|^2 \exp[i(kx_- - \omega_k t_-)]. \tag{3.14}$$

Evaluating x_- along unperturbed orbits, so that

$$x_- = x_{0-} + vt_-, \tag{3.15}$$

and assuming, for convenience, a continuous spectrum of the form

$$|E_k|^2 = \frac{E_0^2}{\Delta k}\left[\left(\frac{k-k_0}{\Delta k}\right)^2 + 1\right]^{-1} \tag{3.16}$$

then allows us to write the correlation function $C(x_-, t_-)$ in the simple, explicit form,

$$C(x_-, \tau) = \int \frac{\mathrm{d}k}{\Delta k} \frac{E_0^2\, \mathrm{e}^{ikx_{0-}} \mathrm{e}^{i(kv-\omega_k)t_-}}{\left[\left(\frac{k-k_0}{\Delta k}\right)^2 + 1\right]}. \tag{3.17}$$

Here $|E_0|^2$ is the spectral intensity, Δk is the spectral width, and k_0 is the centroid of the spectral distribution. Expanding $kv - \omega_k$ as,

$$kv - \omega_k \simeq k_0 v - \omega_{k_0} + \Delta(kv - \omega_k)(k - k_0) + \cdots,$$

the integral in Eq.(3.17) can now easily be performed by residues, yielding,

$$C(x_-, \tau) = 2\pi E_0^2\, \mathrm{e}^{ikx_{0-}} \mathrm{e}^{i(k_0 v - \omega_{k_0})\tau} \times \exp\left[-\Delta|kv - \omega_k|\tau - |\Delta k| x_{0\_}\right]. \tag{3.18}$$

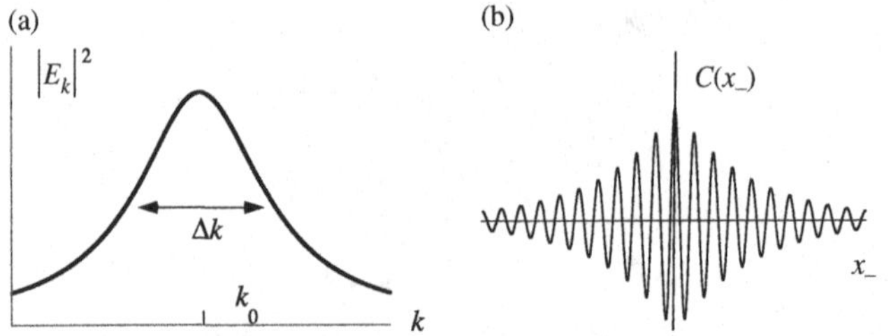

Fig. 3.10. An example of the power spectrum of electric field fluctuation, which is characterized by the peak and width of the wave number (a). Correlation function is given in (b).

As is illustrated in Figure 3.10, Eq.(3.18) is an explicit result for the two-point correlation, constructed using a model spectrum. Equation (3.18) reveals that correlations decay in time according to,

$$C(x_-, \tau) \sim \exp[-\Delta|kv - \omega_k|\tau], \tag{3.19}$$

that is by frequency dispersion $\Delta(\omega_k)$ *and* its interplay with particle streaming, via $\Delta(kv)\tau$. Note that it is, in fact, the *width of the Doppler-shifted frequency which sets the spectral auto-correlation time,* $\tau_{\rm ac}$. Now,

$$1/\tau_{\rm ac} = |\Delta(kv - \omega_k)| = |(v - v_{\rm gr})\Delta k|, \tag{3.20a}$$

so, *for resonant particles* with $v = \omega/k = v_{\rm ph}$,

$$1/\tau_{\rm ac} = |(v_{\rm ph} - v_{\rm gr})\Delta k|, \tag{3.20b}$$

which is identical to the heuristic estimate of the pattern lifetime $\tau_{\rm L}$ given in Eq.(3.7). Thus, we have indeed demonstrated that the dispersion in the Doppler shifted frequency as 'seen' by a resonant particle (moving along an unperturbed orbit) sets the spectral auto-correlation time and thus the lifetime of the field pattern which the particle senses.

We now summarize this discussion by reviewing the basic time-scales characteristic of quasi-linear theory, and the relationships between them which are necessary for the applicability of quasi-linear theory. The basic temporal rates (i.e. inverse time-scales $\sim 1/\tau$) are summarized in Table 3.1. As discussed above, several conditions must be satisfied for quasi-linear theory to be relevant. These are:

$$1/\tau_{\rm b} < 1/\tau_{\rm ac} \tag{3.21a}$$

for the use of unperturbed orbits (linear response theory) to be valid,

$$1/\tau_{\rm relax} \ll 1/\tau_{\rm ac}, \gamma_k \tag{3.21b}$$

Table 3.1. *Basic time-scales of quasi-linear theory*

$1/\tau_{\rm ac} = \lvert(v_{\rm gr} - v_{\rm ph})\Delta k\rvert$	The auto-correlation time or lifetime of the electric field pattern, as sensed by resonant particles.
γ_k	The wave growth or damping rate, as determined by the linear dispersion relation.
$1/\tau_{\rm b} = k\left(\frac{q\phi_{\rm res}}{m}\right)^{1/2}$	The 'bounce' or 'trapping time' for resonant particles in the total packet potential.
$1/\tau_{\rm relax} = \frac{1}{\langle f\rangle}\frac{\partial\langle f\rangle}{\partial t}$	The rate of slow relaxation of the average distribution function.

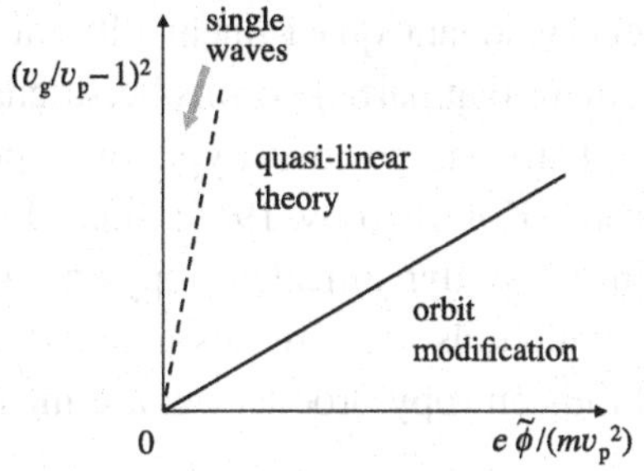

Fig. 3.11. Range of applicability for the quasi-linear theory, Eq.(3.11). Amplitude is normalized by the particle energy at phase velocity on the horizontal axis. The vertical axis shows the magnitude of dispersion, i.e. the difference between the group velocity and phase velocity.

for the closure of the $\langle f\rangle$ equation to be meaningful,

$$1/\tau_{\rm relax} < \gamma_k < 1/\tau_{\rm ac} \tag{3.21c}$$

for the quasi-linear equations to be applicable.

Of course, the irreversibility of resonant quasi-linear diffusion follows from the stochasticity of particle orbits, which in turn requires that the Chirikov overlap criterion (Eq.(3.6)) be met (see Figure 3.11). In retrospect, we see that applicability of the 'trivial' quasi-linear theory naively follows from several rather precise and sometimes even subtle conditions!

3.2.5 Note on entropy production

At this conclusion of our discussion on the origin of irreversibility in quasi-linear theory, it is appropriate to comment briefly on entropy. The Vlasov equation leaves entropy invariant, since entropy

$$S = \int \mathrm{d}v \int \mathrm{d}x\, s(f),$$

and,

$$\frac{\mathrm{d}f}{\mathrm{d}t} = 0$$

in a Vlasov plasma. The quasi-linear equation involves a *coarse graining*, as it describes the evolution of $\langle f \rangle$, not f. Hence, it should be no surprise that quasi-linear relaxation can produce entropy, since such entropy production is intrinsic to phenomena such as resonant particle heating, etc., which occur in the course of the evolution and saturation of plasma turbulence. In this regard, recall that the irreversible quasi-linear evolution of $\langle f \rangle$ requires the onset of chaos, Eq.(3.6). A deeper connection between resonant quasi-linear diffusion and entropy production enters via the requirement that particle orbits be stochastic. Strictly speaking, "stochastic" means that at least one positive Lypunov exponent exists, so the KS (Kolmogorov–Sinai) entropy (Kolmogorov, 1958; Sinai, 1959; Sinai, 1994) is positive, i.e. $h > 0$. Any definition of dynamical entropy entails the definition of some partition of phase space, which also constitutes a coarse graining. Thus, we see that coarse graining, and thus entropy production, are intrinsic to the foundations of quasi-linear theory.

3.3 Energy and momentum balance in quasi-linear theory

3.3.1 Various energy densities

It is no surprise that energy and momentum conservation are non-trivial concerns, since the basic quasi-linear equation for $D(v)$, Eq.(3.4d), makes a clear distinction between resonant and non-resonant particles: Resonant particles, for which,

$$D_{\mathrm{R}}(v) = \sum_k \frac{q^2}{m^2} |E_k|^2 \pi \delta(\omega - kv), \tag{3.22a}$$

exchange energy with waves irreversibly, via Landau resonance. Note that the resonant diffusion coefficient does not depend on the wave growth rate. Non-resonant particles, for which,

$$D_{\mathrm{NR}}(v) \simeq \sum_k \frac{q^2}{m^2} |E_k|^2 \frac{|\gamma_k|}{(\omega - kv)^2}, \tag{3.22b}$$

support the wave by oscillating in it. Their motion is *reversible*, and their quiver velocities increase or decrease with the wave amplitude. Hence, $D_{\mathrm{NR}}(v)$ is explicitly proportional to $|\gamma_k|$, in contrast to D_{R}. It is interesting to note that for $\omega \gg kv$, the non-resonant diffusion reduces to,

$$D_{\rm NR} = \sum_k \frac{q^2}{m^2}\left|\tilde{E}_k\right|^2 \frac{|\gamma_k|}{\omega^2}$$
$$= \left(\frac{1}{n_0 m}\right)\left|\frac{\partial}{\partial t}E_{\rm p}\right|, \tag{3.23a}$$

where $E_{\rm p}$ is the ponderomotive (or quiver) energy density,

$$E_{\rm p} = \sum_k \frac{1}{2}\frac{n_0 q^2}{m^2}\frac{\left|\tilde{E}_k\right|^2}{\omega_k^2}. \tag{3.23b}$$

This observation illustrates that non-resonant diffusion is simply due to reversible quivering of particles in the wave field. Thus, *non-resonant diffusion cannot produce entropy*. Indeed, to understand non-resonant diffusion and energetics in quasi-linear theory, it is important to keep in mind that the familiar quantity W, the total wave energy density,

$$W = \frac{\partial}{\partial \omega_k}(\omega\epsilon)\bigg|_{\omega_k}\frac{|E_k|^2}{8\pi}, \tag{3.24}$$

contains contributions from *both* the electric field energy density ($E^{\rm f}$)

$$E^{\rm f} = |E_k|^2/8\pi, \tag{3.25a}$$

and the non-resonant particle kinetic energy density $E^{\rm nr}_{\rm kin}$. This point is illustrated by considering simple Langmuir oscillations of amplitude E_0 with $\epsilon = 1-\omega_{\rm p}^2/\omega^2$, for which $E^{\rm f} = |E_0|^2/8\pi$, while $W = |E_0|^2/4\pi$. A short calculation reveals that the remaining contribution of $|E_0|^2/8\pi$ is simply the non-resonant particle kinetic energy density ($E^{\rm nr}_{\rm kin}$), which is equal in magnitude to the $E^{\rm f}$ for Langmuir waves. This is easily seen, since $E^{\rm nr}_{\rm kin} = (1/2)nm\,|\tilde{v}|^2$ and $\tilde{v} = q\tilde{E}/\omega m$. Together these give $E^{\rm nr}_{\rm kin} = (1/8)\omega_{\rm p}^2\left|\tilde{E}\right|^2/8\pi\omega^2$, so that for $\omega = \omega_{\rm p}$, the identity $E^{\rm nr}_{\rm kin} = E^{\rm f}$ is clear. Indeed, the thrust of this discussion suggests that since quasi-linear theory divides the particles into *two* classes, namely resonant and non-resonant, there should be *two* ways of balancing the total energy budget. Below, we show that an energy conservation relation can be formulated either as a balance of

resonant particle kinetic energy density	$E^{\rm res}_{\rm kin}$	versus	total wave energy density	W

or of

particle kinetic energy density	E_{kin}	versus	electric field energy density	E^{f}

Momentum balance exhibits similar duality.

3.3.2 Conservation laws

To prove conservation of energy between resonant particles and waves, one must first determine the rate of change of total particle kinetic energy density, E_{kin} by taking the energy moment of the Vlasov equation, i.e.

$$\begin{aligned}\frac{\partial}{\partial t}E_{\text{kin}} &= \frac{\partial}{\partial t}\int \mathrm{d}v\, \frac{mv^2}{2}\langle f\rangle \\ &= \int \mathrm{d}v\, qv\langle \tilde{E}\delta f\rangle,\end{aligned} \tag{3.26}$$

where Eq.(3.2b) is substituted and the partial integration is performed. Because we are studying the balance in the framework of the quasi-linear theory, δf is approximated by the linear Vlasov response, so Eq.(3.26) gives,

$$\frac{\partial}{\partial t}E_{\text{kin}} = -i\int \mathrm{d}v\, \frac{vq^2}{m}\sum_k |E_k|^2\left(\frac{P}{\omega - kv} - i\pi\delta\left(\omega - kv\right)\right)\frac{\partial\langle f\rangle}{\partial v}, \tag{3.27}$$

where P indicates the principal part of the integral and the familiar Plemelj formula has been used to decompose the linear response into resonant and non-resonant pieces. Choosing the resonant piece, we can express the rate of change of resonant particle kinetic energy as

$$\begin{aligned}\frac{\partial}{\partial t}E_{\text{kin}}^{\text{res}} &= -\int \mathrm{d}v\, \frac{\pi q^2}{m}\sum_k \frac{\omega}{k|k|}\delta\left(\frac{\omega}{k} - v\right)\frac{\partial\langle f\rangle}{\partial v}\,|E_k|^2 \\ &= -\frac{\pi q^2}{m}\sum_k \frac{\omega}{k|k|}\left.\frac{\partial\langle f\rangle}{\partial v}\right|_{\omega/k}|E_k|^2.\end{aligned} \tag{3.28}$$

To relate Eq.(3.28) to the change in wave energy density (using Eq.(3.24)), we may straightforwardly write,

$$\begin{aligned}\frac{\partial W}{\partial t} &= \sum_k 2\gamma_k \left.\frac{\partial}{\partial\omega}(\omega\epsilon)\right|_{\omega_k}\frac{|E_k|^2}{8\pi} \\ &= \sum_k 2\gamma_k\omega_k \left.\frac{\partial\epsilon}{\partial\omega}\right|_{\omega_k}\frac{|E_k|^2}{8\pi}.\end{aligned} \tag{3.29}$$

Now, for γ_k:

$$\epsilon = 1 + \frac{\omega_{\mathrm{p}}^2}{k} \int \mathrm{d}v \, \frac{\partial \langle f \rangle / \partial v}{\omega - kv}, \tag{3.30}$$

and,

$$\epsilon_{\mathrm{r}} \left(\omega_k + i\gamma_k \right) + i \mathrm{Im}\, \epsilon = 0, \tag{3.31}$$

so,

$$\gamma_k = - \frac{\mathrm{Im}\, \epsilon}{\left(\partial \epsilon_{\mathrm{r}} / \partial \omega \right)\big|_{\omega_k}}. \tag{3.32}$$

Substituting Eq.(3.32) into Eq.(3.29) gives:

$$\frac{\partial W}{\partial t} = - \sum \omega_k \mathrm{Im}\, \epsilon \left(k, \omega_k \right) \frac{|E_k|^2}{4\pi}. \tag{3.33}$$

However, from Eq.(3.30) we have,

$$\mathrm{Im}\, \epsilon \left(k, \omega_k \right) = - \frac{\pi \omega_{\mathrm{p}}^2}{|k| k} \left. \frac{\partial \langle f \rangle}{\partial v} \right|_{\omega/k}. \tag{3.34}$$

Substituting Eq.(3.34) into Eq.(3.33) then gives,

$$\frac{\partial W}{\partial t} = \frac{\pi q^2}{m} \sum_k \frac{\omega_k}{k|k|} \left. \frac{\partial \langle f \rangle}{\partial v} \right|_{\omega/k} |E_k|^2, \tag{3.35}$$

where the density dependence of $\langle f \rangle$ has been factored out, for convenience. Comparing Eq.(3.28) and Eq.(3.35), we see that, within the scope of quasi-linear theory, we have demonstrated that,

$$\frac{\partial}{\partial t} \left(E_{\mathrm{kin}}^{\mathrm{res}} + W \right) = 0, \tag{3.36}$$

i.e. that energy is conserved between collective modes ("waves") and resonant particles. Equation (3.36) is the fundamental energy conservation relation for quasi-linear theory.

Several comments are in order here. First, the quasi-linear energy conservation relation proved above is a special case of the more general Poynting theorem for plasma waves, which states:

$$\frac{\partial W}{\partial t} + \nabla \cdot S + Q = 0, \tag{3.37}$$

i.e. wave energy density W is conserved against wave radiation ($\nabla \cdot S$, where S is the wave energy density flux) and dissipation ($Q = \langle E \cdot J \rangle$), where E is the electric field and J is the current. For a homogeneous system $\nabla \cdot S = 0$, so the Poynting relation reduces to just $\partial W/\partial t + \langle E \cdot J \rangle = 0$. Computing the plasma current J_k using the linear response $\tilde{f}_k$ then yields an expression identical to Eq.(3.36). The physics here is a simple consequence of the fact that since only resonant particles "see" a DC electric field, so only they can experience a time averaged $\langle E \cdot J \rangle$.

3.3.3 Roles of quasi-particles and particles

A second element of this discussion reveals an alternative form of the energy theorem. As discussed above, the total wave energy density W may be decomposed into pieces corresponding to the field energy density (E^{f}) and the non-resonant particle kinetic energy density ($E_{\mathrm{kin}}^{\mathrm{nr}}$). In these terms, the quasi-linear energy conservation theorem can be written as shown below. We have demonstrated explicitly Eq.(3.36) that,

$$\frac{\partial}{\partial t} W + \frac{\partial}{\partial t} E_{\mathrm{kin}}^{\mathrm{res}} = 0,$$

but also have noted the physically motivated decomposition,

$$W = E^{\mathrm{f}} + E_{\mathrm{kin}}^{\mathrm{nr}},$$

so we have,

$$\frac{\partial}{\partial t}\left(E^{\mathrm{f}} + E_{\mathrm{kin}}^{\mathrm{nr}}\right) + \frac{\partial}{\partial t} E_{\mathrm{kin}}^{\mathrm{res}} = 0.$$

Then, a re-grouping gives,

$$\frac{\partial}{\partial t} E^{\mathrm{f}} + \frac{\partial}{\partial t}\left(E_{\mathrm{kin}}^{\mathrm{res}} + E_{\mathrm{kin}}^{\mathrm{nr}}\right) = 0,$$

where $E_{\mathrm{kin}}^{\mathrm{res}} + E_{\mathrm{kin}}^{\mathrm{nr}} = E_{\mathrm{kin}}$, the total particle kinetic energy density. Thus we arrive at an alternative form of the energy conservation theorem, namely that

$$\frac{\partial}{\partial t}(E^{\mathrm{f}} + E_{\mathrm{kin}}) = 0, \tag{3.38}$$

i.e. electric field energy density E^{f} is conserved against *total* particle kinetic energy density E_{kin}, from Eq.(3.27), *without* the (Plemelj) decomposition of the response into resonant and non-resonant pieces. Returning to Eq.(3.27), we proceed as,

$$\frac{\partial}{\partial t} E_{\mathrm{kin}} = -\sum_k \int \mathrm{d}v \, \frac{\omega_{\mathrm{p}}^2}{k}(kv) \frac{|E_k|^2}{4\pi} \left(\frac{1}{\omega - kv}\right) \frac{\partial \langle f \rangle}{\partial v}. \tag{3.39a}$$

Table 3.2. *Energy balance theorems for quasi-linear theory*

Constituents	Particles	Resonant ($v = \omega/k$) $\rightarrow E_{\mathrm{kin}}^{\mathrm{res}}$ Non-resonant ($v \neq \omega/k$) $\rightarrow E_{\mathrm{kin}}^{\mathrm{nr}}$
	Fields	Electric field energy E^{f} Waves, collective modes $\rightarrow$ total wave energy density (W)
Perspectives	Resonant particles versus waves balance Particles versus fields balance	
Relations and conservation balances	$\frac{\partial}{\partial t}\left(E_{\mathrm{kin}}^{\mathrm{res}} + W\right) \equiv 0 \leftrightarrow$ resonant particles versus waves $\frac{\partial}{\partial t}\left(E_{\mathrm{kin}} + E^{\mathrm{f}}\right) \equiv 0 \leftrightarrow$ total particles versus electric field	

Now, using Eq.(3.30) for $\epsilon(k, \omega)$ we can write,

$$\begin{aligned}\frac{\partial}{\partial t} E_{\mathrm{kin}} &= -i \sum_k \frac{|E_k|^2}{4\pi} \int \mathrm{d}v \, \frac{\omega_{\mathrm{p}}^2}{k} \, (kv - \omega + \omega) \, \frac{1}{(\omega - kv)} \frac{\partial \langle f \rangle}{\partial v} \\ &= -i \sum_k \frac{|E_k|}{4\pi} \int \mathrm{d}v \, \frac{\omega_{\mathrm{p}}^2}{k} \frac{\omega}{\omega - kv} \frac{\partial \langle f \rangle}{\partial v},\end{aligned} \tag{3.39b}$$

as energy is real. Since $\epsilon\,(k, \omega_k) = 0$, by definition of ω_k, we thus obtain,

$$\begin{aligned}\frac{\partial}{\partial t} E_{\mathrm{kin}} &= i \sum_k \frac{|E_k|^2}{4\pi} \omega_k \\ &= -\sum_k \frac{|E_k|^2}{8\pi} \, (2\gamma_k) = -\frac{\partial}{\partial t} E^{\mathrm{f}}.\end{aligned} \tag{3.39c}$$

This completes the explicit proof of the relation $\partial(E_{\mathrm{kin}} + E^{\mathrm{f}})/\partial t = 0$. The energy conservation laws of quasi-linear theory are summarized in Table 3.2. As indicated in the table, the two forms of the quasi-linear energy conservation theorem are a consequence of the two possible conceptual models of a turbulent plasma, namely as an ensemble of either:

a) quasi-particles (waves) and resonant particles, for which $\partial(W + E_{\mathrm{kin}}^{\mathrm{res}})/\partial t = 0$ (Eq.(3.36)) is the appropriate conservation theorem;

or

b) particles (both resonant and non-resonant) and electric fields, for which $\partial(E_{\mathrm{kin}} + E^{\mathrm{f}})/\partial t = 0$ (Eq.(3.38)) is the appropriate conservation theorem.

This distinction is possible since non-resonant diffusion can be counted either as:

a) the sloshing of particles which support the wave energy density;

or as,

b) part of the total particle kinetic energy density.

While both views are viable and valid, we will adopt the former in this book, as it is both appealingly intuitive and physically useful.

Finally, we note in passing that it is straightforward to show that the sum of resonant particle momentum and wave momentum ($P_{\mathrm{W}} = k\,(\partial\epsilon/\partial\omega)_k \times |E_k|^2/8\pi$) is conserved. The proof closely follows the corresponding one for energy, above. No corresponding relation exists for particles and fields, since, of course, purely electrostatic fields have no momentum. In this case, the total particle momentum density is simply a constant. In electromagnetic problems, where the presence of magnetic fields allows a non-zero field momentum density (proportional to the Poynting flux), exchange of momentum between particles and fields is possible, so a second momentum conservation theorem can be derived.

3.4 Applications of quasi-linear theory to bump-on-tail instability

As a complement to the rather general and theoretical discussion thus far, we now discuss two applications of quasi-linear theory. First, to the classic problem of the *bump-on-tail* instability in one dimension and then to transport and relaxation driven by drift wave turbulence in a 3D magnetized plasma. We discuss these two relatively simple examples in considerable depth since they constitute fundamental paradigms, upon which other applications are built.

3.4.1 Bump-on-tail instability

The bump-on-tail instability occurs in the region of positive phase velocities which appears when a gentle beam is driven at high velocities, on the tail of a Maxwellian. The classic configuration of the bump-on-tail is shown in Figure 3.2(a). Based upon our previous discussion, we can immediately write down the set of quasi-linear equations:

$$\epsilon\,(k, \omega_k) = 0, \tag{3.40a}$$

$$\frac{\partial \langle f \rangle}{\partial t} = \frac{\partial}{\partial v} D \frac{\partial \langle f \rangle}{\partial v}, \tag{3.40b}$$

$$D = D_{\mathrm{R}} + D_{\mathrm{NR}} = \sum_k \frac{q^2}{m^2} |E_k|^2 \left\{ \pi\delta(\omega - kv) + \frac{|\gamma_k|}{\omega^2} \right\}, \tag{3.40c}$$

$$\frac{\partial}{\partial t}\left(\frac{|E_k|^2}{8\pi}\right) = 2\gamma_k\left(\frac{|E_k|^2}{8\pi}\right). \tag{3.40d}$$

Initially $\epsilon\,(k, \omega_k)$ should be calculated using the distribution shown in Figure 3.2(a). It is interesting to note that the structure of the bump-on-tail distribution enables us to clearly separate and isolate the regions of resonant and non-resonant diffusion and heating, etc. In particular, since $\partial\langle f\rangle/\partial v > 0$ for a velocity interval on the tail, waves will be resonantly excited in that interval and particles in that region will undergo resonant diffusion. Similarly, since bulk particles are not resonant but do support the underlying Langmuir wave we can expect them to undergo non-resonant diffusion, which can alter their collective kinetic energy but not their entropy.

3.4.2 Zeldovich theorem

Before proceeding with the specific calculation for the bump-on-tail problem, it is useful to discuss the general structure of relaxation in a Vlasov plasma, and to derive a general constraint on the evolution of the mean distribution function $\langle f\rangle$ and on its end state. This constraint is a variant of a theorem first proved by Ya. B. Zeldovich in the context of transport of magnetic potential in 2D MHD turbulence (Zeldovich, 1957). Proceeding, then, the Boltzmann equation states that,

$$\frac{\mathrm{d}}{\mathrm{d}t}(\delta f) = -\frac{q}{m}E\frac{\partial\langle f\rangle}{\partial v} + C\,(\delta f), \tag{3.41}$$

where

$$\frac{\mathrm{d}}{\mathrm{d}t} = \frac{\partial}{\partial t} + v\frac{\partial}{\partial x} + \frac{q}{m}E\frac{\partial}{\partial v},$$

i.e. that fluctuation phase space density is conserved up to collisions (denoted by $C(\delta f)$) and relaxation of the phase space density gradients. Of course, total phase space density is conserved along particle orbits, up to collisions, *only*. Multiplying Eq.(3.41) by δf and averaging then yields,

$$\frac{\mathrm{d}}{\mathrm{d}t}\int \mathrm{d}v\left\langle\delta f^2\right\rangle = \int \mathrm{d}v\left[-\frac{q}{m}\langle E\delta f\rangle\frac{\partial\langle f\rangle}{\partial v} + \langle\delta f C\,(\delta f)\rangle\right]. \tag{3.42}$$

Here, the average implies an integration over space (taken to be periodic), so $\langle\,\rangle = \int \mathrm{d}x$, as well as the explicit integral over velocity. Thus, $\langle \mathrm{d}/\mathrm{d}t\rangle \to \partial/\partial t$. Furthermore, it is useful for physical transparency to represent $C(\delta f)$ using a Crook approximation $C(\delta f) = -\nu(\delta f)$, so that Eq.(3.42) then becomes,

$$\frac{\mathrm{d}}{\mathrm{d}t}\int \mathrm{d}v\left\langle\delta f^2\right\rangle = \int \mathrm{d}v\left[-\frac{q}{m}\langle E\delta f\rangle\frac{\partial\langle f\rangle}{\partial v} - \nu\left\langle\delta f^2\right\rangle\right].$$

Ignoring collisions for the moment, Eq.(3.42) simply states the relation between mean square fluctuation level and the relaxation of the mean distribution fluctuation embodied by the Vlasov equation, i.e.

$$\frac{\mathrm{d}f}{\mathrm{d}t} = 0, \tag{3.43a}$$

and

$$f = \langle f \rangle + \delta f, \tag{3.43b}$$

so,

$$\frac{\mathrm{d}}{\mathrm{d}t}\left(\langle f \rangle + \delta f\right)^2 = 0. \tag{3.43c}$$

Averaging them gives,

$$\frac{\mathrm{d}}{\mathrm{d}t}\int \mathrm{d}v\, \langle \delta f^2 \rangle = -\int \mathrm{d}v\, \langle f \rangle \frac{\partial \langle f \rangle}{\partial t} = \int \mathrm{d}v\, \langle f \rangle \frac{\partial}{\partial v}\left\langle \frac{q}{m} E\delta f \right\rangle, \tag{3.43d}$$

since, of course,

$$\frac{\partial \langle f \rangle}{\partial t} = -\frac{\partial}{\partial v}\left\langle \frac{q}{m} E\delta f \right\rangle. \tag{3.43e}$$

The content of the relation between the left-hand side and right-hand side of Eq.(3.43d) is obvious – relaxation of $\langle f \rangle$ drives $\langle \delta f^2 \rangle$.

Until now, the calculation has been formal, reflecting only the conservative symplectic structure of the Vlasov–Boltzmann equation. Equation (3.42) is a structure relating fluctuation growth to transport ($\sim(q/m)\langle E\delta f \rangle$) and collisional damping. To make contact with quasi-linear theory, we close Eq.(3.42) by taking $\delta f \to f^c$, the coherent linear response, in $\langle E\delta f \rangle$. (The role of incoherent part $\tilde{f}$ in δf, $\delta f = f^c + \tilde{f}$, is explained in Chapter 8.) This gives the *Zeldovich relation*

$$\frac{\partial}{\partial t}\int \mathrm{d}v\, \langle \delta f^2 \rangle = \int \mathrm{d}v\, D\left(\frac{\partial \langle f \rangle}{\partial v}\right)^2 - \int \mathrm{d}v\, \nu \langle \delta f^2 \rangle, \tag{3.44}$$

which connects fluctuation growth to relaxation and collisional damping. Here D is the quasi-linear diffusion coefficient, including both resonant and non-resonant contributions, i.e.

$$D = D_{\mathrm{R}} + D_{\mathrm{NR}}.$$

3.4.3 Stationary states

The point of this exercise becomes apparent when one asks about the nature of a stationary state, i.e. where $\partial\langle\delta f^2\rangle/\partial t = 0$, which one normally associates with instability saturation. In that case, Eq.(3.44) reduces to,

$$\int \mathrm{d}v\, D_\mathrm{R} \left(\frac{\partial\langle f\rangle}{\partial v}\right)^2 = \int \mathrm{d}v\, \nu\langle\delta f^2\rangle, \tag{3.45}$$

which states that fluctuation growth by resonant instability induced relaxation and transport *must* balance collisional damping in a stationary state. This is the Vlasov analogue of the production–dissipation balance generic to the mixing length theory and to turbulent cascades. Notice that *non-resonant diffusion necessarily vanishes at stationarity, since* $D_\mathrm{NR} \sim |\gamma|$, *explicitly*. With the important *proviso that we assume* δf *does not develop singular gradients, then Eq.*(3.45) *states that for a collisionless* ($\nu \to 0$), *stationary plasma,* $\int \mathrm{d}v\, D_\mathrm{R}\, (\partial\langle f\rangle/\partial v)^2$ *must vanish*. Hence either $\partial\langle f\rangle/\partial v \to 0$, so that the mean distribution function flattens (i.e. forms a *plateau*) at resonance, or $D_\mathrm{R} \to 0$, i.e. the saturated electric field spectrum decays and vanishes. These are the two possible end-states of quasi-linear relaxation. Notice also that Eq.(3.45) states that any deviation from the plateau or $D_\mathrm{R} = 0$ state must occur via the action of collisions, alone. We note that accounting for the effect of resonant phase space density granulations introduces dynamical friction, which modifies the structure of this Zeldovich relation. This is discussed further in Chapter 8.

3.4.4 Selection of stationary state

We now proceed to discuss which state (i.e., $D_R = 0$, or $\partial\langle f\rangle/\partial v = 0$) is *actually* selected by the system by explicitly calculating the time dependence of the resonant diffusivity (R. C. Davidson, 1972).

To determine the time evolution of D_R, it is convenient to first re-write it as,

$$D_\mathrm{R} = 16\pi^2\frac{q^2}{m^2}\int_0^\infty \mathrm{d}k\, E^\mathrm{f}(k)\delta(\omega - kv), \tag{3.46}$$

where $E^\mathrm{f}(k) = |E_k|^2/8\pi$. Then we easily see that $\partial D_\mathrm{R}/\partial t$ is given by,

$$\frac{\partial D_\mathrm{R}}{\partial t} = \frac{16\pi^2 q^2}{m^2 v}\left(2\gamma_{\omega_\mathrm{pe}/v}\right) E^\mathrm{f}\left(\frac{\omega_\mathrm{pe}}{v}\right), \tag{3.47a}$$

where $\omega_k = \omega_\mathrm{pe}$. Since $\gamma_k = \gamma_{\omega_\mathrm{pe}/v} = \pi v^2\omega_\mathrm{pe}\,(\partial\langle f\rangle/\partial v)$, a short calculation then yields,

$$D_{\rm R}(v,t) = D_{\rm R}(v,0)\exp\left[\pi\omega_{\rm pe}v^2\int_0^t {\rm d}t'\,\frac{\partial\langle f\rangle}{\partial v}\right]. \tag{3.47b}$$

Using Eq.(3.46) and the expression for γ_k, we also find that,

$$\frac{\partial\langle f\rangle}{\partial t} = \frac{\partial}{\partial t}\frac{\partial}{\partial v}\left[\frac{D_{\rm R}(v,t)}{\pi\omega_{\rm pe}v^2}\right], \tag{3.48a}$$

so

$$\langle f(v,t)\rangle = \langle f(v,0)\rangle + \frac{\partial}{\partial v}\left(\frac{D_{\rm R}(v,t) - D_{\rm R}(v,0)}{\pi\omega_{\rm pe}v^2}\right). \tag{3.48b}$$

Taken together, Eqs.(3.47b) and (3.48b) simply show that quasi-linear saturation must occur via plateau formation. To see this, assume the contrary, i.e. that $D_{\rm R} \to 0$ as $t \to \infty$. In that case, Eq.(3.48b) states that

$$\langle f(v,t)\rangle = \langle f(v,0)\rangle - \frac{\partial}{\partial v}\left[\frac{D_{\rm R}(v,0)}{\pi\omega_{\rm p}v^2}\right]. \tag{3.49}$$

Since $D_{\rm R}(v,0) = 16\pi^2 q^2 E\left(\omega_{\rm p}/v, 0\right)\left(m^2 v\right)^{-1}$, it follows that

$$\langle f(v,t)\rangle = \langle f(v,0)\rangle - \frac{\partial}{\partial v}\frac{2E^{\rm f}\left(\omega_{\rm p}/v,0\right)}{nmv^2/2}, \tag{3.50}$$

so $\langle f(t)\rangle \cong \langle f(0)\rangle$, up to a small correction of O(initial fluctuation energy/bump energy) $\times$ $(n_{\rm b}/n)$, where the bump density $n_{\rm b}$ satisfies $n_{\rm b}/n \ll 1$. Hence $\langle f(v,t)\rangle \cong \langle f(v,0)\rangle$ to excellent approximation. However, if $D_{\rm R} \to 0$ as $t \to \infty$, damped waves require $\partial\langle f\rangle/\partial v < 0$, so $\langle f(v,t)\rangle$ *cannot* equal $\langle f(v,0)\rangle$, and a contradiction has been established. Thus, the time asymptotic state which the system actually selects is one where a plateau forms for which $\partial\langle f\rangle/\partial v \xrightarrow[t\to\infty]{} 0$, in the region of resonance.

To calculate the actual plateau state, it is important to realize that *two* processes are at work, simultaneously. First, resonant particles will be stochastically scattered, so as to drive $\partial f/\partial v \to 0$ by *filling in* lower velocities. This evolution, shown in Figure 3.12, is similar to the propagation of a front of δf from the bump to lower velocities which fall in between the bulk Maxwellian and the bump-on-tail. The end state of the plateau is shown in Figure 3.12(b). Second, the non-resonant bulk particles will experience a *one-sided heating* (for $v > 0$, only) as waves grow during the plateau formation process. It is important to realize that this heating is *fake heating* and does not correspond to an increase in bulk particle entropy,

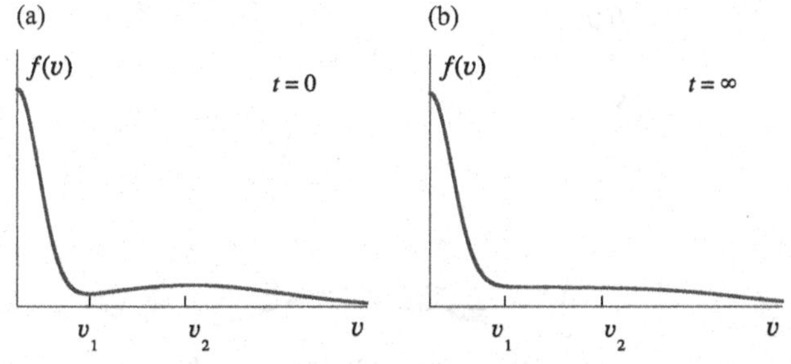

Fig. 3.12. The plateau formation process: initial state (a) and final state (b).

since it originates from non-resonant diffusion. The heating is one sided in order to conserve total momentum between bump-on-tail particles (which slow down) and bulk particles, which therefore must speed up.

To actually calculate the time-asymptotic distribution function and fluctuation saturation level, it is again convenient to separate the evolution into resonant and non-resonant components. The actual saturation level is most expeditiously calculated using the conservation relation $\partial\left(E_{\mathrm{kin}}^{\mathrm{res}} + W\right)/\partial t = 0$. This allows us to equate the change in kinetic energy in the resonant velocity region with the change in the energy of waves in the corresponding region of k values. Thus,

$$\Delta\left(\int_{v_1}^{v_2} \mathrm{d}v\, \frac{mv^2}{2}\langle f\rangle\right) = -2\Delta \int_{k_1}^{k_2} \mathrm{d}k\, E^{\mathrm{f}}(k)\,. \tag{3.51}$$

Here v_1 and v_2 correspond to the lower and upper limits of the range of instability, and, using $k = \omega_{\mathrm{p}}/v$, $k_2 = \omega_{\mathrm{p}}/v_1$, $k_1 = \omega_{\mathrm{p}}/v_2$. The factor of 2 which appears on the right-hand side of Eq.(3.51) reflects the fact that non-resonant particle kinetic energy and field energy (E^{f}) contribute equally to the total wave energy. Then, assuming the fields grow from infinitesimal levels, the total saturated field energy is just

$$\int_{k_1}^{k_2} \mathrm{d}k\, E^{\mathrm{f}}(k) = -\frac{1}{2}\Delta\left(\int_{v_1}^{v_2} \mathrm{d}v\, \frac{mv^2}{2}\langle f\rangle\right). \tag{3.52}$$

To compute the right-hand side explicitly, a graphical, equal area construction is most convenient. Figure 3.13 illustrates this schematically. The idea is that resonant diffusion continues until the upper most of the two rectangles of equal area empties out, toward lower velocity, thus creating a flat spot or plateau between v_1 and v_2. The result of the construction and calculation outlined above gives the saturated field energy and the distortion of the tail.

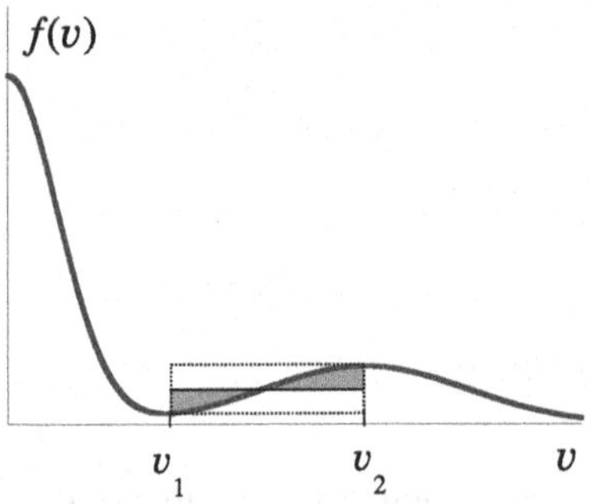

Fig. 3.13. Cartoon of initial and final (plateau) distribution function for resonant region in bump-on-tail instability. Note that quasi-linear diffusion has filled in the initial "hollow" and smoothed out the "bump" centered at v_2.

To determine the change in the bulk distribution function, one must examine the non-resonant diffusion equation. This is,

$$\frac{\partial\langle f\rangle}{\partial t}=\frac{\partial}{\partial t}D_{\rm NR}\frac{\partial\langle f\rangle}{\partial v}\cong\frac{8\pi q^2}{m^2}\int {\rm d}k\, E^{\rm f}(k)\frac{\gamma_k}{\omega_{\rm pe}^2}\frac{\partial^2\langle f\rangle}{\partial v^2}. \tag{3.53a}$$

Here $\gamma_k \geq 0$ for modes in the spectrum, so the absolute value is superfluous. Thus, using the definition of γ_k, we can write the diffusion equation as,

$$\frac{\partial\langle f\rangle}{\partial t}=\left(\frac{1}{nm}\frac{\partial}{\partial t}\int {\rm d}k\, E^{\rm f}(k)\right)\frac{\partial^2\langle f\rangle}{\partial v^2}. \tag{3.53b}$$

Now, defining

$$\tau(t)=\left(\frac{2}{n}\int {\rm d}k\, E^{\rm f}(k,t)\right), \tag{3.54}$$

reduces Eq.(3.53b) to a simple diffusion equation,

$$\frac{\partial\langle f\rangle}{\partial \tau}=\frac{1}{2m}\frac{\partial^2\langle f\rangle}{\partial v^2}, \tag{3.55}$$

with solution (taking the initial bulk distribution to be Maxwellian)

$$\langle f\rangle=\left[\frac{m}{2\pi(T+\tau(t)-\tau(0))}\right]^{1/2}\exp\left[-\frac{mv^2/2}{(T+\tau(t)-\tau(0))}\right]. \tag{3.56}$$

Hence, non-resonant particles of saturation undergo an apparent temperature increase,

$$T\rightarrow T+\frac{2}{n}\int {\rm d}k\left[E^{\rm f}(k,\infty)-E^{\rm f}(k,0)\right], \tag{3.57}$$

so that the bulk electrons appear to be heated by a net increase in field energy. Of course, as is explained in the begining of this subsection, this heating is *fake*, i.e. it does not correspond to an increase in entropy, as it results from non-resonant diffusion. Furthermore it is one sided (i.e. it occurs only for particles with $v > 0$), as a consequence of the need to conserve momentum with beam particles which are slowing down. This result may also be obtained using the conservation relation $\partial \left(E_{\mathrm{kin}} + E^{\mathrm{f}}\right)/\partial t = 0$, and noting that since in this case $\partial \left(E_{\mathrm{kin}}^{\mathrm{res}} + 2E^{\mathrm{f}}\right)/\partial t = 0$, we have,

$$\frac{\partial}{\partial t}\left(E_{\mathrm{kin}}^{\mathrm{nr}} - E^{\mathrm{f}}\right) = 0, \tag{3.58}$$

so $\Delta\left(E_{\mathrm{kin}}^{\mathrm{nr}}\right) = \Delta\left(E^{\mathrm{f}}\right)$, consistent with Eq.(3.57). Note, however, that an explicit computation of $\Delta\left(E^{\mathrm{f}}\right)$ requires an analysis of the distortion of the distribution function in the resonant region. This should not be surprising, since in the bump-on-tail instability, the non-resonant particles are, in some sense, 'slaved' to the resonant particles.

3.5 Application of quasi-linear theory to drift waves

3.5.1 Geometry and drift waves

A second, and very important application of quasi-linear or mean field theory is to drift wave turbulence (Kadomtsev, 1965). A typical geometry is illustrated in Figure 3.14. It is well known that a slab of uniformly magnetized plasma (where $\boldsymbol{B} = B_0\hat{\mathbf{z}}$) which supports cross-field density and/or temperature gradients i.e. $n = n_0(x)$, $T = T_0(x)$, where n_0 and T_0 are the density and temperature profiles, which parameterize the local Maxwellian distribution function, is unstable to low frequency ($\omega < \omega_{ci}$) drift wave instabilities. Such "universal" instabilities,

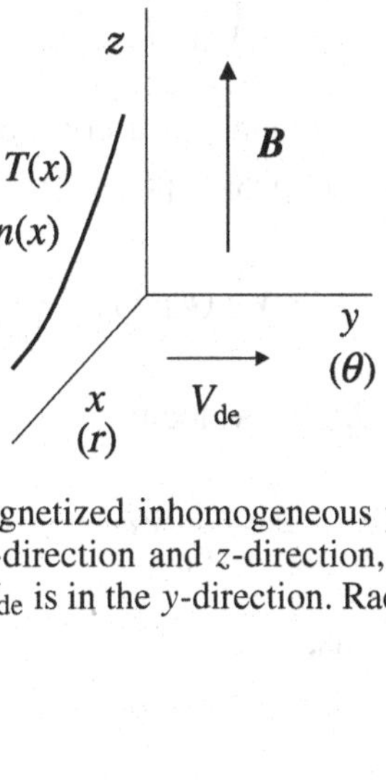

Fig. 3.14. Geometry of magnetized inhomogeneous plasma. The gradients and magnetic field are in the x-direction and z-direction, respectively. The electron diamagnetic drift velocity V_{de} is in the y-direction. Radial and poloidal directions (r, θ) are also illustrated.

which can occur either in collisionless or collisional plasmas, tap expansion free energy stored in radial pressure gradients (i.e. $\partial p/\partial r$) via either collisionless (i.e. wave–particle resonance) or collisional dissipation (Krall and Trivelpiece, 1973; Miyamoto, 1976; Wesson, 1997). Indeed, the collisionless electron drift wave is perhaps the simplest kinetic low frequency instability of the myriad which are known to occur in inhomogeneous plasma. A short primer on the linear properties of drift waves may be found in Appendix 1. Here, we proceed to discuss the quasi-linear dynamics of the collisionless, electron-driven drift instability.

In the collisionless electron drift instability, the ion response is hydrodynamic, while the electrons are described by the drift-kinetic equation,

$$\frac{\partial f}{\partial t} + v_z \frac{\partial f}{\partial z} - \frac{c}{B_0} \nabla \phi \times \hat{z} \cdot \nabla f - \frac{|e|}{m_e} E_z \frac{\partial f}{\partial v_z} = 0. \tag{3.59}$$

Equation (3.59) simply states that phase space density f is conserved (i.e. $\mathrm{d}f/\mathrm{d}t = 0$) along the drift orbits,

$$\frac{\mathrm{d}z}{\mathrm{d}t} = v_z,$$
$$\frac{\mathrm{d}v_z}{\mathrm{d}t} = -\frac{|e|E_z}{m_e},$$
$$\frac{\mathrm{d}\boldsymbol{x}}{\mathrm{d}t} = -\frac{c}{B_0} \nabla \phi \times \hat{z}.$$

These orbits combine Vlasov-like dynamics along the magnetic field with $\boldsymbol{E} \times \boldsymbol{B}$ drift across the field. Note that the phase space flow for drift kinetic dynamics in a straight magnetic field is manifestly incompressible, since,

$$\nabla_\perp \cdot \left(\frac{\mathrm{d}\boldsymbol{x}}{\mathrm{d}t} \right) = 0,$$

and parallel dynamics is Hamiltonian. As a consequence, Eq.(3.59) may be rewritten as a continuity equation in phase space, i.e.

$$\frac{\partial f}{\partial t} + \frac{\partial}{\partial z} v_z f + \nabla \cdot (\boldsymbol{v}_\perp f) + \frac{\partial}{\partial v_z} a_z f = 0, \tag{3.60a}$$

where the perpendicular $\boldsymbol{E} \times \boldsymbol{B}$ flow velocity is,

$$\boldsymbol{v}_\perp = -\frac{c}{B_0} \nabla \phi \times \hat{z}, \tag{3.60b}$$

and the parallel acceleration a_z is,

$$a_z = \frac{|e|}{m_e} \nabla_z \phi. \tag{3.60c}$$

Assuming periodicity in the $\hat{z}$-direction and gradients in the $\hat{x}$-direction, averaging Eqs.(3.60b) then yields the mean field equation for $\langle f \rangle$, i.e.

$$\frac{\partial}{\partial t} \langle f \rangle + \frac{\partial}{\partial x} \left\langle \tilde{v}_x \tilde{f} \right\rangle + \frac{\partial}{\partial v_z} \left\langle \tilde{a}_z \tilde{f} \right\rangle = 0. \tag{3.61}$$

In this example, we see that the quasi-linear dynamics are necessarily two dimensional, and evolve $\langle f \rangle$ in a reduced phase space of $\left(x, v_{\parallel}\right)$, which combines position space (r) and velocity space (v_z) evolution. Thus the quasi-linear evolution involves both a radial flux of particles and energy, as well as heating in parallel velocity, as in the 1D Vlasov example. An energy theorem may be derived by constructing the energy moment of Eq.(3.61), i.e. taking a weighted integral $\left(\int \mathrm{d}^3 v \ \left(m_e v^2/2\right) *\right)$ of the drift-kinetic equation. This gives,

$$\frac{\partial}{\partial t} \langle E_{\mathrm{kin}} \rangle + \frac{\partial}{\partial r} Q_e - \langle E_z J_z \rangle = 0, \tag{3.62a}$$

where

$$\langle E_{\mathrm{kin}} \rangle = \int \mathrm{d}^3 v \, \frac{m_e v^2}{2} \langle f \rangle \tag{3.62b}$$

is the kinetic energy density,

$$Q_e = \int \mathrm{d}^3 v \left\langle \tilde{v}_r \frac{1}{2} m_e v^2 \tilde{f} \right\rangle \tag{3.62c}$$

is the fluctuation-included energy flux and,

$$\begin{aligned} \langle E_z J_z \rangle &= \int \mathrm{d}^3 v \, m_e v_z \left\langle \tilde{a}_z \tilde{f} \right\rangle \\ &= \left\langle \nabla_z \phi \int \mathrm{d}^3 v \, |e| v_z \tilde{f} \right\rangle \end{aligned} \tag{3.62d}$$

is the fluctuation-induced heating. Note that in drift kinetics, the *only* possible heating is parallel heating. In a related vein, the drift wave energy density W_{DW} satisfies a Poynting theorem of the form,

$$\frac{\partial}{\partial t} W_{\mathrm{DW}} + \frac{\partial}{\partial r} S_{\mathrm{r}} = -\left\langle E_{\parallel} J_{\parallel} \right\rangle_{\mathrm{R}} \tag{3.63}$$

where S_{r} is the radial wave energy density flux and $\left\langle E_{\parallel} J_{\parallel} \right\rangle_{\mathrm{R}}$ is the heating by *resonant* particles. Equation (3.63) is seen to be the analogue of the wave energy versus resonant particle energy balance we encountered in 1D, since we can use Eq.(3.62a) to write,

$$\langle E_z J_z \rangle_{\mathrm{R}} = \left(\frac{\partial}{\partial t} \langle E_{\mathrm{kin}} \rangle + \frac{\partial}{\partial r} Q_e \right)_{\mathrm{R}}, \tag{3.64}$$

so that Eq.(3.63) then becomes,

$$\frac{\partial}{\partial t} \left(W_{\mathrm{DW}} + \langle E_{\mathrm{kin}} \rangle_{\mathrm{R}} \right) + \frac{\partial}{\partial r} \left(Q_{e,\mathrm{R}} + S_{\mathrm{r}} \right) = 0. \tag{3.65}$$

Likewise, an energy theorem for the evolution of particle plus field energy may be derived in a similar manner. Interestingly, Eq.(3.65) states that the volume-integrated wave-plus-resonant-particle energy is now conserved only up to losses due to transport and wave radiation through the boundary, i.e.

$$\frac{\partial}{\partial t} \int \mathrm{d}r \left(W_{\mathrm{DW}} + \langle E_{\mathrm{kin}} \rangle_{\mathrm{R}} \right) = - \left(Q_{e,\mathrm{R}} + S_{\mathrm{r}} \right) \Big|_{\mathrm{bndry}}. \tag{3.66}$$

In general, transport exceeds radiation, except where $\tilde{n}/n_0 \to 1$, as at the tokamak edge.

3.5.2 *Quasi-linear equations for drift wave turbulence*

To construct the explicit quasi-linear equation for drift wave turbulence, we substitute the linear response $f_{\boldsymbol{k}}^c$ to $\phi_{\boldsymbol{k}}$ into Eq.(3.61), to obtain a mean field equation for $\langle f \rangle$. Unlike the 1D case, here $f_{\boldsymbol{k}}^c$ responds to drive by both spatial and velocity gradients, so,

$$f_{\boldsymbol{k}}^c = \frac{\phi_{\boldsymbol{k}}}{\omega - k_z v_z} L_{\boldsymbol{k}} \langle f \rangle, \tag{3.67a}$$

where $L_{\boldsymbol{k}}$ is the operator,

$$L_{\boldsymbol{k}} = -\frac{c}{B_0} k_\theta \frac{\partial}{\partial r} + \frac{|e|}{m_e} k_z \frac{\partial}{\partial v_z}. \tag{3.67b}$$

Here it is understood that $\omega = \omega(\boldsymbol{k})$ – i.e. all fluctuations are eigenmodes. Then, the quasi-linear evolution equation for $\langle f \rangle$ can be written as,

$$\frac{\partial}{\partial t} \langle f \rangle = \mathrm{Re} \sum_{\boldsymbol{k}} L_{\boldsymbol{k}} |\phi_{\boldsymbol{k}}|^2 \left(\frac{i}{\omega - k_z v_z} \right) L_{\boldsymbol{k}} \langle f \rangle \tag{3.68a}$$

$$\frac{\partial}{\partial t} \langle f \rangle = \frac{\partial}{\partial r} D_{r,r} \frac{\partial}{\partial r} \langle f \rangle + \frac{\partial}{\partial r} D_{r,v} \frac{\partial}{\partial v_z} \langle f \rangle$$

$$+ \frac{\partial}{\partial v_z} D_{v,r} \frac{\partial}{\partial r} \langle f \rangle + \frac{\partial}{\partial v_z} D_{v,v} \frac{\partial}{\partial v_z} \langle f \rangle, \tag{3.68b}$$

where the four diffusion coefficients describe radial diffusion, i.e.

$$D_{\mathrm{r,r}} = \mathrm{Re} \sum_{k} \frac{e^2}{B_0^2} k_\theta^2 \left|\phi_k\right|^2 \frac{i}{\omega - k_z v_z}, \tag{3.68c}$$

velocity diffusion, i.e.

$$D_{\mathrm{v,v}} = \mathrm{Re} \sum_{k} \frac{e^2}{B_0^2} k_z^2 \left|\phi_k\right|^2 \frac{i}{\omega - k_z v_z}, \tag{3.68d}$$

and two cross-terms,

$$D_{\mathrm{r,v}} = \mathrm{Re} \sum_{k} \frac{c}{B_0} \frac{|e|}{m_e} k_\theta k_z \left|\phi_k\right|^2 \frac{i}{\omega - k_z v_z}, \tag{3.68e}$$

$$D_{\mathrm{r,v}} = \mathrm{Re} \sum_{k} \frac{c}{B_0} \frac{|e|}{m_e} k_\theta k_z \left|\phi_k\right|^2 \frac{i}{\omega - k_z v_z}. \tag{3.68f}$$

In general, some spectral asymmetry i.e. $\langle k_\theta k_z \rangle \neq 0$ (where the bracket implies a spectral average) is required for $D_{\mathrm{r,v}} \neq 0$ and $D_{\mathrm{v,r}} \neq 0$. Equation (3.68b) then, is the quasi-linear equation for $\langle f \rangle$ evolution by drift wave turbulence.

It is interesting to observe that the multi-dimensional structure of wave–particle resonance in, and the structure of the wave dispersion relation for, drift wave turbulence have some interesting implications for the auto-correlation time for stochastic scattering of particles by a turbulent fluctuation field. In general, for drift waves $\omega = \omega\left(k_\theta, k_\parallel\right)$, with stronger dependence of $k_\perp$. Then, modelling,

$$\left|\phi_k\right|^2 = \left|\phi_0\right|^2 \left(\frac{\Delta k_\theta}{\left(k_\theta - k_{\theta_0}\right)^2 + \Delta k_\theta^2}\right) \left(\frac{\Delta k_z}{\left(k_z - k_{z_0}\right)^2 + \Delta k_z^2}\right), \tag{3.69}$$

we see that,

$$\begin{aligned} D_{\mathrm{r,r}} &= \mathrm{Re} \int \mathrm{d}k_\theta \int \mathrm{d}k_\parallel \left|\phi\left(k_\theta, k_\parallel\right)\right|^2 \frac{c^2}{B_0^2} k_\theta^2 \left(\frac{i}{\omega - k_z v_z}\right) \\ &\cong \mathrm{Re} \left|\phi_0\right|^2 \frac{c^2}{B_0^2} k_{\theta_0}^2 \\ &\quad \times i \left\{ \left(\omega\left(k_{\theta_0}, k_{z_0}\right) - k_{z_0} v_z\right) \right. \\ &\quad \left. + i \left| \left(\frac{\mathrm{d}\omega}{\mathrm{d}k_z} - v_z\right) \Delta k_z \right| + i \left| \frac{\mathrm{d}\omega}{\mathrm{d}k_\theta} \Delta k_\theta \right| \right\}^{-1}. \end{aligned} \tag{3.70}$$

Hence, the effective pattern decorrelation rate for a *resonant* particle (i.e. one with $v_z = \omega/k_z$) in drift wave turbulence is,

$$\frac{1}{\tau_{\rm ac}} = \left\{ \left| \left(\frac{{\rm d}\omega}{{\rm d}k_z} - \frac{\omega}{k_z} \right) \Delta k_z \right| + \left| \frac{{\rm d}\omega}{{\rm d}k_\theta} \right| |\Delta k_\theta| \right\}$$
$$\cong \left\{ |v_{\rm T} \Delta k_z| + \left| \frac{{\rm d}\omega}{{\rm d}k_\theta} \Delta k_\theta \right| \right\}. \tag{3.71}$$

The constast with 1D is striking. Since particles do not 'stream' in the θ direction, decorrelation due to poloidal propagation is stronger than that due to parallel propagation, which closely resembles the case of 1D. Usually, the effective turbulence field will decorrelate by simple poloidal propagation at $v_{\rm de}$, and by parallel dispersion at the parallel phase velocity, since ${\rm d}\omega/{\rm d}k_z \cong 0$ for drift waves. Thus, quasi-linear diffusion is, in some sense, more robust for 3D drift wave turbulence than for 1D Vlasov turbulence.

3.5.3 *Saturation via a quasi-linear mechanism*

We can obtain some interesting insights into the mechanisms of saturation of drift wave turbulence by considering the process of 2D plateau formation in the r, v_z phase space for $\langle f \rangle$. Initial contours of constant $\langle f \rangle$ are shown in Figure 3.15(a). Evolved level lines, i.e. contours of $\langle f \rangle$ for which $\partial \langle f \rangle / \partial t = 0$, and which thus define the plateau contours of $\langle f \rangle$ at saturation of the instability, are those of which $L_k \langle f \rangle = 0$. Taking $k_z = \omega_k / v_z$, we see that,

$$L_k \langle f \rangle = \frac{k_\theta}{\Omega_e} \frac{\partial}{\partial r} \langle f \rangle + \frac{\omega_k}{v_z} \frac{\partial}{\partial v_z} \langle f \rangle = 0 \tag{3.72}$$

thus defines the structure of the "plateaued" distribution function. Constant $\langle f \rangle$ curves thus satisfy,

$$\frac{k_\theta}{\omega_{\rm ce}} \frac{\langle f \rangle}{\Delta x} + \frac{\omega}{v_z} \frac{\langle f \rangle}{\Delta v_z} = 0, \tag{3.73}$$

so the level curves of $\langle f \rangle$ are defined by,

$$x - \frac{k_\theta v_z^2}{2\omega_k \omega_{\rm ce}} = \text{const} \tag{3.74}$$

at saturation, The change in level contours is shown in Figure 3.15(b). Note then that any spatial transport which occurs due to the drift wave turbulence is inexorably tied to the concomitant parallel heating. This is no surprise, since the

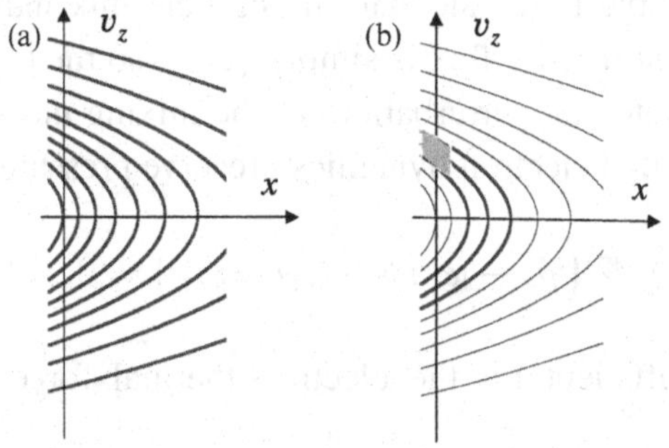

Fig. 3.15. Contour of the mean electron distribution function in the phase space (a). According to the resonance with drift waves, a flattening may occur, and the modification takes place in the level contours of mean distribution, as is illustrated by a shaded region (b).

essence of drift wave instability involves a trade-off between relaxation of density and temperature gradients (which destabilize the waves) and Landau damping (which is stabilizing *but* which also provides the requisite dissipative response in $\langle f \rangle$ to produce instability). In particular, any particle displacement δx from its initial state must be accompanied by a heating (due to Landau damping) δv_z^2, which satisfies,

$$\delta x = \frac{k_\theta}{2\omega_k \omega_{\text{ce}}} \delta v_z^2. \tag{3.75}$$

Since $\omega_k \ll k_\parallel v_{\text{Te}}$, the heating is small i.e. $\delta v_z^2 \sim \alpha v_{\text{Te}}^2$ where $\alpha \ll 1$, so necessarily $\delta x \ll L_n$ – i.e. the maximum displacement is also small, and considerably smaller than the gradient scale length. Hence, the instability quasi-linearly self-saturates at low levels and the resulting particle and/or heat transport is quite modest. To obtain significant steady state transport, the plateau must either be destroyed by collisions or the distribution must be externally "pumped" to maintain it as a Maxwellian.

3.6 Application of quasi-linear theory to ion mixing mode

A third instructive example of quasi-linear theory is that of particle transport due to the ion mixing mode. The ion mixing mode is a type of negative compressibility "ion temperature gradient driven mode" which is likely to occur in collisional plasmas, such as those at the tokamak edge (Rudakov and Sagdeev, 1960; Coppi *et al.*, 1967; Horton, 1999). The mixing mode is driven by ∇T_i, but also transports particles and electron heat. The example of the mixing mode is relevant since it is simple, clear and illustrates:

(a) the application of quasi-linear theory to a purely fluid-like, hydrodynamic instability;
(b) a possible origin of off-diagonal and even counter-gradient transport processes.

The aim of this example is to calculate the particle flux induced by the mixing mode. The quasi-linear density flux is simply $\langle \tilde{v}_r \tilde{n} \rangle$, so the task is to compute the density response to potential perturbation. In the mixing mode, electron inertia is negligible, so the parallel electron dynamics preserve pressure balance, i.e.

$$\nabla \left(\tilde{p}_e - |e| n\phi + \alpha_{\mathrm{T}} n \nabla_{\parallel} \tilde{T}_e \right) = 0, \tag{3.76a}$$

where α_{T} is the coefficient for the electron thermal force (Hinton, 1984), or equivalently,

$$\nabla_{\parallel} \left(\tilde{n}_e + n\tilde{T}_e - |e| n\phi \right) - \alpha_{\mathrm{T}} n \nabla_{\parallel} \tilde{T}_e = 0 \tag{3.76b}$$

$$\frac{\tilde{n}_e}{n} = \frac{|e|\phi}{T_e} - \frac{\tilde{T}_e}{T} (1 + \alpha_{\mathrm{T}}) . \tag{3.76c}$$

To calculate the electron temperature perturbation, we use the temperature evolution equation,

$$\frac{3}{2} n \left(\frac{\partial \tilde{T}_e}{\partial t} + \tilde{v}_e \frac{\mathrm{d}\langle T_e \rangle}{\mathrm{d}x} \right) + nT_e \nabla_{\parallel} \tilde{v}_{\parallel e} = \nabla_{\parallel} n \chi_{\parallel} \nabla_{\parallel} \tilde{T}_e, \tag{3.77a}$$

and the continuity equation,

$$\frac{\partial \tilde{n}}{\partial t} + \tilde{v}_e \frac{\mathrm{d}\langle n \rangle}{\mathrm{d}x} + n \nabla_{\parallel} \tilde{v}_{\parallel e} = 0, \tag{3.77b}$$

to obtain, after a short calculation,

$$\left(\frac{\tilde{T}_e}{T_e} \right)_k = \frac{1}{3\omega/2 + i\chi_{\parallel} k_{\parallel}^2} \left\{ \omega_{*e} \left(\frac{3}{2} \eta_e - 1 \right) \frac{|e|\phi}{T_0} + \omega \frac{\tilde{n}}{n} \right\} . \tag{3.77c}$$

Here $\chi_{\parallel} = v_{\mathrm{Te}}^2 / \nu_e$ is the parallel thermal conductivity, ω_{*e} is the electron diamagnetic frequency and $\eta_e = \mathrm{d} \ln T_e / \mathrm{d} \ln n_e$ represents the temperature gradient parameters. Equations (3.76c) and (3.77c) may then be combined (in the relevant limit of $\chi_{\parallel} k_{\parallel}^2 \gg \omega$) to yield the density perturbation,

$$\begin{aligned} \left(\frac{\tilde{n}_e}{n} \right)_k &= \frac{|e|\phi_k}{T_e} \left\{ 1 + \frac{i(1+\alpha_{\mathrm{T}})}{\chi_{\parallel} k_{\parallel}^2} \left(\omega - \omega_{*e} + \frac{3}{2} \omega_{*e} \eta_e \right) \right\} \\ &\approx \frac{|e|\phi_k}{T_e} \left\{ 1 + \frac{i(1+\alpha_{\mathrm{T}})}{\chi_{\parallel} k_{\parallel}^2} \left(-\omega_{*e} + \frac{3}{2} \omega_{*T_e} \right) \right\} , \end{aligned} \tag{3.78}$$

since the mixing mode has $\omega \approx 0$. Here ω_{*T_e} is just the diamagnetic frequency computed with the electron temperature gradient. Thus, the mixing mode driven particle flux is,

$$\langle \tilde{v}_r \tilde{n}_e \rangle = -D \frac{\partial \langle \tilde{n} \rangle}{\partial x} + V \langle \tilde{n} \rangle, \tag{3.79a}$$

where,

$$D = (1 + \alpha_{\mathrm{T}}) \sum_k \frac{c^2}{B_0^2} \frac{k_\theta^2 \, |\phi_k|^2}{\chi_\parallel k_\parallel^2} \tag{3.79b}$$

$$V = \frac{3}{2} (1 + \alpha_{\mathrm{T}}) \sum_k \frac{c^2}{B_0^2} \frac{k_\theta^2 \, |\phi_k|^2}{\chi_\parallel k_\parallel^2} \frac{1}{\langle T \rangle} \frac{\mathrm{d} \langle T \rangle}{\mathrm{d}x}. \tag{3.79c}$$

Observe that in this example, the quasi-linear particle flux consists of two pieces, the 'usual' Fickian diffusive flux down the density gradient $(-D \times \partial \langle n \rangle / \partial x)$ *and* a convective contribution ($\sim V \langle n \rangle$). It is especially interesting to note that for normal temperature profiles (i.e. $\mathrm{d}\langle T \rangle / \mathrm{d}x < 0$), $V < 0$, so the convective flux is *inward*, and opposite to the diffusive flux! Note that for $\left| \left(1/\langle T \rangle\right) \left(\mathrm{d}\langle T \rangle / \mathrm{d}x\right) \right| > \left| \left(1/\langle n \rangle\right) \left(\mathrm{d}\langle n \rangle / \mathrm{d}x\right) \right|$, the *net* particle flux is consequently inward, and "up" the density gradient. This simple example is typical of a broad class of phenomena manifested in quasi-linear theory which are classified as off-diagonal, gradient-driven fluxes. Off-diagonal inward flows are frequently referred to as a "pinch" (Itoh *et al.*, 1999; Coppi and Spight, 1978; Terry, 1989; Ishichenko *et al.*, 1996; Naulin *et al.*, 1998; Garbet, 2003; Garbet *et al.*, 2005; Estrada-Mila *et al.*, 2005; Angioni and Peeters, 2006; Hahm *et al.*, 2007; Diamond *et al.*, 2008). The temperature gradient-driven pinch described here is sometimes referred to as a thermo-electric pinch. Pinch effects are of great interest in the context of laboratory plasmas, since they offer a possible explanation of profiles which peak on axis, in spite of purely edge fueling. To this end, note that for $V > 0$, the particle flux vanishes for $\left(1/\langle n \rangle\right) \left(\mathrm{d}\langle n \rangle / \mathrm{d}x\right) = V / D$, thus tying the profile scale length to the inward convective velocity. The allusion to 'off-diagonal' refers, of course, to the Onsager matrix which relates the vector of fluxes to the vector of thermodynamic forces. While the diagonal elements of the quasi-linear Onsager matrix are always positive, the off-diagonal elements can be negative, as in this case, and so can drive 'inward' or 'up-gradient' fluxes. Of course, the net entropy production *must* be positive since relaxation occurs. In the case of the ion mixing mode, which is ∇T_i-driven, the entropy produced by ion temperature profile relaxation must exceed the entropy 'destroyed' by the inward particle flux. This requires,

$$\frac{\mathrm{d}S}{\mathrm{d}t} = \int \mathrm{d}r \left\{ \chi_i \left(\frac{1}{\langle T \rangle} \frac{\partial \langle T \rangle}{\partial x} \right)^2 - \langle \tilde{v}_r \tilde{n}_e \rangle \frac{1}{\langle n \rangle^2} \frac{\mathrm{d} \langle n \rangle}{\mathrm{d}x} \right\} > 0, \tag{3.80}$$

where χ_i is the turbulent thermal diffusivity. In practice, satisfaction of this inequality is assured for the ion mixing mode by the ordering $\chi k_{\parallel}^2 \gg \omega_k$, which guarantees that the effective correlation time in χ_i and the ion heat flux is longer than that in the particle flux. We remark that the up-gradient flux is similar to the phenomenon of chemotaxis.

3.7 Nonlinear Landau damping

In this chapter, the quasi-linear response of the perturbation to the mean is explained. The perturbation technique, which is the fundamental element in the procedure, can be extended to higher orders. The nonlinear interactions, which include the higher-order terms, are explained in detail in the next chapters. Before going into a systematic explanation of the interactions in turbulent fluctuations, we briefly describe here the perturbations to higher orders in fluctuation amplitude (Manheimer and Dupree, 1968; Sagdeev and Galeev, 1969; R. C. Davidson, 1972; Krall and Trivelpiece, 1973). The method, which is based on the expansion and truncation of higher-order terms, has limited applicability to turbulence. However, this method can illuminate one essential element in nonlinear interactions, i.e. the Landau resonance of a beat mode. This process is known as 'nonlinear Landau damping', and merits illustration before developing a systematic explanation of coupling in turbulence.

Consider the perturbed electric field (in one-dimensional plasma here for the transparency of the arguments)

$$\frac{\mathrm{d}}{\mathrm{d}t} v = \frac{e}{m} E(x, t) = \frac{e}{m} \sum_k E_k \exp(ikx - i\omega t),$$

where the frequency ω is considered to satisfy the dispersion relation $\omega = \omega_k$. The turbulence is weak, and fluctuations are taken as the sum of linear eigenmodes. (The case of strong turbulence, in which broad band fluctuations are dominantly excited, is not properly treated by these expansions and is explained in the following chapters.) The issue is now to derive the higher-order diffusion coefficient in the velocity space D owing to the fluctuating electric field, by which the mean distribution function evolves $\partial \langle f \rangle / \partial t = \partial / \partial v \left(D \partial \langle f \rangle / \partial v \right)$. The diffusion coefficient in the velocity space is given by the correlation of fluctuating accelerations, i.e.

$$D = \int_0^\infty \mathrm{d}\tau \, \langle F(t + \tau) F(t) \rangle, \qquad F(t) = \frac{e}{m} E(x(t), t). \tag{3.81}$$

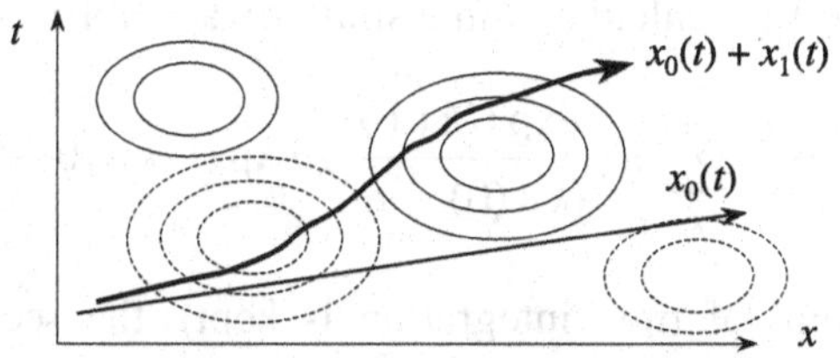

Fig. 3.16. Schematic drawing of the particle orbit in the presence of electric perturbations.

In the method of perturbation expansions, the correlation $\langle F(t+\tau) F(t)\rangle$ is calculated by the successive expansion with respect to the amplitude of electric field perturbation.

The acceleration at time t, $F(t)$, depends on the location of particle $x(t)$, through the space dependence of the electric field $E(x(t), t)$. In the perturbation expansion method, the particle orbit is expanded as,

$$x(t) = x_0(t) + x_1(t) + \cdots, \tag{3.82a}$$

where $x_0(t)$ is the unperturbed orbit, $x_0(t) = x(0) + v(0)t$, and $x_1(t)$ is the first-order correction of the orbit due to the electric field perturbation (as illustrated in Figure 3.16). Associated with this, the net acceleration that particles feel, is given by,

$$\begin{aligned} F(x(t), t) &= F(x_0(t) + x_1(t) + \cdots, t) \\ &= F(x_0(t), t) + x_1(t) \frac{\partial}{\partial x} F(x_0(t), t) + \cdots, \end{aligned} \tag{3.82b}$$

which can be rewritten as $F(t) = F_1(t) + F_2(t) + \cdots$ in a series of electric field amplitude. Note that the expansion (3.82a) is valid as long as the change of orbit occurs in a time that is much shorter than the bounce time of particles in the potential trough, τ_b, i.e., so that,

$$\tau_{ac} << \tau_b,$$

where τ_{ac} is the auto-correlation time that the resonant particles feel, $\tau_{ac}^{-1} = |(\omega/k - \partial\omega/\partial k)\,\Delta k|$, and Δk is the spectral width of $|E_k|^2$. If the bounce time is short, $\tau_{ac} > \tau_b$, the orbit is subject to trapping, and an expansion based on the unperturbed orbit is not valid.

The second-order term with respect to the electric field, $F_2(t)$, is,

$$F_2 = \frac{e}{m} \sum_{k'} ik' x_1(t) E_{k'} \exp\left(ik'x(0)\right) \exp\left(i\left(k'v(0) - \omega'\right)t\right), \tag{3.83}$$

and the third-order term is calculated in a similar way. Noting the fact that,

$$x_1(t) = -\frac{e}{m}\sum_k E_k \frac{\exp(ikx(0))}{(kv(0)-\omega)^2}\exp(i(kv(0)-\omega)t),$$

(where the upper limit of time integration is kept), the second-order term F_2 is the sum of contributions of the beats, $\exp\left[i\left(\left(k \pm k'\right)v(0) - \left(\omega \pm \omega'\right)\right)t\right]$. These beat waves are virtual modes, driven by the nonlinear interaction of primary modes.

The contribution to the diffusion coefficient from the linear response has the phase (which particles feel), $\exp(i(kv(0)-\omega)t)$. Therefore, the quasi-linear contribution (which is second order with respect to the electric field) comes from the resonance,

$$kv(0) - \omega = 0, \tag{3.84a}$$

while the phase of the next-order correction (the fourth order with respect to the electric field) is set by the resonance,

$$\left(k \pm k'\right)v(0) - \left(\omega \pm \omega'\right) = 0. \tag{3.84b}$$

The successive expansion provides that the term, which is $2n$-th order with respect to the electric field, originates from resonances,

$$(k_1 \pm \cdots \pm k_n)v(0) - (\omega_1 \pm \cdots \pm \omega_n) = 0. \tag{3.84c}$$

The fourth-order term in the expansion of the total diffusion coefficient, $D = D_2 + D_4 + \cdots$, is given by,

$$D_4 = \frac{e^4\pi}{m^4}\sum_{k,k'}|E_k|^2|E_{k'}|^2\left(\frac{k-k'}{(kv-\omega)(k'v-\omega')}\right)^2 \delta\left(\left(k-k'\right)v - \left(\omega-\omega'\right)\right), \tag{3.85}$$

where the label of particle velocity $v(0)$ is rewritten as v, and the resonance condition (3.84b) is given in terms of the delta function. The resonance occurs for the particles, which have the velocity at the phase velocity of the beat wave,

$$v = \frac{\omega - \omega'}{k - k'}. \tag{3.86}$$

This scattering process is known as nonlinear Landau damping. The change of kinetic energy $\overline{E}_{\text{kin}}$ associated with the relaxation of the mean distribution function at the fourth order of fluctuating field, $\frac{\partial}{\partial t}\overline{E}^{(2)}_{\text{kin}} = \int \mathrm{d}v \frac{m}{2}v^2 \frac{\partial}{\partial v}D_4\frac{\partial}{\partial v}\langle f\rangle$, gives the additional higher-order damping of wave energy.

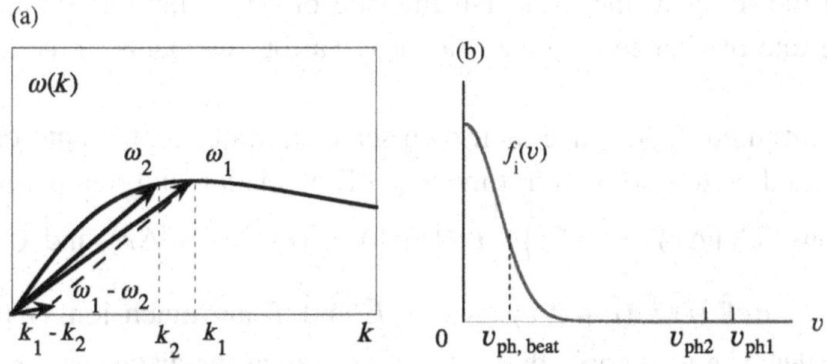

Fig. 3.17. Frequencies of the primary modes (k_1, ω_1, and k_2, ω_2) and the beat mode (a). Primary modes satisfy the dispersion but the beat mode does not. Phase velocities of primary modes and beat mode and the distribution function of ions (b).

This process is effective in connecting electrons and ions via wave excitations. Waves, which are excited by electrons, are often characterized by the phase velocity, $\omega/k \sim v_{\mathrm{T},e}$. For such cases, the phase velocity is too fast to interact with ions. When the wave dispersion is strong, the resonant velocity for the beat, $(\omega - \omega')/(k - k')$, can be much smaller than the phase velocity of primary waves, ω/k and ω'/k'. For the beat mode that satisfies the condition $(\omega - \omega')/(k - k') \sim v_{\mathrm{T},i}$, strong coupling to ions occurs. Figure 3.17 shows schematically the case where beat mode can resonate with ions.

This nonlinear Landau damping is important in the case where waves have strong dispersion. Noting the resonance condition Eq.(3.86), the fourth-order term (3.85) is rewritten as,

$$D_4 = \frac{e^4\pi}{m^4}\sum_{k,k'}|E_k|^2|E_{k'}|^2\frac{(k-k')^6}{(k'\omega - k\omega')^4}\delta\left((k-k')v - (\omega-\omega')\right). \quad (3.87)$$

This result shows that when dispersion is weak so that $\omega/k \simeq \omega'/k'$, the perturbation expansion is invalid. Compared to the first-order term, D_2, the higher-order term D_4 has a multiplicative factor, which is of the order of magnitude, $\sum_{k'}|E_{k'}|^2(k-k')^6(k'\omega - k\omega')^{-4} \propto \tau_{\mathrm{ac}}^4\tau_{\mathrm{b}}^{-4}$. This result shows that the perturbation theory has the expansion parameter $\tau_{\mathrm{ac}}^2\tau_{\mathrm{b}}^{-2}$. The expansion method requires $\tau_{\mathrm{ac}} \ll \tau_{\mathrm{b}}$, as was explained earlier in this chapter.

3.8 Kubo number and trapping

Fluctuations in plasmas can lead to random motion of plasma particles, which may lead to diffusive evolution of the mean distribution. The diffusivity is given by the step size in the jump of the orbit and the rate of the change of orbit. The step size and the rate of change are determined by various elements in the fluctuation spectrum.

First, the turbulent fields have their own scale and rate, i.e. the auto-correlation length, λ_c, and auto-correlation time, τ_c. The spatial and temporal correlation functions, $C_s(\Delta r) = \left\langle \tilde{E}^2 \right\rangle^{-1} L^{-1} \int_0^L \mathrm{d}r \, \tilde{E}(r) \, \tilde{E}(r + \Delta r)$ and $C_t(\Delta\tau) = \left\langle \tilde{E}^2 \right\rangle^{-1} T^{-1} \int_0^T \mathrm{d}t \, \tilde{E}(t) \, \tilde{E}(t + \Delta\tau)$ (where L and T are much longer than characteristic scales of microscopic fluctuations), decay at the distances $\Delta r \sim \lambda_c$ and $\Delta\tau \sim \tau_c$. These correlation length and correlation time are those for 'Eulerian' correlations.

The diffusion is, in reality, determined by the step size (and correlation time) of *particle* motion, and *not* by those of fluctuating field. For the correlation length (and correlation time) of the particle orbit, the Lagrangian correlation is the key, and is not identical to those of the fluctuating field. The Kubo number (sometimes referred to as the Strouhal number) is a key parameter that explains the relation between the Lagrangian correlation of particles and those of the fluctuating field.

Let us consider the $\boldsymbol{E} \times \boldsymbol{B}$ motion of a particle under a strong magnetic field and fluctuating radial electric field, $\tilde{\boldsymbol{E}}$. The equation of motion is written as,

$$\frac{\mathrm{d}}{\mathrm{d}t}\boldsymbol{x} = \boldsymbol{v}(\boldsymbol{x}(t), t), \quad \boldsymbol{v}(\boldsymbol{x}(t), t) = -\frac{1}{B}\tilde{\boldsymbol{E}} \times \boldsymbol{b}, \tag{3.88}$$

where $\boldsymbol{b}$ is a unit vector in the direction of the strong magnetic field. The amplitude of perturbation is characterized by the average of the fluctuating velocity, $\tilde{V} = \sqrt{\langle v^2 \rangle}$. Thus, the fluctuating field is characterized by three parameters, i.e. amplitude $\tilde{V}$, (Eulerian) correlation length and time, λ_c and τ_c respectively. The Kubo number is the ratio of the correlation time τ_c to the eddy circumnavigation time by the $\boldsymbol{E} \times \boldsymbol{B}$ motion $\tau_{cir} = \lambda_c / \tilde{V}$, i.e.

$$\mathcal{K} = \frac{\tau_c}{\tau_{cir}} = \frac{\tau_c \tilde{V}}{\lambda_c}. \tag{3.89}$$

When the Kubo number is much smaller than unity, $\mathcal{K} \ll 1$, the distance of the particle motion during the time period $0 < t < \tau_c$, $\tau_c \tilde{V}$, is much smaller than λ_c. Therefore, particle motion is modelled such that the step size and step time are given by $\tau_c \tilde{V}$ and τ_c. In contrast, for $\mathcal{K} > 1$, the particle motion is decorrelated by moving the distance of the decorrelation length λ_c, not by the decorrelation time of the field τ_c. In this limit, the fluctuation field stays (nearly) unchanged during the period of circumnavigation of particles in the trough of the perturbation potential. Note that $\mathcal{K} \simeq 1$ loosely corresponds to the mixing length fluctuation level.

The transition of transport from the quasi-linear regime to the trapping regime is illustrated briefly here (Vlad *et al.*, 2004; Balescu, 2005). The diffusion coefficient is given by the Lagrangian correlation of fluctuating velocity along the particle orbit,

$$D = \int_0^t \mathrm{d}t' \left\langle v_j\left(\boldsymbol{x}\left(t'\right), t'\right) v_j\left(\boldsymbol{x}(0), 0\right)\right\rangle, \tag{3.90}$$

where $j = x, y$ and coordinates (x, y) are taken perpendicular to the strong magnetic field. In the limit of small Kubo number, $\mathcal{K} \ll 1$, where the decorrelation time of the field is very short, one has $\left\langle v_j\left(\boldsymbol{x}\left(t'\right), t'\right) v_j\left(\boldsymbol{x}(0), 0\right)\right\rangle \sim \left\langle v_j\left(\boldsymbol{x}(0), t'\right) v_j\left(\boldsymbol{x}(0), 0\right)\right\rangle \sim \tilde{V}^2 C_t\left(t'\right)$ for the integrand of Eq.(8.90), and so one has,

$$D \simeq \tilde{V}^2 \tau_{\mathrm{c}} = \frac{\lambda_{\mathrm{c}}^2}{\tau_{\mathrm{c}}} \mathcal{K}^2. \tag{3.91}$$

When the Kubo number becomes larger, the field that particles feel, is decorrelated due to the motion of particles comparable to (longer than) the decorrelation length of the field λ_{c}. Putting the circumnavigation time $\tau_{\mathrm{cir}} = \lambda_{\mathrm{c}}/\tilde{V}$ into the step time, one has,

$$D \simeq \tilde{V}^2 \tau_{\mathrm{cir}} = \tilde{V} \lambda_{\mathrm{c}} = \frac{\lambda_{\mathrm{c}}^2}{\tau_{\mathrm{c}}} \mathcal{K}. \tag{3.92}$$

In this case, the diffusivity is *linearly* proportional to the fluctuation field intensity, provided that λ_{c} and τ_{c} are prescribed. The limit of $\mathcal{K} \gg 1$ is also explained in the literature. The unit $\lambda_{\mathrm{c}}^2 \tau_{\mathrm{c}}^{-1}$ in Eqs.(3.91) and (3.92) is considered to be the limit of complete trapping: in such a limit, particles are bound to the trough of potential, bouncing in space by the length λ_{c}, and the bounce motion is randomized by the time τ_{c}. In reality, the detrapping time for the particle out of the potential trough, τ_{detrap}, and the circumnavigation time τ_{cir} determines the step time (average duration time of coherent motion). The heuristic model for the step time is $\tau_{\mathrm{detrap}} + \tau_{\mathrm{cir}}$,

$$D \simeq \frac{\lambda_{\mathrm{c}}^2}{\tau_{\mathrm{detrap}} + \tau_{\mathrm{cir}}} = \frac{\lambda_{\mathrm{c}}^2}{\tau_{\mathrm{c}}} \mathcal{K} \frac{\tau_{\mathrm{cir}}}{\tau_{\mathrm{detrap}} + \tau_{\mathrm{cir}}}. \tag{3.93}$$

In the large amplitude limit (large $\mathcal{K}$ limit), the circumnavigation time τ_{cir} becomes shorter than the detrapping time, τ_{detrap}. Thus, the ratio $\tau_{\mathrm{cir}}/\left(\tau_{\mathrm{detrap}} + \tau_{\mathrm{cir}}\right)$ is a decreasing function of $\mathcal{K}$, and may be fitted to $\mathcal{K}^{-\gamma}$, where γ is a constant between 0 and 1. The theory based on a percolation in stochastic landscapes has provided an estimate $\gamma = 0.7$ (Isichenko, 1992). The literature (Vlad *et al.*, 2004) reports the result of numerical computations, showing that the power law fitting $\mathcal{K}^{-\gamma}$ holds for the cases $\mathcal{K} \gg 1$, while the exponent depends on the shape of the space-dependence of the Eulerian correlation function of the fluctuating field.

4

Nonlinear wave–particle interaction

流れにうかぶうたかたはかつきえかつむすびて久しくとどまりたるためしなし
Vortices on the flow either annihilate or emerge and have never stayed long.

(Kamo no Chomei, Hojoki)

4.1 Prologue and overview

In this chapter, and those which follow, we introduce and discuss *plasma turbulence theory*. A working theory of plasma turbulence is critical to our understanding of the saturation mechanisms and levels for plasma instabilities, and their associated turbulent transport. Quasi-linear theory alone is not sufficient for these purposes, since fluctuations and turbulence can saturate by coupling to other modes, and ultimately to dissipation, as well as by relaxing the mean distribution function. The energy flow in plasma turbulence is shown schematically in Figure 4.1. In nearly all cases of interest, linear instabilities, driven by externally pumped free energy reservoirs, grow and interact to produce a state of fluctuations and turbulence. As an aside, we note that while the word *turbulence* is used freely here, we emphasize that the state in question is frequently one of spatiotemporal chaos, wave turbulence or weak turbulence, all of which bear little resemblance to the familiar paradigm of high Reynolds number fluid turbulence, with its characteristically broad inertial range. Indeed, it is frequently not even possible to identify an inertial range in plasma turbulence, since sources and sinks are themselves distributed over a wide range of scales. Fundamentally there are two channels by which turbulence can evolve to a saturated state. These are by:

(i) quasi-linear relaxation of the distribution function or profile gradient associated with the free energy reservoir. In this channel $\langle f \rangle$, evolves toward a state where $\gamma_k \to 0$, for all modes k;
(ii) nonlinear interaction of unstable modes with other modes and ultimately with damped modes. In this channel, which resembles the well-known cascade in fluid turbulence,

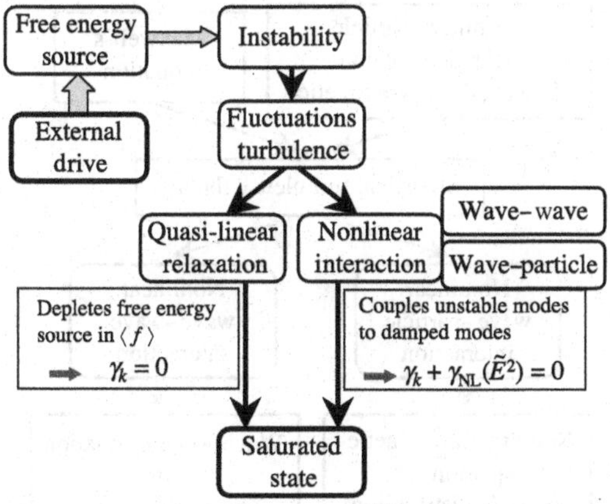

Fig. 4.1. Energy flow in plasma turbulence.

γ_k remains finite ($\gamma_k \neq 0$) but $\gamma_k^{L} + \gamma_k^{NL}\left(|E|^2\right) \to 0$, at each k, so that the *sum* of linear growth or damping and nonlinear transfer allows a finite free energy source and growth rate to be sustained against collisional or Landau damping.

Taken together, routes (i) and (ii) define the pathway from instability to a saturated state of plasma turbulence. Several caveats should be added here. First, the saturated state need not be absolutely stationary, but rather can be cyclic or in bursts, as long as the net intensity does not increase on average in the observed time. Second, the mechanism of nonlinear transfer can work either by wave–wave coupling or by the nonlinear scattering of waves on particles. These two mechanisms subdivide the "nonlinear interaction" channel into two sub-categories referred to, respectively, as "*nonlinear wave–wave interaction*" and "*nonlinear wave–particle interaction*". These two sub-categories for energy transfer are the subjects of this chapter (wave–particle) and the next (wave–wave), and together define the topic of wave turbulence in plasma. The ideas and material of these chapters constitute the essential foundations of the subject of plasma turbulence theory. Finally, we should add that in nearly all cases of practical interest, quasi-linear relaxation co-exists with a variety of nonlinear interaction processes (wave–wave, wave–particle, etc.). Only in rare cases does a single process or transfer channel dominate all the others.

The quasi-linear theory and its structure have already been discussed. The major elements of plasma turbulence theory which we must address are the theory of nonlinear wave–particle and wave–wave interactions. The general structure of plasma turbulence theory is shown schematically in Figure 4.2. Both for particles (via the evolution of f_k) or waves (via the evolution of N, the wave population density), a paradigmatic goal is to derive and understand the physics of the turbulent collision operators C_k^{P} and C_k^{W}. In practice, these two operators are usually strongly coupled. Examples of C_k^{P} include:

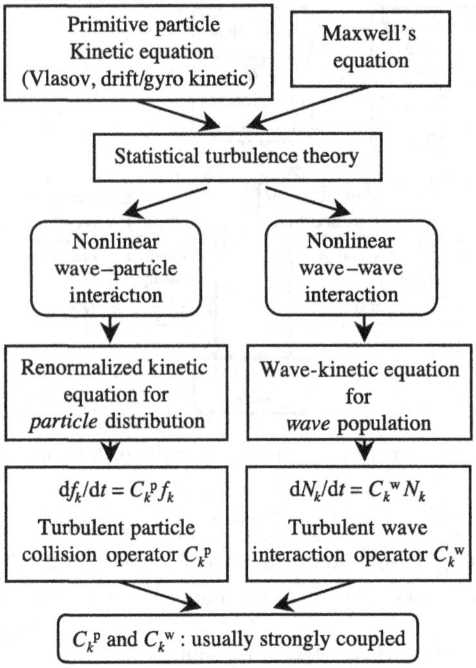

Fig. 4.2. General structure of plasma turbulence theory.

(i) the quasi-linear operator for $\langle f \rangle$ relaxation, discussed in Chapter 3;
(ii) the Balescu–Lenard collision integral, describing the relaxation of $\langle f \rangle$ in a stable plasma near equilibrium, discussed in Chapter 2;
(iii) the Vlasov propagator renormalization,

$$\frac{q}{m}\tilde{E}\frac{\partial \tilde{f}}{\partial v} \rightarrow -\frac{\partial}{\partial v}D_{k,w}\frac{\partial f_{k,\omega}}{\partial v},$$

where $D_{k,\omega} = D\left\{\left|E_{k',\omega'}\right|^2, \tau^{\mathrm{ac}}_{k',\omega'}\right\}$. Here the operator $-\partial/\partial v\, D_{k,\omega}\, \partial/\partial v$ may be viewed as similar to an "eddy viscosity" for Vlasov turbulence, and describes particle scattering by a spectrum of fluctuating electrostatic modes;
(iv) the scattering operator for nonlinear Landau damping $C_k^{\mathrm{P}} f_k \sim N f_k$, which is due to the class of interactions described schematically in Figure 4.3(a). In this case, a nonlinearly generated *beat* or *virtual* mode resonates with particles with their velocity equal to its phase velocity. Mechanisms (iii) and (iv) are discussed in this chapter.

Nonlinear wave–wave interaction processes result from resonant coupling, which is schematically depicted in Figure 4.3(b). The wave–wave collision operator has the generic form $C_k^W N \sim NN$, so the spectral evolution equation takes the form,

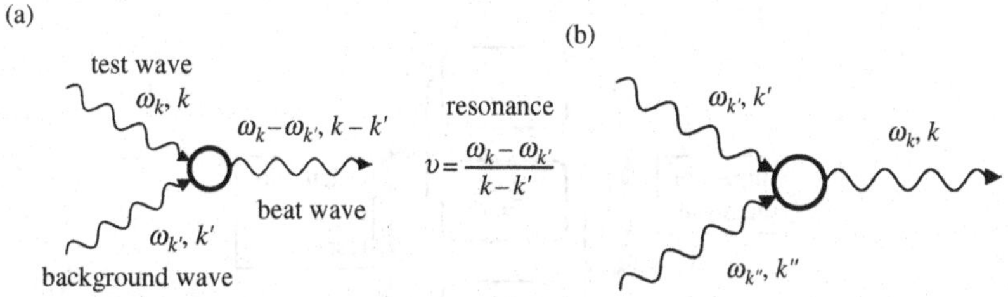

Fig. 4.3. Nonlinear wave–wave interaction process. (a) Nonlinear Landau resonance interaction with beat waves. (b) Nonlinear wave–wave coupling of three resonant modes, where $\boldsymbol{k}+\boldsymbol{k}'+\boldsymbol{k}''=0$ and $\omega_{\boldsymbol{k}}+\omega_{\boldsymbol{k}'}+\omega_{\boldsymbol{k}''}=0$.

$$\frac{\partial\,|E_{\boldsymbol{k}}|^2}{\partial t}-\gamma_k\,|E_{\boldsymbol{k}}|^2+\sum_{\boldsymbol{k}'}C_1\left(\boldsymbol{k},\boldsymbol{k}'\right)|E_{\boldsymbol{k}'}|^2\,|E_{\boldsymbol{k}}|^2=\sum_{\substack{\boldsymbol{k}',\boldsymbol{k}''\\ \boldsymbol{k}'+\boldsymbol{k}''=\boldsymbol{k}}}C_2\left(\boldsymbol{k}',\boldsymbol{k}''\right)|E_{\boldsymbol{k}'}|^2|E_{\boldsymbol{k}''}|^2.$$

Here, the wave kinetics is akin to that of a *birth and death process.* Usually, the incoherent emission term on the right-hand side (so named because it is *not* proportional to $|E_k|^2$) corresponds to *birth,* while the coherent mode coupling term i.e. the third term on the left-hand side (so named because it is proportional to, and coherent with, $|E_{\boldsymbol{k}}|^2$) corresponds to *death* or nonlinear damping. The competition between these two defines the process of nonlinear wave energy transfer, i.e. nonlinear cascade. Generally, $\sum_k$ (incoherent) $=$ $\sum_k$ (coherent), confirming that energy is conserved in the couplings. For weak turbulence, the three-wave resonance function $R_{k,k',k''}$ has negligible width, so Re $R_{k,k',k''}=\pi\delta\left(\omega_k-\omega_{k'}-\omega_{k''}\right)$, while for strong turbulence, $R_{k,k',k''}$ is broadened, and has the form $R_{k,k',k''}=i\big/\left\{\left(\omega_k-\omega_{k'}-\omega_{k''}\right)+i\left(\Delta\omega_k+\Delta\omega_{k'}+\Delta\omega_{k''}\right)\right\}$. The width of $R_{k,k',k''}$ is due to the effects of nonlinear scrambling on the coherence of the three interacting modes. The subject of nonlinear wave–wave interaction is discussed at length in Chapters 4 and 5.

4.2 Resonance broadening theory

4.2.1 Approach via resonance broadening theory

We begin our discussion of nonlinear wave–particle interaction by presenting the theory of resonance broadening (Dupree, 1966). Recall that quasi-linear theory answers the question, "How does $\langle f\rangle$ evolve in the presence of a spectrum of waves, given that the particle orbits are stochastic?". Continuing in that vein, resonance broadening theory answers the question, "How does the plasma distribution

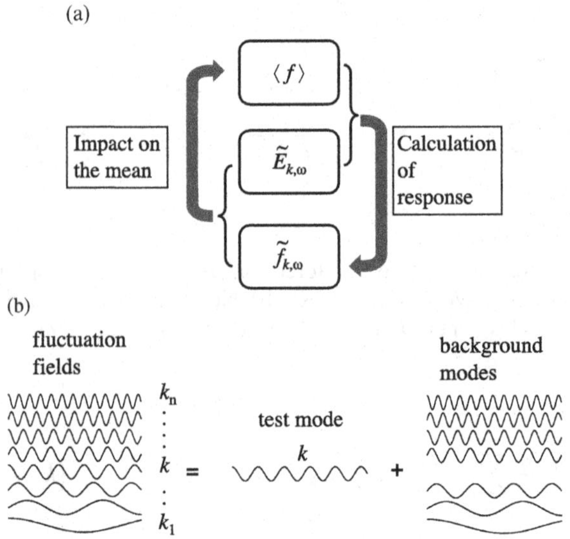

Fig. 4.4. Calculation of the plasma response for given electric field perturbation, and calculation of the evolution of the mean (a). Coupling with test wave and background spectrum is given in (b), depicting the test wave approximation.

function f respond to a test wave $E_{k,\omega}$ at (k, ω), given an existing spectrum of background waves?". This situation is depicted in Figure 4.4.

4.2.1.1 Basic assumptions

It cannot be over-emphasized that *resonance broadening theory rests upon two fundamental assumptions*. First the particle orbits are assumed to be stochastic, so excursions from unperturbed orbits may be treated as a diffusion process. Thus, resonance broadening theory (RBT) is valid only in regions of phase space where the islands around phase space resonances overlap. Resonance broadening theory also tacitly assumes the convergence of the second moment of the scattering step pdf (probability density function) i.e.

$$\langle (\Delta v)^2 \, P \rangle < \infty,$$

for velocity step Δv with pdf P. This allows the scattering to be treated as a diffusion process, via the central limit theorem. Second, the "test wave" approximation is assumed to be valid. The test wave approximation, which appears in various forms in virtually *every* statistical theory of turbulence, envisions the ensemble of interacting modes to be sufficiently large and statistically homogeneous so that *any* one mode may be removed from the ensemble and treated as a *test* wave, without altering the physics of the ensemble of remaining modes. The test wave hypothesis is thus not applicable to problems involving coherent mode coupling, nor to problems involving a few large amplitude or coherent modes interacting with a stochastic bath. In practice, validity of the test wave approximation almost always requires that the number of waves in the ensemble be large, and that the spectral auto-correlation time be short.

4.2.1.2 *Ensemble and path integral*

The essential idea of resonance broadening theory resembles that of the Weiner–Feynman path integral, in that the formal solution of the Vlasov equation, which can be written as an integration over the time history of the exact ("perturbed") orbit, is replaced by an average over a statistical ensemble of excursions from the linear or, "unperturbed", orbit. This concept is shown schematizally in Figure 4.5. To implement this, it is useful to recall that the Vlasov response f_k to an electric field fluctuation E_k may be written as an integration over orbits. So starting from,

$$\frac{\mathrm{d}f_k}{\mathrm{d}t} = -\frac{q}{m}E_k\frac{\partial\langle f\rangle}{\partial v}, \tag{4.1a}$$

when $\mathrm{d}/\mathrm{d}t$ is determined by the characteristic equations (of the Vlasov equation),

$$\frac{\mathrm{d}x}{\mathrm{d}t} = v \qquad \frac{\mathrm{d}v}{\mathrm{d}t} = \frac{q}{m}E, \tag{4.1b}$$

which are also the equations of particle motion, we can write,

$$f_{k,\omega} = -\frac{q}{m}\mathrm{e}^{-ikx}\int_0^\infty \mathrm{d}\tau\, \mathrm{e}^{ik\tau}u(-\tau)\left[\mathrm{e}^{ikx}E_{k,\omega}\frac{\partial\langle f\rangle}{\partial v}\right], \tag{4.2a}$$

where $u(-\tau)$ is the formal, exact orbit propagator, which has the property that,

$$u(-\tau)\mathrm{e}^{ikx} = \mathrm{e}^{ikx(-\tau)}. \tag{4.2b}$$

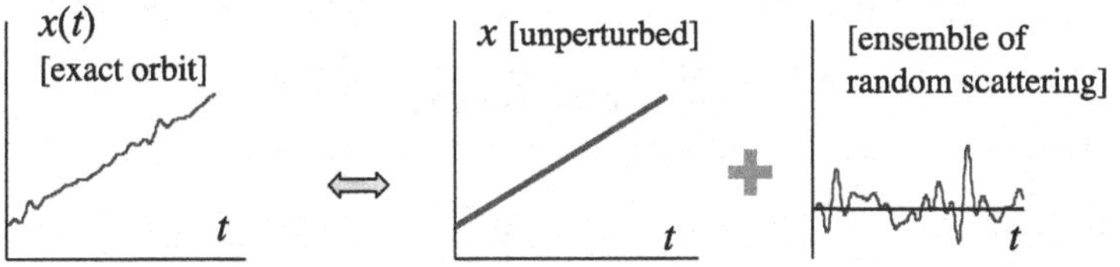

Fig. 4.5. Schematic illustration for decomposition of the particle orbit in resonance broadening theory.

Here, $x(-\tau)$ is the full, perturbed orbit. We can again formally decompose $x(-\tau)$ into the unperturbed piece and a fluctuation around it, as,

$$x(-\tau) = x_0(-\tau) + \delta x(-\tau). \tag{4.2c}$$

Along the unperturbed orbit $x_0(-\tau)$, in this case, $x_0(-\tau) = x - v\tau$. This gives,

$$f_{k,\omega} = -\int_0^\infty \mathrm{d}\tau\, \mathrm{e}^{i(\omega-kv)\tau} \mathrm{e}^{ik\delta x(-\tau)} \frac{q}{m} E_{k,\omega} \frac{\partial \langle f \rangle}{\partial v}. \tag{4.3}$$

Note that here we have assumed that the time scale for orbit scattering τ_s is short compared to the time scale upon which $\langle f \rangle$ varies (i.e. $\tau_s < \tau_{relax}$, where τ_{relax} is the quasi-linear relaxation time in Table 3.1), so that $\langle f \rangle$ may be treated as constant.

4.2.1.3 *Introduction of approximation*

So far all the calculations have been purely formal. We now come to the critical substantive step of resonance broadening theory, which is to *approximate* $f_{k,\omega}$ by its average over a statistical ensemble of orbit perturbations, i.e. to take $f_{k,\omega} \to \langle f_{k,\omega} \rangle_{OE}$, where

$$\langle f_{k,\omega} \rangle_{OE} = -\int_0^\infty \mathrm{d}\tau\, \mathrm{e}^{i(\omega-kv)\tau} \left\langle \mathrm{e}^{ik\delta x(-\tau)} \right\rangle_{OE} \frac{q}{m} E_{k,\omega} \frac{\partial \langle f \rangle}{\partial v}. \tag{4.4}$$

Here the bracket $\langle \quad \rangle_{OE}$ signifies an average over an ensemble of orbits. Note that by employing this ansatz, the orbit perturbation factor appearing in the response time history, i.e. $\exp[ik\delta x(-\tau)]$, which we don't know, is replaced by its ensemble average, which we *can* calculate, by exploiting an assumption concerning the pdf of δx. The approximation, which is used in quasi-linear theory, corresponds to,

$$\left\langle \mathrm{e}^{i\omega\tau - ikx(\tau)} \right\rangle_{QL} \simeq \exp\langle i\omega\tau - ikx(\tau)\rangle = \exp(i\omega\tau - ikv\tau).$$

The mean field theory is employed in evaluating the quasi-linear response.

To calculate $\langle \exp[ik\delta x(-\tau)] \rangle_{OE}$ in RBT, we first note that,

$$\frac{\mathrm{d}x}{\mathrm{d}t} = v = v_0 + \delta v, \tag{4.5a}$$

so

$$\delta x(-\tau) = -\int_0^\tau \mathrm{d}\tau'\, \delta v(-\tau'), \tag{4.5b}$$

and we find,

$$\langle \exp\left[ik\delta x(-\tau)\right]\rangle_{\mathrm{OE}} = \left\langle \exp\left[-ik\int_0^\tau \delta v(-\tau')\mathrm{d}\tau'\right]\right\rangle_{\mathrm{OE}}. \tag{4.5c}$$

Thus the problem has now been reduced to calculating the expectation value in Eq.(4.5c). Now, excursions in velocity from the unperturbed orbit are produced by the fluctuating electric fields of the turbulent wave ensemble, i.e. $\mathrm{d}\delta v/\mathrm{d}t = q\tilde{E}/m$. Consistent with the test wave hypothesis of a statistically homogeneous ensemble of weakly correlated fluctuation, we *assume* a Gaussian pdf of $\tilde{E}$, so that δv behaves diffusively, i.e.

$$\mathrm{pdf}[\delta v] = \frac{1}{\sqrt{\pi D\tau}}\exp\left[-\delta v^2/D\tau\right], \tag{4.6a}$$

and the expectation value of $A(\delta v)$ is just,

$$\langle A\rangle_{\mathrm{OE}} = \int \frac{\mathrm{d}\delta v}{\sqrt{\pi D\tau}}\exp\left[-\delta v^2/D\tau\right]A. \tag{4.6b}$$

Here D is the velocity diffusion coefficient which, like the quasi-linear D, characterizes stochastic scattering of particles by the wave ensemble. In practice, D has the same structure as does the quasi-linear diffusion coefficient. It cannot be over emphasized that the Gaussian statistics of $\tilde{E}$ and the diffusive pdf of δv are *input by assumption*, only. While Gaussian statistics, etc., are often characteristic of nonlinear systems with large numbers of interacting degrees of freedom, there is no a-priori guarantee this will be the case. One well-known example of a dramatic departure from Gaussian behaviour is the Kuramoto transition, in which the phases of an ensemble of N (for $N \gg 1$) strongly coupled oscillators synchronize for coupling parameters above some critical strength (Kuramoto, 1984). Another is the plethora of findings of super-diffusive or sub-diffusive scalings in various studies of the transport of test particles in a turbulent flow. Such non-diffusive behaviours (Yoshizawa *et al.*, 2004), which demand more advanced methods, like SOC models (Bak *et al.*, 1987; Dendy and Helander, 1997; Carreras *et al.*, 1998) and fractional kinetic (Podlubny, 1998; del Castillo-Negrete *et al.*, 2004; Zaslavsky, 2005; Sanchez, 2005), are frequently, but not always associated with the presence of structures in the flow. In spite of these caveats, the Gaussian diffusive assumption is a logical starting point. Furthermore, it is quite plausible that resonance broadening theory can be generalized to treat the fractional kinetics of orbit perturbations, and

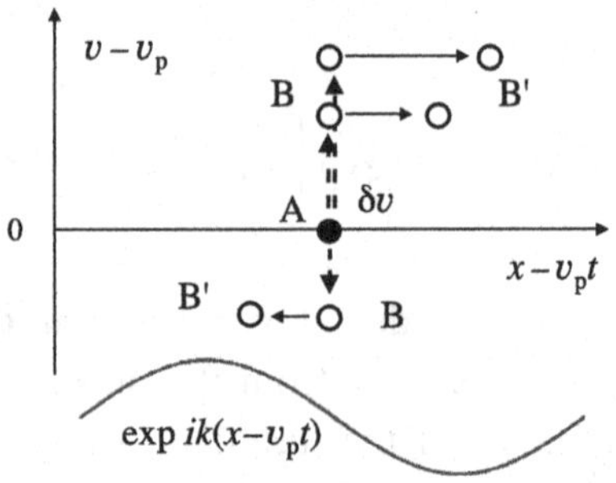

Fig. 4.6. Accelerated decorrelation of resonant particles via diffusion in the velocity space. The deviation of the particle velocity from the phase velocity of the wave enhances the rate of phase change, $\omega - kv$.

so encompass non-diffusive particle motion. Proceeding then, the orbit averaged response factor is given by,

$$\left\langle \exp\left[-ik\int_0^\tau \mathrm{d}\tau'\,\delta v\left(-\tau'\right)\right]\right\rangle_{\mathrm{OE}} = \exp\left[-\frac{k^2 D\tau^3}{6}\right]. \tag{4.7a}$$

The scaling $\langle \delta x^2 \rangle \sim D\tau^3$ is a consequence of the fact that velocity, not position, is directly scattered by electric field fluctuations, so linear streaming can couple to the random walk in velocity to enhance decorrelation. This is shown schematically in Figure 4.6.

4.2.1.4 *Response function and decorrelation rate*

Using Eq.(4.6a), the RBT approximation to the response function is then just,

$$f_{k,\omega} = -\int_0^\infty \mathrm{d}\tau\,\exp\left[i(\omega - kv)\tau - \frac{k^2 D\tau^3}{6}\right]\frac{q}{m}E_{k,\omega}\frac{\partial\langle f\rangle}{\partial v}. \tag{4.7b}$$

If we define the wave–particle correlation time τ_c according to the width in time of the kernel of this response, we have,

$$\frac{1}{\tau_c} = \left(\frac{k^2 D}{6}\right)^{1/3}, \tag{4.7c}$$

so that,

$$f_{k,\omega} = -\int_0^\infty \mathrm{d}\tau\,\exp\left[i(\omega - kv)\tau - \frac{\tau^3}{\tau_c^3}\right]\frac{q}{m}E_{k,\omega}\frac{\partial\langle f\rangle}{\partial v}. \tag{4.7d}$$

Of course, the origin of the name "resonance broadening theory" is now clear, since the effect of scattering by the turbulent spectrum of background waves is to broaden the linear wave–particle resonance, from a delta function of zero width in linear theory to a function of finite width proportional to $1/\tau_c$. In this regard, it is often useful to approximate the result of Eq.(4.7d) by a Lorentzian of width $1/\tau_c$, so,

$$f_{k,\omega} = -\frac{i}{(\omega - kv + i/\tau_c)} \frac{q}{m} E_{k,\omega} \frac{\partial \langle f \rangle}{\partial v}. \tag{4.7e}$$

The principal result of RBT is the identification of $1/\tau_c$ as given by Eq.(4.7c), as the wave–particle decorrelation rate. This is the rate (inverse time) at which a resonant particle scatters a distance of one wavelength ($\lambda = 2\pi/k$) relative to the test wave, and so defines the individual coherence time of a resonant particle with a specific test wave. Of course, τ_c corresponds to τ_s, the particle scattering time referred to earlier but not defined. Since $1/\tau_c$ defines the width of the resonance in time, it also determines a width in velocity, i.e.

$$\frac{1/\tau_c}{(\omega - kv)^2 + 1/\tau_c^2} = \frac{1/\tau_c}{\left\{(\omega/k - v)^2 + 1/k^2\tau_c^2\right\} k^2},$$

so

$$\Delta v_T = \frac{1}{k\tau_c} = \left(\frac{D}{6k}\right)^{1/3} \tag{4.8}$$

is the width in velocity of the broadened resonance. Together, $\Delta x \sim k^{-1}$ and Δv_T define the fundamental scales of an element or *chunk* of turbulent phase space fluid. This fluid element is the analogue for Vlasov turbulence of eddy paradigm, familiar from ordinary fluid turbulence. Note that in contrast to the eddy, with spatial scale independent of amplitude, the velocity scale of a turbulent Vlasov fluid element varies with turbulence intensity via its dependence upon D. In this regard then, $1/\tau_{ck}$ may be viewed as the analogue of the eddy turn-over or decay rate $\Delta\omega_k \sim k\tilde{v}_k$.

To better understand the dependencies and scalings of the RBT parameters τ_c, Δv, etc, and to place these new scales in the context of the fundamental time scales which we encountered in Chapter 3, it is instructive to re-visit the calculation of the quasi-linear diffusion coefficient D_{QL}, now employing the response f_k calculation via RBT. Thus,

$$D = \mathrm{Re}\left[\frac{q^2}{m^2} \sum_k |E_k|^2 \frac{i}{\omega - kv + i/\tau_{ck}}\right]; \tag{4.9a}$$

so employing the Lorentzian spectrum as before,

$$D = \mathrm{Re}\left[\frac{q^2}{m^2}\sum_k \frac{\left|\tilde{E}\right|^2/\Delta k}{1+\{(k-k_0)/\Delta k\}^2}\frac{i}{\omega - kv + i/\tau_{ck}}\right], \tag{4.9b}$$

and performing the spectral summation by contour integration gives,

$$D = \frac{q^2}{m^2}\left|\tilde{E}\right|^2 \mathrm{Re}\left\{i/\left(\omega_{k_0} - k_0 v + i\left|\Delta k\right|\left|v_{\mathrm{gr}} - v\right| + i/\tau_{ck}\right)\right\}. \tag{4.9c}$$

Taking $v = \omega_{k_0}/k_0$ at resonance, we see that D reduces to its quasi-linear antecedent ($D \to D_{\mathrm{QL}}$) for $\left|\Delta k\left(v_{\mathrm{gr}} - v_{\mathrm{ph}}\right)_k\right| > 1/\tau_{ck}$. Thus quasi-linear diffusion is recovered for $\tau_{\mathrm{ac}} < \tau_{\mathrm{c}}$.

4.2.2 Application to various decorrelation processes

4.2.2.1 Scattering in action variable

The structure of the above calculation and the resulting super-diffusive decorrelation are straightforward consequences of the fact that the action variable is scattered (i.e. particle velocity $v \to v + \delta v$), while the response decorrelation is measured by the excursion of the associated angle variable (i.e. $x = \int v\,\mathrm{d}\tau \to x + \delta x$). Moreover drag, a key component in Brownian dynamics, is absent. Thus, we have,

$$\phi \sim \int \omega(J)\,\mathrm{d}\tau, \tag{4.10a}$$

so the scattering induced excursion is,

$$\delta\phi \sim \int \frac{\partial \omega}{\partial J}\delta J\,\mathrm{d}\tau, \tag{4.10b}$$

and

$$\langle \delta\phi \rangle \sim \left(\frac{\partial \omega}{\partial J}\right)^2 D_J \tau^3, \tag{4.10c}$$

so the mean square excursion of the angle variable grows in time $\sim \tau^3$, not $\sim \tau$.

Two particularly important examples of this type of decorrelation process are concerned with:

(a) the decorrelation of an electron by radial scattering in a sheared magnetic field (Fig. 4.7);
(b) the decorrelation of a particle or fluid element by radial scattering in a sheared flow, (Fig. 4.9).

4.2.2.2 Decorrelation in a sheared magnetic field

Regarding electron scattering in the configuration of Figure 4.7, consideration of streaming in the poloidal direction gives,

$$r\frac{\mathrm{d}\theta}{\mathrm{d}t} = v_{\parallel}\frac{B_\theta}{B_\mathrm{T}} \tag{4.11a}$$

(θ: poloidal angle), so

$$\frac{\mathrm{d}\theta}{\mathrm{d}t} = \frac{v_{\parallel}}{Rq\,(r)}. \tag{4.11b}$$

With radial scattering δr, the change in the poloidal angle follows

$$\begin{aligned}\delta\theta &\sim \int \frac{v_{\parallel}}{Rq\,(r+\delta r)}\mathrm{d}\tau \\ &\sim \frac{-v_{\parallel}}{Rq^2}q'\delta r\,\mathrm{d}\tau,\end{aligned} \tag{4.11c}$$

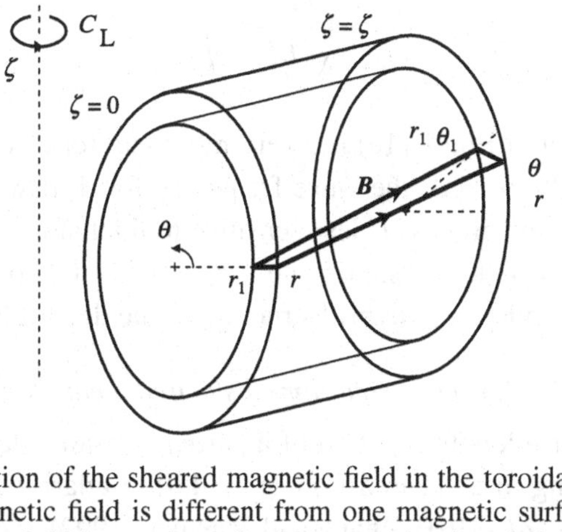

Fig. 4.7. Illustration of the sheared magnetic field in the toroidal plasma. When the pitch of magnetic field is different from one magnetic surface to the other surface, magnetic field has a shear. In this example, the pitch is weaker if the minor radius r increases. Two starting points are at the same poloidal angle at C (the toroidal angle $\zeta = 0$), but $\theta < \theta_1$ holds at C′ following the magnetic field lines.

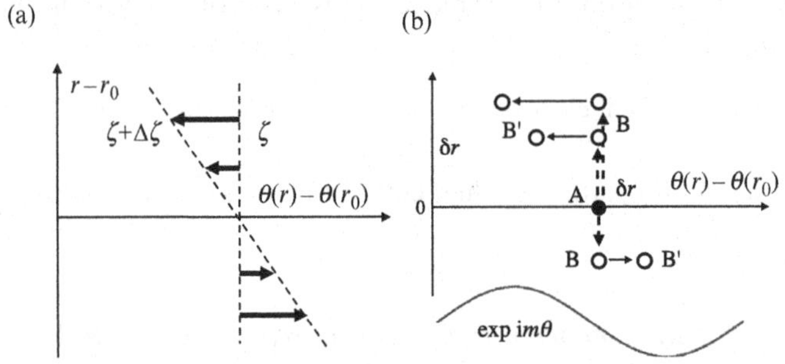

Fig. 4.8. In sheared magnetic field configuration, the poloidal angle of two neighbouring magnetic field lines deviates when the magnetic field lines are followed (a). In this circumstance, accelerated decorrelation (against the wave propagating in the poloidal direction) occurs via diffusion in the velocity space (b).

as is illustrated in Figure 4.8. So, we have (Hirshman and Molvig, 1979),

$$\left\langle \delta\theta^2 \right\rangle \sim \frac{v_\parallel^2}{L_s^2} \frac{D}{r^2} \tau^3. \tag{4.11d}$$

Here $L_s^{-1} = rq'/Rq^2$ is the magnetic shear length and D is the radial diffusion coefficient. As in the 1D velocity scattering case, decorrelation occurs via the synergy of radial scattering with parallel streaming. A useful measure of decorrelation is the time at which $k_\theta^2 r^2 \left\langle \delta\theta^2 \right\rangle \sim 1$. This then defines the decorrelation time τ_c, where,

$$\frac{1}{\tau_c} \sim \left(\frac{k_\theta^2 v_\parallel^2}{L_s^2} D \right)^{1/3}. \tag{4.11e}$$

In practice, the result of Eq.(4.11e) is a very rapid rate for electron scattering, and one which frequently exceeds the wave frequency for drift waves. Note too, that since $1/\tau_c \sim D^{1/3}$, this process is less sensitive to fluctuation levels, etc. than the familiar purely diffusive decorrelation rate $1/\tau_c \sim k_\perp^2 D$. Of course, this is a simple consequence of the underlying hybrid structure of the decorrelation process.

4.2.2.3 Decorrelation in sheared mean flow

Similarly, if one considers the motion of a particle or fluid element which undergoes radial scattering in a sheared flow (Fig. 4.9) (Biglari *et al.*, 1990; Itoh and Itoh, 1990; Shaing *et al.*, 1990; Zhang and Mahajan, 1992), we have,

$$\frac{\mathrm{d}y}{\mathrm{d}t} = V_y(x), \tag{4.12a}$$

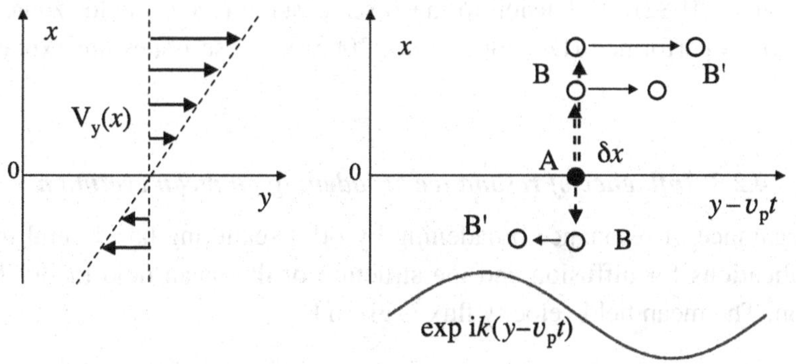

Fig. 4.9. The case of sheared mean flow. Accelerated decorrelation occurs via diffusion in the x-direction. The deviation of the plasma elements in the x-direction enhances the rate of phase change, $\omega - kV$.

so

$$y \sim \int \mathrm{d}\tau \; V_y \left(x + \delta x\right), \tag{4.12b}$$

and

$$\delta y \sim \int \mathrm{d}\tau \left(\frac{\partial V_y}{\partial x}\right) \delta x. \tag{4.12c}$$

The variance in the y-direction evolves as,

$$\left\langle \delta y^2 \right\rangle \sim \left(\frac{\partial V_y}{\partial x}\right)^2 D_x \tau^3, \tag{4.12d}$$

and the decorrelation rate follows as,

$$\frac{1}{\tau_c} \sim \left(\left(\frac{\partial V_y}{\partial x}\right)^2 k_y^2 D_x\right)^{1/3}. \tag{4.12e}$$

Again, note that the hybrid character of the process implies reduced sensitivity to D_x and the fluctuation levels which drive it (Biglari *et al.*, 1990). The interplay of sheared streaming and radial scattering which yields the hybrid decorrelation rate given in Eq.(4.12e) is relevant to the phenomena of suppression of turbulence and transport by a sheared flow (Hahm and Burrell, 1995; Itoh and Itoh, 1996; Terry, 2000). Combined with the theory of electric field bifurcation (Itoh and Itoh, 1988), this turbulence suppression mechanism plays a central role in understanding the phenomenon of confinement improvement (such as the H-mode

(Wagner *et al.*, 1982)). Extension to meso scale radial electric field (zonal flows) has also been performed (Diamond *et al.*, 2005b). These issues are explained in Volume 2.

4.2.3 Influence of resonance broadening on mean evolution

The appearance of resonance broadening by orbit scattering has several interesting implications for diffusion and the structure of the mean field theory for $\langle f \rangle$ evolution. The mean field velocity flux is given by,

$$\begin{aligned}\Gamma_v &= \sum_{\boldsymbol{k},\omega} \frac{q}{m} E_{\substack{\boldsymbol{k}\\-\omega}} f_{\substack{\boldsymbol{k}\\ \omega}} \\ &= -D\frac{\partial \langle f \rangle}{\partial v}, \end{aligned} \tag{4.13a}$$

where by substitution of Eq.(4.7e) for $f_{\substack{k\\ \omega}}$ into Eq.(4.13a), we find,

$$D = \frac{q^2}{m^2} \sum_{k,\omega} \left| E_{\substack{\boldsymbol{k}\\ \omega}} \right|^2 \frac{1/\tau_{c\boldsymbol{k}}}{(\omega - kv)^2 + 1/\tau_{c\boldsymbol{k}}^2}. \tag{4.13b}$$

Noting that $1/\tau_{ck} = \left(k^2 D/6\right)^{1/3}$, we see that D is, in principle, defined as a function of itself in Eq.(4.13). Of course, τ_{ck} constitutes the individual coherence time of distribution perturbations measured relative to a test wave $\boldsymbol{k}$. We also see that τ_{ck} enters the resonance width, and acts to broaden the wave–particle resonance to a finite width.

$$\Delta v_{\mathrm{T}} = \frac{1}{k\tau_{\mathrm{c}}}, \tag{4.14a}$$

so the resonance function now has the broadened form,

$$\frac{1/\tau_{c\boldsymbol{k}}}{(\omega - kv)^2 + 1/\tau_{c\boldsymbol{k}}^2} \to \frac{\Delta v_{\mathrm{T}}/k}{(\omega/k - v)^2 + (1/k\tau_{c\boldsymbol{k}})^2}. \tag{4.14b}$$

Given that the main effect of orbit decorrelation is to broaden the linear, singular resonance function $\sim \delta\,(\omega - kv)$ to one of finite width, one might naturally ask, "How does D really charge," and "is the finite width of the wave–particle resonance at all significant, in the event that the width of the fluctuation spectrum is larger?" As in Chapter 3, we proceed by an ansatz, a simple form of the fluctuation spectrum which facilitates performing the spectrum integrations in order to identify the basic time scales. Assuming D is driven by a Lorentzian spectrum of modes, we have,

$$D = \mathrm{Re}\int \mathrm{d}k\,\frac{q^2}{m^2}\frac{\left|\tilde{E}_0\right|^2}{\left[1+\left(\frac{k-k_0}{\Delta k}\right)^2\right]}\frac{i}{\omega - kv + \frac{i}{\tau_c}}$$

$$\sim \mathrm{Re}\left\{\frac{q^2}{m^2}\left|\tilde{E}_0\right|^2 i \Big/ \left[\omega_{k_0} - k_0 v + i\left|\Delta k\right|\left|\frac{\mathrm{d}\omega}{\mathrm{d}k} - v\right| + \frac{i}{\tau_{ck}}\right]\right\}. \tag{4.15a}$$

Thus we immediately see that if,

$$\left|\Delta k\left(v_{\mathrm{gr}} - \frac{\omega}{k}\right)\right| > \frac{1}{\tau_c}, \tag{4.15b}$$

so that the spectral auto-correlation time $\tau_{\mathrm{ac}} < \tau_c$, the resonance broadening is irrelevant and $D \to D_{\mathrm{QL}}$. Combining Eqs.(4.15a) and (4.7c), Eq.(4.15b) is rewritten as,

$$\left(\frac{\Delta k}{k}\right)^2\left|\frac{v_g}{v_p} - 1\right|^2 > \frac{e\phi}{mv_p^2}.$$

This condition is equivalent to the validity condition for the quasi-linear theory, Eq.(3.11) and Fig. 3.11.

Note that broad spectral width, alone, is *not* sufficient to ensure that D is quasi-linear. Dispersion must be sufficient to ensure that the fluctuation pattern seen by a resonant particle (one with $v = \omega/k$) is short lived in comparison with the correlation time. In this limit of $D \to D_{\mathrm{QL}}$, the particle–wave decorrelation rate scales as,

$$\frac{1}{\tau_{ck}} \sim \left(k^2\left\langle\tilde{a}^2\right\rangle\tau_{\mathrm{ac}}\right)^{1/3} \tag{4.16a}$$

and the resonance width scales as

$$\frac{1}{k\tau_{ck}} \sim \left(\left\langle\tilde{a}^2\right\rangle\frac{\tau_{\mathrm{ac}}}{k}\right)^{1/3}. \tag{4.16b}$$

Here $\left\langle\tilde{a}^2\right\rangle$ is the acceleration fluctuation spectrum and τ_{ac} is the spectral auto-correlation time.

Given the discussion above, the opposite limit of short correlation time naturally arouses one's curiosity. In this limit, for resonant particles $i/(\omega - kv + i/\tau_{ck}) \sim \tau_{ck}$. Hence, ignoring the k-dependence of τ_c, we have,

$$\frac{1}{\tau_c^3} = \frac{k^2 D}{6} = \frac{k^2}{6} \sum_k \frac{q^2}{m^2} |E_k|^2 \tau_c, \tag{4.17a}$$

so

$$\frac{1}{\tau_c^4} \sim \frac{q^2}{m^2} k^4 \left\langle \tilde{\phi}^2 \right\rangle, \tag{4.17b}$$

and

$$\frac{1}{\tau_c} \sim k \left(\frac{q^2}{m^2} \left\langle \tilde{\phi}^2 \right\rangle \right)^{1/4}, \tag{4.17c}$$

and

$$\Delta v_T \sim \left(\frac{q^2}{m^2} \left\langle \tilde{\phi}^2 \right\rangle \right)^{1/4} \sim \left(\frac{q}{m} \left\langle \tilde{\phi}^2 \right\rangle^{1/2} \right)^{1/2}. \tag{4.17d}$$

Not surprisingly, the results for the short-τ_c limit resemble those for a particle interaction with a single wave.

4.3 Renormalization in Vlasov turbulence I: Vlasov response function

4.3.1 Issues in renormalization in Vlasov turbulence

While intuitively appealing in many respects, resonance broadening theory is inherently unsatisfactory, for several reasons. These include, but are not limited to:

1. the theory is intrinsically one of the 'test wave' genre, yet treats E_k as fixed while f_k evolves nonlinearly in response to it – i.e. f_k and E_k are treated asymmetrically. This is especially dubious since the Vlasov nonlinearity consists of the product $E\,\frac{\partial f}{\partial v}$;
2. the resonance broadening theory does not conserve energy, and indeed does not address the issue of energetics. We will elaborate on this point further below, in our discussion of renormalization for drift wave turbulence;
3. resonance broadening theory treats the evolution of f as a Markov process, and neglects memory effects;
4. resonance broadening theory asserts Gaussian statistics a priori for particle orbit scattering statistics.

In view of these limitations, it is natural to explore other, more systematic, approaches to the problem of renormalization. The reader is forewarned, however, that *all* renormalization procedures involve some degree of uncertainty in the accuracy of the approximations they employ. *None* can be fully justified on a rigorous function. None can predict their own errors.

The aim here is to determine the response function relating $f_{k,\omega}$ to the electric field perturbation $E_{k,\omega}$. In formal terms, if one assigns each $E_{k,\omega}$, a multiplicative phase factor $\alpha_{k,\omega} = e^{i\theta_{k,\omega}}$, where $\theta_{k,\omega}$ is the phase of the fluctuation at k, ω, then the Vlasov response function is simply $\delta f_{k,\omega}/\delta\alpha_{k,\omega}$. In other words, the aim here is to extract the portion of the Vlasov equation nonlinearity which is phase coherent with $\alpha_{k,\omega}$. Note that the phase-coherent portion can and will contain pieces proportional to both $f_{k,\omega}$ and $E_{k,\omega}$.

4.3.2 One-dimensional electron plasmas

Proceeding, consider a simple 1D electron plasma with ions responding via a given susceptibility $\chi_i\,(k,\omega)$. Then the Vlasov–Poisson system is just,

$$-i\,(\omega - kv)\, f_{k,\omega} + \frac{\partial}{\partial v}\sum_{k',\omega'}\frac{q}{m}E_{\substack{-k'\\-\omega'}} f_{\substack{k+k'\\ \omega+\omega'}} = -\frac{q}{m}E_{\substack{k\\ \omega}}\frac{\partial\langle f\rangle}{\partial v}, \tag{4.18a}$$

and

$$ikE_{\substack{k\\ \omega}} = 4\pi n_0 q\int \mathrm{d}v\, f_{\substack{k\\ \omega}} - 4\pi n_0 q\,\chi_i\,(k,\omega)\,\frac{q}{T}\phi_{k,\omega}. \tag{4.18b}$$

To obtain the response function for mode k, ω, i.e. $\delta f_{k,\omega}/\delta\phi_{k,\omega}$, we seek to isolate the part of the nonlinearity $(q/m)\,E\,\partial f/\partial v$ which is phase coherent with $E_{k,\omega}$. Here, 'phase coherent' means having the same phase as does $\phi_{k,\omega}$. Note that the philosophy here presumes the utility of a test wave approach, which 'tags' each mode by a phase $\alpha_{k,\omega}$, and assumes that one may examine the phase-coherent response of a given mode in the (dynamic) background of all modes, without altering the statistics and dynamics of the ensemble. Thus, the aim of this procedure is to systematically approximate the nonlinearity,

$$N_{k,\omega} = \frac{\partial}{\partial v}\sum_{k',\omega'}\frac{q}{m}E_{\substack{-k'\\ \omega'}} f_{\substack{k+k'\\ \omega+\omega'}} \tag{4.19a}$$

by a function of the form,

$$N_{k,\omega} = C^{f}_{\mathbf{k},\omega}\, f_{\mathbf{k},\omega} + C^{E}_{k,\omega}\frac{q}{m}E_{\substack{k\\ \omega}}, \tag{4.19b}$$

where $C_{f_{k,\omega}}$ and $C_{E_{k,\omega}}$ are *phase independent* operator function of the fluctuation spectrum, and the response function itself. Note that the form of the renormalized $N_{k,\omega}$ in Eq.(4.19b) suggests that $C_{f_{k,\omega}}$ should reduce to the familiar diffusion operator from resonance broadening theory in certain limits.

4.3.2.1 Renormalization procedure

To answer the obvious question of how one obtains C_f, C_E, we proceed by a perturbative approach. To this end, it is useful to write the Vlasov equation as,

$$-i\,(\omega - kv)\, f_{\substack{k\\ \omega}} + \frac{\partial}{\partial v}\sum_{k',\omega'}\left(\frac{q}{m}E_{\substack{-k'\\ -\omega'}} f^{(2)}_{\substack{k+k'\\ \omega+\omega'}} + f^{(2)}_{\substack{-k'\\ -\omega'}}\frac{q}{m}E_{\substack{k+k'\\ \omega+\omega'}}\right) = -\frac{q}{m}E_{\substack{k\\ \omega}}\frac{\partial\langle f\rangle}{\partial v}. \tag{4.20a}$$

Here, the superscript (2) signifies that the quantity so labelled is driven by the *direct* beating of two modes or fluctuations. Thus,

$$f^{(2)}_{\substack{k+k'\\ \omega+\omega'}} \sim E^{(1)}_{\substack{k'\\ \omega'}} f^{(1)}_{\substack{k\\ \omega}} \propto \exp i\left[\theta_{k',\omega'} + \theta_{k,\omega}\right], \tag{4.20b}$$

and similarly for $E^{(2)}_{\substack{k+k'\\ \omega+\omega'}}$ (here θ is a phase of the mode.) This ensures that the resulting $N_{k,\omega}$ is phase coherent with $\phi_{k,\omega}$. The remaining step of relating $f^{(2)}_{\substack{k+k'\\ \omega+\omega'}}$, $E^{(2)}_{\substack{k+k'\\ \omega+\omega'}}$ to the amplitudes of primary modes is done by perturbation theory, i.e. for $f^{(2)}_{\substack{k+k'\\ \omega+\omega'}}$ we write,

$$\left[-i\left(\omega + \omega' - \left(k + k'\right)v\right) + Cf_{\substack{k+k'\\ \omega+\omega'}}\right] f^{(2)}_{\substack{k+k'\\ \omega+\omega'}} = -\frac{q}{m}\frac{\partial f^{(1)}_{\substack{k'\\ \omega'}}}{\partial v}E^{(1)}_{\substack{k\\ \omega}}, \tag{4.21a}$$

so

$$f^{(2)}_{\substack{k+k'\\ \omega+\omega'}} = L_{\substack{k+k'\\ \omega+\omega'}}\left\{\left(\frac{q}{m}\right)^2 E^{(1)}_{\substack{k'\\ \omega'}}\frac{\partial}{\partial v}L_{\substack{k\\ \omega}}\frac{\partial\langle f\rangle}{\partial v}E^{(1)}_{\substack{k\\ \omega}}\right\}, \tag{4.21b}$$

where the propagator is,

$$L^{-1}_{\substack{k+k'\\ \omega+\omega'}} = -i\left[\omega + \omega' - \left(k + k'\right)v + iC_{\substack{k+k'\\ \omega+\omega'}}\right]. \tag{4.21c}$$

Here we have dropped $E^{(2)}$ contributions since we are concerned with nonlinear wave–particle interaction and the response function for f. Terms from $E^{(2)}$ involve only *moments* of f, and so are not directly relevant to wave–particle interaction. Given this simplification, the subscript 'f' on C_f has also been dropped. Then, substituting Eqs.(4.21b), (4.21c) into Eq.(4.20a) gives the renormalized nonlinearity,

$$N_{k,\omega} = -\frac{\partial}{\partial v}\sum_{k',\omega'}\frac{q^2}{m^2}\left|E_{\substack{k'\\ \omega'}}\right|^2 L_{\substack{k''\\ \omega''}}\frac{\partial}{\partial v}f_{\substack{k\\ \omega}} - \frac{\partial}{\partial v}\sum_{k',\omega'} E_{\substack{-k'\\ \omega'}}\frac{\partial f_{\substack{k\\ \omega}}}{\partial v}L_{\substack{k''\\ \omega''}}\frac{q}{m}E_{\substack{k\\ \omega}}, \quad (4.22a)$$

and the renormalized Vlasov equation thus follows as,

$$-i\left(\omega - kv\right) f_{k,\omega} - \frac{\partial}{\partial v}D_{k,\omega}\frac{\partial}{\partial v}f_{k,\omega} = -\frac{q}{m}E_{k,\omega}\left(\frac{\partial\langle f\rangle}{\partial v} + \frac{\partial}{\partial v}\bar{f}_{k,\omega}\right), \quad (4.22b)$$

where by correspondence with $N_{k,\omega}$,

$$D_{k,\omega} = \sum_{k',\omega'}\frac{q^2}{m^2}\left|E_{k',\omega'}\right|^2 L_{\substack{k''\\ \omega''}} \quad (4.22c)$$

$$\bar{f}_{\substack{k\\ \omega}} = \sum_{k',\omega'}\frac{q}{m}E_{\substack{-k'\\ -\omega'}}\frac{\partial f_{\substack{k'\\ \omega'}}}{\partial v}L_{k'',\omega''}. \quad (4.22d)$$

Cleary, the operator $-\partial/\partial v\, D_{k,\omega}\, \partial/\partial v$ constitutes a propagator dressing or "self-energy" correlation to the bare Vlasov propagator $L_{k,\omega} = i/(\omega - kv)$. We term this a 'self-energy' because it reflects the effect of interactions between the test mode and the ambient spectrum of background waves, just as the self-energy renormalization of the electron propagated in quantum electrodynamics accounts for the interaction of a bare electron with ambient photons induced by vacuum polarization. In a similar vein, $\bar{f}_{k,\omega}$ corresponds to a renormalization of the background or ambient distribution function, and so bears a resemblence to wave function renormalization, familiar from quantum electrodynamics.

4.3.2.2 Non-Markovian property

Although the propagator renormalization derived above has the structure of a diffusion operator, here the diffusion coefficient $D_{k,\omega}$ depends explicitly upon the wave number and frequency of the fluctuation. This important feature reflects the fundamentally *non-Markovian* character of the nonlinear interactions. Recall that a Markovian process, which may be described by a Fokker–Planck equation with a space-time-independent diffusion coefficient, is one with no memory. Thus a Markovian model for f evolution is one with the form,

$$f(t+\tau, v) = f(t, v) + \int \mathrm{d}(\Delta v)\, T(v, \Delta v, \tau)\, f(v - \Delta v, t)\,. \tag{4.23}$$

Here, $T(v, \Delta v, \tau)$ is the transition probability for a step $v - \Delta v \rightarrow v$ in time interval τ. Making a standard Fokker–Planck expansion for small Δv, and taking $\tau \sim \tau_{\mathrm{ac}}$ and $\Delta v \sim \Delta v_{\mathrm{T}}$ recovers the resonance broadening theory result, with $D = D_{\mathrm{QL}}$. Application of this model to the evolution of $f_{k,\omega}$ is sensible *only* if $\partial f_{k,\omega}/\partial t \ll f_{k,\omega}/\tau_{\mathrm{ac}}$, and $|k'| \gg |k|$, $|\omega|' \gg |\omega|$, so that the spectrum of ambient modes appears as a stochastic bath to the test mode in question. Of course, examination of $D_{k,\omega}$ in the $\tau_{\mathrm{ac}} < \tau_{\mathrm{c}}$ limit (so $L_{k'',\omega''}$ may be taken to be bare) reveals that the k, ω dependence of $D_{k,\omega}$, $\bar{f}_{k,\omega}$ is present previously because the test mode at (k, ω) has spatio-temporal scales comparable to, not slower than, the other modes. Indeed, since,

$$D_{k,\omega} = \sum_{k',\omega'} \frac{q^2}{m^2} \left| E_{\substack{k' \\ \omega'}} \right|^2 \mathrm{Re} \left\{ \frac{i}{\omega + \omega' - (k + k')\, v} \right\}, \tag{4.24}$$

we see that $D_{k,\omega} \rightarrow D$ if $k \ll k'$, $\omega \ll \omega'$, which corresponds to the Markovian limit where the random 'kicking' by other modes is so fast that it appears as a sequence of random kicks. Hence, it is apparent that the non-Markovian structure of the theory is a consequence of the fact that the test mode scales are comparable to other scales in the spectrum. However, it should be noted that regardless of the ratio of test wave space-time scales to background scales, the Markovian approximation is always valid for *resonant* particles, for which $\omega = kv$, since in this case,

$$D^{(v)}_{k,\omega} = \sum_{k',\omega'} \frac{q^2}{m^2} \left| E_{\substack{k' \\ \omega'}} \right|^2 \mathrm{Re} \left\{ \frac{i}{\omega' - k' v} \right\} \rightarrow D_{\mathrm{QL}}. \tag{4.25}$$

Thus, the *resonant* particle response *is* amenable to treatment by a Markovian theory.

4.3.2.3 Background distribution renormalization

The other new feature in the theory is the background distribution renormalization $\bar{f}_{k,\omega}$. The function $\bar{f}_{k,\omega}$ accounts for the renormalization or "dressing" of the background distribution function which is necessary for the renormalized response of $f_{k,\omega}$ to reduce to the weak turbulence theory expansion result for $f_{k,\omega}$ in the limit of $kv \ll \omega$ and small fluctuation levels. More generally, $\bar{f}_{k,\omega}$ preserves certain structural properties of $N_{k,\omega}$ which are crucial to energetics and its treatment by the renormalized theory. These features are most readily illustrated in the context of the drift wave dynamics, so it is to this problem we now turn.

4.4 Renormalization in Vlasov turbulence II: drift wave turbulence

4.4.1 Kinetic description of drift wave fluctuations

Recall from Chapter 3 that a simple model for low frequency plasma dynamics in a strongly magnetized plasma is the drift-kinetic equation,

$$\frac{\partial f}{\partial t} + v_z \frac{\partial f}{\partial z} - \frac{c}{B_0} \nabla \phi \times z \cdot \nabla f + \frac{q}{m} E_z \frac{\partial f}{\partial v_z} = 0. \tag{4.26a}$$

The geometry of the plasma is illustrated in Figure 3.14. Indeed, the simplest possible model of drift wave dynamics consists of drift-kinetic ion dynamics, as described by Eq.(4.26a) and a 'nearly Boltzmann' electron response, along with quasi-neutrality, so,

$$\frac{n_{i\,\underset{\omega}{k}}}{n_0} = \int dv\, f_{\underset{\omega}{k}}(v) = \frac{n_{e\,\underset{\omega}{k}}}{n_0} = \left(1 - i\delta_{\underset{\omega}{k}}\right) \frac{|e|\,\phi_{k,\omega}}{T}, \tag{4.26b}$$

where ϕ is the electrostatic potential, so that the electric field in the direction of the main magnetic field, E_z, is given by $-\partial\phi/\partial z$. This simple model can be reduced even further by ignoring the $(qE_z/m)\,\partial f/\partial v_z$ nonlinearity, since $k_z \ll k_\perp$. In that case, the drift-kinetic equation for f simplifies to,

$$\frac{\partial f}{\partial t} + v_z \frac{\partial f}{\partial z} - \frac{cE}{B_0} \nabla f = \frac{c}{B_0} \frac{\partial \phi}{\partial y} \frac{\partial \langle f \rangle}{\partial r} - \frac{q}{m} F_z \frac{\partial \langle f \rangle}{\partial v_z}. \tag{4.26c}$$

Equation (4.26c) has the generic structure explained below,

$$\frac{\partial f}{\partial t} + \underset{①}{v_z \frac{\partial}{\partial z} f} + \underset{②}{\boldsymbol{v}_\perp \cdot \nabla_\perp f} = \underset{③}{\frac{\delta S}{\delta \phi} \phi \langle f \rangle}, \tag{4.27a}$$

where, in Eq.(4.27a), the meaning of the terms is as follows: ①: parallel streaming along $B_0 z$, ②: advection by fluid with $\nabla \cdot \boldsymbol{v} = 0 \rightarrow$ spatial scattering, ③: source-potential perturbation $\langle f \rangle = \langle f(r, v_z) \rangle$. Many variations on this generic form are possible obviously, one is to take $v_\perp \rightarrow 0$ (i.e. $B_0 \rightarrow \infty$), which recovers the structure of the linearized equation.

A second is to take $k_z \rightarrow 0$ and integrate over v, thus recovering,

$$\frac{\partial n}{\partial t} + \boldsymbol{v}_\perp \cdot \nabla_\perp n = \frac{\delta S'}{\delta \phi} \phi, \tag{4.27b}$$

which is similar to the equation for the evolution of a 2D fluid, i.e.

$$\frac{\partial \rho}{\partial t} + \nabla \phi \times z \cdot \nabla \rho - \nu \nabla^2 \rho = 0, \tag{4.27c}$$

where

$$\rho = \nabla^2 \phi. \tag{4.27d}$$

This type of structure appears in the descriptions of 2D fluids, guiding centre plasmas, non-neutral plasmas, etc.

A third variation is found by retaining finite Larmor radius effects, so

$$f_{\boldsymbol{k},\omega} = J_0\,(k\rho)\, f^{\mathrm{gc}}_{\boldsymbol{k},\omega}. \tag{4.27e}$$

Here, $f^{\mathrm{gc}}_{\boldsymbol{k},\omega}$ refers to the guiding centre distribution function and $\rho = v_\perp/\omega_c$. The guiding centre distribution obeys the gyrokinetic equation,

$$\frac{\partial}{\partial t} f^{\mathrm{gc}} + v_z \frac{\partial f^{\mathrm{gc}}}{\partial z} + \langle v_\perp \rangle_\theta \cdot \nabla_\perp f^{\mathrm{gc}} = \left\langle \frac{\partial S}{\partial \phi} \right\rangle_\theta \phi \,\langle f \rangle \,. \tag{4.27f}$$

Here, the averages $\langle\quad\rangle_\theta$ refer to gyro-angle averages. It should be apparent, then, that the structure of Eq.(4.26c) is indeed of general interest and relevant to a wide range of problems.

4.4.2 Coherent nonlinear effect via resonance broadening theory

The simplest approach to the task of obtaining a renormalized response of $f_{k,\omega}$ to $\phi_{k,\omega}$ in Eq.(4.26c) is to employ resonance broadening theory, i.e.

$$\begin{aligned} f_{\substack{k\\ \omega}} &= \mathrm{e}^{ik_z z} \mathrm{e}^{-i\boldsymbol{k}_\perp \cdot \boldsymbol{r}_\perp} \int_0^\infty \mathrm{d}\tau\, \mathrm{e}^{i\omega\tau} \langle u\,(-\tau)\rangle^* \,\mathrm{e}^{ik_z z} \mathrm{e}^{i\boldsymbol{k}_\perp \cdot \boldsymbol{r}_\perp} \frac{\delta S_{k,\omega}}{\delta \phi} \left(-\frac{|e|\,\phi_{\boldsymbol{k},\omega}}{T_i} \right) \langle f \rangle \\ &= \int_0^\infty \mathrm{d}\tau\, \mathrm{e}^{i(\omega - k_z v_z)\tau} \left\langle \mathrm{e}^{i\boldsymbol{k}_\perp \cdot \delta \boldsymbol{r}_\perp (-\tau)} \right\rangle^* \frac{\delta S_{k,\omega}}{\delta \phi} \left(-\frac{|e|\,\phi_{\boldsymbol{k},\omega}}{T_i} \right) \langle f \rangle \,. \end{aligned} \tag{4.28}$$

Here $\delta \boldsymbol{r}$ is the excursion in r induced by random $E \times B$ scattering. Taking the statistical distribution of $\delta \boldsymbol{r}$ to be Gaussian, one then finds,

$$\begin{aligned} \left\langle \mathrm{e}^{i\boldsymbol{k}_\perp \cdot \delta r(-\tau)} \right\rangle &\cong \left\langle 1 + i\boldsymbol{k}_\perp \cdot \delta \boldsymbol{r} - \frac{(k_\perp \delta r)^2}{2} \right\rangle \\ &= \exp\left[-\left(\boldsymbol{k}_\perp \cdot \boldsymbol{D} \cdot \boldsymbol{k}_\perp \right) \tau \right] \\ &= \exp\left[-k_\perp^2 \boldsymbol{D} \tau \right] \end{aligned} \tag{4.29}$$

for isotropic turbulence. Contrary to Eq.(4.7a), a simple exponential decay is recovered. Note that the diffusion is *spatial* here. The precise form of the diffusion tensor $\boldsymbol{D}_\perp$ may be straightforwardly obtained by a quasi-linear calculation on the underlying drift-kinetic equation (i.e. Eq.(4.26c)), yielding,

$$\boldsymbol{k}_\perp \cdot \boldsymbol{D} \cdot \boldsymbol{k}_\perp = \sum_{\boldsymbol{k}'} \frac{c^2}{B_0^2} \left(\boldsymbol{k}_\perp \cdot \boldsymbol{k}'_\perp z\right)^2 \left|\phi_{\boldsymbol{k}'}\right|^2 \pi\delta\left(\omega - k_z v_z\right). \tag{4.30}$$

Thus, for turbulence which is isotropic in $\boldsymbol{k}$ we find the familiar 'classic' form of the renormalized response, as given by resonance broadening theory,

$$f_{k,\omega} = \int_0^\infty \mathrm{d}\tau\, \mathrm{e}^{i\left(\omega - k_z v_z + i k_\perp^2 D\right)\tau} \frac{\delta S_{k,\omega}}{\delta\phi_{k,\omega}} \left(-\frac{|e|}{T_i}\phi_{k,\omega}\right) \langle f \rangle. \tag{4.31}$$

So, for drift-kinetic turbulence, we see that the decorrelation rate for turbulent scattering of a test particle from its unperturbed trajectory scattering is given by,

$$\frac{1}{\tau_{c\boldsymbol{k}}} = k^2 D_\perp. \tag{4.32}$$

This result is, in turn, the underpinning of the classic mixing length theory estimate for the saturated transport associated with drift wave instabilities, namely,

$$D_\perp = \gamma_{\boldsymbol{k}} \big/ k_\perp^2. \tag{4.33}$$

Note that the idea here is simply that the instability saturates when the rate at which a particle is scattered one perpendicular wave length (i.e. $1 / \tau_{c\boldsymbol{k}}$) equals the growth rate $\gamma_{\boldsymbol{k}}$. Of course, coupling to magnetic or flow velocity shear can increase the decorrelation rate, as discussed earlier in this chapter.

4.4.3 Conservation revisited

Note, too, that the essence of resonance broadening theory is simply to replace the nonlinearity of the drift-kinetic equation (i.e. Eq.(4.26c)) by a diffusion operator, so (RHS = right-hand side):

$$\frac{\partial f}{\partial t} + v_z \frac{\partial f}{\partial z} - \frac{c}{B} \nabla\phi \times z \cdot \nabla f = \text{RHS} \tag{4.34a}$$

becomes

$$\frac{\partial f}{\partial t} + v_z \frac{\partial f}{\partial z} - \nabla_\perp \cdot \boldsymbol{D} \cdot \nabla_\perp f = \text{RHS}. \tag{4.34b}$$

Thus, we see that in this application, the result of resonance broadening theory resembles that of a simple 'eddy viscosity' model as used in modelling in fluid turbulence. The apparent direct correspondence between resonance broadening theory and simple eddy viscosity methods then begs the question, "*Is application of resonance broadening theory to the drift-kinetic equation in the vein discussed above correct?*". Two simple observations are quite pertinent to answering this question. One is that since the $\boldsymbol{E} \times \boldsymbol{B}$ nonlinearity is independent of velocity (except for the velocity dependence of f), we can integrate Eq.(4.9c) over velocity ($\int \mathrm{d}^3 v$) to obtain,

$$\frac{\partial n}{\partial t} + \frac{\partial v_z}{\partial z} - \frac{c}{B_0} \nabla \phi \times z \cdot \nabla n = \int \mathrm{d}^3 v \, \mathrm{RHS}, \tag{4.35a}$$

which may then be straightforwardly re-written in the form,

$$\frac{\partial n}{\partial t} + \frac{\partial v_z}{\partial z} + \nabla_\perp \cdot \boldsymbol{J}_\perp = \int \mathrm{d}^3 v \, \mathrm{RHS}, \tag{4.35b}$$

where

$$\boldsymbol{J}_\perp = \boldsymbol{v}_{\boldsymbol{E}\times\boldsymbol{B}} \, n \tag{4.35c}$$

is the perpendicular current carried by $\boldsymbol{E} \times \boldsymbol{B}$ advection of particles. Of course, such a current can not couple to the perpendicular electric field, because $J_\perp$ in Eq.(4.35c) is perpendicular to $E_\perp$ (either to do work or have work done upon it), so we require,

$$\left\langle \boldsymbol{E}_\perp^* \cdot \boldsymbol{J}_\perp \right\rangle = 0, \tag{4.36}$$

where the brackets signify a space-time average. In drift kinetics, the only heating possible is *parallel* heating, so $\langle \boldsymbol{E} \cdot \boldsymbol{J} \rangle = \left\langle E_\parallel J_\parallel \right\rangle$, as in the discussion of the energetics of quasi-linear theory for drift wave turbulence which was presented in Chapter 3. This is seen trivially in real space or in $\boldsymbol{k}$-space, since

$$\left\langle \boldsymbol{E}_\perp^* \cdot \boldsymbol{J}_\perp \right\rangle = \sum_{\boldsymbol{k},\omega} \sum_{\boldsymbol{k}',\omega'} \frac{c}{B_0} \left(\boldsymbol{k} \cdot \boldsymbol{k}', \boldsymbol{\omega}' \times z \right) \phi_{\substack{-\boldsymbol{k} \\ -\omega}} \phi_{\substack{-\boldsymbol{k}' \\ -\omega'}} n_{\substack{\boldsymbol{k}+\boldsymbol{k}' \\ \omega+\omega'}}, \tag{4.37}$$

so that the interchange $(k, \omega) \leftrightarrow \left(k', \omega'\right)$ leaves all in Eq.(4.37) invariant, except for the coupling coefficient $\boldsymbol{k} \cdot \boldsymbol{k}' \times z$, which is anti-symmetry, i.e. $\boldsymbol{k} \cdot \boldsymbol{k}' \times z \rightarrow -\boldsymbol{k} \cdot \boldsymbol{k}' \times z$ so $\left\langle \boldsymbol{E}^* \cdot \boldsymbol{J}_\perp \right\rangle = -\left\langle \boldsymbol{E}^* \cdot \boldsymbol{J}_\perp \right\rangle = 0$. The condition that $\left\langle \boldsymbol{E}^* \cdot \boldsymbol{J}_\perp \right\rangle = 0$ can be re-written as a condition on the nonlinear term $N_{k,\omega}$ since,

$$\left\langle \boldsymbol{E}_\perp^* \cdot \boldsymbol{J}_\perp \right\rangle = \sum_{\boldsymbol{k},\omega} \phi_{\substack{-\boldsymbol{k} \\ -\omega}} N_{\substack{\boldsymbol{k} \\ \omega}} = 0. \tag{4.38}$$

Any renormalization of the nonlinearity $N_{k,\omega}$ must satisfy the condition expressed by Eq.(4.38). From this discussion, we also see that the problem of renormalization in turbulence theory is one of 'representation', i.e. the aim of renormalization theory is to 'represent' the 'bare' nonlinearity by a simpler, more tractable operator which maintains its essential physical properties.

A second, somewhat related property of N is that it annihilates the adiabatic or Boltzmann response, to the lowest order in $1/k_\perp L_\perp$, where $L_\perp$ is the perpendicular scale length of $\langle f \rangle$ variation. In calculations related to drift wave turbulence, it is often useful to write the total fluctuating distribution function as the sum of the Boltzmann response (f_B) plus a non-Boltzmann correction. Thus for ions, we often write,

$$f_{k,\omega} = -\frac{|e|}{T_i}\phi_{k,\omega}\langle f \rangle + g_{k,\omega}. \tag{4.39}$$

Now it is obvious that $\boldsymbol{v}_{E\times\mathrm{B}} \cdot \nabla f_B \to 0$ to the lowest order, since $\boldsymbol{E} \times \boldsymbol{B} \cdot \nabla\phi = 0$. Hence any representation of $N_{k,\omega}$ must preserve the property that $\lim_{f\to f_\mathrm{B}} N_{k,\omega} \to 0$, to the lowest order.

It is painfully clear that resonance broadening theory satisfies *neither* of the constraints discussed above. In particular, in resonance broadening theory, $N_{k,\omega} = k_\perp^2 D f_{k,\omega}$, and rather obviously,

$$\sum_{k,\omega} \phi_{\substack{-k \\ -\omega}} N_{\substack{-k \\ -\omega}} \to \sum_{k,\omega} \phi_{\substack{-k \\ -\omega}} k_\perp^2 D f_{\substack{k \\ \omega}} \neq 0, \tag{4.40a}$$

so that $\langle \boldsymbol{E}_\perp^* \cdot \boldsymbol{J}_\perp \rangle \neq 0$, as it should be. Also, in resonance broadening theory,

$$\lim_{f\to f_\mathrm{B}} N_{\substack{k \\ \omega}} = k^2 D \left(-\frac{|e|}{T}\phi_{\substack{k \\ \omega}} \langle f \rangle \right) \neq 0, \tag{4.40b}$$

so the Boltzmann response is *not* annihilated, either. Hence, resonance broadening theory fails both 'tests' of a successful renormalization. The reason for these shortcomings is the neglect of background renormalization, i.e. $\bar{f}_{k,\omega}$, by the resonance broadening approach.

4.4.4 *Conservative formulations*

This shortcoming can be rectified by the perturbative renormalization procedure presented above in the context of the 1D Vlasov plasma. We now turn to the application of this methodology to drift wave turbulence.

The primitive drift-kinetic equation is, in $\boldsymbol{k}$, ω-space,

$$-i\omega\left(\omega-k_z v_z\right) f_{\substack{\boldsymbol{k}\\ \omega}}+\frac{c}{B_0}\sum_{\boldsymbol{k}'}\left(\boldsymbol{k}\cdot\boldsymbol{k}'\times\boldsymbol{z}\right) f_{\substack{-\boldsymbol{k}'\\ -\omega'}} f_{\substack{\boldsymbol{k}+\boldsymbol{k}'\\ \omega+\omega'}}=\frac{\delta S}{\delta\phi}\left(-\frac{|e|}{T_i}\phi_{\boldsymbol{k},\omega}\right)\langle f\rangle, \tag{4.41a}$$

so that the portion of the nonlinearity which is phase coherent with $\alpha_{\boldsymbol{k},\omega}$, where $\alpha_{\boldsymbol{k},\omega}=\exp i\theta_{\boldsymbol{k}\omega}$ and θ is a phase, may be written as,

$$N_{\substack{\boldsymbol{k}\\ \omega}}\cong\frac{c}{B_0}\sum_{\boldsymbol{k}',\omega}\left(\boldsymbol{k}\cdot\boldsymbol{k}'\times\boldsymbol{z}\right)\left(\phi^{(1)}_{\substack{-\boldsymbol{k}'\\ -\omega'}} f^{(2)}_{\substack{\boldsymbol{k}+\boldsymbol{k}'\\ \omega+\omega'}}-f^{(1)}_{\substack{-\boldsymbol{k}'\\ -\omega'}}\phi^{(2)}_{\substack{\boldsymbol{k}+\boldsymbol{k}'\\ \omega+\omega'}}\right). \tag{4.41b}$$

As before, here we are interested in nonlinear wave–particle interaction, so we ignore $\phi^{(2)}_{\substack{\boldsymbol{k}+\boldsymbol{k}'\\ \omega+\omega'}}$ hereafter. The quantity $f^{(2)}_{\substack{\boldsymbol{k}+\boldsymbol{k}'\\ \omega+\omega'}}$ is given by,

$$\begin{aligned}&-i\left\{\left(\omega+\omega'\right)-\left(k_z+k_z'\right)\right\} f^{(2)}_{\substack{\boldsymbol{k}+\boldsymbol{k}'\\ \omega+\omega'}}+C_{\substack{\boldsymbol{k}+\boldsymbol{k}'\\ \omega+\omega'}} f^{(2)}_{\substack{\boldsymbol{k}+\boldsymbol{k}'\\ \omega+\omega'}}\\ &=\frac{c}{B_0}\left(\boldsymbol{k}\cdot\boldsymbol{k}'\times\boldsymbol{z}\right)\left(\phi_{\substack{\boldsymbol{k}'\\ \omega'}} f_{\substack{\boldsymbol{k}\\ \omega}}-f_{\substack{\boldsymbol{k}'\\ \omega'}}\phi_{\substack{\boldsymbol{k}\\ \omega}}\right),\end{aligned} \tag{4.42a}$$

so

$$f^{(2)}_{\substack{\boldsymbol{k}+\boldsymbol{k}'\\ \omega+\omega'}}=L_{\substack{\boldsymbol{k}+\boldsymbol{k}'\\ \omega+\omega'}}\frac{c}{B_0}\left(\boldsymbol{k}\cdot\boldsymbol{k}'\times\boldsymbol{z}\right)\left(\phi_{\substack{\boldsymbol{k}'\\ \omega'}} f_{\substack{\boldsymbol{k}\\ \omega}}-f_{\substack{\boldsymbol{k}'\\ \omega'}}\phi_{\substack{\boldsymbol{k}\\ \omega}}\right), \tag{4.42b}$$

where

$$L^{-1}_{\substack{\boldsymbol{k}+\boldsymbol{k}'\\ \omega+\omega'}}=-i\left\{\left(\omega+\omega'\right)-\left(k_z+k_z'\right)v_z+iC_{\substack{\boldsymbol{k}+\boldsymbol{k}'\\ \omega+\omega'}}\right\} \tag{4.42c}$$

is the beat wave propagator. Thus, we see that the renormalized nonlinearity has the form,

$$N_{\boldsymbol{k},\omega}=d_{\boldsymbol{k},\omega} f_{\boldsymbol{k},\omega}-\bar{f}_{\boldsymbol{k},\omega}\phi_{\boldsymbol{k},\omega}, \tag{4.43a}$$

where

$$d_{\boldsymbol{k},\omega}=\sum_{\boldsymbol{k}',\omega'}\frac{c^2}{B_0^2}\left(\boldsymbol{k}\cdot\boldsymbol{k}'\times\boldsymbol{z}\right)^2\left|\phi_{\substack{\boldsymbol{k}'\\ \omega'}}\right|^2 L_{\substack{\boldsymbol{k}+\boldsymbol{k}'\\ \omega+\omega'}}\cong k_\perp^2 D_{\boldsymbol{k},\omega} \tag{4.43b}$$

for isotropic turbulence, and

$$\bar{f}_{k,\omega} = \frac{c^2}{B_0^2} \sum_{k',\omega'} \left(k \cdot k' \times z\right)^2 \phi_{\substack{-k' \\ -\omega'}} \phi_{\substack{k' \\ \omega'}} \phi_{\substack{k+k' \\ \omega+\omega'}} . \tag{4.43c}$$

Here, $d_{k,\omega}$ is the rate of test particle scattering and $\bar{f}_{k,\omega}$ is the background distribution renormalization. The term $d_{k,\omega}$ is referred to as the 'test particle scattering rate', since it is different from the actual particle flux, which is necessarily regulated to the non-adiabatic electron response $\delta_{k,\omega}|e|\phi_{k,\omega}/T_e$. Retaining $d_{k,\omega}$, $\bar{f}_{k,\omega}$, we thus see that the renormalized distribution response satisfies the equation

$$-i\left(\omega - k_z v_z\right) f_{\substack{k \\ \omega}} + d_{\substack{k \\ \omega}} f_{\substack{k \\ \omega}} = \left(\frac{\mathrm{d}S}{\mathrm{d}\phi} \langle f \rangle + \bar{f}_{\substack{k \\ \omega}}\right) \phi_{\substack{k \\ \omega}}. \tag{4.44}$$

As in 1D, both resonance broadening and background distribution renormalization are non-Markovian. Finally, note that yet another way to argue for the existence of $\bar{f}_{k,\omega}$ is that both test particles and background particles (somewhat akin to field particles in collision theory) are scattered by the ensemble of fluctuations.

Given the motivation, we first check that the renormalized $N_{k,\omega}$ as given by Eq.(4.43a), satisfies the two properties. First, using Eq.(4.43) we can easily show that,

$$\sum_{k,\omega} \phi_{\substack{-k \\ -\omega}} N_{\substack{k \\ \omega}}$$
$$= \sum_{k,\omega} \sum_{k',\omega'} \frac{c^2}{B_0^2} \left(k \cdot k' \times z\right)^2 \left\{ \left|\phi_{\substack{k' \\ \omega'}}\right|^2 \left(\phi_{\substack{-k \\ -\omega}} f_{\substack{k \\ \omega}}\right) - \left|\phi_{\substack{k \\ \omega}}\right|^2 \left(\phi_{\substack{-k' \\ -\omega'}} f_{\substack{k' \\ \omega'}}\right) \right\} = 0, \tag{4.45}$$

by anti-symmetry under $k, \omega \leftrightarrow k', \omega'$. Thus, we see that the renormalization is consistent with $\left\langle E_\perp^* \cdot J_\perp \right\rangle = 0$. Second, it is straightforward to show that $\lim_{f \to f_B} N_{k,\omega} \to 0$, so the renormalized nonlinearity vanishes in the limit of the Boltzmann response. Hence, we see that the perturbative renormalization procedure respects both properties of $N_{k,\omega}$, as it should. The presence of the background renormalization $\bar{f}_{k,\omega}$ is essential to this outcome! For the sensitive and subtle case of drift-kinetic turbulence, the perturbative renormalization approach, derived from the idea of extracting the piece of $N_{k,\omega}$ phase coherent with $\alpha_{k,\omega}$, clearly is more successful than is resonance broadening theory.

4.4.5 Physics content and predictions

Having addressed some of the questions concerning the formal structure of the renormalized theory of drift wave turbulence, we now turn to more interesting issues of physics content and predictions. Of course, the principal goals of *any* renormalized theory of plasma turbulence, in general, or of drift wave turbulence, in particular, are:

1. to identify and understand nonlinear space-time scales;
2. to identify the relevant nonlinear saturation mechanisms and calculate the corresponding nonlinear damping rates;
3. to identify and predict possible bifurcations in the saturated state.

In the context of the specific example of drift-kinetic turbulence, goals (1) and (2) may be refined further to focus on the specific questions:

1. What is the *physical* meaning of the decorrelation rate and propagator renormalization $d_{\mathbf{k},\omega}$, and how is it related to mixing, transport and heating?
2. What is the rate of nonlinear ion heating? Note that ion heating is required for saturation of drift wave turbulence in order to balance energy input from electron relaxation.

Answering and illuminating these two questions is the task to which we now turn.

4.4.5.1 Propagator renormalization and mixing

The propagator renormalization $d_{\mathbf{k},\omega}$ is a measure of the rate at which the response $f_{\mathbf{k},\omega}$ to a test wave fluctuation $E_{\mathbf{k},\omega}$ is mixed or scrambled by the ensemble of turbulent fluctuations. The term $d_{\mathbf{k},\omega}$ is defined recursively, i.e.

$$d_{\mathbf{k},\omega} = \sum_{\mathbf{k}',\omega'} \frac{c^2}{B_0^2} \left|\phi_{\mathbf{k}',\omega'}\right|^2 \left(\mathbf{k}\cdot\mathbf{k}'\times z\right)^2 \frac{i}{(\omega+\omega') - \left(k_z+k_z'\right)v_z + i d_{\mathbf{k}+\mathbf{k}',\omega+\omega'}}, \tag{4.46}$$

since $d_{\mathbf{k},\omega}$ results from the beat interactions of the test wave with background modes which themselves undergo turbulent decorrelation. (A method based on the recursion is explained in Chapter 6, where the closure model is discussed.) For simplicity, we further specialize to the isotropic turbulence, long wavelength, low frequency limit where,

$$d_{\mathbf{k},\omega} \to k^2 D = \sum_{\mathbf{k},\omega} \frac{c^2}{B_0^2} \left|\phi_{\mathbf{k}',\omega'}\right|^2 \left(\mathbf{k}\cdot\mathbf{k}'\times z\right)^2 \frac{i}{\omega' - k_z' v_z + i d_{\mathbf{k}',\omega'}}. \tag{4.47}$$

This decorrelation rate corresponds to 'resonance broadened' quasi-linear theory. Function $d_{\mathbf{k},\omega}$ behaves rather differently in the weak and strong turbulence regimes, i.e. if $\left|(\mathrm{d}\omega/\mathrm{d}k_\perp)\,\Delta k_\perp\right| + \left|\left(v_{\mathrm{gr}\parallel} - \omega/k_\parallel\right)\Delta k_\parallel\right| > d_{\mathbf{k}',\omega'}$ or $< d_{\mathbf{k}',\omega'}$, respectively. In the first case, which corresponds to weak turbulence theory and resembles the simple quasi-linear prediction,

$$\begin{aligned} \operatorname{Re} d_{\boldsymbol{k},\omega} &= \sum_{\boldsymbol{k}',\omega'} \frac{c^2}{B_0^2} \left|\phi_{\boldsymbol{k}',\omega'}\right|^2 \left(\boldsymbol{k} \cdot \boldsymbol{k}' \times z\right)^2 \pi\delta\left(\omega' - k_z' v_z\right) \\ &= k^2 \left\langle \tilde{v}^2 \right\rangle \tau_{\mathrm{ac}}. \end{aligned} \tag{4.48}$$

Here, the irreversibility inherent to quasi-linear diffusion results from wave–particle resonance and the auto-correlation time τ_{ac} is just $\left(\left|(\mathrm{d}\omega/\mathrm{d}k_\perp)\,\Delta k_\perp\right| + \left|\left(v_{\mathrm{gr}\parallel} - \omega/k_\parallel\right)\Delta k_\parallel\right|\right)^{-1}$, which is determined by the spectrum and the wave dispersion properties. As is usual for such cases, $D \sim \langle \tilde{v}^2 \rangle$, and the physical meaning of $d_{\boldsymbol{k},\omega}$ is simply decorrelation due to resonant diffusion in space. However, on drift wave turbulence k_z is often quite small, so as to avoid the strongly stabilizing effects of ion Landau damping. With this in mind, it is interesting to examine the opposite limit, where $k_z' v_z$ is negligible but propagator broadening is retained. Assuming spectral isotropy, assuming a spectrum of eigenmodes where $\omega = \omega(\boldsymbol{k})$ and ignoring the $\boldsymbol{k}'$, ω' dependence of $d_{\boldsymbol{k}',\omega'}$ then gives,

$$k^2 D \cong \sum_{\boldsymbol{k}'} \frac{c^2}{B_0^2} \left|\tilde{\phi}_{\boldsymbol{k}'}\right|^2 k^2 k'^2 \frac{k'^2 D}{\omega'^2 + \left(k'^2 D\right)^2}, \tag{4.49a}$$

which reduces to,

$$1 \cong \sum_{\boldsymbol{k}'} \frac{c^2}{B_0^2} \left|\tilde{\phi}_{\boldsymbol{k}'}\right|^2 \frac{k'^4}{\omega'^2 + \left(k'^2 D\right)^2}, \tag{4.49b}$$

with understanding a simple scaling as the goal in mind, we throw caution to the winds and boldly pull the right-hand side denominator of Eq.(4.49b) outside the mode summation to obtain,

$$\left(k^2 D\right)^2 + \omega^2 \cong k^2 \left\langle \tilde{v}_{E\times B}^2 \right\rangle, \tag{4.49c}$$

or equivalently,

$$\left(k^2 D\right)^2 \cong k^2 \left\langle \tilde{v}_{E\times B}^2 \right\rangle - \omega^2. \tag{4.49d}$$

Equation (4.49d) finally reveals the physical meaning of D in the limit where $d_{\boldsymbol{k},\omega} > |k_z v_z|$, since it relates the mean squared decorrelation rate $\left(k^2 D\right)^2$ to the *difference* of the mean squared $E \times B$ Doppler shift $\left(k^2 \left\langle \tilde{v}_{E\times B}^2 \right\rangle\right)$ and the mean squared wave frequency $\langle \omega^2 \rangle$. Of course, since the electric field is turbulent, the $E \times B$ Doppler shift is stochastic, and we tacitly presume the second moment $\langle \tilde{v}_{E\times B}^2 \rangle$ is well defined. Thus, Eq.(4.49d) states that there is a critical level of fluctuating $\tilde{v}_{E\times B}$ needed to scatter a test particle *into* resonance, and so render $D \neq 0$. That level is $(\tilde{v}_{E\times B})_{\text{rms}} \sim \omega/k_\perp$, i.e. a stochastic perpendicular velocity which is comparable to the perpendicular phase velocity of the drift wave. The particle in question is called a 'test particle', since nowhere is the field forced to be self-consistent by the imposition of quasi-neutrality.

Given that $(\tilde{v}_{E\times B})_{\text{rms}} \sim \omega/k_\perp$ defines the threshold for stochastization or mixing of a test particle, it is natural to discuss the relationship of this criterion to the familiar "mixing length estimate" for turbulence saturation levels. Note that for drift waves, the stochastic Doppler resonance criterion becomes $\tilde{v}_{E\times B} \sim \omega/k_\perp \sim V_{de}$. Since $\tilde{v}_{E\times B} \sim k_\perp \rho_s c_s (|e|\, \phi/T)$ and $V_{de} = \rho_s c_s/L_n$, $\tilde{v}_{E\times B} \sim V_{de}$ occurs for fluctuation levels of $e\tilde{\phi}/T \sim 1/k_\perp L_n$, which, noting that $\tilde{n}/n \sim e\tilde{\phi}/T$ for electron drift waves, is *precisely* the traditional mixing length estimate of the saturation level. This occurrence is not entirely coincidental, as we now discuss.

Basically, all of the standard drift wave type plasma instabilities are *gradient driven*, and (in the absence of external drive) tend to radially mix or transport the driving gradient, and so *relax* or *flatten* the gradient thus turning off the gradient drive. Thus, electron drift waves tend to mix density n or electron temperature T_e and so to relax ∇n and ∇T_e, ion temperature gradient driven modes tend to mix T_i and relax ∇T_i, etc. The essence of the mixing length estimate is that the growth of an instability driven by a local gradient will cease when the 'mixing term' or nonlinearity grows to a size which is comparable to the gradient drive. Thus, if one considers advection of density in the context of a drift wave, the density fluctuation would satisfy an equation with the generic structure,

$$\frac{\partial n}{\partial t} + \tilde{v}_{E\times B} \cdot \nabla n + \cdots = -v_{E\times B,r} \frac{\partial \langle n \rangle}{\partial r}. \tag{4.50a}$$

A concrete example of such a balance would be that between the linear term $(\tilde{v}_r\, \partial \langle n \rangle / \partial r)$ and the nonlinear term $(\tilde{v}_r\, \partial \tilde{n}/\partial r)$ of Eq.(4.50a), which yields,

$$\frac{\tilde{n}}{\langle n \rangle} \sim \frac{l_\perp}{L_n}, \tag{4.50b}$$

which is the conventional mixing length 'estimate' of the saturation level. Note that Eq.(4.50b) relates the density fluctuation level $\tilde{n}/n$ to the ratio of two length scales, namely $l_\perp$, the "mixing length" and L_n, the density gradient scale length. This then begs the question of just what precisely *is* the mixing length $l_\perp$. Intuitively speaking, it is the length over which a fluid or plasma element is scattered by instability-induced fluctuations. The mixing length, $l_\perp$ is often thought to correspond to the width of a typical convection cell, and motivated by concerns of calculation, is frequently estimated by the radial wavelength of the underlying linear instability. We emphasize that this is *purely* an approximation of convenience, and that there is absolutely *no* reason why $l_\perp$ for a state of fully developed turbulence should be tied to the scale of the original linear instability. In general, $l_\perp$ is unknown, and the accuracy to which it can be calculated varies closely with the depth of one's understanding of the fundamental nonlinear dynamics. For example, in the core of Prandtl mixing length theory discussed in Chapter 2, the choice of the distance to the wall as the mixing length most likely was motivated by an appreciation of the importance of self-similarity of the mean velocity gradient and the need to fit empirically determined flow profiles. Thus, mixing length theory should be considered *only* as a guideline for estimation, and practitioners of mixing length theory should keep in mind the old adage that "mixing length theory is always correct, *if* one knows the mixing length". Finally, we should add that local mixing length estimation, of the form described above, is also based upon the tacit presumption that $l_\perp \ll L_n$, so there are many cells within a gradient scale length. In this sense, the system is taken to be more like a sandpile than Rayleigh–Benard convection in a box, which is dominated by a single big convection cell, so $l_\perp \sim L_n$. Non-local mixing length models have been developed, and resemble in structure those of flux-limited transport, where $l_{\text{mfp}} \sim L_n$. However, applications of such models has been limited in scope. (The multiple-scale problem is discussed in Chapter 7 and in Volume 2. The interested reader should also see the discussion in Frisch (1995).)

4.4.5.2 *Nonlinear heating and saturation mechanism*

We now turn to the second physics issue, namely that of nonlinear heating and the saturation mechanism. In this regard, it is useful to recall the key points of the previous discussion, which were:

(a) while turbulent test particle scattering and decorrelation occur at a rate given by $k_\perp^2 D$, where for $k_z \to 0$, $D \neq 0$ requires that a threshold in intensity be exceeded, i.e. $k^2 \langle \tilde{v}_{\text{E}\times\text{B}}^2 \rangle \gtrsim \langle \omega^2 \rangle$;

(b) actual *heating* occurs only via *parallel* $\boldsymbol{E}\cdot\boldsymbol{J}$ work, i.e. $\langle E_\parallel J_\parallel \rangle$, since the energy density satisfies the conservation equation (compare Eq.(3.62a)),

$$\frac{\partial E}{\partial t} + \frac{\partial Q}{\partial x} + \langle E_z J_z \rangle = 0,$$

this is a consequence of the fact that $\langle E_\perp \cdot J_\perp \rangle = 0$ for drift kinetics.

Thus, *any nonlinear heating which leads to saturation* ***must*** *be in proportion to a power of* k_z, since $\langle E_z J_z \rangle \to 0$ for $k_z \to 0$. Equivalently, any 'action' from the nonlinearity to saturate the turbulence occurs via k_z.

4.4.5.3 Description by moments of the drift-kinetic equation

Given this important observation, and the fact that $k_z v_{\mathrm{Ti}}/\omega < 1$ for drift wave turbulence, it is useful to work from moments of the drift-kinetic equation, i.e. Eq.(4.26c). Assuming a Maxwellian $\langle f \rangle$, the drift-kinetic equation may be written as,

$$\frac{\partial f}{\partial t} + v_z \frac{\partial f}{\partial z} - \frac{c}{B_0} \nabla \phi \times z \cdot \nabla f = -i \left(k_z v_z - \omega_{*i} \right) \frac{|e|\phi}{T_{\mathrm{i}}} \langle f \rangle . \qquad (4.51)$$

Then the relevant moments are:

$$n = \int \mathrm{d}^3 v \, f \qquad \text{: density} \qquad (4.52a)$$

$$J = |e| \int \mathrm{d}^3 v \, v_z f \qquad \text{: parallel current} \qquad (4.52b)$$

$$p = m \int \mathrm{d}^3 v \, v^2 f \qquad \text{: energy} \qquad (4.52c)$$

and satisfy the fluid equations,

$$\frac{\partial n}{\partial t} + \frac{1}{|e|} \frac{\partial J}{\partial z} - \frac{c}{B_0} \nabla \phi \times z \cdot \nabla n = \frac{c}{B_0} \nabla \phi \times z \cdot \nabla n_0 \qquad (4.53a)$$

$$\frac{\partial J}{\partial t} + \frac{\partial}{\partial z} \frac{|e|\, p}{m} + \frac{e^2}{m} n \frac{\partial \phi}{\partial z} - \frac{c}{B_0} \nabla \phi \times z \cdot \nabla J = 0 \qquad (4.53b)$$

$$\frac{\partial p}{\partial t} + \frac{\partial}{\partial z} \overline{n m v_z v^2} - E_z J_z - \frac{c}{B_0} \nabla \phi \times z \cdot \nabla p = 0. \qquad (4.53c)$$

Equations (4.53a)–(4.53c) may be further simplified by noting that,

$$\frac{\tilde{n}}{n} \simeq \frac{|e|\phi}{T_{\mathrm{e}}}, \qquad (4.54a)$$

$$p = m \frac{v_{\mathrm{Ti}}^2}{2} n, \qquad (4.54b)$$

so

$$-\frac{c}{B_0} \nabla \phi \times z \cdot \nabla n \to 0 \qquad (4.54c)$$

$$-\frac{c}{B_0} \nabla \phi \times z \cdot \nabla p \to 0. \qquad (4.54d)$$

In this limit, the density equation is *strictly linear*, so the system of fluid equations reduces to

$$(\omega - \omega_{*e}) \frac{|e|\,\phi_{\boldsymbol{k}}}{T} = \frac{k_z}{n_0\,|e|} J_{z\,\boldsymbol{k},\omega} \tag{4.55a}$$

and

$$J_{z\,\substack{\boldsymbol{k}\\ \omega}} = \frac{k_z v_{\mathrm{Ti}}^2}{\omega} \frac{n_0\,|e|^2}{T} \phi_{\boldsymbol{k},\omega} + \frac{-i}{\omega} \sum_{\boldsymbol{k}',\omega'} \frac{c}{B_0} (k \cdot k' \times z)\, \phi_{-k'-\omega'} J_{\substack{\boldsymbol{k}+\boldsymbol{k}'\\ \omega+\omega'}}. \tag{4.55b}$$

Linear theory tells us that the waves here are drift-accoustic modes, with

$$\omega = \omega_{*e} + \frac{k_z^2 v_{\mathrm{Ti}}^2}{\omega}. \tag{4.56}$$

This structure, and the fact that ion heating requires finite k_z, together, strongly suggest that *shear viscosity of the parallel flow will provide the requisite damping.* Since turbulent shear viscosity results from $\boldsymbol{E} \times \boldsymbol{B}$ advection of J, we now focus on the renormalization of the current equation.

In $\boldsymbol{k}$, ω space, the equation for parallel flow or current is,

$$-i\omega J_{\substack{\boldsymbol{k}\\ \omega}} + \frac{c}{B_0} \sum_{\boldsymbol{k}',\omega'} (k \cdot k' \times z) \left\{ \phi_{\substack{-\boldsymbol{k}'\\ -\omega'}} J^{(2)}_{\substack{\boldsymbol{k}+\boldsymbol{k}'\\ \omega+\omega'}} - \phi^{(2)}_{\substack{\boldsymbol{k}+\boldsymbol{k}'\\ \omega+\omega'}} J_{\substack{-\boldsymbol{k}'\\ -\omega'}} \right\} = -i \frac{k_z}{\omega} v_{\mathrm{Ti}}^2 n_0 \frac{|e|^2}{T} \phi_{\substack{\boldsymbol{k}\\ \omega}}. \tag{4.57}$$

As before, since we are concerned with heating, we focus on nonlinear wave–particle interaction, and so neglect the $\phi^{(2)}_{\substack{\boldsymbol{k}+\boldsymbol{k}'\\ \omega+\omega'}}$ contribution. For $J^{(2)}_{\substack{\boldsymbol{k}+\boldsymbol{k}'\\ \omega+\omega'}}$ we can then immediately write,

$$\left(-i\,(\omega+\omega') + d_{\substack{\boldsymbol{k}''\\ \omega''}}\right) J^{(2)}_{\substack{\boldsymbol{k}+\boldsymbol{k}'\\ \omega+\omega'}} = \frac{c}{B_0} (k \cdot k' \times z) \left(\phi_{\substack{\boldsymbol{k}'\\ \omega'}} J_{\substack{\boldsymbol{k}\\ \omega}} - J_{\substack{\boldsymbol{k}'\\ \omega'}} \phi_{\substack{\boldsymbol{k}\\ \omega}} \right), \tag{4.58a}$$

so

$$J^{(2)}_{\substack{\boldsymbol{k}+\boldsymbol{k}'\\ \omega+\omega'}} = L_{\substack{\boldsymbol{k}+\boldsymbol{k}'\\ \omega+\omega'}} \frac{c}{B_0} (k \cdot k' \times z) \left(\phi_{\substack{\boldsymbol{k}'\\ \omega'}} J_{\substack{\boldsymbol{k}\\ \omega}} - J_{\substack{\boldsymbol{k}'\\ \omega'}} \phi_{\substack{\boldsymbol{k}\\ \omega}} \right), \tag{4.58b}$$

where

$$L^{-1}_{\substack{\boldsymbol{k}+\boldsymbol{k}'\\ \omega+\omega'}} = -i\,(\omega+\omega') + d_{\substack{\boldsymbol{k}+\boldsymbol{k}'\\ \omega+\omega'}} \tag{4.58c}$$

is the propagator. Substitution of Eq.(4.58b) into Eq.(4.57) then gives the renormalized parallel flow or current equations as,

$$\left(-i\omega + d_{\boldsymbol{k},\omega}\right) J_{\substack{\boldsymbol{k}\\ \omega}} = i\frac{k_z}{\omega} v_{\mathrm{Ti}}^2 n_0 \frac{|e|^2}{T} \phi_{\boldsymbol{k},\omega} + \beta_{\boldsymbol{k},\omega}\phi_{\boldsymbol{k},\omega} \tag{4.59a}$$

$$d_{\boldsymbol{k},\omega} = \sum_{\boldsymbol{k}',\omega'} \left(k \cdot k' \times z\right)^2 \frac{c^2}{B_0^2} \left|\phi_{\substack{\boldsymbol{k}'\\ \omega'}}\right|^2 L_{\substack{\boldsymbol{k}+\boldsymbol{k}'\\ \omega+\omega'}} \tag{4.59b}$$

$$\beta_{\boldsymbol{k},\omega} = \sum_{\boldsymbol{k}',\omega'} \left(\boldsymbol{k} \cdot \boldsymbol{k}' \times z\right)^2 \frac{c^2}{B_0^2} \phi_{\substack{-\boldsymbol{k}'\\ -\omega'}} J_{\substack{\boldsymbol{k}'\\ \omega'}} L_{\substack{\boldsymbol{k}+\boldsymbol{k}'\\ \omega+\omega'}}. \tag{4.59c}$$

Furthermore, since both linear response theory and more general considerations of energetics suggest that $J_{z\,\boldsymbol{k}} \sim k_z\phi_{\boldsymbol{k}}$, we have,

$$\begin{aligned}\beta_{\boldsymbol{k},\omega} &\approx \sum_{\boldsymbol{k}',\omega'} \left(\boldsymbol{k}' \cdot \boldsymbol{k}' \times z\right)^2 \frac{c^2}{B_0^2} \left|\phi_{\boldsymbol{k}',\omega'}\right|^2 k_z L_{\substack{\boldsymbol{k}+\boldsymbol{k}''\\ \omega+\omega''}} \\ &\to 0,\end{aligned} \tag{4.60}$$

since the integrand is odd in k_z. Thus, we see that Eq.(4.59a) simplifies to,

$$\left(-i\omega + d_{\boldsymbol{k},\omega}\right) J_{\boldsymbol{k},\omega} = -ik_z v_{\mathrm{Ti}}^2 n_0 \frac{|e|^2}{T} \phi_{\boldsymbol{k},\omega}. \tag{4.61}$$

Note this simply states that in a system of drift wave turbulence, the response of the parallel flow is renormalized by a shear viscosity, and that this is the leading-order nonlinear effect. Combining Eq.(4.61) and Eq.(4.55a) then gives,

$$\omega - \omega_{*e} = \frac{k_z^2 v_{\mathrm{Ti}}^2}{\omega + d_{\boldsymbol{k},\omega}}, \tag{4.62a}$$

so for low or moderate fluctuation levels we find,

$$\omega \cong \omega_{*e} + \frac{k_z^2}{\omega} v_{\mathrm{Ti}}^2 d_{\boldsymbol{k},\omega}, \tag{4.62b}$$

which says that turbulent dissipation is set by the product of the shear viscous mixing rate and the hydrodynamic factor $k_z^2 v_{\mathrm{Ti}}^2/\omega^2$. Note that nonlinear damping enters in proportion to the non-resonance factor $k_z^2 v_{\mathrm{Ti}}^2/\omega^2$, since the only heating which can occur is *parallel heating*, $\sim \langle E_z J_z \rangle$. Also note that since $k_z^2 v_{\mathrm{Ti}}^2/\omega^2 < 1$, the size nonlinear damping rate $\gamma_{\boldsymbol{k}}^{\mathrm{NL}}$ is $\left|\gamma_{\boldsymbol{k}}^{\mathrm{NL}}\right| < d_{\boldsymbol{k},\omega}$. Thus, the nonlinear damping rate is reduced relative to naive expectations by the factor $k_z^2 v_{\mathrm{Ti}}^2/\omega^2$. Finally for $k_z \to 0$, $\gamma_{\boldsymbol{k}}^{\mathrm{NL}} \to 0$, as it must.

This simple case of weakly resonant drift-kinetic turbulence is a good example of the subtleties of renormalized turbulence theory, energetics, symmetry, etc. Further application of related techniques to other propagator renormalization problems may be found in (Kim and Dubrulle, 2001; Diamond and Malkov, 2007). The moral of this story is clearly one that physical insight and careful consideration of dynamical constraints are essential elements of the application of any renormalized theory, no matter how formally appealing it may be.

5

Kinetics of nonlinear wave–wave interaction

A wave is never found alone, but is mingled with as many other waves as there are uneven places in the object where said wave is produced. At one and the same time there will be moving over the greatest wave of a sea innumerable other waves, proceeding in different directions.

(Leonardo da Vinci, "Codice Atlantico")

5.1 Introduction and overview

5.1.1 Central issues and scope

After our discussions of quasi-linear theory and nonlinear wave–particle interaction, it is appropriate to pause, to review where we've been, and to survey where we're going. Stepping back, one can say that the central issues which plasma turbulence theory must address may be classified as:

(1) mean field relaxation – how the mean distribution function evolves in the presence of turbulence, and what sort of heating, cross-field transport, etc. results from that evolution. Though much maligned, mean field theory forms the backbone of most approaches to turbulence. Chapter 3 deals with the most basic approach to the mean field theory of relaxation, namely quasi-linear theory – based upon closure using the linear response. Future chapters will discuss more advanced approaches to describing relaxation in a turbulent, collisionless plasma;

(2) response – how the distribution function evolves in response to a test perturbation in a turbulent collisionless plasma. Problems of response including nonlinear Landau damping, resonance broadening theory, propagator renormalization, etc. are discussed in Chapters 3 and 4, on the kinetics of nonlinear wave–particle interaction;

(3) spectra and excitation – how a system of nonlinearly interacting waves or modes couples energy or 'wave quanta density' across a range of space-time scales, given a distribution of forcing, growth and dissipation. In essence, the quest to understand the mechanism of spectral energy transfer and distribution *defines* the "problem of turbulence" – the classic Kolmogorov 1941 (K41) theory of 3D Navier–Stokes

turbulence being the most familiar and compelling example. Hence, the discussion of this chapter, now turns to the problems of spectra and excitation.

We approach the problem of spectral transfer dynamics by examining a sequence of illustrative paradigms in wave–wave interaction theory. This sequence begins with coherent wave–wave interactions, proceeds to wave turbulence theory and its methodology, and then addresses simple scaling models of cascades. Future chapters will discuss extensions and related topics, such as closure methods for strong turbulence, methods for the reduction of multi-scale problems, disparate scale interactions, etc. The list of familiar examples discussed includes, but is not limited to:

- resonant wave interactions
- wave kinetics
- decay and modulational processes
- scaling theory of turbulent cascade

Familiar conceptual issues encountered along the way include:

- the implication of conservation laws for transfer processes
- basic time-scale orderings and their relevance to the unavoidable (yet unjustifiable!) *assumption* of the applicability of statistical methods
- restrictions on the wave kinetic equation
- the type of interactions and couplings (i.e. local and non-local) possible. Anticipating subsequent discussions of structure formation, we give special attention to *non-local* wave–wave interactions
- spectral structure and its sensitivity to the distribution of sources and sinks

Both lists are long and together leave no doubt that the subject of wave–wave interaction is a vast and formidable one!

5.1.2 Hierarchical progression in discussion

Given the scope of the challenge, we structure our discussion as a hierarchical progression through three sections. These are:

(a) The integrable dynamics of three coupled modes.
These simple, "toy model" studies reveal the basic elements of the dynamics of mode–mode interactions (such as the Manley–Rowe relations) and illustrate the crucial constraints which conservation laws impose on nonlinear transfer processes. Though simple, these models constitute the essential foundation of the theory of wave–wave interaction. One appealing feature of these basic paradigms is that their dynamics can be described using easily visualizable geometrical constructions, familiar from classical mechanics.

(b) The physical kinetics of multi-wave interactions and wave turbulence.
The culmination of this discussion is the derivation of a wave-kinetic equation, similar to the Boltzmann equation, for the exciton or wave population density $N(\boldsymbol{k}, \omega_{\boldsymbol{k}}, t)$ and its evolution. The wave population density $N(\boldsymbol{x}, \boldsymbol{k}, t)$ may be thought of as a distribution function for quasi-particles in the phase space $(\boldsymbol{x}, \boldsymbol{k})$. Thus, wave population dynamics resembles quasi-particle dynamics. Like the Boltzmann equation for particles, the wave-kinetic equation is *fundamentally statistical*, rests on assumptions of weak correlation and microscopic chaos, and takes the generic form,

$$\frac{\partial N}{\partial t} + \left(v_{\mathrm{gr}} + v\right) \cdot \nabla N - \frac{\partial}{\partial \boldsymbol{x}} \left(\omega + \boldsymbol{k} \cdot v\right) \cdot \frac{\partial N}{\partial \boldsymbol{k}} = C(N)$$

$$C(N) = \sum_{\substack{\boldsymbol{k}',\boldsymbol{k}'' \\ \boldsymbol{k}'+\boldsymbol{k}''=\boldsymbol{k}}} \delta\left(\omega_{\boldsymbol{k}} - \omega_{\boldsymbol{k}'} - \omega_{\boldsymbol{k}''}\right) \times \left\{C_{\mathrm{s}}\left(\boldsymbol{k}', \boldsymbol{k}''\right) N_{\boldsymbol{k}'} N_{\boldsymbol{k}''} - C_{\mathrm{s}}\left(\boldsymbol{k}', \boldsymbol{k}\right) N_{\boldsymbol{k}'} N_{\boldsymbol{k}}\right\}.$$

Wave kinetics first appeared in the theory of the statistical mechanics of lattice vibrational modes – i.e. as in the Einstein and Debye theory of solids, etc. We emphasize, though, that those early applications were concerned with systems of thermal fluctuations *near* equilibrium, while wave turbulence deals with strongly excited systems *far* from equilibrium. One quantity which distinguishes 'equilibrium' from 'non-equilibrium' solutions of the wave-kinetic equation is the exciton density flux in $\boldsymbol{k}$, i.e.

– near equilibrium, the flux of exciton density (i.e. energy) from mode to mode is weak, so the spectrum is determined by a scale-by-scale balance of emission and absorption, consistent with the fluctuation–dissipation theorem (Fig. 2.1).

– far from equilibrium the scale-to-scale flux (which can be either local or non-local in $\boldsymbol{k}$!) dominates local emission and absorption, as in the inertial range cascade in fluid turbulence. The spectrum is determined by the condition of flux stationarity, which entails the solution of a differential equation with appropriate boundary conditions.

The wave-kinetic equation can either be solved to obtain the spectrum or can be integrated to derive (moment) equations for net wave energy and momentum density. The latter may be used to calculate macroscopic consequences of wave interaction, such as wave-induced stresses and wave energy flux, by analogy with the theory of radiation hydrodynamics.

(c) The scaling theory of cascades in wave turbulence.
The classic example of a cascade scaling theory is the K41 model of Navier–Stokes turbulence. Though they may appear simple, or even crude, to a casual observer, scaling theories can be subtle and frequently are the only viable approach to problems

of wave turbulence. Scaling theories and wave kinetics are often used synergistically when approaching complex problems in wave turbulence.

(d) Non-local interaction in wave turbulence.
Wave turbulence can transfer energy both locally (i.e. between modes with neighboring $\boldsymbol{k}$s) and non-locally (i.e. between modes of very different scale). In this respect, wave turbulence is much richer and more complex than high R_e fluid turbulence (as usually thought of), despite the fact that wave turbulence is usually weaker, with a lower 'effective Reynolds number'. Non-local interaction is important to the dynamics of structure formation, as a possible origin of intermittency in wave turbulence, and as an energy flow channel which cannot be ignored.

Note that as we progress through the sequence of sections (a) → (b) → (c), the level of rigour with which we treat the dynamics degrades, in return for access to a broader regime of applicability – i.e. to systems with more degrees of freedom or higher wave intensity levels. Thus, in (a), the analysis is *exact* but restricted to resonant interaction of only three modes. In (b), the description is statistical and cast in terms of a Boltzmann-like wave kinetic equation. As for the Boltzmann equation in the kinetic theory of gases (KTG) Chapman and Cowling (1952), wave kinetics rests upon assumptions of weak interaction (akin to weak correlations in KTG) and microscopic chaos – i.e. the random phase approximation (akin to the principle of molecular chaos in KTG). However, wave kinetics *does* capture anisotropy and scale dependency in the coupling coefficients and in the selection rules which arise from the need for resonant matching of frequencies. In contrast to the toy models discussed in (a), the wave kinetics of (b) can describe the evolution of broad spectra with many interacting waves, although this gain comes with the loss of all phase information. However, wave kinetics cannot address wave breaking (i.e. fluctuation levels at or beyond the mixing length level) or other instances where nonlinear rates exceed linear wave frequencies. A systematic treatment of breaking, wave resonance broadening and other strongly nonlinear stochastic phenomena requires the use of renormalized closure theory, which is the subject of Chapter 6. At the crudest level, as discussed in c), one can proceed in the spirit of the Kolmogorov and Richardson model of the energy cascade in Navier-Stokes turbulence and construct scaling theories for spectral evolution. Scaling models, which are basically zero-dimensional, constitute one further step along the path of simplification, in that they release *all* phase and memory constraints in return for ease of applicability to systems with greater complexity. In scaling models, scale-dependent couplings are represented by simple multiplicative factors, anisotropy is often (though not always) ignored and spectral transfer is assumed to be local in scale. Nevertheless, scaling arguments are often the only tractable way to deal with problems of strong wave turbulence, and so merit discussion in this chapter.

5.2 The integrable dynamics of three coupled modes

"When shall we three meet again?"

(Shakespeare, "Macbeth")

In this section, we discuss the simple and fundamental paradigm of three resonantly coupled waves. This discussion forms the foundation for much of our later treatment of wave kinetics and so is of some considerable importance.

5.2.1 *Free asymmetric top (FAT)*

Surely the *simplest* example of a system with three resonantly coupled degrees of freedom is the free asymmetric top (FAT), familiar from elementary mechanics Landau and Lifshitz (1976). The FAT satisfies Euler's equation,

$$\frac{\mathrm{d}\boldsymbol{L}}{\mathrm{d}t} + \boldsymbol{\Omega} \times \boldsymbol{L} = 0, \tag{5.1a}$$

which may be written component-by-component as,

$$\frac{\mathrm{d}\Omega_1}{\mathrm{d}t} + \left(\frac{I_3 - I_2}{I_1}\right)\Omega_2\Omega_3 = 0, \tag{5.1b}$$

$$\frac{\mathrm{d}\Omega_2}{\mathrm{d}t} + \left(\frac{I_1 - I_3}{I_2}\right)\Omega_3\Omega_1 = 0, \tag{5.1c}$$

$$\frac{\mathrm{d}\Omega_3}{\mathrm{d}t} + \left(\frac{I_2 - I_1}{I_3}\right)\Omega_1\Omega_2 = 0. \tag{5.1d}$$

Here the three components of the angular momentum $\boldsymbol{\Omega}$ may be thought of as analogous to evolving coupled mode amplitudes. The inertia tensor is diagonal, with principal axes 1, 2, 3 and $I_3 > I_2 > I_1$. Of course $\boldsymbol{L} = \boldsymbol{I} \cdot \boldsymbol{\Omega}$, where $\boldsymbol{I}$ is the moment of inertia. Equation (5.1) can be solved straightforwardly in terms of elliptic functions, but it is far more illuminating to 'solve' the FAT geometrically, via the elegant Poinsot construction. The essence of the Poinsot construction exploits the fact that, the FAT has *two* quadratic integrals of the motion, namely $L^2 = \boldsymbol{L} \cdot \boldsymbol{L}$ – the magnitude of the angular momentum vector, and the energy. These invariants follow from Euler's equation. Thus, since,

$$L^2 = L_1^2 + L_2^2 + L_3^2 = L_0^2, \tag{5.2a}$$

$$E = \frac{L_1^2}{2I_1} + \frac{L_2^2}{2I_2} + \frac{L_3^2}{2I_3} = E_0, \tag{5.2b}$$

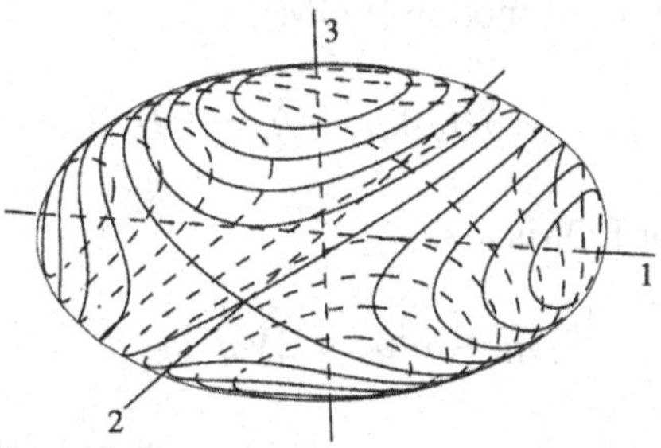

Fig. 5.1. Poinsot construction for the free asymmetric top. Note that trajectories originating at the poles of the "2" axis encircle the ellipsoid.

$\boldsymbol{L}(t) = \boldsymbol{I} \cdot \boldsymbol{\Omega}(t)$ is *simultaneously* constrained to evolve on:

(i) the surface of a sphere of constant radius L_0;
(ii) the surface of an ellipsoid, with semi-major axes of length $\left(2E_0 I_{1,2,3}\right)^{1/2}$.

The set of curves which trace out the possible intersections of the sphere and ellipsoid also trace the possible trajectories of FAT motion. This constitutes the Poinsot construction, and is shown in Figure 5.1.

Several features of the FAT motion can be deduced simply by inspection of the Poinsot construction. First, all trajectories are *closed*, so the dynamics are *reversible*. There is *no* intrinsic tendency for energy to accumulate in any one degree of freedom. Second, trajectories originating near the "1" or "3" axes (corresponding to I_1 or I_3) are closed and *localized* to the vicinity of those axes, while trajectories initialized near the "2" axis wrap around the body of the ellipsoid, and so are *not localized*. Thus, an initial condition starting near the "2" axis is linearly unstable, as is well known from rigid body stability theory. However, the fact that all trajectories are closed tells us that the linear solution breaks down as instability grows and then saturates, and $\boldsymbol{L}$ ultimately returns to its point of origin.

5.2.2 Geometrical construction of three coupled modes

Interestingly, a similar geometrical construction, which captures the essential dynamics of resonant 3-mode coupling, may be derived from the resonant mode coupling equations and their conservation properties. Most generally, resonant coupling dynamics arises in the perturbative solution to the interaction of three nonlinearly coupled harmonic oscillators, with Hamiltonian,

$$H = \sum_i \left(\frac{p_i^2}{2m} + \frac{\omega_i^2 q_i^2}{2} + 2V q_1 q_2 q_3 \right), \tag{5.3a}$$

and Hamiltonian equations of motion (EOM),

$$\dot{p}_i = -\partial H/\partial q_i, \quad \dot{q}_i = \partial H/\partial p_i, \tag{5.3b}$$

so the coupled oscillator EOMs are,

$$\ddot{q}_i + \omega_i^2 q_i = 2V q_j q_k. \tag{5.3c}$$

The essence of the perturbative, *weak coupling* approximation applied here is to restrict amplitudes to the limit where nonlinearity is small relative to the wave frequency (i.e. $\left|Vq \ll \omega^2\right|$), so the time for nonlinear energy transfer is slower than a wave oscillation time. Thus,

$$q_i(t) = a_i(t)\, \mathrm{e}^{-i\omega_i t} + a_i^*(t)\, \mathrm{e}^{i\omega_i t}, \tag{5.4}$$

where the time variation of the phase factor $\exp(\pm i\omega_i t)$ accounts for the fast oscillation frequency and that of the amplitude $a_i(t)$ accounts for slow variation due to nonlinear interaction. The basic ordering is $|\dot{a}_i/a| \ll \omega_i$. Substitution of Eq.(5.4) into Eq.(5.3c) (for $i = 1$) then yields, after some rearrangement,

$$\frac{\mathrm{d}^2 a_1}{\mathrm{d}t^2} - 2i\omega_1 \frac{\mathrm{d}a_1(t)}{\mathrm{d}t} = \mathrm{e}^{2i\omega_1 t}\left(\frac{\mathrm{d}^2 a_1^*}{\mathrm{d}t^2} + 2i\omega_1 \frac{\mathrm{d}a_1^*}{\mathrm{d}t}\right)$$

$$- 2V\Big(a_2 a_3 \exp[i(\omega_1 - \omega_2 - \omega_3)t] + a_2 a_3^* \exp[i(\omega_1 - \omega_2 + \omega_3)t]$$

$$+ a_2^* a_3 \exp[i(\omega_1 + \omega_2 - \omega_3)t] + a_2^* a_3^* \exp[i(\omega_1 + \omega_2 + \omega_3)t]\Big). \tag{5.5}$$

Since the question of concern here is the nature of energy transfer among oscillators on time scales that are *long* compared to the wave period $2\pi/\omega_1$, we aim to describe secular evolution of $a_1(t)$, which can occur only if the right-hand side of Eq.(5.5) does *not* oscillate rapidly in time. Hence, we arrive at the *resonance* or *frequency matching condition* or '*selection rule*' which is,

$$\omega_1 \pm \omega_2 \pm \omega_3 = 0. \tag{5.6}$$

Satisfying this 3-wave resonance condition ensures the secular drive of each mode by the other two. Proceeding without loss of generality by taking $\omega_3 = \omega_1 + \omega_2$, we obtain,

$$i\omega_1 \frac{\mathrm{d}a_1(t)}{\mathrm{d}t} = V a_2^* a_3 + \text{oscillatory terms},$$

which for weak interaction as $t \to \infty$ reduces to,

$$i\omega_1 \frac{da_1(t)}{dt} = V a_2^* a_3. \tag{5.7}$$

Implementing similar expansions for $i = 2, 3$ yields the *resonant 3-wave coupling equations*,

$$i\omega_1 \frac{da_1(t)}{dt} = V a_2^*(t)\, a_3(t), \tag{5.8a}$$

$$i\omega_2 \frac{da_2(t)}{dt} = V a_1^*(t)\, a_3(t), \tag{5.8b}$$

$$i\omega_3 \frac{da_3(t)}{dt} = V a_1(t)\, a_2(t). \tag{5.8c}$$

In general, the coupling coefficient V may be complex and may depend on the parameters of waves 1, 2, 3. For example, the resonant coupling equations for three interacting Rossby waves (as given by Pedlosky (1987), after Longuet-Higgins and Gill (Longuet-Higgins *et al.*, 1967)) are,

$$\frac{da_1}{dt} + \frac{B(\boldsymbol{k}_2, \boldsymbol{k}_3)}{k_1^2 + F} a_2 a_3 = 0,$$

$$\frac{da_2}{dt} + \frac{B(\boldsymbol{k}_3, \boldsymbol{k}_1)}{k_2^2 + F} a_3 a_1 = 0,$$

$$\frac{da_3}{dt} + \frac{B(\boldsymbol{k}_1, \boldsymbol{k}_2)}{k_3^2 + F} a_1 a_2 = 0. \tag{5.9a}$$

See Appendix 1 for explanation of the equations for Rossby waves. Here the coupling coefficient is,

$$B(\boldsymbol{k}_m, \boldsymbol{k}_n) = \frac{1}{2}\left(k_m^2 - k_n^2\right)\left(\boldsymbol{k}_m \times \boldsymbol{k}_n \cdot \hat{z}\right). \tag{5.9b}$$

The parameter F indicates the scale length that dictates the wave dispersion, and the resonance conditions are,

$$\omega_{mn} \pm \omega_m \pm \omega_n = 0,$$

$$\boldsymbol{k}_{mn} \pm \boldsymbol{k}_m \pm \boldsymbol{k}_n = 0. \tag{5.9c}$$

Certainly, the resonant coupling equations for model amplitudes $a_1(t)$, $a_2(t)$, $a_3(t)$ bear a close resemblance to the Euler equations for $\Omega_1(t)$, $\Omega_2(t)$, $\Omega_3(t)$ in the FAT. Analogy with the top suggests we should immediately identify the integrals of the motion (IOMs) for the resonant coupling equations. It is no surprise that one quantity conserved by Eq.(5.8) is the total energy of the system, i.e.

$$E = \frac{1}{2}\omega_1^2 |a_1(t)|^2 + \frac{1}{2}\omega_2^2 |a_2(t)|^2 + \frac{1}{2}\omega_3^2 |a_3(t)|^2, \tag{5.10}$$

which is also derived from Eq.(5.8) by noting the relation $\omega_3 = \omega_1 + \omega_2$. Conservation of energy, i.e. $\mathrm{d}E/\mathrm{d}t = 0$, is demonstrated straightforwardly using Eqs.(5.8a)–(5.8c), their complex conjugates and the resonance condition $\omega_3 - \omega_1 - \omega_2 = 0$. A (somewhat) less obvious conservation relation may be derived from the mode amplitude equations (Eq.(5.8)) by observing,

$$\begin{aligned}\frac{\mathrm{d}}{\mathrm{d}t}\left(\omega_1 |a_1|^2\right) = \frac{\mathrm{d}}{\mathrm{d}t}\left(\frac{\omega_1^2 |a_1^2|}{\omega_1}\right) &= V\,\mathrm{Re}\left(-ia_1^* a_2^* a_3\right) \\ &= V\,\mathrm{Im}\left(a_1^* a_2^* a_3\right),\end{aligned} \tag{5.11a}$$

and similarly,

$$\frac{\mathrm{d}}{\mathrm{d}t}\left(\frac{\omega_2^2 |a_2|^2}{\omega_2}\right) = V\,\mathrm{Im}\left(a_1^* a_2^* a_3\right), \tag{5.11b}$$

$$\begin{aligned}\frac{\mathrm{d}}{\mathrm{d}t}\left(\frac{\omega_3^2 |a_3|^2}{\omega_3}\right) &= V\,\mathrm{Im}\left(a_3^* a_1 a_2\right) \\ &= -V\mathrm{Im}\left(a_1^* a_2^* a_3\right),\end{aligned} \tag{5.11c}$$

since $\mathrm{Im}\, a^* = -\mathrm{Im}\, a$. Taken together, Eqs.(5.11a), (5.11b), (5.11c) state that,

$$\frac{\mathrm{d}}{\mathrm{d}t}(E_1/\omega_1) = \frac{\mathrm{d}}{\mathrm{d}t}(E_2/\omega_2) = -\frac{\mathrm{d}}{\mathrm{d}t}(E_3/\omega_3), \tag{5.12}$$

where $E_j = \omega_j^2 |a_j|^2$ ($j = 1, 2, 3$). We remind the reader that here, the selection or frequency match rule is $\omega_3 = \omega_2 + \omega_1$. That is, ω_3 is the highest frequency.

5.2.3 *Manley–Rowe relation*

Equations (5.11) and (5.12) form a geometrical construction for three (resonantly) coupled modes. Equation (5.12) is a particular statement of an important and

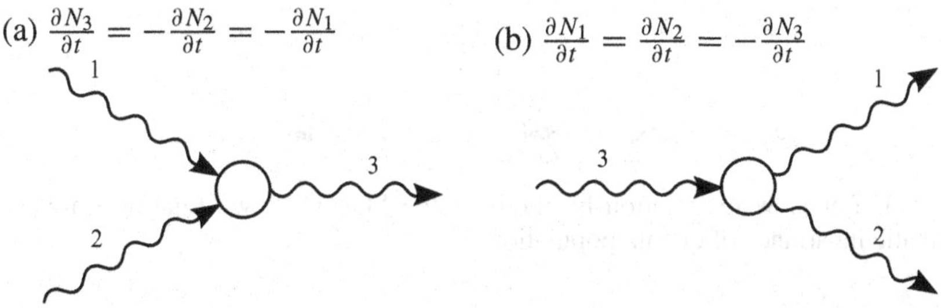

Fig. 5.2. Equivalent statements of the Manley–Rowe relations for three interacting waves.

general identity called the Manley–Rowe (M–R) relation. This relation is best understand by observing that E_i/ω_i has dimensions of action (i.e. energy $*$ time) and so may be thought of as a *mode action*,

$$N_i = E_i/\omega_i, \quad E_i = N_i\omega_i$$

which also reminds us of the familiar semi-classical formula $E = N\omega$, which relates N, the number of quanta, to the energy E and frequency ω (for $N \gg 1$). Hence N may also be usefully thought of as the 'number' of wave quanta. The significance of the M–R relation then emerges as an input-out balance for wave quanta! Specifically, the M–R relation which (most generally) requires that if,

$$\omega_{k_\alpha} + \omega_{k_\beta} = \omega_{k_\gamma}, \tag{5.13a}$$

then

$$\frac{\mathrm{d}N(k_\alpha)}{\mathrm{d}t} = \frac{\mathrm{d}N\left(k_\beta\right)}{\mathrm{d}t} = -\frac{\mathrm{d}N\left(k_\gamma\right)}{\mathrm{d}t}, \tag{5.13b}$$

which effectively states that should modes α, β beat together to drive mode γ, then for every exciton or wave 'created' in modes γ, one quantum *each* must be lost from modes α, β. The M–R relation is also reversible, so that if instead one quantum of mode γ is destroyed, then one quantum each for modes α, β must be created as a result of the interaction process. The M–R relation is depicted graphically in Figure 5.2. The M–R relation has interesting implications for the often relevant limit where one mode has frequency much lower than the other two, i.e. $\omega_1 + \omega_2 = \omega_3$, but $\omega_1 \ll \omega_2, \omega_3$. Such instances of slow modulation are relevant to problems of drift wave–zonal flow interaction, Langmuir turbulence and other important applications involving structure formation, discussed in Chapter 6. In this case, where $\omega_2 \cong \omega_3$, the behaviour of modes 2 and 3 must be virtually identical, but the M–R relation forces $\mathrm{d}N_2/\mathrm{d}t = -\mathrm{d}N_3/\mathrm{d}t$!? This paradox is resolved

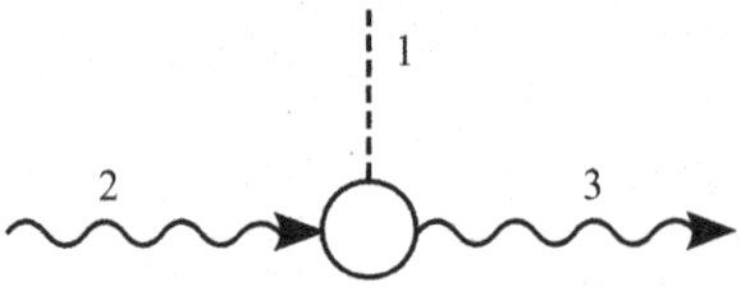

Fig. 5.3. For slow modulation by mode 1, the Manley–Rowe relation implies adiabatic invariance of quanta populations.

by requiring,

$$\frac{dN_2}{dt} = \frac{dN_3}{dt} = 0,$$

so that the number of quanta N is conserved in the interaction (i.e. $dN/dt = 0$). Thus, *in the case where one mode is a slow modulator of the other two, the Manley–Rowe relation is equivalent to the statement of adiabatic invariance of the wave quanta population.* This is sketched in Figure 5.3.

Taken together, then, energy conservation and the M–R relations, i.e.

$$E_0 = \left(\omega_1^2 a_1^2 + \omega_2^2 a_2^2 + \omega_3^2 a_3^2\right), \tag{5.14}$$

and

$$\omega_1 a_1^2 + \omega_3 a_3^2 = N_1(0) + N_3(0), \tag{5.15a}$$

or equivalently,

$$\omega_2 a_2^2 + \omega_3 a_3^2 = N_2(0) + N_3(0), \tag{5.15b}$$

specify *two* constraints on the interacting mode amplitudes in a resonant triad. Here, the number of quanta of mode i, $N_i(0)$, refers to the initial quanta number of mode i, and so, $E_0 = \omega_1 N_1(0) + \omega_2 N_2(0) + \omega_3 N_3(0)$. Consideration of Eqs.(5.14), (5.15b) reveals that the system's trajectories in the phase space $\left(\sqrt{\omega_1}a_1, \sqrt{\omega_2}a_2, \sqrt{\omega_3}a_3\right)$ are given by the curves of intersection between:

(i) the ellipsoid of constant energy, with semi-major axes $(E_0/\omega_1)^{1/2}$, $(E_0/\omega_2)^{1/2}$, $(E_0/\omega_3)^{1/2}$, and
(ii) Manley–Rowe cylinders, oriented parallel to the a_1 and a_2 axes, with radii $(N_2(0) + N_3(0))^{1/2}$, $(N_1(0) + N_3(0))^{1/2}$.

This construction clearly resembles, but is subtly different from, the Poinsot construction for the FAT trajectories. From Figure 5.4, we can immediately see that:

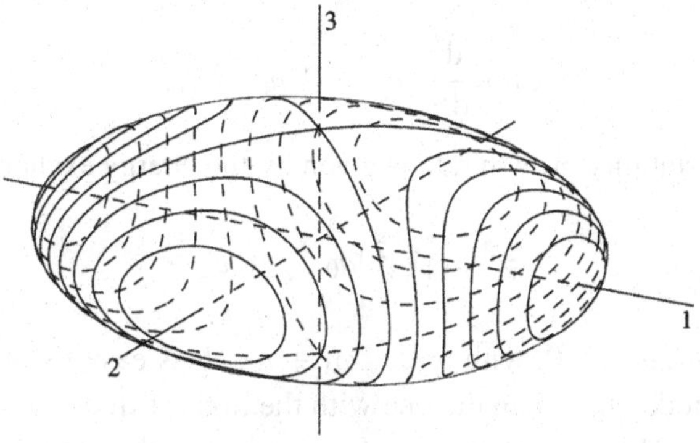

Fig. 5.4. Poinsot construction for three interacting modes. Note that trajectories originating at the "3" axis encircle the ellipsoid.

(i) as for the FAT, all trajectories are closed curves, so all *motion is reversible and periodic*

but,

(ii) since $\omega_3 > \omega_2 > \omega_1$, the intersections of the Manley–Rowe cylinders and the energy ellipsoid always encircle the ellipsoid if $N_3(0) \gg N_2(0), N_1(0)$, while trajectories with initial conditions for which $N_2(0) \gg N_1(0), N_3(0)$ remain localized near the poles at the a_2 or a_1 axes.

5.2.4 Decay instability

This tells us that the *highest* frequency mode is subject to *decay instability* if it is initialized with the largest population or externally driven. Recall for contrast that in the FAT, decay instability occurred for initialization near the Ω_2 axis, i.e. corresponding to the mode with intermediate moment of inertia ($I_3 > I_2 > I_1$). In both cases, however, the nonlinear motion is periodic and all trajectories close on themselves.

The prediction of decay instability may be verified by linearizing Eqs.(5.8a)–(5.8c) about the state $a_3(t) \cong a_0$, $a_{1,2}(t) = \delta a_{1,2} \ll a_0$. This gives,

$$i\omega_1 \frac{\mathrm{d}\delta a_1}{\mathrm{d}t} = V\delta a_2^* a_0 \tag{5.16a}$$

$$i\omega_2 \frac{\mathrm{d}\delta a_2}{\mathrm{d}t} = V\delta a_1^* a_0 \tag{5.16b}$$

$$i\omega_3 \frac{\mathrm{d}a_3}{\mathrm{d}t} = V\delta a_1 \delta a_2 \simeq 0, \tag{5.16c}$$

so

$$\omega_1\omega_2\frac{\mathrm{d}^2}{\mathrm{d}t^2}\delta a_1 = |Va_0|^2\,\delta a_1, \tag{5.17}$$

and the decay instability growth rate is given by the energy exchange rate,

$$\gamma_{\mathrm{E}}^2 = |V|^2\,|a_0|^2\,/\omega_1\omega_2. \tag{5.18}$$

Here, we took $\omega_1, \omega_2 > 0$, with $\omega_3 = \omega_1 + \omega_2$. It is easy to verify that should the pump be a mode other than the one with the highest frequency, then the decay process is stable. The time evolutions for decay unstable and decay stable processes are shown in Figure 5.4. If the initial state is very close to the X-point (i.e. the intersection with the third axis), the trajectory deviates from the X-point, and the energy of the mode-3 is converted to those of mode-1 and mode-2. In contrast, if the initial state is close to the O-point (e.g., the intersection with the second axis), the trajectory encircles the O-point, and the deviation does not grow. Observe that the important limit of the *parametric subharmonic instability*, with $\omega_1 \sim \omega_2$ and $\omega_3 \sim 2\omega_1$, is a particular case of the decay instability described here. (See Mima and Nishikawa (1984) for detailed explanation of parametric instabilities.)

The decay instability growth rate γ_{E} is a fundamental time scale. Equation (5.18) sets the rate of coherent energy transfer out of a strongly populated or 'pumped' mode. The rate γ_{E} is one of a few characteristic rates, the ordering of which determines which theoretical description is appropriate for the system under study. For example, whether a coherent or stochastic mode coupling approach is relevant depends on the relative size of the spectral auto-correlation (self-coherence) rate and the decay instability rate of Eq.(5.16). This comparison is discussed further in the next section.

5.2.5 Example – drift–Rossby waves

As drift–Rossby waves and drift wave turbulence are critically important to this discussion of plasma turbulence and self-organization, the problem of three interacting drift–Rossby waves merits special discussion here. The alert reader may already have noted that Eqs.(5.9a) – the mode amplitude equations for three interacting Rossby waves – don't have quite the same structure as Eqs.(5.8) – the usual, generic amplitude equations for three coupled nonlinear oscillators, and that no analogue of Eq.(5.12) is apparent. However, given that the Hasegawa–Mima or quasi-geostrophic equation, Hasegawa and Mima (1978), conserves *both* energy and potential enstrophy (see Appendix 1), it is evident that Eq.(5.9a) must also conserve these, so we can simply state that,

$$E_0 = \sum_i \left(k_i^2 + F\right) a_i^2/2, \tag{5.19a}$$

and

$$\Omega_0 = \sum_i \left(k_i^2 + F\right)^2 a_i^2/2 \tag{5.19b}$$

are both integral of motions (IOMs). Interestingly, the structure of these IOMs – and, more generally, the structure of the problem of 3-mode interaction for drift waves – are virtually *identical* to their counterparts for the FAT. Thus, the correspondence rules,

$$\left(k_i^2 + F\right) a_i^2 \rightarrow L_i^2$$

$$\left(k_i^2 + F\right) \rightarrow 1/I_i, \tag{5.20}$$

effectively map the problem of 3 interacting drift–Rossby waves to the familiar example of the FAT. This isomorphism enables us to again utilize geometrical intuition gained from experience with the Poinsot construction. To this end, we can immediately note that:

- the ordering of moments of inertia $I_1 < I_2 < I_3$ maps to the wave number ordering $k_1^2 > k_2^2 > k_3^2$;
- then for three interacting drift–Rossby waves, the correspondence to Figure 5.1 indicates that the wave with the *intermediate* value of k^2 (i.e. k_2) should be unstable to decay, if it is strongly excited;
- this expectation is supported by a "calculation by correspondence", i.e. since for the FAT,

$$\gamma_{\text{decay}}^2 = \left(\frac{(I_3 - I_2)(I_2 - I_1)}{I_1 I_3}\right) \Omega_2^2(0), \tag{5.21}$$

so using the correspondence 'rules' gives,

$$\rightarrow \frac{\left[\left(\frac{1}{(k_3^2+F)} - \frac{1}{(k_2^2+F)}\right)\left(\frac{1}{(k_2^2+F)} - \frac{1}{(k_1^2+F)}\right)\right] A_0^2}{\left(1/\left(k_1^2 + F\right)\right)\left(1/\left(k_3^2 + F\right)\right)}$$

so

$$\gamma_{\text{decay}}^2 \sim \frac{\left(k_2^2 - k_3^2\right)\left(k_1^2 - k_2^2\right)}{\left(k_2^2 + F\right)^2} A_0^2. \tag{5.22}$$

Here $A_0^2 \sim a_2(0)^2$. This calculation by correspondence confirms our expectations that the intermediate wave-number mode is the one which can be decay unstable. These results may also be obtained from a straightforward linear analysis of Eqs.(5.9);
- again, the three wave interaction dynamics is intrinsically *periodic* and *reversible*. There is no a-priori tendency for energy to accumulate in any single mode.

Two comments are in order here. First, the decay instability of the intermediate wave-number mode pumps both the shorter and longer wave length modes, and so may be thought of as a "dual decay process". This has sometimes been invoked as a harbinger of the familiar dual cascade of 2D turbulence. Although both the dual decay and dual cascade have their origins in the simultaneous conservation of energy and enstrophy, the direct relevance of dual decay to dual cascade is dubious since:

- the decay process is one of resonant 3-wave coupling – i.e. frequency match is required;
- the 3-wave interaction process is time reversible, while the cascade is not.

Second, care must be taken to recognize that the decay of intermediate wave number does not translate trivially to frequency. Since $\omega_k = k_y V_{de}/\left(F + k^2\right)$, where V_{de} is the diamagnetic velocity of electrons for drift waves, the decay instability criterion,

$$\left(k_3^2 - k_2^2\right)\left(k_2^2 - k_1^2\right) > 0 \tag{5.23a}$$

can be re-expressed as,

$$\left(\frac{k_{y2}}{\omega_2} - \frac{k_{y1}}{\omega_1}\right)\left(\frac{k_{y3}}{\omega_3} - \frac{k_{y2}}{\omega_2}\right) > 0. \tag{5.23b}$$

This implies that

$$\frac{\left(\omega_3 k_{y2} - \omega_2 k_{y3}\right)^2}{\omega_1 \omega_3 \omega_2^2} > 0 \tag{5.24}$$

for instability. Hence, instability requires $\omega_1\omega_3 > 0$, but the triad resonance condition also requires $\omega_2 = -\left(\omega_1 + \omega_3\right)$. These two requirements are reconcilable *only* if the unstable wave has the *highest frequency* in the triad. Thus, we see that in the resonant interaction of drift–Rossby waves, the *highest frequency* wave is the unstable one. Note that the resemblance of this result to that of Eq.(5.18) is somewhat coincidental. Here, an explicit form of the drift wave dispersion relation was used to relate $\boldsymbol{k}$ to $\omega_{\boldsymbol{k}}$, while in the discussion leading up to Eq.(5.18), we considered generic nonlinear oscillators.

5.2.6 Example – unstable modes in a family of drift waves

Another illustrative example of the three coupled mode is constructed for unstable modes in a family of drift waves. Analysis of the ion-temperature-gradient (ITG) mode (Rudakov and Sagdeev, 1960; Coppi *et al.*, 1967; Mikailovski, 1992; Horton, 1999; Weiland, 2000); instability is briefly shown here following the discussion in Lee and Tang (1988), Parker *et al.* (1994) and Watanabe *et al.* (2000).

A simple inhomogeneous plasma slab is chosen (main magnetic field is in the z-direction, and the density and ion temperature have gradients in the x-direction, the scale lengths of gradients of these are given by L_n and $L_{\mathrm{T_i}}$, respectively). Unstable waves propagate in the direction of the diamagnetic drift (y-direction). In order to construct a three-mode model, one unstable mode with $\left(k_x, k_y\right) = (\pm k, \pm k)$ is kept, which has a linear growth rate γ_{L}, and the second harmonics with $\left(k_x.k_y\right) = (\pm 2k, 0)$, which works for the nonlinear stabilization of the linearly unstable mode, is taken into account as well. The wave-number component in the direction of the magnetic field is given by $k_z = \theta \rho_i L_n^{-1} k_y$, where θ is a fixed parameter here. The Vlasov equation, which is truncated at these two components, takes a form,

$$\left(\frac{\partial}{\partial t} + ik\theta v\right) f_{1,1} + 2ik^2\phi \mathrm{Im}\, f_{2,0} = -ik\phi\left(1 + \frac{2L_n}{L_{\mathrm{T_i}}}\left(v^2 - 1\right) + \theta v\right) F_{\mathrm{M}}, \tag{5.25a}$$

$$\frac{\partial}{\partial t}\mathrm{Im}\, f_{2,0} = 4k^2 \mathrm{Im}\left(\phi f_{1,1}\right), \tag{5.25b}$$

where ϕ is the normalized electrostatic potential perturbation for the unstable mode, v is the parallel velocity, F_{M} is the local Maxwellian distribution, the suffixes 1, 1 and 2, 0 denote unstable and stable modes, respectively, and length and velocity are normalized to the ion gyroradius ρ_i and ion thermal velocity, respectively. In the linear response, the growth rate is determined by the eigenvalue equation $\int \mathrm{d}v\, f_{1,1} = \phi$,

$$\int \mathrm{d}v \frac{1 + 2L_n L_{T_i}^{-1}\left(v^2 - 1\right) + \theta v}{\omega - k\theta v + i\gamma_{\mathrm{L}}} k F_{\mathrm{M}} = 1.$$

When the gradient ratio $L_n L_{\mathrm{T_i}}^{-1}$ is large, strong instability can occur.

The eigenfunction of the linearly unstable mode for the perturbed distribution function, $f_{\mathrm{L}}(v) = f_{\mathrm{L,r}}(v) + i f_{\mathrm{L,i}}(v)$, is employed and the perturbed distribution function of the unstable mode is set as,

$$f_{1,1}(v,t) = \left\{a(t)\, f_{\mathrm{L,r}}(v) + ib(t)\, f_{\mathrm{L,i}}(v)\right\}\exp(-i\omega t), \tag{5.26a}$$

where ω is the real frequency, and $a(t)$ and $b(t)$ indicate the amplitude. The imaginary part of the second harmonics has the same functional form as $f_{\mathrm{L,i}}(v)$, and

$$\mathrm{Im}\, f_{2,0}(v,t) = c(t)\, f_{\mathrm{L,i}}(v). \tag{5.26b}$$

Substituting Eq.(5.26) into Eq.(5.25), with the help of charge neutrality condition, $\phi(t) = \int \mathrm{d}v\, f_{1,1}(v,t)$, a set of coupled equations for amplitudes (a, b, c) is obtained as,

$$\frac{\mathrm{d}}{\mathrm{d}t} a = \gamma_{\mathrm{L}} b, \tag{5.27a}$$

$$\frac{\mathrm{d}}{\mathrm{d}t} b = \gamma_{\mathrm{L}} a - 2k^2 ac, \tag{5.27b}$$

$$\frac{\mathrm{d}}{\mathrm{d}t} c = 4k^2 ac. \tag{5.27c}$$

The nonlinear coupling terms are quadratic, although not identical to those in Eq.(5.8).

This set of equations shows an exponential growth when the amplitude is small, $a, b, c \to 0$. In addition, this set has two types of stationary solutions, $\left(a_0, 0, \gamma_{\mathrm{L}} k^{-2}/2\right)$ and $(0, 0, c_0)$, where a_0 and c_0 are arbitrary constants. The integrals of motions are deduced from Eqs.(5.27) as,

$$a^2 + b^2 + \frac{1}{2}c^2 - \frac{\gamma_{\mathrm{L}}}{k^2} c = E_0, \tag{5.28a}$$

$$b^2 + \frac{1}{2}c^2 - \frac{\gamma_{\mathrm{L}}}{2k^2} c = E_1, \tag{5.28b}$$

$$a^2 - \frac{\gamma_{\mathrm{L}}}{2k^2} c = E_2. \tag{5.28c}$$

Similar geometrical constructions, using an ellipsoid, a cylinder and a parabola, are thus derived. (Note that one integral of motion is deduced from the other two in Eq.(5.28).) Figure 5.5 illustrates an example of the construction. Orbits are shown to be periodic. A typical orbit is illustrated by a solid curve in Figure 5.5.

5.3 The physical kinetics of wave turbulence

5.3.1 Key concepts

We now leap boldly from the *terra firma* of deterministic, integrable systems of 3 interacting modal degrees of freedom to the *terra nova* of wave

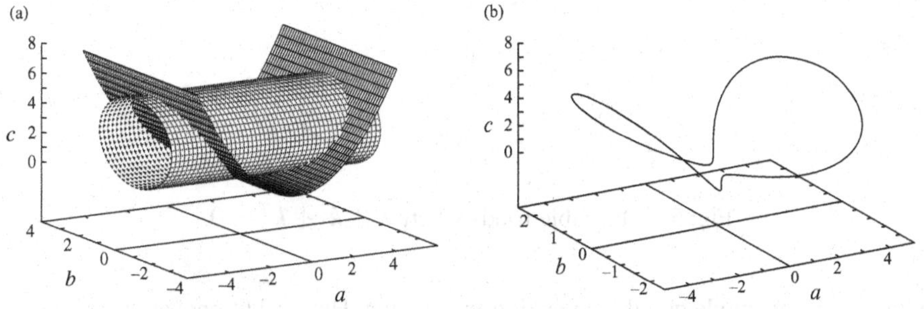

Fig. 5.5. Three-wave model for the case that includes an unstable wave. Integrals of motion are illustrated in the (a, b, c) space (a). A trajectory (the initial value of which is characterized by a small amplitude b and $a = c = 0$) is shown in (b).

turbulence – systems of N interacting waves, where $N \gg 1$. Statistical methods are required to treat such problems which involve the nonlinear interaction of many degrees of freedom. *The fundamental idea of the statistical theory of wave or weak turbulence is that energy transfer occurs by a random walk of mode couplings in the space of possible resonant interactions.* Each coupling event persists for a coherence time, which is short in comparison with the spectral evolution time. A net mode population density (i.e. energy) evolution then occurs via the accumulation of many of these short kicks or energy transfer events which add incoherently, as in a diffusion process. We remind the reader that the familiar theory of random walks and diffusion is based on:

(1) two disparate time scales τ_{ac}, τ_D such that $\tau_{ac} \ll \tau_D$. These are:
  (a) the spectral auto-correlation time τ_{ac} – which corresponds to an inverse bandwidth – and sets the duration of *one* random kick or step time;
  (b) the diffusion time $\tau_D \sim \Delta v^2 / D_v$ – the time to diffuse some finite interval in velocity Δv. *Many* steps occur during one τ_D.
(2) an evolution equation involving a fluctuating force, i.e. $\mathrm{d}v/\mathrm{d}t = q\tilde{E}/m$ so $\delta v \sim (q/m)\, \tilde{E} \tau_{ac}/m$ sets the step *size*.

Thus, $D \sim \langle \delta v^2 \rangle / \tau_{ac} \sim (q/m)^2 \langle E^2 \rangle \tau_{ac}$ is the diffusion coefficient which gives the rate of evolution. By analogy, we can say that the statistical theory of wave turbulence is based on:

(1) having two disparate time scales τ_{T_c} and τ_E, such that $\tau_{T_c} \ll \tau_E$. Here, the fundamental time scales correspond to:
  (a) the coupling or triad coherence time τ_{T_c} – which is the duration time of any specific three-wave coupling. Possible triad structures are shown in Figure 5.6. The term τ_{T_c} is set by the inverse bandwidth (i.e. net dispersion) of the frequency mismatch;
  (b) the energy transfer time, τ_E which is analogous to γ_E^{-1} for the case of coherent coupling (γ_E: defined as Eq.(5.18)).

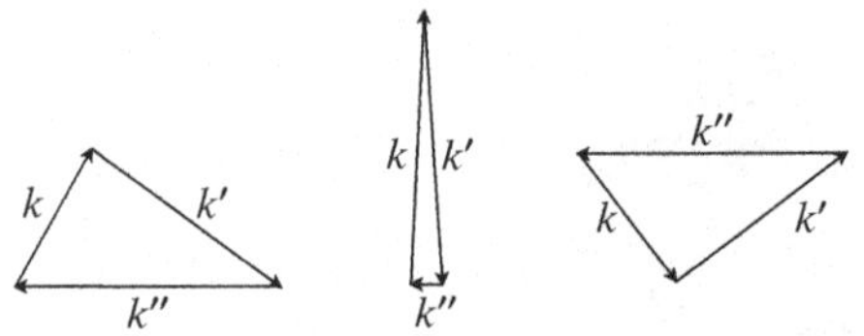

Fig. 5.6. Possible triads where $\boldsymbol{k} + \boldsymbol{k}' + \boldsymbol{k}'' = 0$.

(2) the stochastic mode population evolution equation. Here, additional assumptions such as the random phase approximations are required for closure, since stochastic amplitudes mean that the noise is multiplicative, not additive, as in the case of Brownian motion.

We should add here that wave turbulence or weak turbulence differs from fully developed or 'strong' turbulence since, for the latter, linear frequencies are completely washed out, so the triad coherence and energy transfer times are not distinguishable. For wave turbulence, standard perturbation theory based on linear wave response is possible, while for strong turbulence, renormalization – an uncontrolled approximation which effectively sums some portion of perturbation theory to all orders – is required.

Given that $N \gg 1$, even wave turbulence theory is highly non-trivial, and several assumptions are needed to make progress. The central element of the statistical theory of wave turbulence is the *wave-kinetic equation* (WKE), which is effectively a Boltzmann equation for the wave population density $N(\boldsymbol{x}, \boldsymbol{k}, t)$. Since N is proportional to the wave intensity, all phase information is lost en route to the WKE, which is derived by utilizing a 'random phase' or 'weak coupling' approximation. Attempts at justifying the random phase approximation often invoke notions of "many modes" (i.e. $N \gg 1$) or "broad spectra", but it must be said that these criteria are rather clearly inadequate and can indeed be misleading. For example, the Kuramoto model of synchronization involves $N \gg 1$ coupled nonlinear oscillators, yet exhibits states of synchronization – perfect phase coherence in the $N \to \infty$ limit – diametrically *opposite* behaviour to that of a randomly phased ensemble of waves! Simply having a large number of degrees of freedom does not – in and of itself – ensure stochasticity! A more plausible rationale for a statistical approach is to appeal to the possibility that the triad coherence time τ_{T_c} is shorter than the coherent energy transfer time τ_{E}, i.e. $\tau_{T_c} \ll \tau_{\mathrm{E}}$. In this case, since a particular mode $\boldsymbol{k}$ will participate in many uncorrelated triad couplings prior to significant change of its population via nonlinear energy transfer, such dynamics are amenable to description as a random walk in the space of possible resonant interactions. Having $N \gg 1$ modes suggests that $M > 1$ resonant couplings or kicks can occur in the course of spectral evolution, thus permitting us to invoke the

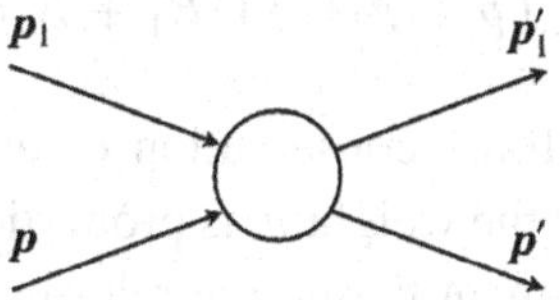

Fig. 5.7. Two-particle collision for gas kinetics.

central limit theorem to justify a statistical approach. Truth in advertising compels us to admit that this is little more than a physically appealing plausibility argument though, since we have no a-priori knowledge of the density of resonant triads in $\boldsymbol{k}$-space or the statistical distribution of triad coherence times. In this regard, we remark that a tail on the coherence time pdf – due to long lived triads – could be one possible indication of intermittency in wave turbulence.

5.3.2 Structure of a wave kinetic equation

It is useful to heuristically survey the theory of wave turbulence prior to delivering into the technical exposition – both to see the 'big picture' and to identify key time and space scales. Since much of the structure of wave turbulence theory is analogous to that of the kinetic theory of gases (KTG), it is advantageous to discuss the theory by comparison and contrast with the KTG. Just as the Boltzmann equation (BE) evolves the one-body distribution function $f(1)$ via single particle orbits and collisions, i.e.

$$\frac{\partial f}{\partial t} + Lf = C(f), \tag{5.29}$$

the wave kinetic equation (WKE) evolves the wave quanta density – usually the wave action density, given by $E_{\boldsymbol{k}}/\omega_{\boldsymbol{k}}$, according to,

$$\frac{\partial N_{\boldsymbol{k}}}{\partial t} + L_{\boldsymbol{k}} N_{\boldsymbol{k}} = C\{N\}. \tag{5.30}$$

Here, L is the linear evolution operator, and $L_{\boldsymbol{k}}$ generates the evolution of N along ray trajectories. The Boltzmann collision integral (the centrepiece of the KTG) has the structure,

$$C(f) = \int \mathrm{d}\Gamma_1 \mathrm{d}\Gamma' \mathrm{d}\Gamma'_1 \, w_{\mathrm{T}} \left(f' f'_1 - f f_1\right), \tag{5.31a}$$

where w_{T} is the transition probability for an individual scattering event (sketched in Figure 5.7) and has the structure,

$$w_{\mathrm{T}} \sim w\delta\left(\left(\boldsymbol{p}_1' + \boldsymbol{p}'\right) - \left(\boldsymbol{p} + \boldsymbol{p}_1\right)\right)\delta\left(\left(E_1' + E'\right) - \left(E + E_1\right)\right). \tag{5.31b}$$

Here, the delta functions enforce conservation of energy and momentum in an individual collision event and the weight w is proportioned to the collisional cross-section σ. It is useful to comment that in the relevant case of number-conserving interactions with small momentum transfer (i.e. $\boldsymbol{p}' = \boldsymbol{p} - \boldsymbol{q}$, where $|\boldsymbol{q}| \ll |\boldsymbol{p}|$), $C(f)$ can be written in the form of a divergence of a flux, i.e.

$$C(f) = -\nabla_{p} \cdot \boldsymbol{S}_{p}, \tag{5.32a}$$

where the phase-space flux $\boldsymbol{S}_{p_\alpha}$ is,

$$\boldsymbol{S}_{p_\alpha} = \int_{q_\alpha > 0} \mathrm{d}^3p'\mathrm{d}^3q\, \boldsymbol{w}\left(\boldsymbol{p}, \boldsymbol{p}', \boldsymbol{q}\right)\left[\boldsymbol{f}(\boldsymbol{p})\frac{\partial f'(\boldsymbol{p}')}{\partial \boldsymbol{p}_\beta'} - f'(\boldsymbol{p}')\frac{\partial f(\boldsymbol{p})}{\partial \boldsymbol{p}_\beta}\right] q_\alpha q_\beta. \tag{5.32b}$$

The reader should not be surprised to discover that the collision integral in the wave kinetic equation often takes the form,

$$C\{N\} = \int \mathrm{d}^3k'\left[\left|V_{\boldsymbol{k}',\boldsymbol{k}-\boldsymbol{k}'}\right|^2 N_{\boldsymbol{k}'}N_{\boldsymbol{k}-\boldsymbol{k}'} - \left|V_{\boldsymbol{k},\boldsymbol{k}'}\right|^2 N_{\boldsymbol{k}'}N_{\boldsymbol{k}}\right]$$
$$\times\, \delta\left(\boldsymbol{k}'' - \boldsymbol{k} - \boldsymbol{k}'\right)\delta\left(\omega_{\boldsymbol{k}+\boldsymbol{k}'} - \omega_{\boldsymbol{k}} - \omega_{\boldsymbol{k}'}\right), \tag{5.33}$$

since the theory of wave kinetics also models pdf evolution as a sequence of many weak interactions which add incoherently. Here the transition probability w and the coupling functions $V_{\boldsymbol{k},\boldsymbol{k}',\boldsymbol{k}''}$ parametrize the basic interaction strengths, $f\left(\boldsymbol{p}'\right)$ and $N_{\boldsymbol{k}'}$ account for the distribution of 'field particles' (which scatter a given test particle) or the background mode population (which scatters a given test mode), and the factor of $\delta\left(\boldsymbol{k}'' - \boldsymbol{k} - \boldsymbol{k}'\right) \times \delta\left(\omega_{\boldsymbol{k}''} - \omega_{\boldsymbol{k}} - \omega_{\boldsymbol{k}'}\right)$ enforces momentum and energy conservation in an elementary interaction. In the case where couplings result in small increments of the test mode wave vector, $C\{N\}$ can also be simplified to the convection–diffusion form,

$$C\{N\} = -\nabla_{\boldsymbol{k}} \cdot \boldsymbol{S}_{\boldsymbol{k}}, \tag{5.34a}$$

$$S_{k_\alpha} = V_\alpha \left\langle N(\boldsymbol{k})\right\rangle - D_{\alpha\beta}\frac{\partial \langle N\rangle}{\partial k_\beta}. \tag{5.34b}$$

Here the convection velocity V_α is usually associated with local interactions between comparable scales, and the diffusion $D_{\alpha\beta}$ (called induced diffusion) is usually due to random straining or refraction of small scales by larger ones. Both

$C(f)$ and $C\{N\}$ appear as the difference of two competing terms, since both model evolution by a succession of inputs and outputs, or emissions and absorptions. Both $C(f)$ and $C\{N\}$ are derived from an *assumption* of microscopic chaos – in the case of KTG, the "principle of molecular chaos" is used to justify the factorization,

$$f(1,2) = f(1)\, f(2). \tag{5.35}$$

For wave turbulence, the random phase approximation (RPA) – which approximates all modal phases as random variables (i.e. $\Phi_k = A_{\boldsymbol{k}} e^{-i\alpha_k}$, with $\alpha_{\boldsymbol{k}}$ random) with Gaussian distribution allows,

$$\begin{aligned}\left\langle \Phi_{k_1}\Phi_{k_2}\Phi_{k_3}\Phi_{k_4}\right\rangle &= \left|\Phi_{k_1}\right|^2 \left|\Phi_{k_3}\right|^2 \delta\left(\boldsymbol{k}_1-\boldsymbol{k}_2\right)\delta\left(\boldsymbol{k}_3-\boldsymbol{k}_4\right) + \text{symmetric term}\\ &\sim N\left(\boldsymbol{k}_1\right) N\left(\boldsymbol{k}_3\right)\delta\left(\boldsymbol{k}_1-\boldsymbol{k}_2\right)\delta\left(\boldsymbol{k}_3-\boldsymbol{k}_4\right) + \text{s.t.}\end{aligned} \tag{5.36}$$

In a related vein, $C(f)$ is derived from an assumption of diluteness or weak correlation, while $C\{N\}$ is based on a test mode hypothesis, which assumes that the statistics and other properties of all modes are similar. Table 5.1 summarizes the comparison and contrast of the KTG and wave kinetics.

One advantage of the 'preview' of wave turbulence theory given above is that we can identify the basic time scales and explore the implications of their ordering. Inspection of Eq.(5.19b) reveals the basic temporal rates:

- the mismatch frequency,

$$\omega_{\mathrm{MM}} = \omega_{\boldsymbol{k}''} - \omega_{\boldsymbol{k}} - \omega_{\boldsymbol{k}'}, \tag{5.37}$$

  which gives the net oscillation rate for any given triad. Obviously, $\omega_{\mathrm{MM}} \to 0$ for resonant triads. The number density of resonant triads in a given range of wave vectors is central to quantifying the efficiency of resonant interactions and determining whether they are stochastic or coherent. In practice, this number density is set by the range of wave vectors, the dissipative cut-off, and the structure of the dispersion relation.
- the rate of dispersion of ω_{MM},

$$\Delta\omega_{\mathrm{T}} = \left|\frac{\mathrm{d}\omega_{\mathrm{MM}}}{\mathrm{d}\boldsymbol{k}}\Delta\boldsymbol{k}'\right| \cong \left|\left(\frac{\mathrm{d}\omega_{\boldsymbol{k}''}}{\mathrm{d}\boldsymbol{k}''} - \frac{\mathrm{d}\omega_{\boldsymbol{k}'}}{\mathrm{d}\boldsymbol{k}'}\right)\Delta\boldsymbol{k}'\right| \tag{5.38}$$

  which gives the rate at which a particular triad disperses due to wave propagation. The function $\Delta\omega_{\mathrm{T}}^{-1}$ is a plausible estimate for the dispersion-induced triad coherence time $\tau_{T_{\mathrm{c}}}$. It is enlightening to observe that the triad decoherence rate $\Delta\omega_{\mathrm{T}} \sim \left|\left(\boldsymbol{v}_g\left(\boldsymbol{k}'\right) - \boldsymbol{v}_g\left(\boldsymbol{k}''\right)\right)\cdot\Delta\boldsymbol{k}'\right|$ i.e. the typical rate at which two interacting wave packets disperse at their different group speeds. Thus, the triad coherence which enters the lifetime of mode $\boldsymbol{k}$ is set simply by the rate at which the interacting packets stream away from each other at their respective group velocities. Note also that $\Delta\omega_{\mathrm{T}} \sim$

Table 5.1. *Comparison and contrast of the kinetic theory of gases and wave kinetics*

Kinetic theory of gases	Wave kinetics
Time scales	
collision frequency ν_c	triad de-coherence rate $(1/\tau_{T_c})$
relaxation time	spectral evolution time
Structure	
particle Liouvillian L	eikonal Liouvillian $L_{\boldsymbol{k}}$
Boltzmann $C(f)$	wave–wave interaction operator $C\{N_{\boldsymbol{k}}\}$
Landau collision operator	wave Fokker–Planck operator
cross-section σ	coupling coefficients $\left\|V_{\boldsymbol{k},\boldsymbol{k}',\boldsymbol{k}''}\right\|^2$
energy and momentum conservation factors i.e. $\delta\left(\sum \boldsymbol{P}_{\text{in}} - \sum \boldsymbol{P}_{\text{out}}\right)\delta\left(\sum E_{\text{in}} - E_{\text{out}}\right)$	selection rules for $\boldsymbol{k}$, ω matching i.e. $\delta\left(\boldsymbol{k}'' - \boldsymbol{k} - \boldsymbol{k}'\right)\delta\left(\omega_{\boldsymbol{k}''} - \omega_{\boldsymbol{k}} - \omega_{\boldsymbol{k}'}\right)$
field particles	background, ambient waves
test particle	test wave
Irreversibility	
principle of molecular chaos	random phase approximation
micro-reversibility due detailed balance	micro-reversibility due selection rules, conservation laws
coarse graining → macro irreversibility → H-theorem	coarse graining → macro irreversibility → H-theorem
uniform Maxwellian equilibrium solution	Bose–Einstein, zero-flux equilibrium solution
transport → finite flux in $\boldsymbol{x}$	cascade solution → finite flux in $\boldsymbol{k}$

$\left|\left(\partial^2\omega/\partial k_\alpha \partial k_\beta\right)\Delta k_\alpha \Delta k_\beta\right|$ is related to the strength of diffraction in the waves, and is sensitive to anisotropy in dispersion.

- the energy transfer rate γ_{E}. For *coherent* interactions, $\gamma_{\text{E}}^{\text{coh}}$ is similar to that given by Eq.(5.16), i.e.

$$\gamma_{\text{E}}^{\text{coh}} \sim \left(|V|^2 |a|^2 / \omega_1 \omega_2\right)^{1/2}.$$

For *stochastic* interactions in wave kinetics, we will soon see that,

$$\gamma_{\text{E}}^{\text{stoch}} \sim \sum_{\boldsymbol{k}'} V_{\boldsymbol{k},\boldsymbol{k}',\boldsymbol{k}-\boldsymbol{k}'} N_{\boldsymbol{k}'} \tau_{T_c} \sim \left(\gamma_{\text{E}}^{\text{coh}}\right)^2 \tau_{T_c}. \tag{5.39}$$

Expressing the results for these two limiting cases in a consistent notation,

$$\gamma^{\text{stoch}}/\gamma^{\text{coh}} \sim \tau_{T_c}\gamma^{\text{coh}}, \tag{5.40}$$

indicates that for comparable intensity levels, energy transfer is *slower* in wave kinetics than for coherent interaction. This is because in wave kinetics transfer occurs via a random walk of step duration τ_{T_c}, where $\tau_{T_c} < \gamma_{\text{E}}^{-1}$, so many steps are required to stochastically transfer an amount of energy comparable to that transferred in a single coherent interaction. Validity of wave kinetics requires $\Delta\omega_T < \gamma_E^{coh}$ *and* $\Delta\omega_T < \gamma_E^{stoch}$. We again emphasize that, as in quasi-linear theory, *wave dispersion* is crucial to the applicability of perturbative, weak turbulence methodology. A broad spectrum (large $|\Delta \boldsymbol{k}|$) alone is *not* sufficient for validity of wave kinetics, since in the absence of dispersion, resonant triads in that spectrum remain correlated for dynamically long times, forcing $\tau_{T_c}\gamma^E \rightarrow 1$. This is suggestive of either nonlinear structure formation (i.e. shocks, developed by steeping), or the onset of strong turbulence.

5.3.3 'Collision' integral

Moving beyond generality, we now turn to the concrete task of constructing the collision integral for the wave kinetic equation. As with coherent interaction, this is best done via two complementary examples:

(i) a general calculation for a 'generic' model evolution equation, assuming random transitions. This parallels the treatment for coherent interaction;
(ii) a calculation for the specific and relevant example of the Hasegawa–Mima equation for drift–Rossby waves.

5.3.3.1 *Model dynamical equation*

A generic form for the nonlinear oscillator equation is,

$$\frac{d^2 \mathrm{a}_k}{dt^2} + \omega_k^2 \mathrm{a}_k = \sum_{k'} V_{k,k',k-k'} \mathrm{a}_{k'} \mathrm{a}_{k-k'}. \tag{5.41}$$

Here, a_k is the field variable of mode $\boldsymbol{k}$, ω_k is the linear wave frequency and $V_{k,k',k-k'}$ is the coupling function which ordinarily is $\boldsymbol{k}$-dependent. Here $V_{k,k',k-k'}$ has the symmetries,

$$V_{k,k',k-k'} = V_{k,k-k',k'} = V_{-k,-k',-k+k'}, \tag{5.42a}$$

$$V_{k,k',k-k'} = V_{k-k',-k',k} \, \mathrm{sgn}\left(\omega_k \omega_{k-k'}\right). \tag{5.42b}$$

Extracting the fast oscillation, i.e.

$$\mathrm{a}_k(t) = a_k(t)\, e^{-i\omega_k t},$$

where $a_k(t)$ indicates the wave amplitude, gives

$$\frac{d}{dt} a_k(t) = i \sum_{k'} V_{k,k',k-k'} a_{k'}(t)\, a_{k-k'}(t) \exp\left[-i\left(\omega_{k'} + \omega_{k-k'} - \omega_k\right) t\right], \tag{5.43}$$

which may be thought of as the multi-mode counterpart of Eqs.(5.8), the modal amplitude equation. A factor of ω has been absorbed in the coupling coefficient. Here the occupation density or number of quanta for a particular mode $\boldsymbol{k}$ is $N_{\boldsymbol{k}} = |a_{\boldsymbol{k}}|^2$. Since ultimately we seek the collision operator for the evolution of $N_{\boldsymbol{k}}(t)$, we proceed by standard time-dependent perturbation theory. As the change in occupation $\Delta N_{\boldsymbol{k}}(t)$ is given by,

$$\Delta N_{\boldsymbol{k}}(t) = \left\langle |a_{\boldsymbol{k}}(t)|^2 \right\rangle - \left\langle |a_{\boldsymbol{k}}(0)|^2 \right\rangle, \tag{5.44a}$$

working to second order in δa, $a_{\boldsymbol{k}}(t) - a_{\boldsymbol{k}}(0) = \delta a_{\boldsymbol{k}}^{(1)} + \delta a_{\boldsymbol{k}}^{(2)} + \cdots$, gives,

$$\Delta N_{\boldsymbol{k}}(t) = \left\langle \left| \delta a_{\boldsymbol{k}}^{(1)}(t) \right|^2 \right\rangle + \left\langle a_{\boldsymbol{k}}^*(0)\, \delta a_{\boldsymbol{k}}^{(2)}(t) \right\rangle + \left\langle \delta a_{\boldsymbol{k}}^{(2)*} a_{\boldsymbol{k}}(0) \right\rangle. \tag{5.44b}$$

Expanding in a straightforward manner gives,

$$\frac{\mathrm{d}}{\mathrm{d}t}\left(\delta a_{\boldsymbol{k}}^{(1)}(t) + \delta a_{\boldsymbol{k}}^{(2)}(t) + \cdots\right) = i \sum_{\boldsymbol{k}'} V_{\boldsymbol{k},\boldsymbol{k}',\boldsymbol{k}-\boldsymbol{k}'} \left(a_{\boldsymbol{k}'}(0) + \delta a_{\boldsymbol{k}}^{(1)}\right)$$
$$\times \left(a_{\boldsymbol{k}-\boldsymbol{k}'}(0) + \delta a_{\boldsymbol{k}-\boldsymbol{k}'}^{(1)}\right) \exp\left[-i\left(\omega_{\boldsymbol{k}'} + \omega_{\boldsymbol{k}-\boldsymbol{k}'} - \omega_{\boldsymbol{k}}\right)t\right], \tag{5.45a}$$

so

$$\delta a_{\boldsymbol{k}}^{(1)}(t) = i \sum_{\boldsymbol{k}',\boldsymbol{k}''} a_{\boldsymbol{k}'}(0)\, a_{\boldsymbol{k}''}(0) \int_0^t \mathrm{d}t'\, \hat{V}_{\boldsymbol{k},\boldsymbol{k}',\boldsymbol{k}''}(t'), \tag{5.45b}$$

where,

$$\hat{V}_{\boldsymbol{k},\boldsymbol{k}',\boldsymbol{k}''}(t) = V_{\boldsymbol{k},\boldsymbol{k}',\boldsymbol{k}''} \exp\left[-i\left(\omega_{\boldsymbol{k}'} - \omega_{\boldsymbol{k}''} - \omega_{\boldsymbol{k}}\right)t\right], \tag{5.45c}$$

and we understand that $\boldsymbol{k}'' = \boldsymbol{k} - \boldsymbol{k}'$ here. Similar straightforward calculations give for $\delta a^{(2)}$,

$$\delta a_{\boldsymbol{k}}^{(2)} = -\sum_{\substack{\boldsymbol{k}',\boldsymbol{k}'' \\ \boldsymbol{q}',\boldsymbol{q}''}} \left\{ a_{\boldsymbol{k}'}(0)\, a_{\boldsymbol{q}'}(0)\, a_{\boldsymbol{q}''}(0) \int_0^t \mathrm{d}t' \int_0^{t'} \mathrm{d}t''\, \hat{V}_{\boldsymbol{k},\boldsymbol{k}',\boldsymbol{k}''}(t')\, \hat{V}_{\boldsymbol{k}',\boldsymbol{q}'\boldsymbol{q}''}(t'') \right.$$
$$\left. + a_{\boldsymbol{k}''}(0)\, a_{\boldsymbol{q}'}(0)\, a_{\boldsymbol{q}''}(0) \int_0^t \mathrm{d}t' \int_0^{t'} \mathrm{d}t''\, \hat{V}_{\boldsymbol{k},\boldsymbol{k}',\boldsymbol{k}''}(t')\, \hat{V}_{\boldsymbol{k}.\boldsymbol{q}',\boldsymbol{q}''}(t'') \right\}. \tag{5.45d}$$

In the first term on the right-hand side of Eq.(5.45d), $\boldsymbol{q}' + \boldsymbol{q}'' = \boldsymbol{k} - \boldsymbol{k}'$, while in the second $\boldsymbol{q}' + \boldsymbol{q}'' = \boldsymbol{k}'$.

5.3.3.2 Extraction of response and closure

To close the calculation of $\Delta N_{\mathbf{k}}(t)$ – the change in occupation number, correlators such as $\langle \delta a^{(1)} \delta a^{(1)} \rangle$ and $\langle \delta a \delta a^{(2)} \rangle$ are simplified by the random phase approximation (RPA). The essence of the RPA is that if the duration of phase correlations is *shorter than any other time scale in the problem*, then the phases of the modal amplitude may be taken as random, i.e.

$$a_{\mathbf{k}} \rightarrow \hat{a}_{\mathbf{k}} \mathrm{e}^{i\theta_{\mathbf{k}}},$$

with $\theta_{\mathbf{k}}$ random. Then, for

$$\langle \quad \rangle = \langle \quad \rangle_{\text{ensemble}} = \int \mathrm{d}\theta \, P(\theta),$$

where $P(\theta)$ is the pdf of phase $\theta_{\mathbf{k}}$,

$$\begin{aligned} \langle a_{\mathbf{k}} a_{\mathbf{k}'} \rangle &= \left\langle \hat{a}_{\mathbf{k}} e^{i\theta_{\mathbf{k}}} \hat{a}_{\mathbf{k}'} e^{i\theta_{\mathbf{k}'}} \right\rangle \\ &= \left| \hat{a}_{\mathbf{k}} \right|^2 \delta_{\mathbf{k},-\mathbf{k}'}, \end{aligned} \tag{5.46a}$$

and

$$N_{\mathbf{k}} = \left| \hat{a}_{\mathbf{k}} \right|^2. \tag{5.46b}$$

We should comment that:

(i) the RPA should be considered as arising from the need to close the moment hierarchy. Truly random phases of the physical modal amplitudes would preclude any energy transfer, since all triad couplings would necessarily vanish. Rather, the RPA states that modal correlations are in some sense weak, and induced only via nonlinear interaction of resonant triads;
(ii) the RPA is an uncontrolled approximation. It lacks rigorous justification and cannot predict its own error;
(iii) the RPA is seemingly relevant to a system with short triad coherence time (i.e. $\gamma_{\mathrm{E}} \tau_{T_c} \ll 1$), and so it should be most applicable to ensembles of waves where many strongly dispersive waves interact resonantly, though this connection has not been firmly established.

Having stated all those caveats, we must add that there is no way to make even the crudest, most minimal progress on wave turbulence without utilizing the RPA. It is the *only* game in town.

Closure modelling is explained in detail in the next chapter, and we discuss here the extraction of interaction time briefly. Proceeding, the increment in occupation

ΔN driven by weak coupling can thus be written as,

$$\Delta N_k(t) =$$

$$\left\langle \mathrm{Re} \sum_{\substack{k',k'' \\ q',q''}} \left\{ a_{k'}(0)\, a_{k''}(0)\, a^*_{q'}(0)\, a^*_{q''}(0) \int_0^t \mathrm{d}t'\, \hat{V}_{k,k',k''}(t') \int_0^t \mathrm{d}t''\, \hat{V}^*_{k,q',q''}(t'') \right.\right.$$

$$- a_k(0)\, a_{k'}(0)\, a^*_{q'}(0)\, a^*_{q''}(0) \int_0^t \mathrm{d}t'\, \hat{V}_{k,k',k''}(t') \int_0^t \mathrm{d}t''\, \hat{V}^*_{k'',q',q''}(t'')$$

$$\left.\left. - a_k(0)\, a_{k''}(0)\, a^*_{q'}(0)\, a^*_{q''}(0) \int_0^t \mathrm{d}t'\, \hat{V}_{k,k',k''}(t') \int_0^t \mathrm{d}t''\, \hat{V}^*_{k',q',q''}(t'') \right\} \right\rangle. \tag{5.47}$$

Since the bracket refers to an average over phase, we can factorize and reduce the averages of quartic products by the requirement of phase matching. For example, the first term on the right-hand side of Eq.(5.47) may be written as:

$$\Delta N_1 = \sum_{\substack{k',k'' \\ q',q''}} \left\langle a_{k'}(0)\, a_{k''}(0)\, a^*_{q'} a^*_{q''} \right\rangle \int_0^t \mathrm{d}t'\, \hat{V}_{k,k',k''}(t') \int_0^t \mathrm{d}t''\, \hat{V}^*_{k,q',q''}(t''), \tag{5.48a}$$

where possible factorizations of the quartic product are given by,

$$\left\langle a_{\mathbf{k}'}(0)\, a_{\mathbf{k}''}(0)\, a^*_{\mathbf{q}'} a^*_{\mathbf{q}''} \right\rangle \rightarrow \left\langle \overset{1}{a_{\mathbf{k}'}}(0)\, \overset{2}{a_{\mathbf{k}''}}(0)\, \overset{3}{a^*_{\mathbf{q}'}} \overset{4}{a^*_{\mathbf{q}''}} \right\rangle$$

$$= \left|\hat{a}_{k'}\right|^2 \delta_{k',q'} \left|\hat{a}_{k''}\right|^2 \delta_{k'',q''} + \left|\hat{a}_{k'}\right|^2 \delta_{k',q''} \left|\hat{a}_{k''}\right|^2 \delta_{k'',q'}.$$

Thus

$$\Delta N_1 = \sum_{k',k''} N_{k'} N_{k''} \int_0^t \mathrm{d}t'\, \hat{V}_{k,k',k''}(t') \int_0^t \mathrm{d}t''\, \hat{V}^*_{k,k',k''}(t'')$$

$$+ \sum_{k',k''} N_{k'} N_{k''} \int_0^t \mathrm{d}t'\, \hat{V}_{k,k',k''}(t') \int_0^t \mathrm{d}t''\, \hat{V}^*_{k,k'',k'}(t''). \tag{5.48b}$$

Using the coupling function symmetries as given by Eq.(5.42), simple but tedious manipulation then gives,

$$\Delta N_1 = 2 \sum_{k',k''} N_{k'} N_{k''} \left[\int_0^t \mathrm{d}t'\, \hat{V}^*_{k,k',k''}(t') \int_0^t \mathrm{d}t''\, \hat{V}_{k,k',k''}(t'') \right]. \tag{5.49}$$

To perform the time integration, as in the case of Brownian motion it is convenient to transform to relative $(\tau = (t' - t'')/2)$ and average $(T = (t' + t'')/2)$ time variables and then symmetrize to obtain,

$$\left[\int_0^t dt' \hat{V}^*_{k,k',k''}(t') \int_0^t dt'' \hat{V}_{k,k',k''}(t'')\right]$$
$$= \left|V_{k,k',k''}\right|^2 2\int_0^t dT \int_0^T d\tau \exp\left[i\left(\omega_k - \omega_{k'} - \omega_{k''}\right)\tau\right]$$
$$= \left|V_{k,k',k''}\right|^2 2\int_0^t dT \frac{\left(\exp\left(i\left(\omega_k - \omega_{k'} - \omega_{k''}\right)T\right) - 1\right)}{i\left(\omega_k - \omega_{k'} - \omega_{k''}\right)}, \tag{5.50a}$$

so taking $T \gg \omega_{\mathrm{MM}}^{-1}, \Delta\omega_{\mathrm{MM}}^{-1}$ then gives,

$$= 2\pi t \left|V_{k,k',k''}\right|^2 \delta\left(\omega_k - \omega_{k'} - \omega_{k''}\right). \tag{5.50b}$$

Equation (5.50) has the classic structure of a transition probability element, as given by the Fermi golden rule for incoherent couplings induced by time-dependent perturbations. Here the time dependency arises from the limited duration of the triad coherence set by the dispersion in ω_{MM}. The term ΔN is in proportion to t. Since t is, by construction, large in comparison to any other time scale in the problem, as in Fokker–Planck theory we can write,

$$\frac{\partial N_1}{\partial t} = 4\pi \sum_{k',k''} N_{k'} N_{k''} \left|V_{k,k',k''}\right|^2 \delta\left(\omega_k - \omega_{k'} - \omega_{k''}\right). \tag{5.51}$$

A similar set of calculations for the second and third terms on the right-hand side then gives the total collision integral for population evolution due to stochastic wave–wave interaction with short coherence time as,

$$C\{N_k\} = 4\pi \sum_{k',k''} \left(V_{k,k',k''}\right)^2 \delta_{k,k',k''} \delta\left(\omega_k - \omega_{k'} - \omega_{k''}\right)$$
$$\times \left[N_{k'} N_{k''} - \left(\operatorname{sgn}\left(\omega_k \omega_{k''}\right) N_{k'} + \operatorname{sgn}\left(\omega_k \omega_{k'}\right) N_{k''}\right) N_k\right]. \tag{5.52a}$$

A related form of $C\{N_k\}$ with more general symmetry properties is,

$$C\{N_k\} = \pi \sum_{k',k''} \left\{\left|V_{k,k',k''}\right|^2 \left(N_{k'} N_{k''} - \left(N_{k'} + N_{k''}\right) N_k\right)\right.$$
$$\times \delta\left(\mathbf{k} - \mathbf{k}' - \mathbf{k}''\right) \delta\left(\omega_k - \omega_{k'} - \omega_{k''}\right)$$
$$+ 2\left|V_{k',k,k''}\right|^2 \left(N_{k''} N_k - N_{k'}\left(N_{k''} + N_k\right)\right)$$
$$\left.\times \delta\left(\mathbf{k}' - \mathbf{k} - \mathbf{k}''\right) \delta\left(\omega_{k'} - \omega_k - \omega_{k''}\right)\right\}. \tag{5.52b}$$

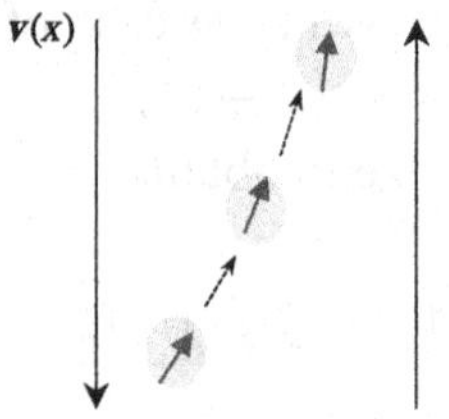

Fig. 5.8. Wave packet (which is denoted by the shade) with wave vector $\boldsymbol{k}$ (thick solid arrow) moves in the presence of a large-scale strain field $\boldsymbol{v}(x)$. Thin dotted arrows show the motion of a packet in real space.

The factor of 2 arises from the arbitrary choice of $\boldsymbol{k}'$ to interchange with $\boldsymbol{k}$ in the second term. The full wave kinetic equation is then,

$$\frac{\partial N}{\partial t} + \left(\boldsymbol{v}_{\mathrm{gr}} + \boldsymbol{v}\right) \cdot \nabla N - \frac{\partial}{\partial \boldsymbol{x}} \left(\omega + \boldsymbol{k} \cdot \boldsymbol{v}\right) \cdot \frac{\partial N}{\partial \boldsymbol{k}} = 2\gamma_{\boldsymbol{k}} N_{\boldsymbol{k}} + C\left\{N_{\boldsymbol{k}}\right\}. \qquad (5.53)$$

Here $\boldsymbol{v}$ is an ambient large-scale shear flow which advects the interacting wave population and $\gamma_{\boldsymbol{k}}$ is the linear growth or damping rate for the wave population. The left-hand side of Eq.(5.53) is conservative – i.e. can be written in the form $\mathrm{d}N/\mathrm{d}t$ – and describes evolving N along Hamiltonian ray trajectories (see Fig. 5.8),

$$\frac{\mathrm{d}\boldsymbol{x}}{\mathrm{d}t} = \boldsymbol{v}_{\mathrm{gr}} + \boldsymbol{v}, \qquad \frac{\mathrm{d}\boldsymbol{k}}{\mathrm{d}t} = -\frac{\partial\left(\omega + \boldsymbol{k} \cdot \boldsymbol{v}\right)}{\partial \boldsymbol{x}}.$$

Equation (5.52) gives the collision integral for stochastic wave–wave interaction by random triad couplings of short duration, and constitutes a central result in the theory of wave–wave interactions.

Equations (5.52), (5.53) certainly merit discussion in detail.

- $C\{N\}$ has the characteristic structure indicating evolution of the population $N_{\boldsymbol{k}}$ at a given wave vector $\boldsymbol{k}$ via a competition between input by (incoherent) noise and relaxation by nonlinear couplings which produce outflow to other $\boldsymbol{k}$s, i.e.

$$\frac{\partial N}{\partial t} \sim \sum |V|^2 \{ \quad \underbrace{N_{\boldsymbol{k}'} N_{\boldsymbol{k}''}} \quad - \quad \underbrace{N_{\boldsymbol{k}'} N_{\boldsymbol{k}}} \quad \}.$$

— noise/incoherent emission INTO $\boldsymbol{k}$ from $\boldsymbol{k}'$, $\boldsymbol{k}''$ interaction

— not $\sim N_{\boldsymbol{k}}$

— relaxation OUTFLOW from $\boldsymbol{k}$ via nonlinear interaction

— $\sim N_{\boldsymbol{k}}$

This structure is common to virtually *all* wave kinetic equations.

- The population outflow or damping term in Eq.(5.53) identifies the characteristic nonlinear relaxation rate of a test mode $\boldsymbol{k}$ in wave turbulence theory, as $\partial N_{\boldsymbol{k}}/\partial t \sim -N_{\boldsymbol{k}}/\tau_{\mathrm{R}\boldsymbol{k}}$, where the relaxation rate $1/\tau_{\mathrm{R}k}$ is

$$1/\tau_{\mathrm{R}k} \sim \sum_{k',k'} \left|V_{k,k',k''}\right|^2 \delta_{k,k',k''}\delta\left(\omega_{k''}-\omega_{k'}-\omega_k\right) N_{k'}$$

$$\sim \sum_{\substack{k' \\ \text{resonant}}} \left|V_{k,k'}\right|^2 \tau_{T_c k} N_{k'}.$$

Notice that the relaxation rate is set by:
– the resonance condition and the number of resonant triads involving the test mode $\boldsymbol{k}$
– the coherence time $\tau_{T_c k}$ of triads involving the test mode $\boldsymbol{k}$
– the mean square coupling strength and the ambient mode intensity.

The relaxation rate $1/\tau_{\mathrm{R}k}$ is the stochastic counterpart of the coherent energy decay rate $\gamma_{\mathrm{E}}^{\mathrm{coh}}$. Most estimates of solution levels and transport in weak turbulence theory are derived by the balance of some linear growth rate $\gamma_{\mathrm{L},k}$ with $1/\tau_{\mathrm{R}k}$. This gives a generic scaling of the form for a fluctuation level,

$$N \sim \left(\gamma_{\mathrm{L}}/\tau_{T_c}\right)\left|V_{k,k'}\right|^{-2}$$

- $C\{N\}$ conserves the spectrum integrated wave energy $\mathcal{E}$ ($\mathcal{E} = N\omega$) and momentum $\boldsymbol{P}$ ($\boldsymbol{P} = N\boldsymbol{k}$) densities since the resonance conditions enforce these conservation laws in the microscopic interactions. In the case where the scattering increment ($\Delta \boldsymbol{k}$) of the test mode is small, so $C\{N\} \to -\nabla_k \cdot \boldsymbol{S}_k$, $C\{N\}$ also conserves excitation *number*. In general, however, the Manley–Rowe relations tell us that total exciton number need not be conserved in three-mode interactions – i.e. two waves in, one out (or the reverse) – though it is for the case of resonant four-wave processes (i.e. two in → two out).
- As should be apparent from the triad resonance conditions, $C\{N\}$ supports several types of wave–wave interaction processes, depending on dispersion, coupling strength and behaviour, etc. Interactions can be local in $\boldsymbol{k}$, in which case $C\{N\}$ takes the generic form,

$$C\{N\} = -\frac{\partial}{\partial \boldsymbol{k}}\left(\boldsymbol{V}(\boldsymbol{k}, N) N_k\right),$$

as in the Leith model of turbulence. Here $\boldsymbol{V}(\boldsymbol{k}, N)$ represents a flux or flow of quanta density in wave vector. Interactions can be non-local in $\boldsymbol{k}$ but proceed via small $\Delta \boldsymbol{k}$ increments, in which case $\boldsymbol{S}_k$ is diffusive, i.e.

$$\boldsymbol{S}_k = -D_k \frac{\partial N}{\partial \boldsymbol{k}},$$

so

$$C\{N\} = \frac{\partial}{\partial \boldsymbol{k}} D_k \frac{\partial N_k}{\partial \boldsymbol{k}}.$$

Such interactions are referred to as *induced diffusion*. The physics of certain generic classes of non-local wave interactions including induced diffusion, will be discussed later in this chapter.

- Given the parallel development of wave kinetics and the kinetic theory of gas, it is no surprise that one can construct and prove an H-theorem for $C\{N\}$ in analogy with the Boltzmann H-theorem. Here, the entropy is,

$$\mathcal{S} = \int \mathrm{d}\boldsymbol{k} \, \ln(N_{\boldsymbol{k}}), \tag{5.54}$$

and the distribution for which $\mathrm{d}\mathcal{S}/\mathrm{d}t = 0$ corresponds to a Rayleigh–Jeans-type distribution, $N_{\boldsymbol{k}} = T/\omega_{\boldsymbol{k}}$. This implies that equipartition of energy is one stationary population distribution $N_{\boldsymbol{k}}$ in the limit $\omega_{\boldsymbol{k}} \ll T$. This equilibrium distribution corresponds to a state of zero spectral flux or energy dissipation rate. The theory of entropy production in wave turbulence is developed further in the monograph by Zakharov *et al.* (1992), and discussed further in the next chapter.

5.3.4 Application to drift–Rossby wave

Proceeding as in our discussion of coherent wave–wave interactions, we now present the theory of wave kinetics for the very relevant case of drift–Rossby wave turbulence in its simplest incarnation, namely the Hasegawa–Mima equation. There are several reasons for explicit consideration of this example, which include:

- the relevance of drift–Rossby wave turbulence to confinement physics
- the impact of the dual conservation of energy and potential enstrophy – both quadratic invariant of moments – on the wave spectrum evolution
- the consequent appearance of 'negative viscosity phenomena' – i.e. the tendency of wave energy to be scattered toward large scale
- the relation of wave interaction to potential vorticity transport
- the breaking of scale invariance in the coupling factors, on account of $k_\perp \rho_s$ independence.

5.3.4.1 Model

The quasi-geostrophic or Hasegawa–Mima equation for drift–Rossby waves is,

$$\frac{\mathrm{d}}{\mathrm{d}t}\left(F\phi - \nabla^2\phi\right) + V_{\mathrm{de}}\frac{\partial \phi}{\partial y} = 0, \tag{5.55a}$$

where advection is by $\boldsymbol{E} \times \boldsymbol{B}$ velocity,

$$\frac{\mathrm{d}}{\mathrm{d}t} = \frac{\partial}{\partial t} + \nabla\phi \times \hat{z} \cdot \nabla. \tag{5.55b}$$

So for mode $\boldsymbol{k}$, we have the generic amplitude evolution equation,

$$\frac{\partial \phi_{k}(t)}{\partial t}+i\omega_{k}\phi_{k}(t)=\sum_{k'+k''=k} V_{k,k',k''}\phi_{k'}(t)\,\phi_{k''}(t), \tag{5.56a}$$

where the coupling coefficient is,

$$V_{k,k',k''}=\frac{1}{2}\frac{\left(k'\cdot k''\times\hat{z}\right)\left(k''^{2}-k'^{2}\right)}{F+k^{2}}, \tag{5.56b}$$

and the drift wave frequency is just,

$$\omega_{k}=k_{y}V_{\mathrm{de}}/\left(F+k_{\perp}^{2}\right). \tag{5.57}$$

Derivation and explanation for the model equation (5.55) are given in Appendix 1. Equation (5.56a) is the basic equation for the evolution of modal amplitudes $\phi_k(t)$. The long time evolution of the wave intensity $|\phi_k|^2$ due to nonlinear couplings is given by the triad correlator i.e.

$$\frac{\partial}{\partial t}\left|\phi_{k}(t)\right|^{2}=\sum_{k'+k''=k} V_{k,k',k''}\left\langle\phi_{k}^{*}(t)\,\phi_{k'}(t)\,\phi_{k''}(t)\right\rangle=T_{k}. \tag{5.58}$$

The triad correlator T_k is non-vanishing due to test wave couplings which survive the ensemble average over random phases denoted by the brackets $\langle\ \ \rangle$, and which satisfy the resonance condition. Hence,

$$T_{k}=\sum_{k'+k''=k} V_{k,k',k''}\Big[\underbrace{\left\langle\phi_{k'}(t)\,\phi_{k''}(t)\,\delta\phi_{k}^{*(2)}(t)\right\rangle}_{\text{incoherent emission } T_{1}}$$
$$+\underbrace{\left\langle\phi_{k}(t)\,\delta\phi_{k''}{}^{(2)}\phi_{k}^{*}(t)\right\rangle}_{\text{relaxation } T_{2}}+\underbrace{\left\langle\delta\phi_{k'}{}^{(2)}(t)\,\phi_{k''}(t)\,\phi_{k}^{*}(t)\right\rangle}_{\text{relaxation } T_{2}}\Big], \tag{5.59}$$

where in the first term, which ultimately represents nonlinear noise or incoherent emission,

$$\delta\phi_{k}^{(2)}\sim\phi_{k'}\phi_{k''},\quad \text{so}\quad T_{1k}\sim\left|\phi_{k'}\right|^{2}\left|\phi_{k-k'}\right|^{2},$$

and the second two ultimately represent nonlinear relaxation i.e.

$$\delta\phi_{k''}^{(2)}\sim\phi_{k'}\phi_{k},\quad \text{so}\quad T_{2k}\sim\left|\phi_{k'}\right|^{2}\left|\phi_{k}\right|^{2}.$$

To complement our derivation of Eq.(5.52), here we will explicitly calculate T_2 – the nonlinear relaxation response – and simply state the result for the incoherent contribution to T_k. For the nonlinear response contribution T_2, after symmetrization, etc. we straightforwardly obtain,

$$T_{2k} = \sum_{k'} \left(\frac{\mathbf{k}' \cdot \mathbf{k} \times \hat{z}}{F + k^2} \right) \left(k''^2 - k'^2 \right) \left\langle \phi_{-k}(t)\, \phi_{-k'}(t)\, \delta\phi^{(2)}_{k+k'}(t) \right\rangle, \qquad (5.60a)$$

and

$$\frac{\partial}{\partial t} \delta\phi^{(2)}_{k+k'} + i\omega_{k+k'}\delta\phi^{(2)}_{k+k'} = \left(\frac{\mathbf{k}' \cdot \mathbf{k} \times \hat{z}}{F + k''^2} \right) \left(k'^2 - k^2 \right) \phi_{k'}(t)\, \phi_k(t), \qquad (5.60b)$$

so

$$\delta\phi^{(2)}_{k+k'} = \int_{-\infty}^{t} \mathrm{d}t' \exp\left[i\omega_{k+k'} \left(t - t' \right) \right] \left(\frac{\mathbf{k}' \cdot \mathbf{k} \times \hat{z}}{F + k''^2} \right) \left(k'^2 - k^2 \right) \phi_{k'}(t')\, \phi_k(t'). \qquad (5.60c)$$

Causality and $t > t'$ together imply that $\omega_{k+k} \rightarrow \omega + i\epsilon$, so ϕ^2 is damped at $t \rightarrow -\infty$. Combining all this gives the T_2 contribution,

$$T_{2k} = \sum_{k'} \frac{\left(\mathbf{k}' \cdot \mathbf{k} \times \hat{z} \right)^2}{\left(F + k^2 \right) \left(F + k''^2 \right)} \left(k''^2 - k'^2 \right) \left(k'^2 - k^2 \right) \times \int_{-\infty}^{t} \mathrm{d}t' \exp\left[i\omega_{k+k'} \right] \left\langle \phi_{-k}(t)\, \phi_{-k'}(t)\, \phi_{k'}(t)\, \phi_k(t') \right\rangle. \qquad (5.61)$$

We further take the two time correlator as set by wave frequency, nonlinear dynamics and causality, alone. This allows the two time scale ansatz,

$$\left\langle \phi_k^*(t)\, \phi_{k'}(t') \right\rangle = \left| \phi_k(t') \right|^2 \delta_{k+k',0} \exp\left[-i\omega_k \left(t - t' \right) \right], \qquad (5.62)$$

since $t' > t$, $\omega \rightarrow \omega + i\epsilon$ guarantees that correlation decays as $t \rightarrow \infty$. It is understood that the amplitude varies slowly relative to the phase. Using Eq.(5.62) in Eq.(5.61) then gives,

$$T_{2k} = \sum_{k'} \frac{\left(\mathbf{k} \cdot \mathbf{k}' \times \hat{z} \right)^2}{\left(F + k^2 \right) \left(F + k'^2 \right)} \left(k''^2 - k'^2 \right) \left(k'^2 - k^2 \right) \left| \phi_{k'} \right|^2 \left| \phi_k \right|^2 \Theta_{k,k',k''}, \qquad (5.63a)$$

where Θ, the triad coherence time, is,

$$\Theta_{\boldsymbol{k},\boldsymbol{k}',\boldsymbol{k}''} = \int_{-\infty}^{t} \mathrm{d}t' \, \exp\left[-i\left(\omega_{\boldsymbol{k}} + \omega_{\boldsymbol{k}'} - \omega_{\boldsymbol{k}''}\right)\left(t - t'\right)\right], \tag{5.63b}$$

and $\boldsymbol{k}'' = \boldsymbol{k} + \boldsymbol{k}'$ taking the real part finally gives (for weak nonlinear interaction),

$$\mathrm{Re}\,\Theta_{\boldsymbol{k},\boldsymbol{k}',\boldsymbol{k}''} = \pi\delta\left(\omega_{\boldsymbol{k}} + \omega_{\boldsymbol{k}'} - \omega_{\boldsymbol{k}''}\right), \tag{5.63c}$$

from where appears the resonant coupling condition. Hence, the nonlinear response contribution to T is,

$$T_{2\boldsymbol{k}} = \sum_{\boldsymbol{k}'} \frac{\left(\boldsymbol{k}\cdot\boldsymbol{k}'\times\hat{z}\right)^2}{\left(F + k^2\right)\left(F + k''^2\right)}\left(k''^2 - k'^2\right)\left(k'^2 - k^2\right)$$
$$\times \left|\phi_{\boldsymbol{k}'}\right|^2 \left|\phi_{\boldsymbol{k}}\right|^2 \pi\delta\left(\omega_{\boldsymbol{k}} + \omega_{\boldsymbol{k}'} - \omega_{\boldsymbol{k}''}\right). \tag{5.64}$$

T_1 – the noise emission part of the spectral transfer – can similarly be shown to be,

$$T_{1\boldsymbol{k}} = \sum_{\substack{\boldsymbol{p},\boldsymbol{q} \\ \boldsymbol{p}+\boldsymbol{q}=\boldsymbol{k}}} \frac{\left(\boldsymbol{p}\cdot\boldsymbol{q}\times\hat{z}\right)^2}{\left(F + k^2\right)^2}\left(q^2 - p^2\right)\left(q^2 - p^2\right)$$
$$\times \left|\phi_{\boldsymbol{p}}\right|^2 \left|\phi_{\boldsymbol{q}}\right|^2 \pi\delta\left(\omega_{\boldsymbol{k}} - \omega_{\boldsymbol{p}} - \omega_{\boldsymbol{q}}\right), \tag{5.65}$$

so

$$T_{\boldsymbol{k}} = T_{1\boldsymbol{k}} + T_{2\boldsymbol{k}}. \tag{5.66}$$

Several aspects of Eqs.(5.64) and (5.65) merit detailed discussion.

- The relaxation time $\tau_{\mathrm{R},\boldsymbol{k}}$ can be read off directly from Eq.(5.64) as,

$$1/\tau_{\mathrm{R},\boldsymbol{k}} = \sum_{\boldsymbol{k}'} \frac{\left(\boldsymbol{k}\cdot\boldsymbol{k}'\times\hat{z}\right)^2}{\left(1 + k^2\right)\left(1 + k''^2\right)}\left(k''^2 - k'^2\right)\left(k^2 - k'^2\right)$$
$$\times \left|\phi_{\boldsymbol{k}'}\right|^2 \pi\delta\left(\omega_{\boldsymbol{k}} + \omega_{\boldsymbol{k}'} - \omega_{\boldsymbol{k}''}\right). \tag{5.67}$$

For $|\boldsymbol{k}| \ll \left|\boldsymbol{k}'\right|$, $1/\tau_{\mathrm{R},\boldsymbol{k}} < 0$, while for $|\boldsymbol{k}| \gg \left|\boldsymbol{k}'\right|$ as usual, $1/\tau_{\mathrm{R},\boldsymbol{k}} > 0$. This is suggestive of 'inverse transfer' or a 'negative viscosity' phenomenon, whereby intensity is scattered to large scales from smaller scales. This is a recurring theme in 2D and drift–Rossby turbulence and follows from the dual conservation of energy and potential enstrophy.

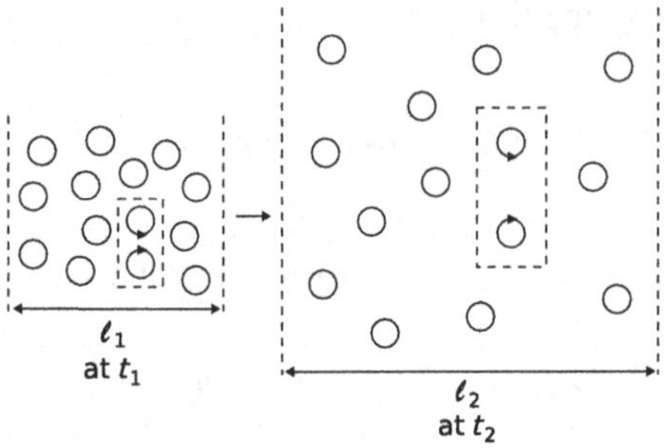

Fig. 5.9. Broadening or spreading ($l_2 > l_1$, $t_2 > t_1$) of a patch of 2D turbulence by mutual induction of vortex motion.

Geostrophic and 2D turbulence dynamics is discussed in Chapter 2, while geostrophic turbulence is discussed further in Chapter 6 and in Volume 2. The reader should be cautioned here that "negative viscosity" refers only to the tendency of the system to scatter energy to large *scale*. A patch of turbulence in such systems tends to broaden and spread itself in space by scattering via mutual induction of the interacting vortices. (See Figure 5.9 for illustration.) Also, while the dynamics of negative viscosity phenomena in drift–Rossby wave turbulence resembles that of the inverse cascade, familiar from 2D fluids, we stress that these two are *not* identical. Wave turbulence dynamics depends sensitively on triad resonance and thus on dispersion, etc. and transfer of energy need not be local in $\boldsymbol{k}$. The inverse cascade is a local process in $\boldsymbol{k}$, and is insensitive to the details of wave dynamics. Further discussion is given in Chapter 7.

- The conservation property for the transfer function T_k must be noted. One can straightforwardly show that,

$$\sum_k \left(F + k^2\right) T_k = 0$$

$$\sum_k \left(F + k^2\right)^2 T_k = 0,$$

so both energy E and potential enstrophy Ω,

$$E = \sum_k \left(1 + k^2\right) |\phi_k|^2$$

$$\Omega = \sum_k \left(1 + k^2\right)^2 |\phi_k|^2$$

are conserved by the wave coupling process. (See also the explanation of conservation in Appendix 1.) Note that proper treatment of both noise and nonlinear response is required for balance of the energy and potential enstrophy budgets. In particular, potential enstrophy conservation is manifested in wave kinetics from the direct correlation between nonlinear noise and response terms. Conservation is a simple consequence of symmetrization of the perturbation theory and is not an especially discriminating test of a turbulence theory.

- In general, drift wave dispersion is quite strong, so resonant triads are, in the same sense, rather 'special'. This is due at least in part, to the gyro/Rossby radius dependence (i.e. $k_\perp^2 \rho_i^2$ factor (ρ_i: ion gyro radius)) in the dispersion relation, which breaks scale invariance. An exception to this occurs at long wavelength, i.e. $k^2 \ll F$ (i.e. $k_\perp^2 \rho_i^2 < 1$), where $\omega_{\boldsymbol{k}} \sim (k_y V_{\mathrm{de}}/F)(1 - k_\perp^2/F)$ (i.e. $\omega_{\boldsymbol{k}} \sim k_y V_{\mathrm{de}}(1 - k_\perp^2 \rho_i^2)$), so the waves are nearly dispersion free. There $\tau_{T_c} \sim O(F/k^2)$ (i.e. $\sim O(1/k^2\rho_i^2)$), so the triad coherence diverges and renormalization is definitely required. The general structure of resonant triads for Rossby wave turbulence was considered by Longuet-Higgins and Gill (Longuet-Higgins *et al.*, 1967), and we refer the reader to that original source for further discussion. We do remark, though, that if one of the interacting modes is a low frequency or zero frequency shear flow (i.e. a zonal flow or GAM (Diamond *et al.*, 2005b)), resonance occurs much more easily. This is one reason for the dominant role of zonal flows in drift wave turbulence. This topic will be extensively discussed in Volume 2.

5.3.5 Issues to be considered

After this introduction to the derivation and structure of the theory of wave kinetics, it is appropriate to pause to take stock of the situation and to reflect on what *has* and *has not* been accomplished. So far, we have developed a perturbative, statistical theory of wave population evolution dynamics. Indeed, by construction the structure of wave kinetics closely resembles the theory for incoherent emission and absorption of photons in atomic transitions, and the theory of vibrational mode interactions in a solid. Wave kinetics is built upon the dual assumptions of negligible mode–mode coherence (i.e. the RPA) and short triad lifetime ($\tau_{T_c}\gamma_{\mathrm{E}} < 1$), and so is limited in applicability to systems with a large number of dispersive waves. Wave kinetics *does*:

- provide a framework within which we may identify and assess nonlinear interaction mechanisms and with which to calculate mode population evolution
- preserve relevant energy, momentum etc. conservation properties of the primitive equations. We emphasize that this is not an especially rigorous test of the theory, however
- provide a collision operator $C\{N\}$ for the wave kinetic equation which enables the construction of radiation hydrodynamic equations for the wave momentum and energy fields, etc.
- enable the identification of relevant time and space scales. However, from the trivial solution of energy equipartition, we have *not* yet:

- demonstrated the actual existence of any solutions to the wave kinetic equation which annihilate $C\{N\}$
- characterized the class of possible local and non-local interactions and their impact on spectral evolution
- considered the stability of possible solutions.

We now turn to these issues.

5.4 The scaling theory of local wave cascades

5.4.1 Basic ideas

We now turn from the formal development of wave interaction theory to discuss:

- how one might actually *use* wave kinetics to calculate the wave spectrum.

We discuss several examples of wave cascades:

- solutions of the WKE with finite spectral flux
- the relation of local cascades in wave turbulence to the K41 theory discussed in Chapter 2.

Truth in advertising compels us to admit the wave kinetic equation is rarely solved outright, except in a few very simple and rather academic cases. Given its complexity, this should be no surprise. Instead, usually its structure is analyzed to determine which types of coupling mechanisms are dominant, and how to construct a *simpler* dynamical model of spectral evolution for that particular case. In this section, we present some instructive 'studies' in this approach. For these, wave kinetics plays an important role as a general structure within which to determine relaxation time. Sadly though, there is no universal solution. Rather, the form of the dispersion relation and coupling coefficients make each case a challenge and opportunity.

5.4.1.1 Fokker–Planck approach

We begin by examining processes which produce a small increment in the wave vector $\boldsymbol{k}$. Here, *small increment* refers to couplings between roughly comparable scales or $\boldsymbol{k}$ values, which result in a slight shift $\Delta\boldsymbol{k}$ in the test mode wave vector. The Kolmogorov (K41) cascade is a classic example of a *small increment process*

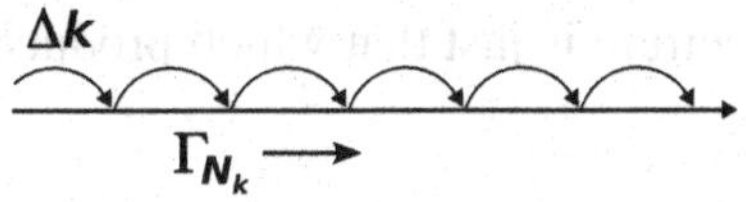

Fig. 5.10. Flux of N_k induced by a series of small increments in steps $\Delta \boldsymbol{k}$.

in scale space. For small increments, one might think of N (N: action density in k space) evolving as in a Fokker–Planck process, i.e.

$$N\,(t+\Delta t, \boldsymbol{k}) = \int \mathrm{d}\Delta \boldsymbol{k}\, N\,(t, \boldsymbol{k}-\Delta \boldsymbol{k})\, T\,(\Delta \boldsymbol{k}, \Delta t)\,, \tag{5.68a}$$

where $T\,(\Delta \boldsymbol{k}, \Delta t)$ is the transition probability for a step or $\boldsymbol{k}$-increment of size $\Delta \boldsymbol{k}$ in a time interval Δt. The usual algebra then gives,

$$\begin{aligned}\frac{\partial N_k}{\partial t} &= -\frac{\partial}{\partial \boldsymbol{k}} \cdot \left\{\left\langle \frac{\mathrm{d}\boldsymbol{k}}{\mathrm{d}t}\right\rangle N - \frac{\partial}{\partial \boldsymbol{k}} \cdot \left\langle \frac{\Delta \boldsymbol{k} \Delta \boldsymbol{k}}{2\Delta t}\right\rangle N\right\} \\ &= -\frac{\partial}{\partial \boldsymbol{k}} \cdot \left\{(\boldsymbol{V}_k N) - \frac{\partial}{\partial \boldsymbol{k}} \cdot (\mathsf{D}_k N)\right\},\end{aligned} \tag{5.68b}$$

where:

- $\boldsymbol{V}_k = \langle \mathrm{d}\boldsymbol{k}/\mathrm{d}t\rangle$ is the mean 'flow velocity' in $\boldsymbol{k}$. This flux in $\boldsymbol{k}$ results from a series of small increments $\Delta \boldsymbol{k}$, as shown in Figure 5.10
- $\mathsf{D}_k = \langle \Delta \boldsymbol{k} \Delta \boldsymbol{k}/2\Delta t\rangle$ is the $\boldsymbol{k}$-space diffusivity which describes evolution of the $\boldsymbol{k}$-variance of N.

Since the flow and diffusion in $\boldsymbol{k}$ are produced by mode–mode interactions, obviously $\boldsymbol{V}_k = \boldsymbol{V}_k\,\{N\}$, $\mathsf{D}_k = \mathsf{D}_k\,\{N\}$. Sources (i.e. growth) and sinks (i.e. damping) may also be added to Eq.(5.41). While $\boldsymbol{k}$-space diffusion is a small increment process, it is in general *not* a local one. Rather, the strain field driving D_k is usually localized at larger and slower scales than those of the test wave. For local interactions producing a net mean exciton flux in $\boldsymbol{k}$-space, the lowest-order description of a small increment process simply neglects diffusion and, in the absence of sources and sinks, reduces to,

$$\frac{\partial N_k}{\partial t} + \frac{\partial}{\partial \boldsymbol{k}} \cdot (\boldsymbol{V}_k N_k) = 0. \tag{5.69a}$$

Here, the stationary spectrum is just that which produces a divergence-free flux in $\boldsymbol{k}$ – i.e. that for which

$$\frac{\partial}{\partial \boldsymbol{k}} \cdot (V_{\boldsymbol{k}} N_{\boldsymbol{k}}) = 0. \tag{5.69b}$$

5.4.1.2 Leith model

Frequently, it is useful to formulate the flux in a 1D (i.e. scalar) $\boldsymbol{k}$-space, or scale (i.e. l) space. In such cases, the relaxation rate $\sim 1/\tau_{\mathrm{R}k}$ is often easier to construct than a flow velocity. When the spectrum is isotropic, N_k depends only on $k = |\boldsymbol{k}|$, the one-dimensional density $N(k) = 4\pi k^2 N_{\boldsymbol{k}}$ is used, and a measure of the wave-number quanta within the wave number k,

$$\mathcal{n}(k) = kN(k),$$

is introduced. The velocity in the k-space is also specified by a scalar variable $V(k)$, which denotes the velocity in the $|\boldsymbol{k}|$-direction. In the next step, the velocity in the k-space $V(k)$ is rewritten in terms of the relaxation rate as,

$$\frac{1}{k} V(k) = \frac{\mathrm{d}}{\mathrm{d}t} \ln k \sim \frac{1}{\tau_{\mathrm{R}k}}. \tag{5.70a}$$

Rewriting the divergence operator $(\partial/\partial \boldsymbol{k}) \cdot$ in one-dimensional form, $k^{-2}(\mathrm{d}/\mathrm{d}k)\, k^2$, Eq.(5.69a) is rewritten as,

$$\frac{\partial}{\partial t} \mathcal{n}(k) + \frac{\mathrm{d}}{\mathrm{d} \ln k} \left(\frac{1}{\tau_{\mathrm{R}k}} \mathcal{n}(k) \right) = 0. \tag{5.70b}$$

The stationary spectrum satisfies,

$$\frac{\mathrm{d}}{\mathrm{d}k} \left(\frac{1}{\tau_{\mathrm{R}k}} \mathcal{n}(k) \right) = 0. \tag{5.71}$$

This is the idea underpinning the Leith model, a useful and generalizable approach to modelling cascades and local processes. The Leith model, which is motivated by analogy with radiation and neutron transport theory, aims to represent the cascade's flux of energy in $\boldsymbol{k}$-space as a simple, local nonlinear diffusion process. Such a representation is extremely useful for applications to multi-scale modelling, transport, wave radiation hydrodynamics, etc. Of course, other assumptions or physics information is required to determine V_k or $1/\tau_{\mathrm{R}k}$ and relate them to $N_{\boldsymbol{k}}$. Wave kinetic theory provides a framework with which to determine these quantities.

Like most things in turbulence theory, the Leith model is motivated by the K41 theory of Navier–Stokes turbulence, which balances local energy flux through scale l at the rate $v(l)/l$ with a constant dissipation rate ϵ, taken to be independent of scale and viscosity. Thus, we recall that,

$$\epsilon = \frac{E(l)}{\tau(l)} = \frac{v^3(l)}{l},$$

or equivalently in k-space,

$$\epsilon = [kE(k)]\left[k^3 E(k)\right]^{1/2}.$$

Here $E(k)$ is in the 1D spectrum, so $kE(k)$ is a measure of the energy in wave number k and $\left[k^3 E(k)\right]^{1/2} = k[kE(k)]^{1/2} = 1/\tau_{Rk}$ is just the eddy turn-over rate. In essence, k appears as a density-of-states factor. Hence, $E(k) \sim \epsilon^{2/3} k^{-5/3}$, so we recover the familiar K41 spectrum, which has finite, constant spectral energy flux ϵ. Now in the Leith model of fluid turbulence, the effective quanta density $n(k)$ is just $kE(k)$, the relaxation rate is $1/\tau_{Rk} = \left[k^3 E(k)\right]^{1/2}$ and so Eq.(5.71) is just,

$$\frac{\partial}{\partial \ln k}\left(k\,(kE(k))^{3/2}\right) = \frac{\partial}{\partial \ln k}\left(k^{5/2} E(k)^{3/2}\right) = 0, \tag{5.72}$$

the solution of which recovers the K41 spectrum.

5.4.1.3 Leith model with dissipation

It is instructive to discuss a slightly more complicated example of the Leith model, in order to get a feel for spectral flow constructions in a familiar context. Retaining viscous damping as an explicit high-k sink changes the Leith model spectral flow continuity equation to,

$$\frac{\partial}{\partial t} n(k) + \frac{\partial}{\partial \ln k}\left(k n(k)^{3/2}\right) + \nu k^2 n(k) = 0. \tag{5.73}$$

Here $n(k) = kE(k)$. The stationary state spectrum must then balance spectral flux and dissipation to satisfy,

$$\frac{\partial}{\partial \ln k}\left(k n(k)^{3/2}\right) + \nu k^2 n(k) = 0,$$

with an initial condition (influx condition) that at k_0, the stirring or input wave number, $n(k_0) = v_0^2$. Then taking $U = kn(k)^{3/2}$ (note U is just the energy flow rate!) transforms the flow equation to,

$$\frac{dU(k)}{dk} + \nu k^{1/3} U(k)^{2/3} = 0, \tag{5.74}$$

so

$$U(k) = \epsilon - \nu k^{4/3}/4. \tag{5.75}$$

In the inertial range, $\epsilon \gg \nu k^{4/3}/4$, so $U(k) = \epsilon$, $n(k) = (\epsilon/k)^{2/3}$ and $E(k) = \epsilon^{2/3}k^{-5/3}$, the familiar Kolmogorov spectrum. Observe that in the Leith model, the energy dissipation rate ϵ appears as a constant of integration. Matching the boundary condition at k_0 requires $U(k_0) = k_0 n(k_0)^{3/2} = k_0 v_0^3 \sim v_0^3/l_0$. This gives the integration constant ϵ as,

$$\begin{aligned} \epsilon &= U(k_0) + \nu k_0^{4/3}/4 \\ &\cong k_0 v_0^3. \end{aligned} \tag{5.76}$$

The second term on the right-hand side is just a negligible $O(1/R_e)$ correction to the familiar formula which relates the dissipation rate to the stirring scale parameters. (R_e: Reynolds number. Illustration is given in Figure 5.11.)

From this simple example, we see that the essence of the Leith model approach, which is useful and applicable *only* for local, steady small increment processes, is to:

- identify an exciton density which 'flows' in k or scale space by local increments. The relevent quantity is usually suggested by some indication of self-similarity (i.e. a power-law spectrum over a range of scales)
- use physics input or insight to identify a flow rate or relaxation time and, in particular, its dependence on N_k. Wave kinetics is very helpful here
- impose stationarity to determine the spectrum. A quanta source and sink must be identified, and the necessary constant of integration is usually related to the net flow rate.

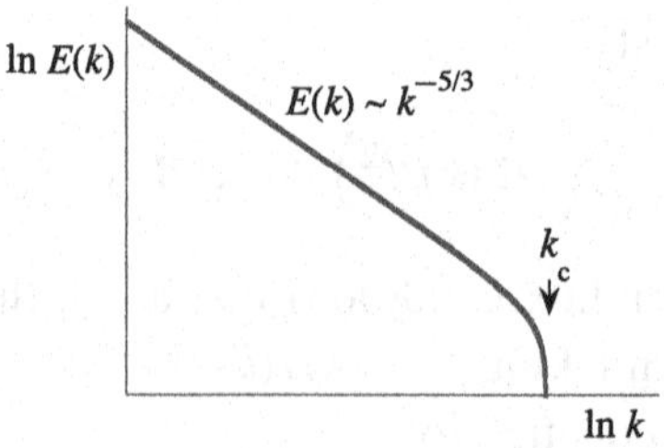

Fig. 5.11. Leith model with small but finite molecular viscosity ν. The spectrum is cut at k_c, which satisfies the relation $k_c = (4\epsilon/\nu)^{3/4}$.

5.4.2 Gravity waves

Another instructive application of wave kinetics is to the spectrum of surface gravity waves (Lighthill, 1978). Indeed, three of the earliest and best known studies of wave kinetics are the pioneering works of Phillips (1966), Hasselmann (1962; 1968), and Zakharov and Filonenko (1967), all of which dealt with surface wave turbulence. Ocean surface gravity waves are excited by the wind, and continuously fill a range of scales with wave number $k_{\mathrm{w}} < k < k_{\mathrm{cap}}$. Here k_{w} is the wave number of the wind wave $k_{\mathrm{w}} = g/v_{\mathrm{w}}^2$, and k_{cap} is set by the small scale where surface tension becomes important, i.e. $k_{\mathrm{cap}} \sim (\rho g/\sigma)^{1/2}$ (where σ is the coefficient of surface tension, g is the gravitational acceleration, and ρ is the mass density). The surface wave displacement spectrum is a power law over this range, and asymptotes to a quasi-universal form $\left|\tilde{\xi}\right|^2 \sim k^{-4}$, at the upper end of the gravity wave range (ξ: displacement of surface). This universal spectrum is referred to as the 'Phillips spectrum,' after O. M. Phillips (1966), who first proposed it in 1955. The basic idea of the Phillips model is that the waves are saturated, i.e. sitting just at the threshold of breaking, so the wave slope $\tilde{s}$ is discontinuous, i.e. $\mathrm{d}\tilde{s}/\mathrm{d}x = \delta\left(x - x'\right)$ at a wave crest (see Figure 5.12). Since $\tilde{s} = k\tilde{\xi}$, this implies a displacement spectrum of $\left|\tilde{\xi}\right|^2 \sim 1/k^4$, the essence of the Phillips model. Intensive studies by a variety of sensing and analysis techniques all indicate that:

- the excitations on the ocean surface are very well approximated by an ensemble of surface waves, with a power law spectrum of wave slopes;
- the Phillips spectrum is a good fit, at least at the upper end of the gravity wave range.

The scale invariance of gravity waves on the interval $k_{\mathrm{w}} < k < k_{\mathrm{cap}}$, the universality of the Phillips spectrum and the strong excitation of gravity waves by even modest wind speeds have all combined to motivate application of the theory of wave kinetics to the problem of the gravity wave spectrum, in the hope of developing the first principles of a theory. Gravity waves have several interesting features which distinguish them from Alfvén or drift waves, and which also make this example especially instructive. In particular:

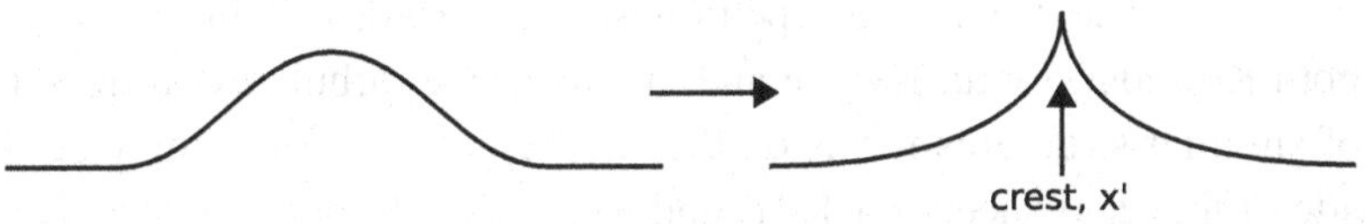

Fig. 5.12. Formation of wave slope discontinuity at crest of breaking wave. $\mathrm{d}\tilde{s}/\mathrm{d}x = \delta\left(x - x'\right)$

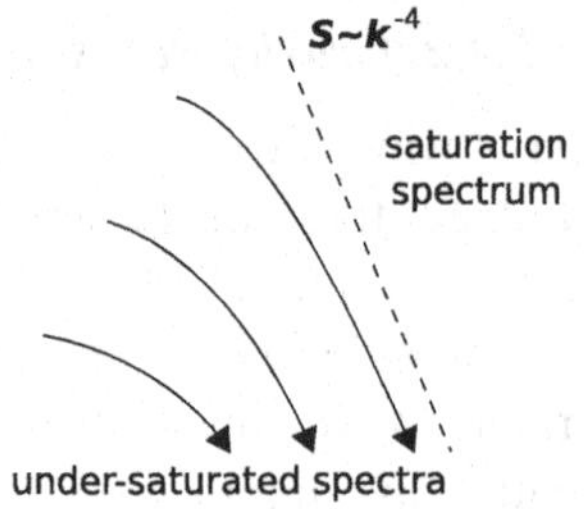

Fig. 5.13. Spectral steepening as waves approach saturation at the Phillips spectrum $S \sim k^{-4}$.

- on account of the gravity wave dispersion relation $\omega = \sqrt{gk}$, there are no three-wave resonant couplings among gravity waves. Rather, the fundamental resonant interaction is *four-wave coupling*. Resonant three-wave coupling *is* possible among two gravity waves and a very small gravity-capillary wave. Indeed, the absence of the three-wave resonance among gravity waves is one likely reason why they are seemingly such a good model for ocean surface excitations;
- unlike Alfvén waves in incompressible MHD, gravity waves can break, so applicability of weak turbulence theory requires that wave displacement $\tilde{\xi}$ be small enough so that the wave slope is subcritical to breaking (i.e. $k\tilde{\xi} < 1$). This imposes a limit on the wave amplitudes and energies which are compatible with wave kinetics. More generally, it is plausible to expect the weak turbulence spectrum to be 'softer' than the Phillips spectrum (i.e. $\sim k^{-\alpha}$, $\alpha < 4$), since resonant four-wave coupling is not as efficient as wave breaking in disposing of wave energy. (Fig. 5.13).

Given the self-similar power law structure of the gravity wave spectrum, it is natural to try to model gravity wave interaction as a local energy cascade from large (wind-wave scale) to small (gravity-capillary wave scale) scales. In this sense, the gravity wave energy cascade resembles the Kolmogorov cascade. An important difference between these two cascades is the nature of the effective dissipation which terminates them. Instead of viscosity as for ordinary Navier–Stokes turbulence, the gravity wave cascade terminates by some combination of wave crest instability and wave breaking, which involves the combined effects of surface tension, vorticity at the surface layer, and the dynamics of air–sea interaction (i.e. white capping, foam and bubble formation, etc.). The dissipative dynamics of the gravity wave cascade remains an open question. Indeed, it is interesting to note that in both Kolmogorov turbulence and gravity wave turbulence, a fractal distribution of singular structures forms on the smallest scales. These may be thought of as vortex tubes and sheets for K41, and to marginally breaking wave crests or a foam of small bubbles for ocean waves. Also, in neither case does a rigorous understanding of the dissipation rate exist at present. That said, motivated by the empirical self-similarity of ocean wave turbulence, we will plunge ahead boldly and impose constant energy flux as,

$$\epsilon = \frac{kE(k)}{\tau_{Rk}}.$$

Here, it is understood that the system is 2D, so $kE(k)$ is energy/area. The relaxation time is determined by four considerations, namely:

(i) that the dynamics are scale invariant and self-similar;
(ii) that the characteristic transfer rate is proportional to the surface wave frequency $\omega_k = \sqrt{gk}$ – the only temporal rate in the gravity wave range;
(iii) that the fundamental interaction is four-wave coupling, so $1/\tau_{Rk} = [kE(k)]^2$ – i.e. an extra power of energy appears, as compared to three-wave coupling;
(iv) that the waves must be subcritical to breaking, i.e. fluid parcel velocities v_k should be less than the wave phase ω/k ($v_k < \omega/k$), so $kE(k) < \rho_w(\omega^2/k^2)k$. Nonlinear transfer must thus be small in the ratio $E(k)(\omega/k)^{-2}\rho_w^{-1}$. Here ρ_w is the density of water and the additional factor of k^{-1} appears since $kE(k)$ has dimensions of energy/area. The particular association of length with wavelength follows from the fact that ocean wave perturbations decay exponentially with depth as $\sim e^{-|kz|}$.

Assembling these pieces enables us to construct the spectral flow equation (with ϵ constant),

$$\epsilon = \frac{kE(k)}{\tau_{Rk}}, \tag{5.77a}$$

where,

$$\frac{1}{\tau_{Rk}} = \left(\frac{kE(k)}{\rho_w\left(\omega^2/k^2\right)k}\right)^2 \omega_k, \tag{5.77b}$$

i.e. the relaxation rate is the wave frequency, multiplied by *two* powers (four-wave interaction!) of intensity, normalized to the breaking threshold. Thus, the spectral transfer equation is,

$$\epsilon = \omega_k \left(\frac{kE(k)}{\rho_w\left(\omega^2/k^2\right)k}\right)^2 (kE(k)). \tag{5.78}$$

Subsequently using the dispersion relation $\omega = \sqrt{gk}$ gives $E(k) \sim k^{-5/2}$ and a surface displacement spectrum $\left|\tilde{\xi}(k)\right|^2 \sim k^{-7/2}$. This does *not* agree with the scaling of the Phillips spectrum, nor should it, since the latter is based on a hypothesis of saturation by wave breaking, which is outside the 'event horizon' of weak

Table 5.2. *Elements of local wave cascade models*

Constituent	
Alfvén wave	gravity wave
$\omega = k_{\parallel} v_A$	$\omega = \sqrt{kg}$
incompressible MHD	surface of ocean
Interaction	
3-wave	4-wave
Basic rate	
$\dfrac{v^2(x_\perp)}{l_\perp^2(\Delta k_\parallel v_A)}$	$\omega_k \left(\dfrac{kE(k)}{\rho_w(\omega^2/k^2)k}\right)^2$
Constraint limit	
critical balance	sub-critical to breaking
$\Delta k_\parallel v_A \sim v(l_\perp)/l_\perp$	$kE(k) < \rho_w(\omega^2/k^2)k$
Spectral balance	
$\epsilon = \dfrac{v^2(l_\perp)v(l_\perp)}{l_\perp^2(\Delta k_\parallel v_A)}$	$\alpha \equiv \omega(k)\left(\dfrac{kE(k)}{\rho_w(\omega^2/k^2)k}\right)^2 kE(k)$
Limiting result	
$k_\parallel \sim \dfrac{\epsilon^{1/3}k_\perp^{2/3}}{v_A}$	$\|\tilde{\varepsilon}(k)\|^2 \sim k^{-7/2}$
$E(k)$ as K41	softer than Phillips

turbulence theory! It is indeed reassuring to see that the weak turbulence gravity wave spectrum is softer than the Phillips spectrum (i.e. α=7/2<4), as we expect. The self-similar gravity wave spectrum $\left|\tilde{\xi}(k)\right|^2 \sim k^{-7/2}$, sometimes referred to as the 'Kolmogorov spectrum' for gravity waves by Zakharov and collaborators (Zakharov *et al.*, 1992), might be relevant to regimes where waves are driven weakly, slightly above the wind excitation threshold. Certainly, it cannot properly describe the process of wave spectrum saturation at high wind speed. Indeed, no attempt to connect the Phillips spectrum to perturbation theory or wave turbulence theory has yet succeeded, although recent efforts by Newell and Zakharov (2008) appear promising in this respect. We include this example here as an illustration of how to apply wave turbulence theory to obtain results for a more complex problem. The applications of wave kinetics to Alfvén and gravity wave turbulence are summarized in Table 5.2. (See Chapter 9 for detailed explanation of Alfvén wave turbulence.)

5.5 Non-local interaction in wave turbulence

No doubt the reader who has persevered this far is thinking, "Surely not *all* wave interaction processes are simply local cascades!?" Such scepticism is now to be rewarded as we turn to the important and often-neglected subject of non-local interaction in wave turbulence. Non-local interaction in wave turbulence refers to the resonant interaction processes of three waves in which the magnitudes of the three frequencies and/or wave vectors are not comparable. These are contrasted with local interactions in the diagram in Figure 5.6 (left). Crudely put, in local interactions, the triangles defined by resonant triad $\boldsymbol{k}$ vectors are nearly equilateral, while those corresponding to non-local interactions deviate markedly from equilateral structure, as shown in Figure 5.6 (centre). Local interactions transfer exciton density (i.e. energy) between neighboring $\boldsymbol{k}$, while non-local interaction can transfer energy between quite disparate scales – as occurs when large-scale shears strain smaller scales, for example. However, non-local interaction *is* compatible with the notion of a "small-increment process," as discussed earlier. Indeed, we shall see that stochastic shearing by large scales produces a random walk in $\boldsymbol{k}$ space at small scales. This is a classic example of non-local interaction resulting in a small increment, diffusive scattering of the population density.

5.5.1 Elements in disparate scale interaction

We have already encoutered one generic example of a non-local wave interaction process, namely the decay or parametric subharmonic instability in a resonant triad, discussed in Section 5.2. There, a populated high-frequency mode at $\omega \sim 2\Omega$ decays to two daughter waves at $\omega \sim \Omega$. Note that this mechanism requires a *population inversion* – the occupation density of the pump, or high frequency mode, must exceed those of the daughters in order for decay to occur. Another generic type of interaction is *induced diffusion* – an interaction where large-scale, low-frequency waves strain smaller, higher-frequency waves. Induced diffusion arises naturally when one leg of the resonant triad is much shorter than the other two, indicative of a near self-beat interaction of two short-wavelength, high-frequency waves with one long-wavelength, low-frequency wave. Thus, the interaction triad is a thin, nearly isosceles triangle, as shown in Figure 5.6 (centre). The "diffusion" in induced diffusion occurs in $\boldsymbol{k}$-space, and is a consequence of the spatio-temporal scale disparity between the interacting waves. This allows an eikonal theory description of the interaction between the long-wavelength strain field and the short-wavelength mode with wave vector $\boldsymbol{k}$, which undergoes refractive scattering. Thus since,

$$\frac{\mathrm{d}\boldsymbol{k}}{\mathrm{d}t} = -\nabla\left(\omega + \boldsymbol{k}\cdot\boldsymbol{v}\right), \tag{5.79a}$$

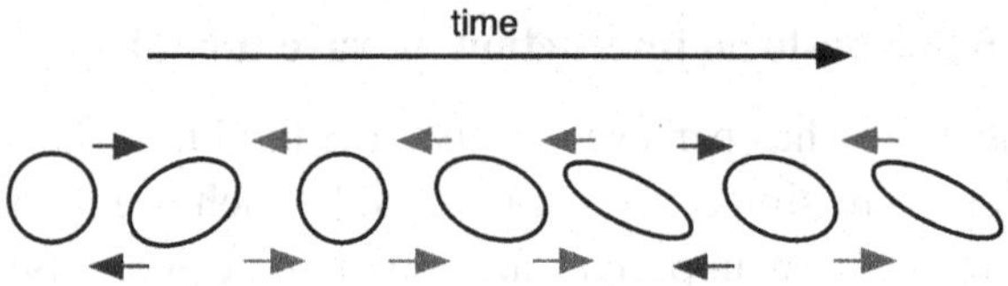

Fig. 5.14. Distortion of a small-scale perturbation (shown by circular or elliptical vortex) by an ambient large-scale sheared flow. The large-scale flow changes its sign so that the wave vector of small-scale perturbation is subject to diffusive change.

$\delta \boldsymbol{k}$, the excursion in $\boldsymbol{k}$ due to inhomogeneities in $\omega_{\boldsymbol{k}}$ and v, is,

$$\frac{\mathrm{d}}{\mathrm{d}t}\delta \boldsymbol{k} = -\nabla\left(\tilde{\omega} + \boldsymbol{k}\cdot\tilde{\boldsymbol{v}}\right), \tag{5.79b}$$

(see Figure 5.14 for illustration (Diamond *et al.*, 2005b)).

$$\mathsf{D}_k = \frac{\mathrm{d}}{\mathrm{d}t}\left\langle \delta k^2 \right\rangle = \sum_q \boldsymbol{q}\boldsymbol{q}\left|\left(\tilde{\omega} + \boldsymbol{k}\cdot\tilde{\boldsymbol{v}}\right)_{\boldsymbol{q}}\right|^2 \tau_{\boldsymbol{k},\boldsymbol{q}}, \tag{5.80}$$

where $\tau_{\boldsymbol{k},\boldsymbol{q}}$ is the coherence time of the scattering field $\boldsymbol{q}$ with the scattered ray $\boldsymbol{k}$. Insight into the physics of $\tau_{\boldsymbol{k},\boldsymbol{q}}$ follows from consideration of the triad resonance function in the limit where $|\boldsymbol{q}| \ll \boldsymbol{k}, \boldsymbol{k}'$, i.e.

$$\begin{aligned}\Theta_{\boldsymbol{k},\boldsymbol{k}',\boldsymbol{q}} &= \frac{i}{\omega_{\boldsymbol{q}} + \omega_{\boldsymbol{k}'} - \omega_{\boldsymbol{k}'+\boldsymbol{q}}} \\ &\simeq \frac{i}{\omega_{\boldsymbol{q}} + \omega_{\boldsymbol{k}'} - \omega_{\boldsymbol{k}'} - \boldsymbol{q}\cdot\left(\partial\omega_{\boldsymbol{k}'}/\partial\boldsymbol{k}'\right)} \\ &= \frac{i}{\omega_{\boldsymbol{q}} - \boldsymbol{q}\cdot\boldsymbol{v}_{\mathrm{gr}}(\boldsymbol{k})},\end{aligned} \tag{5.81a}$$

so

$$\mathrm{Re}\,\Theta_{\boldsymbol{k},\boldsymbol{k}',\boldsymbol{q}} \cong \pi\delta\left(\omega_{\boldsymbol{q}} - \boldsymbol{q}\cdot\boldsymbol{v}_{\mathrm{gr}}(\boldsymbol{k})\right). \tag{5.81b}$$

Hence, the coherence time $\tau_{\boldsymbol{k},\boldsymbol{q}}$ is set by the larger of:

– the dispersion rate of the strain field spectrum (i.e. the qs), as "seen" by a wave packet with group velocity $\boldsymbol{v}_{\mathrm{gr}}(k)$, i.e.

$$\frac{1}{\tau_{\mathrm{ac},\boldsymbol{q},\boldsymbol{k}}} = \left|\left(\frac{\mathrm{d}\omega_{\boldsymbol{q}}}{\mathrm{d}\boldsymbol{q}} - \boldsymbol{v}_{\mathrm{gr}}(\boldsymbol{k})\right)\cdot\Delta\boldsymbol{q}\right|. \tag{5.82}$$

For resonant packets, this is proportional to the difference of straining wave group and phase speeds, reminiscent of what we encountered in our discussion of quasi-linear theory. This time scale is relevant to 'pure' weak turbulence theory;
- with resonance broadening – the self-decorrelation rate of the straining wave or test wave. These are nonlinear timescales related to wave dynamics and enter when resonance broadening is considered. For example, the test wave lifetime is often of the order of magnitude of the inverse growth rate $|\gamma_{\boldsymbol{k}}|$.

Like the parametric subharmonic process, the effect of induced diffusion also is sensitive to the exciton population profile. Thus, for energy density $E=N\omega$, diffusion of N means the net change of the short-wave energy is given by,

$$\frac{\mathrm{d}E_{\mathrm{sw}}}{\mathrm{d}t} = \frac{\mathrm{d}}{\mathrm{d}t}\int \mathrm{d}\boldsymbol{k}\, E(\boldsymbol{k},t) = -\int \mathrm{d}\boldsymbol{k}\, \frac{\mathrm{d}\omega}{\mathrm{d}\boldsymbol{k}} \cdot \mathsf{D} \cdot \frac{\mathrm{d}N}{\mathrm{d}\boldsymbol{k}}. \tag{5.83}$$

Hence, the poulation profile gradient $\mathrm{d}N/\mathrm{d}\boldsymbol{k}$ and the group velocity $\mathrm{d}\omega/\mathrm{d}\boldsymbol{k}$ together determine $\mathrm{d}E_{\mathrm{sw}}/\mathrm{d}t$. For $\mathrm{d}\omega/\mathrm{d}\boldsymbol{k} > 0$, $\mathrm{d}N/\mathrm{d}\boldsymbol{k} \lessgtr 0$ implies $\mathrm{d}E_{\mathrm{sw}}/\mathrm{d}t \gtrless 0$, with correspondingly opposite results for $\mathrm{d}\boldsymbol{\omega}/\mathrm{d}\boldsymbol{k} < 0$. Of course, $\mathrm{d}E_{\mathrm{sw}}/\mathrm{d}t>0$ means the short waves are *gaining* energy from the straining waves, while $\mathrm{d}E_{\mathrm{sw}}/\mathrm{d}t < 0$ means that they are *losing* energy to the longer wavelengths.

It is interesting to note that induced diffusion is one limit where the fundamental origin of irreversibility in wave turbulence is rigorously clear. In eikonal theory, $\boldsymbol{k} = \nabla\phi$, where ϕ is the wave phase function. Then, integrating the eikonal equation for $\boldsymbol{k}$ gives,

$$\frac{\mathrm{d}}{\mathrm{d}t}\delta\phi = -\left(\tilde{\omega} + \boldsymbol{k}\cdot\tilde{\boldsymbol{v}}\right), \tag{5.84a}$$

so

$$\left\langle \delta\phi^2 \right\rangle = D_\phi t, \tag{5.84b}$$

where the phase diffusion coefficient is,

$$D_\phi = \sum_{\boldsymbol{q}} \left|\left(\tilde{\omega} + \boldsymbol{k}\cdot\tilde{\boldsymbol{v}}\right)_{\boldsymbol{q}}\right|^2 \tau_{\boldsymbol{k},\boldsymbol{q}}, \tag{5.85a}$$

and in weak wave turbulence,

$$\tau_{\boldsymbol{k},\boldsymbol{q}} \to \pi\delta\left(\omega_{\boldsymbol{q}} - \boldsymbol{q}\cdot\boldsymbol{v}_{\mathrm{gr}}(\boldsymbol{k})\right). \tag{5.85b}$$

Thus, we see that if resonances between the strain field phase velocity and the wave packet group velocity overlap (in the sense of island overlap, as in Chirikov, the wave phase will evolve diffusively, consistent with the notion of a random phase. Here then, *ray stochasticity*, emerges as the dynamical underpinning of irreversibility for induced diffusion. More generally, since a range of large-scale waves or straining flows is a ubiquitous element in fluctuation spectra, it is tempting to speculate that the utility of RPA-based techniques may be rooted in phase stochasticity driven by induced diffusion. This is especially likely when the straining field and the energy-containing regions of the spectrum coincide. This speculation could be explored by comparing the coherent energy transfer rate in a resonant triad (i.e. γ_E) with the rate at which the frequency wanders due to phase stochastization by straining during that time period, i.e. $\Delta\omega_{\text{phase}} \sim (v_{\text{gr}}^2 q^2 D_\phi/\gamma_E)^{1/2}$.

5.5.2 Effects of large/meso scale modes on micro fluctuations

Non-local interactions have received far less attention than local, 'Kolmogorov' cascades in wave turbulence. Nevertheless, non-local interactions are of great importance since they:

- are useful as a framework for describing large-scale structure formation in turbulence;
- describe and account for interactions which break scale invariance and so induce intermittency.

We now briefly discuss these two important roles of non-local interactions. One frequently utilized approach to the problem of structure formation is to consider when and how an ambient spectrum of waves and turbulence is, in some sense, unstable to the growth of a large-scale seed perturbation. For example, in mean field dynamo theory one considers the stability of a spectrum of turbulence to a large-scale magnetic field. The induced diffusion interaction, introduced here and developed much further in Chapter 7, is especially useful for this type of consideration, since it naturally describes the exchange of energy between an ambient short-wavelength wave spectrum and a seed spectrum of large-scale excitations. The theory can be extended to address saturation of structure formation by a variety of processes, including self-consistent alteration or evolution of the ambient wave spectrum. This picture of coupled evolution of the large-scale strain field and smaller-scale wave field (which exerts a stress on the former) leads naturally to a 'predator–prey' type model of the self-consistent interaction of the two components, as discussed further in Volume 2. Parametric subharmonic decay interaction is also interesting as a means for non-local transfer of energy in frequency – from a pump to lower frequency waves. This mechanism is exploited in some scenarios of low-frequency structure formation.

Non-local interactions are of interest as a possible origin of intermittency. Of course, intermittency has many forms and many manifestations. One frequently invoked definition of intermittency is that of "a process which breaks scale similarity by inducing an explicit memory of one class of scales in another," the connection of which is driven by a cascade. For example, the β-model and multifractal models of inertial range intermittency in K41 turbulence all invoke a notion of embedded turbulence, where a 'footprint' of the stirring scale l_0 survives in the inertial range via explicit spectral dependence on l_0 (Note, here 'explicit' means dependence on (l_0) not only via the dissipation rate ϵ). Non-local interactions clearly can produce such multi-scale memory – random straining (as occurs in induced diffusion) will surely leave an imprint of a large-scale, energetic strain field on smaller wave scales. Of course, the size and strength of this effect will to some extent depend upon the relative sizes of the large-scale induced distortion or strain rate and the rate of local energy transfer among smaller scales. In this regard it is worth noting that a recent interesting line of research (Laval *et al.*, 2001) on the dynamics of intermittency in 3D Navier–Stokes turbulence has suggested a picture where:

- like-scale interactions are self-similar and non-intermittent, and are responsible for most of the energy transfer. These are well described by conventional scaling arguments;
- disparate scale straining is the origin of scale symmetry breaking and intermittency, and is well described by rapid distortion theory, which is closely related to induced diffusion.

Apart from its intrinsic interest, this approach to the problem of intermittency is noteworthy since it is entirely compatible with a simple statistical or weak turbulence model, and does not require invoking more exotic theoretical concepts such as multi-fractality, coherent structures, etc.

5.5.3 Induced diffusion equation for internal waves

As a case study in non-local interaction, we focus on the interaction of oceanic internal waves. Internal waves (IWs) have the dispersion relation,

$$\omega^2 = k_{\mathrm{H}}^2 N_{\mathrm{BV}}^2 / k^2, \tag{5.86a}$$

where N_{BV} is the Brunt–Väisälä (BV) buoyancy frequency, i.e.

$$N_{\mathrm{BV}}^2 = +\frac{g}{\rho_0}\frac{\mathrm{d}\rho_0}{\mathrm{d}z}, \tag{5.86b}$$

where z grows with depth, down from the surface, so IWs may be thought of as the stable counterpart to the Rayleigh–Taylor instability for the case of a continuous density profile (i.e. no interface). Here k_{H} refers to the horizontal wavenumber and k_V is the vertical wavenumber, so $k^2 = k_H^2 + k_V^2$. Internal waves are excited at mesoscales by Rossby waves, the interaction of large currents with bottom topography, and large storms. Internal wave interaction generates a broad spectrum of IWs which is ultimately limited by breaking and over-turning at small scales. A phenomenological model of the IW spectrum, called the Garret–Munk (GM) model (Garret and Munk, 1975), provides a reasonable fit to the measured IW spectrum. The GM spectrum density of IWs is peaked at large scales.

In an edifying and broadly relevant study, McComas and Bretherton (1977) identified three types of non-local interactions which occur in IW turbulence. These are:

- induced diffusion
- parametric subharmonic instability
- elastic scattering

In all cases, the origin of the generic type of interaction can be traced to the wave dispersion relation structure and the basic wave–wave coupling equations. Here, we outline the key points of this instructive analysis. In this context, the wave population density is denoted by $A(\boldsymbol{k})$ (to avoid confusion with BV frequency N) and the wave–wave collision integral has the generic form following Eq.(5.33), i.e.

$$\begin{aligned}\frac{\mathrm{d}}{\mathrm{d}t}A(\boldsymbol{k}) = F\{A\} = \int \mathrm{d}\boldsymbol{k}' \int \mathrm{d}\boldsymbol{k}'' \big\{ & D^{+}\delta\left(\boldsymbol{k}'+\boldsymbol{k}''-\boldsymbol{k}\right)\delta\left(\omega'+\omega''-\omega\right) \\ & \times \left[A(\boldsymbol{k}')A(\boldsymbol{k}'') - A(\boldsymbol{k}')A(\boldsymbol{k}) - A(\boldsymbol{k}'')A(\boldsymbol{k})\right] \\ & + 2D^{-}\delta\left(\boldsymbol{k}'-\boldsymbol{k}''-\boldsymbol{k}\right)\delta\left(\omega'-\omega''-\omega\right) \\ & \times \left[A(\boldsymbol{k}')A(\boldsymbol{k}'') + A(\boldsymbol{k}')A(\boldsymbol{k}) - A(\boldsymbol{k}'')A(\boldsymbol{k})\right]\big\}. \end{aligned} \quad (5.87)$$

Here D^{+} and D^{-} are coupling coefficients, and the notation is obvious.

To extract induced diffusion from $F\{A\}$, it is useful to divide the spectrum into two pieces, as shown in Fig. 5.15. In that representation:

- $\boldsymbol{k}_2$ refers to the straining waves at large scale, with spectral density $B(\boldsymbol{k}_2)$;
- $\boldsymbol{k}_1$ and $\boldsymbol{k}_3$ are the short-wavelength waves, with spectral density $F(\boldsymbol{k}_{1,3})$.

Thus,

$$A(\boldsymbol{k}) = B(\boldsymbol{k}) + F(\boldsymbol{k}).$$

We assume B and F do not overlap, so $B(\boldsymbol{k}_3) = B(\boldsymbol{k}_1) = 0$. Since short-wavelength evolution due to large-scale wave effects is of interest here, we seek $\partial F(\boldsymbol{k}_3)/\partial t$, and can re-write Eq.(5.87) as,

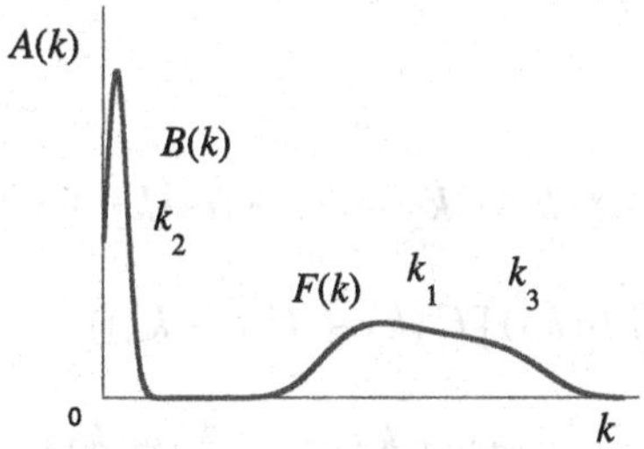

Fig. 5.15. Representation of long wave–short wave interactions. The spectral function $A(k)$ is composed of $B(k)$ and $F(k)$.

$$\frac{\partial}{\partial t}F(\boldsymbol{k}_3) = \int \mathrm{d}\boldsymbol{k}' \int \mathrm{d}\boldsymbol{k}'' \left\{D^{+}\delta\left(\boldsymbol{k}' + \boldsymbol{k}'' - \boldsymbol{k}_3\right)\delta\left(\omega' + \omega'' - \omega_3\right)\right.$$
$$\times \left[B(\boldsymbol{k}')\left(F(\boldsymbol{k}'') - F(\boldsymbol{k}_3)\right) + B(\boldsymbol{k}'')\left(F(\boldsymbol{k}') - F(\boldsymbol{k}_3)\right) + F(\boldsymbol{k}')F(\boldsymbol{k}'')\right.$$
$$\left. - F(\boldsymbol{k}_3)F(\boldsymbol{k}') - F(\boldsymbol{k}_3)F(\boldsymbol{k}'')\right] + 2D^{-}\delta\left(\boldsymbol{k}' - \boldsymbol{k}'' - \boldsymbol{k}_3\right)\delta\left(\omega' - \omega'' - \omega_3\right)$$
$$\left.\times \left[B(\boldsymbol{k}'')\left(F(\boldsymbol{k}') - F(\boldsymbol{k}_3)\right) + F(\boldsymbol{k}_3)F(\boldsymbol{k}') - F(\boldsymbol{k}_3)F(\boldsymbol{k}) - F(\boldsymbol{k}_3)F(\boldsymbol{k}'')\right]\right\}. \tag{5.88}$$

The point here is to isolate the influence of the large-scale waves on the smaller scales. Hence, we decompose $\partial F(\boldsymbol{k}_3)/\partial t$ into,

$$\frac{\partial F(\boldsymbol{k}_3)}{\partial t} = \left.\frac{\partial F(\boldsymbol{k}_3)}{\partial t}\right|_{\text{local}} + \left.\frac{\partial F(\boldsymbol{k}_3)}{\partial t}\right|_{\text{non-local}} \tag{5.89a}$$

and can read off $\partial F(\boldsymbol{k}_3)/\partial t|_{\text{non-local}}$ from Eq.(5.88) as,

$$\left.\frac{\partial}{\partial t}F(\boldsymbol{k}_3)\right|_{\text{non-local}} = 2\int \mathrm{d}\boldsymbol{k}_1 \int \mathrm{d}\boldsymbol{k}_2$$
$$\left\{D^{+}\delta\left(\boldsymbol{k}_1 + \boldsymbol{k}_2 - \boldsymbol{k}_3\right)\delta\left(\omega_1 + \omega_2 - \omega_3\right)B(\boldsymbol{k}_2)\left[F(\boldsymbol{k}_1) - F(\boldsymbol{k}_3)\right]\right.$$
$$\left. + D^{-}\delta\left(\boldsymbol{k}_1 - \boldsymbol{k}_2 - \boldsymbol{k}_3\right)\delta\left(\omega_1 - \omega_2 - \omega_3\right)B(\boldsymbol{k}_2)\left[F(\boldsymbol{k}_1) - F(\boldsymbol{k}_3)\right]\right\}. \tag{5.89b}$$

Here the spectral energy is contained predominantly in $B(\boldsymbol{k}_2)$. Now expanding about $\boldsymbol{k}_3$ in the first term in brackets while expanding about $\boldsymbol{k}_1$ in the second term in brackets, and noting that D^{+} and D^{-} are real and symmetric in indices gives,

$$\left.\frac{\partial}{\partial t}F(\boldsymbol{k}_3)\right|_{\text{non-local}} =$$

$$= 2\int \mathrm{d}\boldsymbol{k}_1 \int \mathrm{d}\boldsymbol{k}_2\, D^{+}\delta\left(\boldsymbol{k}_1+\boldsymbol{k}_2-\boldsymbol{k}_3\right)\delta\left(\omega\left(\boldsymbol{k}_3+\left(\boldsymbol{k}_1-\boldsymbol{k}_3\right)\right)\right.$$
$$\left. -\,\omega(\boldsymbol{k}_3)+\omega(\boldsymbol{k}_2)\right) B(\boldsymbol{k}_2)\left[F\left(\boldsymbol{k}_3+\left(\boldsymbol{k}_1-\boldsymbol{k}_3\right)\right)-F(\boldsymbol{k}_3)\right]$$
$$+\,2\int \mathrm{d}\boldsymbol{k}_1 \int \mathrm{d}\boldsymbol{k}_2\, D^{-}\delta\left(\boldsymbol{k}_1-\boldsymbol{k}_2-\boldsymbol{k}_3\right)\delta\left(\omega(\boldsymbol{k}_1)-\omega(\boldsymbol{k}_2)\right.$$
$$\left. -\,\omega\left(\boldsymbol{k}_1+\left(\boldsymbol{k}_3-\boldsymbol{k}_1\right)\right)\right) B(\boldsymbol{k}_2)\left[F(\boldsymbol{k}_1)-F\left(\boldsymbol{k}_1+\left(\boldsymbol{k}_3-\boldsymbol{k}_1\right)\right)\right]$$
$$= 2\int \mathrm{d}\boldsymbol{k}_1 \int \mathrm{d}\boldsymbol{k}_2\, H(\boldsymbol{k})B(\boldsymbol{k}_2)\left(\boldsymbol{k}_1-\boldsymbol{k}_3\right)\cdot\left[\frac{\partial}{\partial \boldsymbol{k}}F(\boldsymbol{k}_1)-\frac{\partial}{\partial \boldsymbol{k}}F(\boldsymbol{k}_3)\right], \quad (5.89c)$$

where,

$$H(\boldsymbol{k}) = D^{+}\delta\left(\boldsymbol{k}_1+\boldsymbol{k}_2-\boldsymbol{k}_3\right)\delta\left(\omega_1+\omega_2-\omega_3\right). \quad (5.89d)$$

Again expanding about $\boldsymbol{k}_3$ finally yields,

$$\frac{\partial}{\partial t}F(\boldsymbol{k}_3) = \frac{\partial}{\partial \boldsymbol{k}_3}\cdot \mathsf{D}_k \cdot \frac{\partial}{\partial \boldsymbol{k}_3}F(\boldsymbol{k}_3). \quad (5.90a)$$

Here the $\boldsymbol{k}$-space diffusion tensor D_k is given by,

$$\mathsf{D}_k = 2\int \mathrm{d}\boldsymbol{k}_1 \int \mathrm{d}\boldsymbol{k}_2\, D^{+}B(\boldsymbol{k}_2)$$
$$\times\left[\left(\boldsymbol{k}_3-\boldsymbol{k}_1\right)\left(\boldsymbol{k}_3-\boldsymbol{k}_1\right)\right]\delta\left(\boldsymbol{k}_1+\boldsymbol{k}_2-\boldsymbol{k}_3\right)\delta\left(\omega_2-\boldsymbol{k}_2\cdot\boldsymbol{v}_{\mathrm{gr}}(\boldsymbol{k}_3)\right), \quad (5.90b)$$

and the same procedure as in Eqs.(5.81a), (5.81b) has been used to simplify the frequency matching condition.

The brief calculation sketched above shows that induced diffusion can be recovered from a systematic expansion of $C\{N\}$, and so is more robust than suggested by the heuristic, 'back-of-the-envelope' argument using the eikonal equations. Clearly, induced diffusion corresponds to adiabatic modulation of short waves by long ones, and so should conserve the total number of wave quanta, as required by the Manley–Rowe relations (Eq.(5.13)). Equation (5.89) clearly *does* satisfy quanta conservation. In view of its foundation in adiabatic theory, it is not surprising then, that induced diffusion can also be derived from mean field theory for the collisionless wave kinetic equation. Treating the refractive term as multiplicative modulation induced by large-scale perturbations, and neglecting $C\{N\}$, the wave kinetic equation is,

$$\frac{\partial N}{\partial t} + v_{\mathrm{gr}} \cdot \nabla N - \frac{\partial}{\partial \boldsymbol{x}} (\tilde{\omega} + \boldsymbol{k} \cdot \tilde{\boldsymbol{v}}) \cdot \frac{\partial N}{\partial \boldsymbol{k}} = 0, \tag{5.91a}$$

so the mean field equation for a spatially homogeneous (or slowly varying) mean population $\langle N \rangle$ is,

$$\frac{\partial \langle N \rangle}{\partial t} = \frac{\partial}{\partial \boldsymbol{k}} \cdot \left\langle \left(\frac{\partial}{\partial \boldsymbol{x}} (\tilde{\omega} + \boldsymbol{k} \cdot \tilde{\boldsymbol{v}}) \right) \tilde{N} \right\rangle. \tag{5.91b}$$

Then, writing,

$$\begin{pmatrix} \tilde{\omega} \\ \tilde{v} \end{pmatrix} = \sum_{q,\Omega} \begin{pmatrix} \tilde{\omega}_q \\ \tilde{v}_q \end{pmatrix} \exp\left(i\left(\boldsymbol{q} \cdot \boldsymbol{x} - \Omega t\right)\right), \tag{5.91c}$$

where $|\boldsymbol{q}| \ll |\boldsymbol{k}|$ and $\Omega \ll \omega$ and computing the linear response $\tilde{N}$ of the population to the modulation field gives,

$$\tilde{N}_{q,\Omega} = \frac{-\boldsymbol{q}\, (\tilde{\omega} + \boldsymbol{k} \cdot \tilde{\boldsymbol{v}})_{q,\Omega}}{\left(\Omega - \boldsymbol{q} \cdot v_{\mathrm{gr}}\right)} \cdot \frac{\partial \langle N \rangle}{\partial \boldsymbol{k}}. \tag{5.92}$$

A simple quasi-linear closure of Eq.(5.91b) finally gives the induced diffusion equation,

$$\frac{\partial \langle N \rangle}{\partial \boldsymbol{k}} = \frac{\partial}{\partial \boldsymbol{k}} \cdot \mathsf{D}_k \cdot \frac{\partial \langle N \rangle}{\partial \boldsymbol{k}}, \tag{5.93a}$$

where the diffusion tensor is,

$$\mathsf{D}_k = \sum_{q} \boldsymbol{q}\boldsymbol{q} \left|(\tilde{\omega} + \boldsymbol{k} \cdot \tilde{\boldsymbol{v}})_q\right|^2 \pi \delta\left(\Omega - \boldsymbol{q} \cdot v_{\mathrm{gr}}(\boldsymbol{k})\right). \tag{5.93b}$$

The correspondence between Eq.(5.93b) and Eq.(5.90b) is obvious. Induced diffusion and its role in the self-consistent description of disparate scale interaction processes will be discussed in much more depth in Chapter 7 and in Volume 2.

5.5.4 Parametric interactions revisited

Parametric–subharmonic interaction is best approached first from the viewpoint of coherent interaction, and then subsequently discussed in the context of wave kinetics. At its roots, parametric–subharmonic interaction occurs due to parametric variation in wave oscillation frequency. Thus for linear internal waves (IWs), where $\omega^2 = k_H^2 N_{BV}^2 / k^2$, a fluid element will oscillate vertically according to,

$$\frac{d^2 z}{dt^2} + \frac{k_H^2}{k^2} N_{BV}^2 z = 0.$$

If parametric variation is induced in the BV frequency (Eq.(5.86b)) at some frequency Ω, so N_{BV}^2 becomes time-dependent, i.e.

$$N_{BV}^2 = \frac{k_H^2}{k^2} N_{BV,0}^2 \left(1 + \delta \cos\left(\Omega t\right)\right),$$

then the motion of the fluid element may exhibit *parametric growth* according to the solution of the Mathieu equation,

$$\frac{d^2 z}{dt^2} + \frac{k_H^2}{k^2} N_{BV,0}^2 \left(1 + \delta \cos\left(\Omega t\right)\right) z = 0.$$

In particular, it is well known that parametric instability will occur for $\Omega \sim 2\omega_{IW} \sim 2\left(k_H^2 N_{BV,0}^2 / k^2\right)^{1/2}$. Of course, this simple argument completely ignores spatial dependence. In the context of IWs (or any other waves), *both* wave number and frequency matching conditions must be satisfied, so that,

$$\boldsymbol{k}_3 = \boldsymbol{k}_1 + \boldsymbol{k}_2,$$

$$\omega_3 = \omega_1 + \omega_2.$$

Now, it is interesting to observe that one can ‘arrange’ a high-frequency, but spatially quasi-homogeneous variation in the BV frequency by a three-wave interaction where $\boldsymbol{k}_1 = -\boldsymbol{k}_2 + \epsilon \boldsymbol{k}$ (here $\epsilon \ll 1$) and $\omega_1 \sim \omega_2$. For this pair of counter-propagating waves with comparable frequencies, we have $\boldsymbol{k}_3 = \boldsymbol{k}_1 + \boldsymbol{k}_2 = \epsilon \boldsymbol{k}$ and $\omega_3 \sim 2\omega$, which is precisely the sought-after situation of spatially uniform parametric variation. Of course, coherent resonant three-wave interactions are reversible, so we can view this traid as one consisting of:

- a ‘pump,’ at $\omega_3 \sim 2\omega$, with $|\boldsymbol{k}_3| \sim O(\epsilon)$;
- two ‘daughter’ waves at $\boldsymbol{k}_1$, ω_1 and $\boldsymbol{k}_2 = -\boldsymbol{k}_1 + \epsilon \boldsymbol{k}$, ω_1.

In this light, we easily recognize the parametric–subharmonic instability as a variant of the decay instability, discussed earlier in Section 5.2. Thus, taking

$a_3 \sim$ const. as the pump, Eqs.(5.16) can immediately be applied, so (noting that here, the coupling coefficient is D),

$$i\omega \frac{\mathrm{d}a_1}{\mathrm{d}t} = D a_2^* a_3, \tag{5.94a}$$

$$i\omega \frac{\mathrm{d}a_2}{\mathrm{d}t} = D a_1^* a_3, \tag{5.94b}$$

so the parametric–subharmonic growth rate is just,

$$\gamma_{\mathrm{PS}}^2 = \frac{D^2}{\omega^2} |a_3|^2 . \tag{5.94c}$$

To see how parametric–subharmonic instability emerges in wave kinetics, it is convenient to take $\boldsymbol{k}_2$ as the pump, so $A(\boldsymbol{k}_2) \gg A(\boldsymbol{k}_1), A(\boldsymbol{k}_3)$, and $|\boldsymbol{k}_2| \ll |\boldsymbol{k}_1|, |\boldsymbol{k}_3|$, $\omega_2 \approx 2\omega_1$. In this limit, we can neglect contributions to the wave kinetic equation for $A(\boldsymbol{k}_2)$, which are proportional to the product $A(\boldsymbol{k}_1)A(\boldsymbol{k}_2)$, just as we neglected terms of $O(a_1, a_2)$ in the coherent equations. This gives,

$$\begin{aligned}\frac{\partial A(\boldsymbol{k}_2)}{\partial t} \cong & -\int \mathrm{d}\boldsymbol{k}_1 \int \mathrm{d}\boldsymbol{k}_3 \, D\delta\left(\boldsymbol{k}_1 + \boldsymbol{k}_3 - \boldsymbol{k}_2\right) \delta\left(\omega_1 + \omega_3 - \omega_2\right) \\ & \times \left[\left(A(\boldsymbol{k}_1) + A(\boldsymbol{k}_3)\right) A(\boldsymbol{k}_2)\right],\end{aligned} \tag{5.95}$$

which describes the depletion in the pump energy due to parametric–subharmonic coupling to $\boldsymbol{k}_1$ and $\boldsymbol{k}_3$. Related expressions for the growth of $A(\boldsymbol{k}_1)$ and $A(\boldsymbol{k}_3)$ are easily obtained from the general expression for $C\{A\}$, given in Eq.(5.87). Energy transfer by parametric–subharmonic interaction will continue until the pump is depleted, i.e. until $A(\boldsymbol{k}_2) \sim A(\boldsymbol{k}_1), A(\boldsymbol{k}_3)$. Equation (5.95) is also consistent with our earlier observation concerning the ratio of the growth rates for stochastic and coherent decay processes. Once again, for parametric–subharmonic (PS) interaction, we have $\gamma_{\mathrm{PS}}^{\mathrm{stoch}}/\gamma_{\mathrm{PS}}^{\mathrm{coh}} \sim \tau_{T_c}\gamma_{\mathrm{E}}^{\mathrm{coh}}$. Thus in wave kinetics, the interaction growth is reduced in proportion to the ratio of the triad coherence time to the coherent growth rate interaction.

Induced diffusion and parametric–subharmonic interaction both involve the interaction of nearly counter-propagating waves with a low $\boldsymbol{k}$ wave. In one case (induced diffusion), the frequencies nearly cancel too, while in the other (parametric–subharmonic) the frequencies add. These triads are shown in Figure 5.16. The third type of non-local interaction, called *elastic scattering*, is complementary to the other two, in that the magnitudes of the interacting $\boldsymbol{k}$s are *comparable* and only the frequencies are *disparate*. For elastic scattering, we consider a triad of $\boldsymbol{k}_1, \boldsymbol{k}_2, \boldsymbol{k}_3$ with:

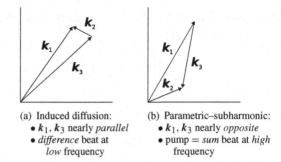

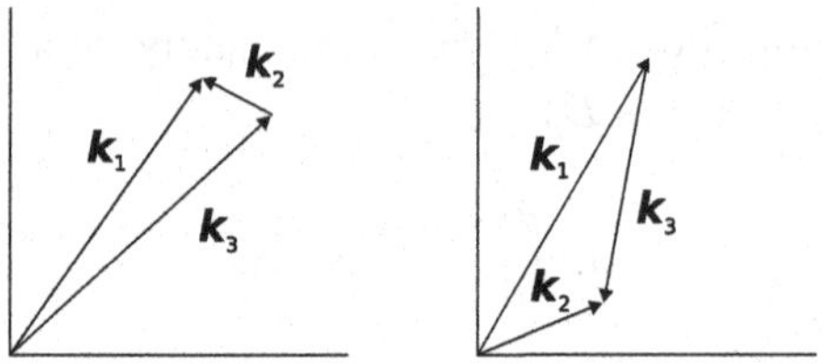

(a) Induced diffusion:
• $\boldsymbol{k}_1$, $\boldsymbol{k}_3$ nearly *parallel*
• *difference* beat at *low* frequency

(b) Parametric–subharmonic:
• $\boldsymbol{k}_1$, $\boldsymbol{k}_3$ nearly *opposite*
• pump = *sum* beat at *high* frequency

Fig. 5.16. Triad structure for non-local interactions of induced diffusion, parametric–subharmonic types.

- $k_{1\mathrm{H}} \sim k_{3\mathrm{H}}$, $k_{1,\mathrm{V}} \sim -k_{3,\mathrm{V}}$ with $k_{1,\mathrm{V}} < 0$;
- $k_{\mathrm{V}_1} \sim 2k_{\mathrm{V}_2}$, so $k_{1,\mathrm{V}} + k_{2,\mathrm{V}} = k_{3,\mathrm{V}}$;
- $\omega_2 < \omega_1, \omega_3$.

The coherent interaction equations for the amplitudes c_1, c_2, c_3 are,

$$i\omega \frac{\mathrm{d}c_1}{\mathrm{d}t} = Dc_2^* c_3, \tag{5.96a}$$

$$i\omega \frac{\mathrm{d}c_2}{\mathrm{d}t} = Dc_1^* c_3, \tag{5.96b}$$

$$i\omega \frac{\mathrm{d}c_3}{\mathrm{d}t} = Dc_1 c_2, \tag{5.96c}$$

so the energies $E_i = \omega_i^2 |c_i|^2$, $i = 1, 2, 3$ evolve according to,

$$\frac{\partial E_1}{\partial t} = -\omega_1 R, \tag{5.97a}$$

$$\frac{\partial E_2}{\partial t} = \omega_2 R, \tag{5.97b}$$

$$\frac{\partial E_3}{\partial t} = \omega_3 R, \tag{5.97c}$$

where $R = -D\,\mathrm{Im}\left\{c_1^* c_2^* c_3\right\}$. Since $\omega_2 \ll \omega_1, \omega_3$, E_2 is nearly constant, as compared to E_1 and E_3, so the essential dynamics are described by Eqs.(5.97a), (5.97c), i.e.

$$\frac{\partial E_1}{\partial t} = -\omega_1 R \quad ; \quad \frac{\partial E_3}{\partial t} = \omega_3 R.$$

Table 5.3. *Summary of non-local internal wave interactions*

Induced diffusion

$$\begin{cases} \Omega_2 \sim \omega_1 - \omega_3 \ll \omega_1, \omega_3 \\ |\boldsymbol{k}_2| \sim |\boldsymbol{k}_1 - \boldsymbol{k}_3| \ll |\boldsymbol{k}_1|, |\boldsymbol{k}_3| \end{cases}$$

slow, stochastic straining by low frequency, large scale diffusion in $\boldsymbol{k}$

Parametric–subharmonic interaction

$$\begin{cases} \Omega_2 \sim \omega_1 + \omega_3 > \omega_1, \omega_3 \\ |\boldsymbol{k}_2| \sim |\boldsymbol{k}_1 + \boldsymbol{k}_3| < |\boldsymbol{k}_1|, |\boldsymbol{k}_3| \end{cases}$$

pumping by high frequency
decay instability

Elastic scattering

$$\begin{cases} |\boldsymbol{k}_1| \sim |\boldsymbol{k}_2| \sim |\boldsymbol{k}_3| \\ \omega_2 < \omega_1, \omega_3 \end{cases}$$

wave packets 1, 3 on static scattering field of 2
elastic scattering, as in Bragg

Thus, we see that in this system, energy is exchanged between modes 1 and 3, moving through the static, low-frequency field of mode 2. The low-frequency scattering field is essentially unaffected by the scattering process – hence the name "elastic scattering." As in the familiar case of Bragg scattering, waves 1 and 3 are back-scattered by the component of the background with half the vertical wavelength of the scattered wave. Elastic scattering is instructive, as it illustrates the rich variety of non-local interactions possible among nonlinearly interacting dispersive waves in anisotropic media. The types of non-local interactions at work in internal wave turbulence are summarized in Table 5.3.

Some advanced topics in wave interactions such as the weak turbulence of filamentary structures (Dyachenko *et al.*, 1992) and limited domains of resonance overlap (Kartashova, 2007) are discussed in the research literatures.

6
Closure theory

In general, commanding a large number is like commanding a few. It is a question of dividing up the numbers. Fighting with a large number is like fighting with a few. It is a question on configuration and design.

(Sun Tzu, "The Art of War")

6.1 Concepts in closure

This chapter discusses closure theory, which is often referred to as strong turbulence theory. The motivation for turbulence theory arises naturally from consideration of weak turbulence theory in the limit where the triad coherence time becomes very long or divergent (i.e. $\tau_{R,k}$ or $\Theta_{\mathbf{k},\mathbf{k}',\mathbf{k}''}$ in Chapter 5 goes to infinity), so that the nonlinear transfer is so large as to predict negative spectra. Of course, strong turbulence theory has its own identity as the theoretical description of high Reynolds number Navier–Stokes turbulence and other problems with strong excitation of many degrees of freedom. This suggests that physics other than wave dispersion or (weak) dissipation, upon which we focused in Chapter 5, is controlling the triad coherence time. Intuition suggests that the physics is *nonlinear scrambling*, which may be thought of as the turbulent mixing of a test mode by advection by the ensemble of turbulence. Such turbulent mixing occurs via an effective eddy viscosity, which is a functional of the turbulent velocity field. The goal of closure theory is to realize this intuition from a systematic, deductive mathematical procedure, in which all approximations are clear *ab initio*.

Closure theory may be modified and approached from at least two directions. One – from a narrow perspective – is as a remedy for the ills of wave kinetics and weak turbulence theory. The other – more general in nature – is as an answer to the question of how to reduce the dimension or number of degrees of freedom of a very complex problem, such as fluid turbulence at high Reynolds number, or phase transition dynamics near criticality. The general class of answers to this question is the set of *renormalization group* (RG) *transformations*. These transformations encompass renormalization and closure, as these constitute the first step of the RG transformation, which then continues with some statement which demands invariance to changing the renormalization point (i.e. the boundary between relevant, or resolved, and irrelevant, or unresolved variables). While RG theory is beyond the scope of this monograph, we do think it important to at least set our discussion of closures in this context. Readers interested in RG theory are directed to the following excellent books (Ma, 1976; Barenblatt, 1979; Collins, 1984; Goldenfeld, 1992), and others which treat this subject in depth. Most RG and closure theories have the common elements of:

(i) a statistical formulation, since information concerning eliminated degrees of freedom is necessarily incomplete;
(ii) a structure where evolution is determined by a competition between effective memory or dressed response, and an effective noise. This structure is *constructed* by obvious analogy with Brownian motion, and is motivated by the necessary statistical formulation of the problem.

Since closure theory deals directly with the construction of the effective response and memory, and of the effective noise, it is a necessary prerequisite for the study of RG theory, and so readers should find this chapter useful for that purpose. Readers who are especially interested in problem reduction should take special note of Section 2 of this chapter, which deals with the Mori–Zwanzig theory, of closure and scale elimination.

At first, concepts in closure theory are introduced briefly. The object of the closure of nonlinear statistical equations is to derive equations for observables of interest. Observables of interest often include response functions (such as the ratio of a quantity to the external force, transport coefficients, time scales of interactions), spectra of fluctuations, or lower-order moments of a spectrum, etc. Note that these quantities correspond to *moments* of the full probability density function (PDF). The latter is of interest but beyond the scope of standard closure theory approaches.

An equation that is closed within a small number of observables can be obtained only by approximations. Thus, the relevance of approximations must be understood before using the closed set of equations. Understanding is composed of two aspects. One is physics insight that lets the researcher employ a particular approximation, driven by the objective of solving a problem. The other is concerned with mathematical structure and pinpoints where and what approximations are introduced.

In the first part of this chapter (Section 6.1), we highlight the physics elements, which are introduced in the closure theories. In the later parts of this chapter (Sections 6.2 to 6.5), the systematology of some of the prototypical closure models is explained. An explanation is made in Section 6.5 on the deduction of a fluid moment equation for kinetic dynamics in the case of rare collisions. This is another example of the reduction of variables. At the end, a brief discussion is given on the outlook.

6.1.1 Issues in closure theory

Let us take, as an example, a nonlinear equation for a variable v with quadratic nonlinearity (three-wave coupling), which is written in a Fourier space, as,

$$\frac{\partial}{\partial t} v_k + i\omega_k v_k + \frac{1}{2} \sum_{k'} N_{k,k'} v_{-k'} v_{k+k'} = F_k^{\text{ext}},$$

where ω_k, $N_{k,k'}$ and F_k^{ext} represent the linear dispersion relation, nonlinear coupling coefficient and the external forcing, respectively. Multiplying v_{-k} by this equation, we have,

$$\frac{\partial}{\partial t} |v_k|^2 + \text{Re} \sum_{k'} N_{k,k'} v_{-k} v_{-k'} v_{k+k'} = 2\text{Re} \left(v_{-k} F_k^{\text{ext}} \right).$$

Noting the fact that the external forcing includes the random excitation (which is controlled externally), one takes a statistical average,

$$\frac{\partial}{\partial t} \left\langle |v_k|^2 \right\rangle + \text{Re} \sum_{k'} \left\langle N_{k,k'} v_{-k} v_{-k'} v_{k+k'} \right\rangle = 2\text{Re} \left\langle v_{-k} F_k^{\text{ext}} \right\rangle.$$

The equation for the quadratic moment (spectrum) is governed by the third-order moments. The third-order moments are controlled by higher-order moments, e.g.,

$$\frac{\partial}{\partial t} \left\langle v_{-k'} v_{-k} v_{k+k'} \right\rangle \sim - \sum_{k''} \left\langle v_{-k'} v_{-k} N_{k+k',k''} v_{-k''} v_{k+k'+k''} \right\rangle$$
$$+ \text{ symmetric terms.}$$

Thus, a chain of coupled equations for moments is deduced. The coupled moment equations have their origin in the original equation, and one must somehow close the hierarchy.

The essence in the modelling of the closure theory is not to truncate the higher-order moments, but to approximately evaluate the fourth-order moments by lower-order moments, e.g.

$$\langle vvNNvv \rangle \rightarrow \tau_c \langle vv \rangle \langle vv \rangle .$$

This symbolic expression illustrates the following issues: How are the fourth-order moments decoupled as products of quadratic moments? What is the relevant time scale that limits the interaction between the three waves that couple nonlinearly via the quadratic nonlinearity?

Of course, the molecular viscosity gives rise to a molecular dissipation rate. However, this rate is very slow in the high temperature plasmas and invicid fluids, which gives a long triad coherence time. This, in turn, results in unpredictably strong physical, nonlinear interactions and possibly negative spectra. Thus, some nonlinear timescale must be deduced to limit triad coherence so as to maintain realizability, i.e. to obtain results which fulfill our expectations for a physically meaningful system. In particular, the triad coherence time must be regulated so as to avoid negative spectra. We discuss realizability in more detail at the end of this section. In addition, there are various candidate time scales for the modal interactions. The list of candidates includes:

(t1) νk^2: rate of damping by the molecular viscosity ν;
(t2) γ_{NL}: nonlinear damping rate (the rate at which energy is taken from the mode);
(t3) $\left|\left(\frac{\omega}{k} - \frac{\partial\omega}{\partial k}\right)\Delta k\right|$: the auto-correlation rate of resonant particle and wave, (which is relevant for strong kinetic turbulence in phase space);
(t4) $|\Delta\omega_{\mathrm{MM}}|$: auto-correlation rate of wave-wave interaction that is set by the mis-match (MM) of the frequency of the beat component from the dispersion, $\Delta\omega_{\mathrm{MM}} = \omega_{k+k'} - \omega_k - \omega_{k'}$;
(t5) $\Delta\omega_k$: nonlinear scrambling rate (the nonlinear Doppler shift, e.g. due to the convective nonlinearity).

Illustrations of the auto-correlation rate of resonant particles and the frequency mismatch are given in Figure 3.8. As noted in Chapter 5, $\left|\omega_{k+k'} - \omega_k - \omega_{k'}\right| \sim \left|\left(\mathrm{d}\omega_{k+k'}/\mathrm{d}k' - \mathrm{d}\omega_k/\mathrm{d}k\right) \cdot \Delta k'\right| \left|\Delta k \cdot \frac{d^2\omega_k}{\mathrm{d}k\mathrm{d}k} \cdot \Delta k\right|$, and so $\Delta\omega_{\mathrm{MM}}$ is very sensitive to the structure of the wave dispersion relation. This effect is particularly important in problems of wave turbulence. Note that $\Delta\omega_{\mathrm{MM}}$ is the rate at which two rays separated by $\Delta k'$ disperse in space. Careful consideration of the physics and a systematic procedure are necessary to identify which time scale is chosen to close the higher-order moments in terms of lower-order moments. For example, weak turbulence theory (which is explained in, e.g. Chapters 3 and 4) assumes that (t4) is much smaller than (t2) and (t5) (thus only a set of linear modes are treated). In addition, (t3) is assumed larger than (t2), (t5) and the rate of evolution of the mean plasma variables in calculating wave–particle interactions. The ansatz may work in the cases where matching conditions for linear eigenmodes are satisfied and where linear growth rates (or linear damping rates) are large. Thus, weak turbulence theory has shown some success in understanding plasma turbulence. For broader circumstances, such as strong turbulence discussed in Chapter 4, more advanced methodology is necessary.

As is explained above, key issues are the consequences of truncation, decomposition of higher-order moments, extraction of an appropriate time scale of nonlinear interaction, satisfying the conservation laws which the original equations obey and realizability. These issues are illustrated in the following example. First, the problem of truncation is explained. Because effecting the statistical average requires a long-time average, a procedure to treat the long-time behaviour is described in the context of the stochastic oscillator. Then, the process of extracting the appropriate time scale of interaction is explained. Although the original equation is Markovian, the elimination of irrelevant variables (i.e. the degrees of freedom which we are not interested in) introduces the non-Markovian property in the closed equations. In this procedure, the concepts of the test mode and renormalization are explained. The nonlinear term is divided into the 'coherent' term which determines the nonlinear interaction and the residual part. Then the residual part is approximated in such a way that the equation for the spectral function which we construct satisfies the energy conservation relation. The illustration of the method for extraction of the nonlinear interaction time (nonlinear decorrelation time) and for the formation of the spectral equation is given by the example of the Driven Burgers/KPZ Equation. A short, heuristic explanation is given after this.

6.1.2 Illustration: the random oscillator

An essential motivation in the theory of closure is the modelling of the long-time behaviour of correlations in the presence of randomly varying fluctuations. This requires a method to capture long-time behaviour. A simple method, such as a perturbation in a time series or a truncation of moments, fails for such a purpose. The issue is illustrated by employing an example of a (single) random oscillator.

A random oscillator obeys an equation,

$$\frac{\mathrm{d}q}{\mathrm{d}t} + ibq = 0, \tag{6.1}$$

where $q(t)$ denotes the displacement of the oscillator (either the velocity or spatial location) and b stands for a random frequency variable (constant in time) which is prescribed by the probability density function (PDF) $P(b)$. (The symbol b is used here in analogy to the Brownian motion.) This is a linear equation with multiplicative noise. This equation is an idealized limit of the turbulence problem, which is seen as follows. In the dynamical equation with convective nonlinearity, $\boldsymbol{v} \cdot \nabla \boldsymbol{v}$, a test mode (with frequency ω_0) is subject to a Doppler shift by other fluctuations, ω_{NL},

$$\frac{\partial \boldsymbol{v}}{\partial t} + i\omega_0 \boldsymbol{v} + i\omega_{\mathrm{NL}} \boldsymbol{v} + \mathrm{Res} = 0 \tag{6.2}$$

where Res indicates the residual impact of nonlinear interactions. Owing to the complicated evolution of a large number of interacting modes, ω_{NL} appears as a

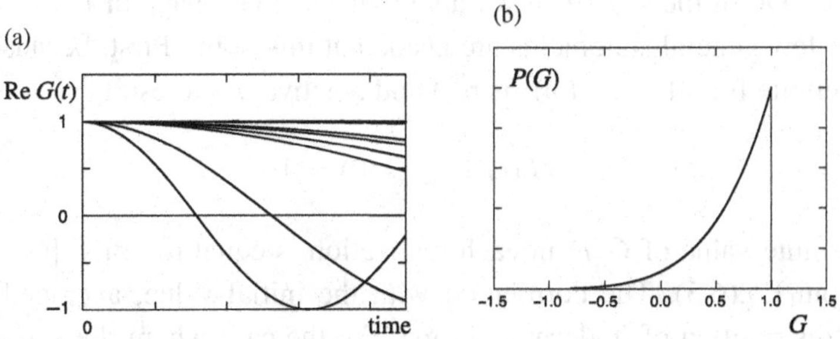

Fig. 6.1. Illustrations of a few realizations $G(t)$, (a), and $P(G(t))$ at some time slice, (b).

rapidly changing statistical variable. When one is interested in the influence of larger-scale perturbations on the test mode, this effect ω_{NL} is idealized as a statistical variable which is constant in time. Separating the unperturbed frequency as $\boldsymbol{v}(t) = q(t) \exp(-i\omega_0 t)\, \boldsymbol{v}(0)$ (and neglecting the residual force), the random oscillator equation is recovered. Thus, the random oscillator represents one of the prototypical processes in turbulence.

The long-time behaviour is described by the statistical average of the response function $G(t)$,

$$\frac{\mathrm{d}G}{\mathrm{d}t} + ibG = 0, \text{ with } G(0) = 1. \tag{6.3}$$

The statistical average of $G(t)$ is given by,

$$\langle G(t) \rangle \equiv \int G(t)\, P(b)\, \mathrm{d}b. \tag{6.4}$$

In this model case, Eq.(6.3) is solved as,

$$G(t) = \exp(-ibt). \tag{6.5}$$

Figure 6.1 illustrates the time evolution of $G(t)$ in various realizations, together with $P(b)$ and the PDF of $G(t)$ at a certain time slice.

Equation (6.5) shows that the Fourier transform of $\langle G(t) \rangle$ is given by that of $P(b)$,

$$\begin{aligned} \langle G(\omega) \rangle &= \left\langle \frac{1}{2\pi} \int G(t)\, \mathrm{e}^{i\omega t} \mathrm{d}t \right\rangle \\ &= \frac{1}{2\pi} \int_{-\infty}^{\infty} \mathrm{d}t \int_{-\infty}^{\infty} \mathrm{e}^{i(\omega - b)t} P(b)\, \mathrm{d}b = P(\omega). \end{aligned} \tag{6.6}$$

That is, the PDF of the statistical variable b gives the propagator in the frequency domain. A few general statements are needed at this point. First, because $P(b)$ is positive definite for all b, $\langle G(\omega)\rangle$ is real and positive. As a result,

$$|\langle G(t)\rangle| \le G(0) = 1. \tag{6.7}$$

The absolute value of $G(t)$ in each realization is equal to unity, $|G(t)| = 1$, as is seen from Eq.(6.5). The correlation with the initial value, averaged over the statistical distribution of b, decays in time. For the case where the variable b has Gaussian statistics,

$$P(b) = \frac{1}{\sigma\sqrt{\pi}} \exp\left(-\frac{b^2}{\sigma^2}\right), \tag{6.8}$$

where $\sigma^2 = 2\langle b^2\rangle$, one has the statistical average

$$\langle G(t)\rangle = \exp\left(-\frac{1}{2}\left\langle b^2\right\rangle t^2\right). \tag{6.9}$$

In this case, memory of the initial condition is lost in a time of the order $\langle b^2\rangle^{-1/2}$.

In real problems, the PDF $P(b)$ is not known, but is specified by some of the series of moments, $\langle b^n\rangle$,

$$\left\langle b^n\right\rangle = \int_{-\infty}^{\infty} \mathrm{d}b P(b)\, b^n. \tag{6.10}$$

The task that one usually faces is to construct approximate (but relevant, we hope) solutions using the given moments $\langle b^n\rangle$. Standard methods are:

(i) naive perturbation theory;
(ii) truncation of higher-order moments.

It is useful to see how these methods succeed and how they fail.

In naive perturbation theory, one may expand the solution in time series in the vicinity of $t = 0$. One has a Taylor expansion of the solution as,

$$G(t) = 1 - ibt - \frac{b^2}{2}t^2 + \cdots \tag{6.11}$$

Figure 6.2(a) shows the approximation by use of the truncation of time series (dashed line) and the exact solution of $\langle G(t)\rangle$. If one continues the expansion, one obtains the result,

$$\langle G(t)\rangle = 1 + \sum_{n=1}^{\infty} (-i)^n \frac{\langle b^n\rangle}{n!} t^n. \tag{6.12}$$

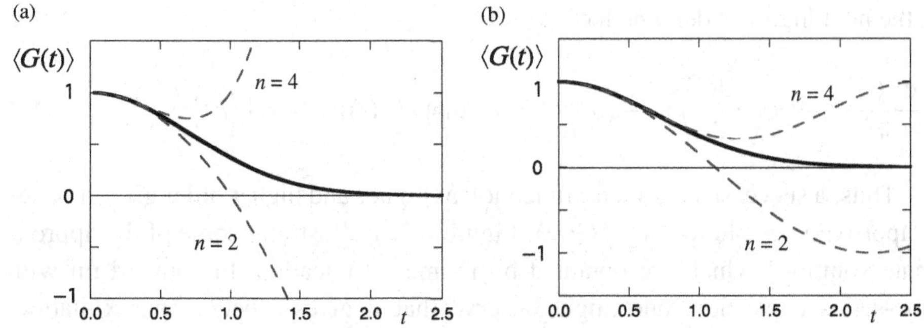

Fig. 6.2. Comparison of $\langle G(t)\rangle$ (solid line) and its truncated approximation (dashed line) for the case of time-series expansion (a). Approximation by use of the moment truncation is shown in (b) by the dashed lines (for the case of $\langle b^2\rangle = 2$).

This result provides $\langle G(t)\rangle = \exp\left(-\frac{1}{2}\langle b^2\rangle t^2\right)$ for the case of Gaussian statistics, i.e. $\langle b^n\rangle = 0$ for odd-n and $\langle b^n\rangle = \left(\langle b^2\rangle/2\right)^n 2n!/n!$. The perturbation expansion, if performed to all orders, may reproduce the exact solution. The expansion form Eq.(6.12) is absolutely convergent for all t (in this example), but convergence is slow at long times, and requires more and more terms at longer times. If any *truncation* is employed, convergence ultimately fails. The approximation with any truncation at finite-n leads to $|\langle G(t)\rangle| \to \infty$ at $t \to \infty$. So, for no finite truncation, we obtain a uniformly valid approximation for $G(t)$. Hence the statistical evaluation which is given by the long-time average cannot be obtained from the truncation of the series expansion.

In the approach where moments are discarded in some way (e.g. the random phase approximation, RPA), one starts from a hierarchy of coupled equations for Eq.(6.3) as,

$$\frac{\mathrm{d}\langle G\rangle}{\mathrm{d}t} = -i\langle bG\rangle \tag{6.13a}$$

$$\frac{\mathrm{d}\langle bG\rangle}{\mathrm{d}t} = -i\left\langle b^2 G\right\rangle, \tag{6.13b}$$

and so on. This moment hierarchy Eq.(6.13) is treated with the help of modelling of the moments $\langle b^n\rangle$. For (the most ideal) case of Gaussian statistics, one can simplify, via $\langle bG\rangle = \langle b\rangle\langle G\rangle = 0$, $\left\langle b^2 G\right\rangle = \left\langle b^2\right\rangle\langle G\rangle$, $\left\langle b^3 G\right\rangle = \left\langle b^2\right\rangle\langle bG\rangle = 0$, $\left\langle b^4 G\right\rangle = 3\left\langle b^2\right\rangle^2\langle G\rangle$, etc., so as to obtain the series of equations. In lowest order,

$$\frac{\mathrm{d}\langle G\rangle}{\mathrm{d}t} = -i\langle bG\rangle = 0, \text{ so that } \langle G\rangle = 1. \tag{6.14a}$$

In the next higher order, one has,

$$\frac{\mathrm{d}^2 \langle G \rangle}{\mathrm{d}t^2} = -i \left\langle b \frac{\mathrm{d}G}{\mathrm{d}t} \right\rangle = -\left\langle b^2 \right\rangle \langle G \rangle, \text{ so that } \langle G(t) \rangle = \cos\left(\left\langle b^2 \right\rangle^{1/2} t \right), \quad (6.14b)$$

etc. Thus, a successive moment truncation at higher and higher order gives a series of approximate solutions for $\langle G(t) \rangle$. Figure 6.2(b) illustrates some of the approximate solutions which are obtained by moment truncation. In comparison with time-series expansion, one might observe that it gives a better approximation, because the condition $|\langle G(t) \rangle| \le 1$ is satisfied. However, the approximate form of $|\langle G(t) \rangle|$ given in this method does not vanish in the limit of $t \to \infty$. Another observation on this point is that the approximation implies $\langle G(\omega) \rangle$ is a sum of delta-functions (i.e. *not* a smooth function), while the true solution of $\langle G(\omega) \rangle$ must be smooth, since it follows from averaging.

From these observations, we see that both perturbation theory and the moment truncation approach are *inadequate* for predicting statistical properties, even for a simple random oscillator problem. Both do not reproduce key elements (e.g. the necessary conditions $|\langle G(t) \rangle| \le 1$, or $\langle G(t) \rangle \to 0$ at $t \to \infty$). Therefore, one should explore alternative approaches.

6.1.3 Illustration by use of the driven-Burgers/KPZ equation (1)

6.1.3.1 Model

Next, the issue of real-space advection due to nonlinearity is illuminated. This process can induce turbulent diffusivity. The driven-Burgers equation is a case study which illustrates one-dimensional, nonlinear advective dynamics (in the (x, t)-coordinates). This is a very special case for the study of turbulence, because the one-dimensional problem cannot describe (nonlinear) deformations of vortices and waves such as stretching, shearing and tearing by background fluctuations. Nevertheless, this system of nonlinear equations highlights a nonlinear mechanism that generates smaller scales. At a very small scale, where molecular viscosity becomes important, fluctuation energy is dissipated. Therefore, enhanced energy dissipation rate due to nonlinearity is described here. That is set by nonlinearity and dissipation, the interplay of which is a key element in turbulence. Thus, this simple model is useful in considering the problem of closures. The driven-Burgers equation takes a form,

$$\frac{\partial v}{\partial t} + v \frac{\partial v}{\partial x} - \nu \frac{\partial^2 v}{\partial x^2} = F^{\mathrm{ext}} \quad (6.15)$$

where v is the velocity in the x-direction, ν is the collisional viscosity (molecular viscosity) and F^{ext} is the external (stochastic) forcing.

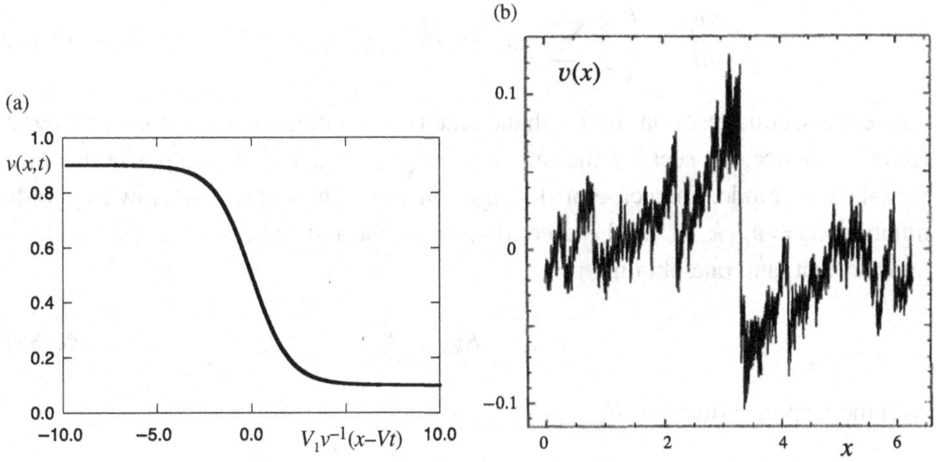

Fig. 6.3. A shock-type solution of the Burgers equation (a). An example of snapshot illustration of $v(x,t)$ for the random-force driven-Burgers equation (b) (see Chekhlov and Yakhot (1995)).

The determination of the response function, i.e. the statistical evaluation of the response of v to the external forcing F^{ext}, is a major goal of this study. For this, a naïve perturbative approach easily fails. For instance, the shock formation in the Burgers equation for small value of viscosity ν, is not captured in a perturbation expansion which is truncated at a finite order. In the absence of external forcing, the Burgers equation has solutions of the shock type, as is shown in Figure 6.3(a),

$$v(x,t) = V + V_1 \frac{1 - \exp(\alpha(x - Vt))}{1 + \exp(\alpha(x - Vt))}, \tag{6.16}$$

where the parameter which determines the sharpness of the shock is given by $\alpha = V_1/\nu$. The existence of shock solutions to the Burgers equation is a consequence of the exact solvability of that equation using the Hopf–Cole transformation. In the presence of external forcing, a large number of steep gradients (the sharpness of which depend on the height) are generated at various locations and move with speeds proportional to the size. An example is illustrated in Figure 6.3(b). The response of a particular mode (spatio-temporal structure), in the presence of background fluctuations is studied in the discussion.

6.1.3.2 Renormalized memory function

The response function is most conveniently evaluated by use of the Fourier representation of Eq.(6.15),

$$\frac{\partial v_k}{\partial t} + \left(\frac{ik}{2} \sum_{k'} v_{-k'} v_{k+k'} \right) + \nu k^2 v_k = F_k^{\text{ext}}, \tag{6.17}$$

where the second term on the left-hand side (i.e. the nonlinearity) is symmetrized for convenience. By seeking the ratio $\delta v_k / \delta F_k^{\text{ext}}$, we obtain the response function for k-Fourier mode. For a case of slow motion and high viscosity, i.e. low Reynolds number, $R_e = v/\nu k \ll 1$, the second term on the left-hand side of Eq.(6.17) is unimportant, and one obtains,

$$v_{k,\omega} = R_{k,\omega} F_{k,\omega}^{\text{ext}}, \tag{6.18a}$$

with the response function $R_{k,\omega}$ as,

$$R_{k,\omega} = \frac{1}{-i\omega + \nu k^2}. \tag{6.18b}$$

In this formula, the time scale for the coupling is set by the viscosity. The result of Eq.(6.18b) cannot be straightforwardly applied to general cases with a finite value of R_e. A perturbative calculation of the effect of the nonlinearity does not work. For instance, if one substitutes Eq.(6.18) into the second term on the left-hand side of Eq.(6.17), one may derive a next-order 'correction' by the nonlinearity as,

$$v_{k,\omega} = R_{k,\omega} F_{k,\omega} - \frac{ik}{2} \sum_{k',\omega'} R_{k,\omega} R_{-k',-\omega'} R_{k+k',\omega+\omega'} F_{-k',-\omega'} F_{k+k',\omega+\omega'} + \cdots .$$

It eventually gives an infinitely large response in the limit of $R_e \gg 1$. Instead, one must *extract* effective time scale from the nonlinearity. The physics of the time scale to be determined is the nonlinear scrambling/coupling, which is derived from the nonlinear response. This nonlinear scrambling process may be thought of as mixing by an effective eddy viscosity, resulting from nonlinear mode coupling.

Extraction of the effective time scale may be performed by an implicit procedure. Assume there exists an effective time scale that limits, through nonlinearity, the coupling between F_k^{ext} and v_k. Then we may write Eq.(6.17) symbolically as,

$$\frac{\partial v_k}{\partial t} + \Gamma_k v_k + \text{Res} + \nu k^2 v_k = F_k^{\text{ext}}, \tag{6.19}$$

where the nonlinear term is represented by an effective damping rate Γ_k, which limits the response of v_k to F_k^{ext} in the limit of $R_e \gg 1$, and the other nonlinear interactions (which is denoted Res on the left-hand side of Eq.(6.19)). That is, the response of the test mode is sought by considering its interaction with the rest of the turbulence. In this subsection, the role of effective damping is illustrated, and the influence of the residual term (Res), related to the energy conservation in closure modelling, is explained in the next subsection. By this focus, the response function is now given from Eq.(6.19) as,

$$R_{k,\omega} = \frac{1}{-i\omega + \nu k^2 + \Gamma_{k,\omega}}. \tag{6.20}$$

Of course, the term $\Gamma_k v_k$ reflects the physics of the nonlinear term $\frac{ik}{2}\sum_{k'} v_{-k'}v_{k+k'}$. The term $\Gamma_k v_k$ is phase-coherent with respect to v_k, so that, in the lowest order, Γ_k is proportional to amplitude but phase independent. That is, the term $\Gamma_k v_k$ comes from a particular combination in $v_{-k'}v_{k+k'}$, which is induced by the nonlinear coupling between v_k and other modes (we call $v_{k'}^{(c)}$ the mode driven by the direct *beat* between v_k and $v_{-k+k'}$),

$$\Gamma_{k,\omega}v_{k,\omega} = \frac{ik}{2}\sum_{k',\omega'}\left(v_{-k'}^{(c)}v_{k+k'} + v_{-k'}v_{k+k'}^{(c)}\right) = ik\sum_{k',\omega'} v_{-k',-\omega'}v_{k+k',\omega+\omega'}^{(c)}. \tag{6.21}$$

The direct beat mode $v_{k+k'}^{(c)}$ is generated by the test mode and background fluctuations, i.e. so that the nonlinear term, including v_k, is separated from others as,

$$\frac{\partial v_{k+k'}^{(c)}}{\partial t} + \left(\frac{i\left(k+k'\right)}{2}\sum_{k''\neq -k} v_{-k''}v_{k+k'+k''}\right) + \nu\left(k+k'\right)^2 v_{k+k'}^{(c)}$$
$$= -i\left(k+k'\right)v_k v_{k'}.$$

We introduce here an approximation that the second term on the left-hand side of this equation has the same role as that in Eq.(6.17) (with the replacement $k \to (k + k')$), although one term (the right-hand side) is extracted from the original nonlinear terms. This approximation is based on the idea that one term (the right-hand side) constitutes for only a small portion of the total nonlinear effect if a large number of Fourier modes are excited, so that isolating it does *not* change the character of the response. This ansatz is often referred to as the "test wave hypothesis". Just as the test particle hypothesis (discussed in Chapter 2), which is based on the idea that we can treat the response to one test particle as identical to the total plasma response (i.e. so isolating one particle makes no difference), the test wave hypothesis is based on the idea that isolating one test wave leaves the net response of the system unchanged. Both rely on the idea that since $N \gg 1$ (where N is the number of particles or modes, respectively), the dynamics of $N - 1$ elements will be identical to that of N elements. *Both* ignore the possibility of strong phase correlations among subsets of interacting degrees of freedom. With this ansatz, the direct beat mode $v^{(c)}_{k+k'}$ is also expressed by use of the screened response function (6.20) and the source $-i\left(k+k'\right)v_{k',\omega'}v_{k,\omega}$, so,

$$v^{(c)}_{k+k',\omega+\omega'}=-i\left(k+k'\right)R_{k+k',\omega+\omega'}v_{-k',-\omega'}v_{k,\omega}. \tag{6.22}$$

Then, self-consistency requires,

$$\Gamma_{k,\omega}v_{k,\omega}=\left(k^2\sum_{k',\omega'}\left|v_{k',\omega'}\right|^2\left(1+\frac{k'}{k}\right)R_{k+k',\omega+\omega'}\right)v_{k,\omega}. \tag{6.23}$$

For a stationary value of v_k, the effective damping rate (extracted from the nonlinearity) must satisfy the relation,

$$\Gamma_{k,\omega}=\nu_{t,k}k^2\equiv k^2\sum_{k',\omega'}\left|v_{k',\omega'}\right|^2\left(1+\frac{k'}{k}\right)R_{k+k',\omega+\omega'}. \tag{6.24}$$

Here $\nu_{t,k}$ indicates the renormalized turbulent viscosity, which represents the turbulent scrambling rate of the test mode by background fluctuations. It is recursively defined, as,

$$\Gamma_{k,\omega}=\sum_{k',\omega'}\frac{\left|v_{k',\omega'}\right|^2k\left(k+k'\right)}{-i\left(\omega+\ \omega'\right)+\nu(k+k')^2+\Gamma_{k+k',\omega+\omega'}}, \tag{6.25}$$

as is seen by combining Eq.(6.20) and Eq.(6.24). This is because nonlinear scrambling, represented by $\Gamma_{k+k',\omega+\omega'}$, limits the time history of the propagator within $\Gamma_{k,\omega}$.

6.1.3.3 Non-Markovian property and nonlocal interaction

The effective damping rate $\Gamma_{k,\omega}$ includes dependence on both the wave number k and frequency ω. That is, if Eq.(6.19) is rewritten in terms of real space and time variables $v(x,t)$, one has the relation,

$$\Gamma_{k,\omega}v_{k,\omega}\rightarrow\int\mathrm{d}x'\int\mathrm{d}t'K\left(x-x';t-t'\right)v\left(x',t'\right). \tag{6.26}$$

The convolution kernel $K\left(x-x';t-t'\right)$ represents nonlocal interaction in space and the memory effect. We see that the causality in the response function determines that the time integral over t' comes from the 'past', i.e., $t'<t$. The nonlocal kernel $K\left(x-x';t-t'\right)$ may be rewritten as follows.

Let us write, after Eq.(6.24), $\Gamma_{k,\omega}=k^2\nu_t\left(k,\omega\right)$, where the dependence of $\nu_t(k,\omega)$ on (k,ω) indicates the deviation from a simple diffusion process. (If ν_t

is a constant, it is interpreted as a diffusion process.) The coherent damping force in real space is given as,

$$\Gamma(x,t)\,v(x,t)=\int \mathrm{d}k \int \mathrm{d}\omega \Gamma_{k,\omega} v_{k,\omega} \exp(ikx-i\omega t)\,, \tag{6.27a}$$

so that,

$$\Gamma(x,t)\,v(x,t)=\frac{1}{4\pi^2}\int \mathrm{d}x' \int \mathrm{d}t' \int \mathrm{d}k \int \mathrm{d}\omega \times k^2 \nu_t(k,\omega)\exp\left(ik\left(x-x'\right)-i\omega\left(t-t'\right)\right) v\left(x',t'\right). \tag{6.27b}$$

Noting that $k^2 \exp\left[ik\left(x-x'\right)-i\omega\left(t-t'\right)\right]=-\partial^2/\partial x'^2 \exp\left[ik\left(x-x'\right)-i\omega\left(t-t'\right)\right]$, and performing the partial integral over x' twice, one has the coherent drag force in non-Markovian form,

$$\Gamma(x,t)\,v(x,t)=-\iint \mathrm{d}x'\mathrm{d}t' K\left(x-x';t-t'\right)\nabla^2_{x'} v\left(x',t'\right), \tag{6.28}$$

with the nonlocal interaction kernel,

$$K\left(x-x';t-t'\right)=\frac{1}{4\pi^2}\iint \mathrm{d}k\mathrm{d}\omega \nu_t(k,\omega)\exp\left(ik\left(x-x'\right)-i\omega\left(t-t'\right)\right). \tag{6.29}$$

This result shows that the dependence of $\nu_t(k,\omega)$ on (k,ω) dictates the range of nonlocal and non-Markovian interactions. If $\nu_t(k,\omega)$ is a constant – i.e. is white – $K\left(x-x';t-t'\right)$ reduces to a delta-function and the local diffusion limit is recovered. While the wave-number-dependence of $\nu_t(k,\omega)$ shows that it is reduced for $|k|>k_0$, $K\left(x-x';t-t'\right)$ remains finite in the range of $\left|x-x'\right|<k_0^{-1}$. Figure 6.4 illustrates the nonlocal kernel.

The convolution kernel is finite in time with a scale which is determined by the nonlinear decorrelation process. We stress again that the non-Markovian property

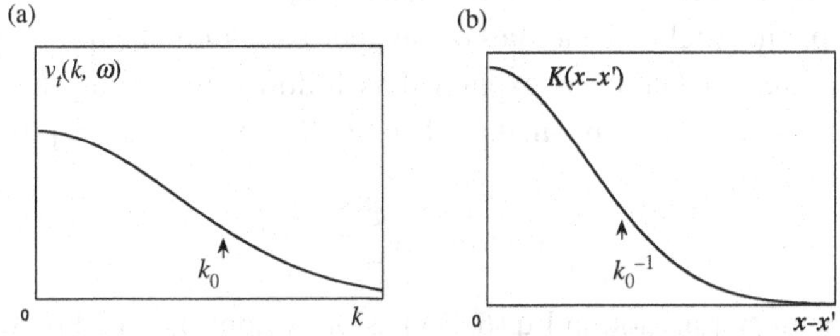

Fig. 6.4. When the response $\nu_t(k,\omega)$ is limited in Fourier space, a nonlocal interaction appears in the coherent drag force.

(Eq.(6.26)) arises, even though the original equation is Markovian, since the irrelevant variables (i.e. the variables which are not of interest and so are averaged) are eliminated. For the same value of a relevant variable of interest at any given time, the eliminated degrees of freedom may have a variety of values, so that statistical averaging is necessary for meaningful elimination. Thus, the time evolution of the relevant variable of interest cannot be determined using only the values of the variables of interest at that time. Equivalently, the memory of the relevant variable reflects the history of the dynamics of the eliminated variables. As a result of this, the future evolution of a variable is determined by *both* its instantaneous value as well as its memory. It is interesting to relate the discussion of closure to that of the memory function, so we now turn to explicating that connection.

The appearance of the memory or non-Markovian character due to the elimination of irrelevant variables can be illustrated by a simple example. Take the case of coupled oscillators,

$$\frac{\mathrm{d}}{\mathrm{d}t}x + i\omega x = -\varepsilon y, \tag{6.30a}$$

$$\frac{\mathrm{d}}{\mathrm{d}t}y + \gamma y = x. \tag{6.30b}$$

In this model, an oscillator $x(t)$ is chosen as observable, and couples with the damped motion $y(t)$ (which is chosen as an unobserved degree of freedom). The driven damped motion $y(t)$ is expressed in terms of observable variable $x(t)$ as $y(t) = \int_{-\infty}^{t} \mathrm{d}t' \exp\left(-\gamma t + \gamma t'\right) x\left(t'\right)$, so that the dynamical equation with elimination of y is given from Eq.(6.30a) as,

$$\frac{\mathrm{d}}{\mathrm{d}t}x + i\omega x + \varepsilon \int_0^\infty \mathrm{d}\tau \exp\left(-\gamma\tau\right) x\left(t-\tau\right) = 0. \tag{6.31}$$

Here the integration variable is redefined as $t' = t - \tau$. Equation (6.31) in this case takes a form in which the memory function is a convolution in time. The dynamic equation for $x(t)$ is now non-Markovian. This is an illustration that reduction of an observed variable causes non-Markovian nature.

At this step, one might take a Markovian approximation of Eq.(6.31). The basis of Markovian approximation is explained as follows. Assuming that the rate of temporal variation of $x(t)$ is smaller than γ, $|\partial x/\partial t| \ll \gamma x$, Eq.(6.31) can be Markovian made as,

$$\frac{\mathrm{d}}{\mathrm{d}t}x + i\omega x + \frac{\varepsilon}{\gamma}x = 0, \tag{6.32}$$

so that the memory function in Eq.(6.31) is simply approximated by the damping rate ϵ/γ. The validity of this Markovian approximation is supported by the primitive equations (6.30a) and (6.30b). From Eq.(6.30b), we see that the approximate solution, $y \simeq x/\gamma$, is valid as long as the temporal evolution of the variable x is slow, i.e. $|\omega| \ll \gamma$. This is an adiabatic response of the variable y; i.e. the variable y reaches equilibrium solution during dynamic evolution of the observed (relevant) variable x. This analogy is extended to the case where more than one degree of freedom is unobservable (i.e. irrelevant). A more systematic explanation is given in Section 6.2.

6.1.3.4 Limit of an effective transport coefficient

The recurrent form of Eq.(6.25) is a generalization of the quasi-linear formula of the turbulent transport coefficient. In the case that the test mode has much longer wavelength and oscillation period than background fluctuations, $k \ll k'$, and $\omega \ll \omega'$, $\nu_{t,k',\omega'}$ reduces to the turbulent transport coefficient,

$$\nu_{t,k} \rightarrow \nu_t \cong \sum_{k',\omega'} \left|v_{k',\omega'}\right|^2 R_{k',\omega'} = \sum_{k',\omega'} \left|v_{k',\omega'}\right|^2 \frac{\nu_{t,k',\omega'} k'^2}{\omega'^2 + \left(\nu_{t,k',\omega'} k'^2\right)^2}. \tag{6.33}$$

(Note that the sum of terms in proportion to kk' vanishes, in the limit of $k \ll k'$, because $\left|v_{k',\omega'}\right|^2$ is an even function of k'.) In such a case (i.e. the test mode varies slowly in space and time, while background fluctuations change very rapidly), the interaction with background fluctuations appears as a process of random kicks without memory, as in Brownian motion. (In the opposite limit, where the test mode changes rapidly in space and time, while the background fluctuations change slowly but with random phases, the model of a random oscillator, as is explained in the preceding subsection, illustrates the response.) It is important to note that, in the formula of ν_t, the term $\nu_{t,k',\omega'}$ appears as renormalized. Therefore, we still need a method to calculate $\nu_{t,k',\omega'}$ in a consistent manner. The term $\nu_{t,k',\omega'}$ is governed by the mutual interactions between components which may have a similar scale.

6.1.3.5 Ballistic and diffusive dynamics

Equation (6.33) tells us that the test mode is subject to diffusive damping due to the background turbulence. This implies that the statistical average of the spatial deviation δx increases as,

$$\left\langle (\delta x)^2 \right\rangle \sim \nu_t t.$$

This might be puzzling, if we recall the fact that the nonlinear interaction leads to ballistic motion, where $\delta x \sim Vt$ holds as is explained in Eq.(6.16). The possible ballistic dynamics (supported by the nonlinearity) and the apparent diffusive response must be reconciled. This point is understood since the turbulent diffusivity ν_t is a function of turbulent motion δx and turbulent velocity. For a dimensional argument, one takes ν_t as a product of deviation $\left\langle(\delta x)^2\right\rangle^{1/2}$ and fluctuating velocity $\tilde{V}$ as,

$$\nu_t \sim \sqrt{\left\langle(\delta x)^2\right\rangle}\tilde{V}.$$

Substitution of this mixing length estimate into the dispersion owing to the diffusion, $\left\langle(\delta x)^2\right\rangle \sim \nu_t t$, one has,

$$\left\langle(\delta x)^2\right\rangle \sim \tilde{V}^2 t^2.$$

This recovers ballistic scaling of the perturbed motion. Readers should note that this simple argument can be deceptive and its validity is restricted to the one-dimensional case of Burgers turbulence. In this idealized model of Burgers turbulence, a large number of small-scale, finite-lifetime perturbations, which move ballistically, are excited. A large number of short-range ballistic perturbations are considered (as nearly) independent kicks, and result in the diffusive decay of a large-scale test mode. Figure 6.3(b) shows an example of direct numerical solution of the randomly forced Burgers equation. A large number of small and spiky fluctuations are induced by random forcing and convective nonlinearity.

6.1.3.6 Irreversibility

The other issue in Eq.(6.24) is that irreversibility (i.e. $\Gamma_{k,\omega}$ is positive and finite) is induced by the nonlinear interactions. The memory of the initial condition is lost in a time scale of $\Gamma_{k,\omega}^{-1}$. The irreversibility of the test mode originates from mixing in the inertial range, and the dissipation that consequently occurs at very small scale (at the scale determined by the molecular viscosity ν). This is in contrast to resonant wave–particle nonlinear interaction. In the latter case, as is explained in Chapter 3, irreversibility is induced by orbit chaos resulting from overlap of resonances where $\omega = kv$, (v is the particle velocity), so that,

$$D = \frac{\pi q^2}{m^2} \sum_k \left|\tilde{E}_k\right|^2 \delta\left(\omega - kv\right),$$

where q and m are the charge and mass of a particle, respectively, $\tilde{E}_k$ is the fluctuating electric field, and $\delta(\omega - kv)$ is Dirac's delta function. In collisionless plasmas, the decorrelation of plasma response from the imposed field (in the presence of background fluctuations) can occur either through mixing by fluid motion or through the kinetic resonance of plasma particles with accelerating fields. In considering real problems, one must keep both possibilities in mind when determining the nonlinear decorrelation rate.

6.1.4 Illustration by use of the driven-Burgers/KPZ equation (2)

We now illustrate the issue of the spectral equation, putting emphasis on energy conservation in the closure model. The damping through turbulent viscosity in Eq.(6.19) (the second term on the left-hand side) shows that the intensity of the mode decays owing to this term. The *residual term* in Eq.(6.19) plays an important role in the evolution of the spectrum. Combining the coherent memory function and residual term, the spectral equation is deduced.

6.1.4.1 Excitation of the test mode by a nonlinear fluctuating force

Multiplying $\overset{*}{v_k}(=v_{-k})$ to Eq.(6.19) and taking an average over the statistical distribution of the random forcing, one has,

$$\frac{\partial}{\partial t}\left\langle |v_k|^2\right\rangle + \Gamma_k\left\langle |v_k|^2\right\rangle + \left\langle \overset{*}{v_k}\mathrm{Res}\right\rangle + \nu k^2\left\langle |v_k|^2\right\rangle = \left\langle \overset{*}{v_k}F_k^{\mathrm{ext}}\right\rangle, \tag{6.34}$$

where the right-hand side indicates the source of fluctuation energy by forcing. In this expression, the second term on the left-hand side leads to decay of the total intensity $\langle \tilde{v}^2\rangle = \sum_k \langle |v_k|^2\rangle$, as does a molecular viscosity (the fourth term on the left-hand side). This observation demonstrates the importance of the residual term. The original equation (6.15) shows that the total intensity is conserved by nonlinear interaction. Multiplying v and integrating it over space, one has,

$$\frac{\partial}{\partial t}\int \mathrm{d}x\, v^2 + \int \mathrm{d}x\, v^2\frac{\partial v}{\partial x} + \int \mathrm{d}x\, \nu\left(\frac{\partial v}{\partial x}\right)^2 = \int \mathrm{d}x\, vF^{\mathrm{ext}}.$$

The second term on the left-hand side, the nonlinear coupling term, vanishes, because it is a total derivative of $v^3/3$. Therefore, the total energy $\int \mathrm{d}x\, v^2$ does not decrease by nonlinear interactions. Thus, the modelling of the nonlinear damping term simultaneously requires evaluation of the residual term such that together, they conserve energy through nonlinear interactions.

Here we explain briefly modelling of the terms $\Gamma_k\left\langle |v_k|^2\right\rangle + \left\langle \overset{*}{v_k}\mathrm{Res}\right\rangle$ in Eq.(6.34). Returning to the original equation, and using a Fourier representation, the evolution equation of $\langle |v_k|^2\rangle$ is rewritten as,

$$\frac{\partial}{\partial t}\left\langle |v_k|^2\right\rangle = \left\langle \overset{*}{v_k}F_k\right\rangle + \nu k^2\left\langle |v_k|^2\right\rangle - T_k, \tag{6.35a}$$

$$T_k = \frac{1}{3}\left\langle \left(\frac{\partial}{\partial x}v^3\right)_k\right\rangle, \tag{6.35b}$$

where T_k indicates the nonlinear transfer of energy between different modes through nonlinear interactions. The conservation of total energy through nonlinear interactions implies that

$$\sum_k T_k = 0 \tag{6.36}$$

holds. Thus T_k must either be a sum of two terms, which cancel upon summation, or be anti-symmetric with respect to k. The argument that leads to Eq.(6.21) is applied to all nonlinear coupling terms, so,

$$T_k = 2i \sum_{k'} \left(k + k'\right) v_{-k} v_{-k'} v_{k+k'}^{(c)} - 2i \sum_{p+q=k} k v_{-q} v_{-p} v_{p+q}^{(c)}, \tag{6.37}$$

where the suffix denoting the frequency is suppressed for simplicity of the expression. The first term is the *coherent mode coupling* with the test mode, so that,

$$T_k^{(C)} \equiv 2i \sum_{k'} \left(k + k'\right) v_{-k} v_{-k'} v_{k+k'}^{(c)} \simeq \Gamma_k \left\langle |v_k|^2 \right\rangle. \tag{6.38a}$$

This denotes the dissipation of $\left\langle |v_k|^2 \right\rangle$ due to turbulent viscosity. The second term on the right-hand side of Eq.(6.37) is the *incoherent excitation* from other modes,

$$T_k^{(I)} \equiv -2i \sum_{p+q=k} k v_{-q} v_{q-k} v_{p+q}^{(c)} \propto -\left\langle |v_p|^2 \right\rangle \left\langle |v_q|^2 \right\rangle. \tag{6.38b}$$

The modes v_p and v_q are excited independently. The beat of these generates v_k if the condition $p + q = k$ is satisfied. Therefore, the term Eq.(6.38b) is considered as emission. This incoherent term denotes the birth of $\left\langle |v_k|^2 \right\rangle$ by nonlinear noise emission into the k-mode. Figure 6.5 gives a schematic drawing of the transfer function T_k and the evolution of the spectrum $E_k = \left\langle v_k^2 \right\rangle$. Steady state is illustrated in Figure 6.5(b). In the region where the source from external stirring

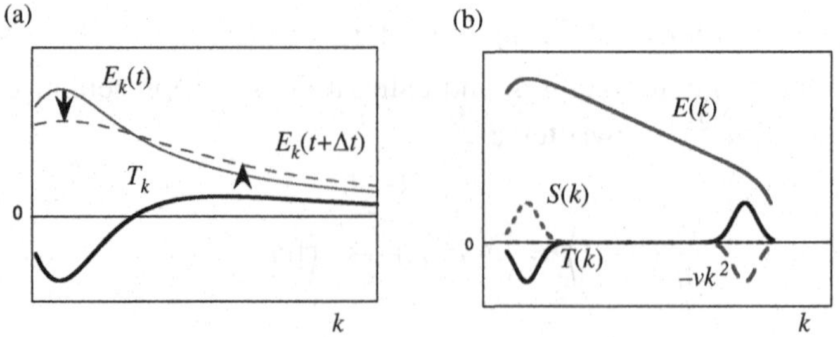

Fig. 6.5. Schematic drawing of the transfer function T_k and the evolution of the spectrum $E_k = \langle v_k^2 \rangle$ (a). Schematic drawing of the stationary state (b).

$S(k) = \langle v_{-k} F_k \rangle$ exists, the transfer function T_k is negative. In the short-wavelength regime, where the molecular viscosity induces dissipation, T_k is positive. The source input is carried to small scale by turbulence and is dissipated at high-k.

6.1.4.2 *Derivation of the transfer function*

One can formulate the coherent term and incoherent term simultaneously so as to obtain Eq.(6.25). Writing the time dependence of Eq.(6.22) explicitly,

$$v_{k+k'}^{(c)}(t) = -i\left(k+k'\right)\int_{-\infty}^{t} dt' L_{k+k'}\left(t-t'\right) v_{k'}\left(t'\right) v_k\left(t'\right), \tag{6.39a}$$

where the propagator includes the effects of molecular viscosity and the damping rate via nonlinear mixing in the time response history,

$$L_{k+k'}(t) = \exp\left(-\left(\nu\left(k+k'\right)^2 + \Gamma_{k+k'}\right)t\right). \tag{6.39b}$$

In a similar manner, we have,

$$v_{p+q}^{(c)}(t) = -i\left(p+q\right)\int_{-\infty}^{t} dt' L_{p+q}\left(t-t'\right) v_p\left(t'\right) v_q\left(t'\right). \tag{6.39c}$$

Upon substituting Eq.(6.39a) into $v_{k+k'}^{(c)}$ of Eq.(6.38a), one derives the coherent part of the transfer function as,

$$T_k^{(C)}(t) = 2\sum_{k'}\left(k+k'\right)^2\left\langle v_{-k}(t)\, v_{-k'}(t)\int_0^{\infty} d\tau\, L_{k+k'}(\tau)\, v_k(t-\tau)\, v_{k'}(t-\tau)\right\rangle. \tag{6.40}$$

In this expression, the time variable is replaced by τ using the definition $t' = t - \tau$. In writing this, the memory effect in $L_{k+k'}(\tau)$ is explicitly shown, connecting the transfer function $T_k^{(C)}$ at the time of t to the fluctuating field at $t-\tau$. In a similar way, the incoherent part (source for the test mode from incoherent background fluctuations) is evaluated as,

$$T_k^{(I)}(t) = 2\sum_{p+q=k}(p+q)^2\left\langle v_{-p}(t)\, v_{-q}(t)\int_0^{\infty} d\tau\, L_{p+q}(\tau)\, v_p(t-\tau)\, v_q(t-\tau)\right\rangle. \tag{6.41}$$

Now, a model of temporal self-coherence, $\langle v_{-k}(t)\, v_k(t-\tau)\rangle$, is necessary. The coherent damping by nonlinearity was extracted as Eq.(6.19). Therefore, we adopt a *model* that the self-correlation decays at the rate given by the response time (i.e. the sum of nonlinear damping rate and molecular-viscosity damping) as,

$$\langle v_{-k}(t)\, v_k(t-\tau)\rangle = \left\langle |v_k|^2(t)\right\rangle \exp\left(-\left(\Gamma_k + \nu k^2\right)\tau\right), \tag{6.42}$$

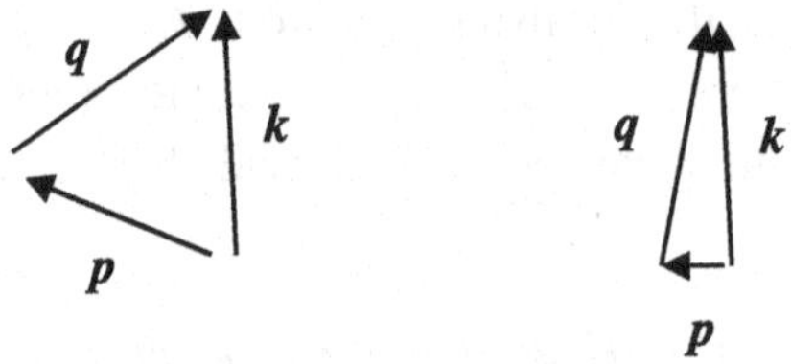

Fig. 6.6. Examples of triad pairs. Interaction among like-scale perturbations (left) and that among disparate scales (right).

where $\left(\Gamma_k + \nu k^2\right)$ is the rate at which memory is lost, and $\left\langle |v_k|^2 (t)\right\rangle$ captures the slow time evolution of the intensity of the k-mode. (The basis of this assumption (6.42) is revisited in Section 6.4.) By use of this self-consistent modelling of the correlation function, Eq.(6.40) is,

$$T_k^{(C)}(t) = 2\sum_{k'} \left(k + k'\right)^2 \Theta_{k,k',k+k'} \left\langle |v_{k'}|^2 \right\rangle \left\langle |v_k|^2 \right\rangle, \tag{6.43}$$

$$\Theta_{k,k',k+k'} \equiv \int_0^\infty \mathrm{d}\tau \exp\left(-\left(\Gamma_k + \Gamma_{k'} + \Gamma_{k+k'} + \nu k^2 + \nu k'^2 + \nu\left(k+k'\right)^2\right)\tau\right), \tag{6.44}$$

where $\Theta_{k,k',k+k'}$ is the triad coherence time, i.e. the time that a particular mode triad keeps coherence so as to effect energy transfer. (Note that we use the symbol $\Theta_{k,k',k+k'}$ here, instead of τ in Chapter 5, in order to illustrate explicitly that the nonlinear scrambling effect is included in the determination of the interaction time through closure.) A triad (see an example in Figure 6.6) keeps mutual coherence only if all three interacting modes keep coherence (relative to the initial time when the accumulation of interaction starts). (The well-known problem of the consistency of this form of $\Theta_{k,k',k+k'}$ with Galilean invariance has been intensively studied. We give a brief discussion of this issue at the end of this section.) Therefore, the interaction time is limited by the sum of the decay rates of the individual components in the triad. Similarly, the expression for the incoherent emission term is given as,

$$T_k^{(I)}(t) = -2 \sum_{p+q=k} (p+q)^2 \Theta_{p,q,k} \left\langle |v_p|^2 \right\rangle \left\langle |v_q|^2 \right\rangle. \tag{6.45}$$

6.1.4.3 Spectral equation

Now, observe that the transfer function is expressed in terms of the spectral function and the nonlinear interaction time, while the nonlinear interaction time is given in terms of the spectral function. Therefore, the original dynamical equation, from which a hierarchy of moment-equations follows, is now 'closed'. The spectral equation takes the form,

$$\frac{\partial}{\partial t}\left\langle |v_k|^2\right\rangle + \nu k^2\left\langle |v_k|^2\right\rangle + 2\sum_{k'}\left(k+k'\right)^2\Theta_{k,k',k+k'}\left\langle |v_{k'}|^2\right\rangle\left\langle |v_k|^2\right\rangle$$
$$= \langle v_{-k}F_k\rangle + 2\sum_{p+q=k}(p+q)^2\Theta_{p,q,k}\left\langle |v_p|^2\right\rangle\left\langle |v_q|^2\right\rangle, \tag{6.46}$$

where the second and third terms on the left-hand side denote damping by molecular viscosity and by turbulent viscosity (induced by nonlinear decay), respectively, and the first and second terms on the right-hand side stand for stirring by external force and nonlinear noise excitation (which is also induced by the nonlinear interaction), respectively. The expression for the triad interaction time (lifetime of interaction), given by Eqs.(6.25) and (6.39), supplements the spectral equation.

The closed spectral equation (6.46) has the structure of a Langevin equation, in which the noise and drag are both renormalized. In the invicid limit, (ν is small), the molecular dissipation appears only at very short wavelength. Forcing (either by external forcing or by some instability mechanism) may be limited to a particular scale length. Thus, there exists a range of wave numbers in between the excitation and molecular damping, where the second term on the left-hand side and the first term on the right-hand side are negligibly small in Eq.(6.46).

It is noted that Eqs.(6.43) and (6.45) form a set of turbulent damping and excitation terms, which satisfies the energy conservation relation. The total summation for the transfer functions is,

$$\sum_k T_k(t) = \sum_k T_k^{(C)}(t) + \sum_k T_k^{(I)}(t)$$
$$= 2\sum_k\sum_{k'}\left(k+k'\right)^2\Theta_{k,k',k+k'}\left\langle |v_{k'}|^2\right\rangle\left\langle |v_k|^2\right\rangle$$
$$-2\sum_k\sum_{p+q=k}(p+q)^2\Theta_{p,q,k}\left\langle |v_p|^2\right\rangle\left\langle |v_q|^2\right\rangle. \tag{6.47}$$

One finds that the right-hand side of Eq.(6.47) vanishes by re-labelling. That is, the sum of coherent damping equals the sum of incoherent emission. If one mode is chosen as a test mode k, this mode is subject to a coherent damping by background turbulence. The energy extracted from this test mode k is distributed to incoherent sources of other modes, so that the energy conserving property of the nonlinear interaction in the original equation is maintained by this closure model,

$$\frac{\partial}{\partial t}\left\langle |v_k|^2\right\rangle + \nu_{\text{turb}}k^2\left\langle |v_k|^2\right\rangle = -T_k^{(I)}, \tag{6.48}$$

where $-T_k^{(I)}$ indicates the nonlinear noise source for $|v_k|^2$.

6.1.5 Short summary of elements in closure theory

Examining these three examples in the preceding subsections, one observes the necessity to evaluate the statistical average of nonlinear interaction without naïve truncation at a particular order. Thus, a method *to close* the hierarchy, *not to truncate* it, must be introduced. This is the essence of the closure theory of strong turbulence. Closure theory is based upon considerations of the response function and decorrelation time. That is, the first step is to analyze the response,

$$\delta v / \delta F^{\mathrm{ext}}.$$

This step is performed in a perturbative manner, but information from turbulent decorrelation (i.e. renormalization) is introduced in an implicit manner. The nonlinear decorrelation rate is extracted from nonlinear interaction terms. During this extraction process, the response $\delta v/\delta F$ is utilized. A test mode and its interactions are identified in the large number of ambient (random) fluctuating components. The beat between the test mode and background fluctuations (i.e. the 'polarization of the background by the test mode) is calculated. By use of this decorrelation time, the spectral equation (dynamical equation for $\langle |v_k|^2 \rangle$) is expressed in terms of the spectral function $\langle |v_{k'}|^2 \rangle$ and the decorrelation rate. Thus the closed set of equations includes both the spectral equation and the equation that determines the correlation time (or propagator).

The closure models are consistent with the conservation laws of the original equation. The triad interaction time $\Theta_{k,k',k+k'}$ is limited by the sum of the decorrelation rates of elements in the triad,

$$\frac{1}{\Theta_{k,k',k+k'}} \sim \Gamma_k + \Gamma_{k'} + \Gamma_{k+k'}.$$

Usually, no restriction on the shape of interacting triads is imposed. (See Figure 6.6, for examples.) Therefore, a distinction between the sweeping and stretching, etc., is not made. (Figure 6.7 illustrates sweeping and stretching of a

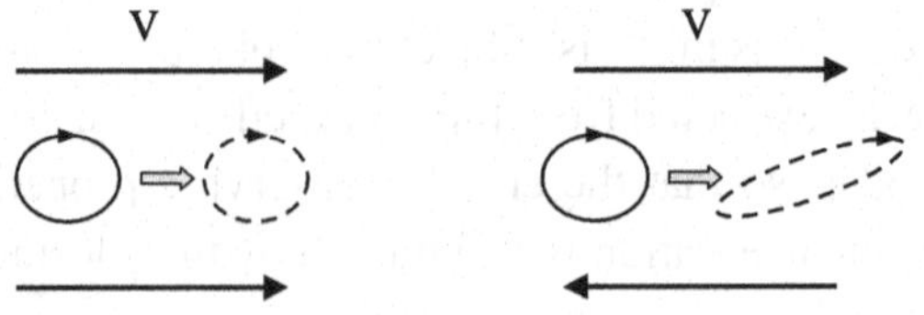

Fig. 6.7. Sweeping (left) and stretching (right) of a test vortex by ambient velocity fields. In the case of sweeping, a vortex (shown by a small circle) moves in space (dashed circle), keeping its identity. In the case of stretching, a test vortex (solid circle) is deformed as time goes on (dashed line).

test vortex.) This point might be important, depending on the problems under study (e.g. when a strongly anisotropic mode is considered, or uniform advection, etc.). In the extraction of a nonlinear timescale from nonlinearities, one must introduce physical insight that pinpoints the deduction of the response $\delta v/\delta F$ in a renormalized but perturbative manner. In this place, the central assumption (thus the main limitation of the model) is introduced. Depending on the physics insights, several different but related closure methodologies have been developed. As famous and successful methods, one may count eddy viscosity models, eddy-damped quasi-normal Markovian (EDQNM) approximation (Orszag, 1970), the DIA closure model, and the Mori–Zwanzig method.

6.1.6 On realizability

We conclude this introduction to closure theory with a brief discussion of two 'special topics' in renormalized turbulence theory, namely, realizability and the random coupling mode, and Galilean invariance (Kraichnan, 1961; Kraichnan, 1970; Krommes, 1984). We now briefly discuss these two topics.

Realizability is concerned with the consistency of closure theory predictions with our expectations for a real, physical system. Of course, avoidance of negative spectra is one aspect of realizability. To this end, we think it illuminating (following Kraichnan) to address realizability from the perspective of the question, "Given that the closure equations are *approximate* solutions to full fluid equations, *to what physical system are they exact solutions*?" Such a system is a physical realization of the closure model. Kraichnan answered this question by identifying the realization as the system of *randomly coupled stochastic oscillators* (Kraichnan, 1961), with the dynamic equation,

$$\frac{\mathrm{d}}{\mathrm{d}t} q_\alpha = \frac{-i}{\sqrt{M}} \sum_\beta \psi_{\alpha,\beta,\alpha-\beta} b_\beta q_{\alpha-\beta}, \tag{6.49a}$$

where the coupling coefficient $\psi_{\alpha,\beta,\alpha-\beta}$ has the form,

$$\psi_{\alpha,\beta,\alpha-\beta} = \exp\left(i\theta_{\alpha,\beta,\alpha-\beta}\right). \tag{6.49b}$$

Here $\theta_{\alpha,\beta,\alpha-\beta}$ is random on $[0, 2\pi]$ and $\theta_{\alpha,\beta,\alpha-\beta} = -\theta_{\beta,\alpha,\alpha-\beta}$. The random coupling model is exactly solvable, with a temporal response function $G(t)$ determined by,

$$\frac{\mathrm{d}}{\mathrm{d}t} G(t) + \left\langle b^2 \right\rangle \int_0^t \mathrm{d}s G(t-s) G(s) = 0, \tag{6.50a}$$

so,

$$G(t) = J_1(2b_* t)/b_* t. \tag{6.50b}$$

Here J_1 is the first-order Bessel function of the first kind and b_* is the root square mean of b. The term $G(t)$ is identical to the temporal response predicted for Navier–Stokes turbulence by the direct interaction approximation (DIA) (McComb, 1990; P. A. Davidson, 2004). Thus, the random coupling model propagator exactly solves the set of equations which the DIA closure uses to approximate the full dynamics of the Navier–Stokes equations.

It is interesting to note that the essence of the DIA-type closure is revealed to be one of *random coupling*, rather than *random phase*, as is often quoted. This is consistent with our intuition that the test wave hypothesis is valid only for regimes of weak phase correlation and limited triad coherence, and suggests that the duration of phase correlation must be *short*, in order to apply stochastic renormalization techniques. It is also interesting to comment that the limit of the random coupling model is precisely the diametrically opposite limit from that of the Kuramoto oscillator model, where though $N \to \infty$, phase synchronization can occur due to strong nonlinear oscillator coherence (Kuramoto, 1984). In this regard, it is interesting to define a phase Kubo number,

$$\mathcal{K}_\theta = \frac{\dot{\theta}\tau_\theta}{\Delta\theta},$$

(where $\dot{\theta}$ is the phase evolution rate, τ_θ is the phase coherency time and $\Delta\theta$ is a typical phase separation of correlation in the ensemble) by analogy to the usual Kubo number or Strouhal number for particle or fluid element displacement, i.e.

$$\mathcal{K} = \frac{\tilde{V}\tau_c}{l_c}.$$

Then we see that the random coupling model and synchronized Kuramoto model correspond to the limits of $\mathcal{K}_\theta \to 0$ and $\mathcal{K}_\theta \to \infty$, respectively. We remark that a systematic exploration of the transition from $\mathcal{K}_\theta \ll 1$ to $\mathcal{K}_\theta \gg 1$ could be an interesting approach to understanding the development of intermittency.

Galilean invariance and how the closure theory handles the distinction between "sweeping" and "straining' are other issues which merit comment. It is generic to DIA-type closure theory that decorrelation rates *add*, so $\Theta^{-1}_{k,k',k+k'} \sim \Gamma_k + \Gamma_{k'} + \Gamma_{k+k'}$. Here Γ_k refers to a decorrelation rate. As originally pointed out by Kraichnan, this structure misrepresents the effect of a random Galilean transformation (i.e. three modes are Doppler-shifted uniformly), which must necessarily leave coherent times of the triad invariant, as it only 'sweeps' along interacting eddies, but does not affect their Lagrangian correlation time. This is depicted in Figure 6.8. For the case of Navier–Stokes turbulence, unphysical sweeping contributions to eddy lifetimes can be eliminated by restricting the lower cut-off of the spectrum sum in Γ_k. A more fundamental, but quite labour intensive, alternative is

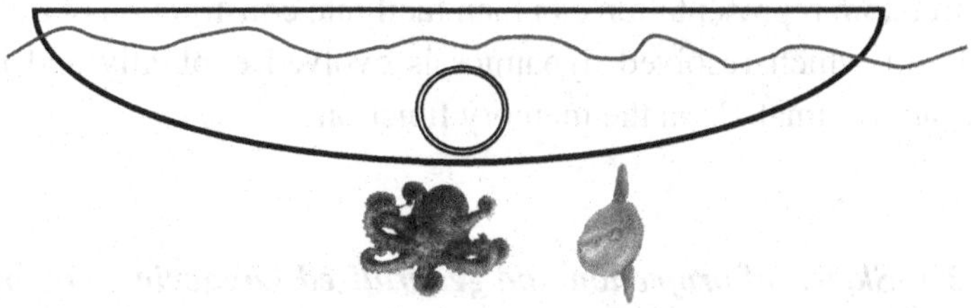

Fig. 6.8. Correlation of observed (Eulerian) signals is affected by sweeping. When a mola mola and an octopus are swept by the sea current, after the elapse of a short time, a fish (observed from a porthole of a ship) seems to metamorphose into an octopus (after Yoshizawa (1999)).

the Lagrangian history direct interaction approximation (LHDIA) closure. Further discussion of this fascinating topic is beyond the scope of this work, and readers are referred to the original research literature (Kraichnan, 1977; Kaneda, 1981; McComb, 1990; P. A. Davidson, 2004).

6.2 Mori–Zwanzig theory and adiabatic elimination

As illustrated in Section 6.1, closure theories aim to extract the dynamics of 'relevant' (observable, i.e. of interest) variables out of a large number of degrees of freedom. Systematic elimination or reduction in representation of "small or fast" degrees of freedom is a goal of many approximation procedures, including renormalization group theory, adiabatic theory and other familiar methods. A systematic procedure has been postulated using an approach involving a projection operator. Mori–Zwanzig theory (M–Z theory) is a prototypical methodology for this task (Mori, 1965; Zwanzig, 2001). The M–Z theory is concerned with systems with multiple-interacting degrees of freedom, for which:

(i) Some elements of noise and/or irreversibility and damping (e.g. molecular viscosity at very short scale) are present;
(ii) A broad range of relaxation rates exist, such that some modes are 'fast', and come quickly to equilibrium and so can be integrated out. Some another modes are 'slow' (and of interest) and need to be resolved on time scales of interest.

For this system, the method provides a route to:

(iii) Systematically project the dynamics onto that of the slow modes;
(vi) Extend Markovian Fokker–Planck theory to describe slow mode dynamics.

We shall see that the M–Z theory has several elements in common with the method of quasi-particles, which is developed in conjunction with the adiabatic theory of Langmuir turbulence. Disparate-scale interaction, in the context of the example of Langmuir turbulence, is discussed in Chapter 7. These methods presume a scale disparity between the modes of interest and ambient components. The reduced degrees of freedom representation can then facilitate construction of tractable subgrid models, in which resolved dynamics is evolved explicitly and unresolved dynamics is approximated via the memory function.

6.2.1 Sketch of projection and generalized Langevin equation

6.2.1.1 Elimination of irrelevant variables

Consider a system with N variables $\boldsymbol{A} = (a_1, a_2, \cdots, a_N)$, which obeys the dynamic equation,

$$\frac{\mathrm{d}}{\mathrm{d}t} a_i = h_i(\boldsymbol{A}), i = 1, 2, \cdots, N. \tag{6.51}$$

In order to illuminate the difference in the damping rate, we change variables, $\xi = \xi(\boldsymbol{A})$, so that,

$$\frac{\mathrm{d}}{\mathrm{d}t} \xi_i + \gamma_i \xi_i = h_i(\boldsymbol{\xi}), i = 1, 2, \cdots, N. \tag{6.52}$$

Here γ_i is the linear damping rate (relaxation rate) of the mode, some of which are slow and some fast. This distinction requires some sort of relative time scale separation. The modes are divided into two classes:

(i) $s = 1, 2, \cdots, M$, for which dynamics is slow, and
(ii) $i = M + 1, \cdots, N$ for fast modes.

The fast variables are obtained first, and then plugged into the evolution equation for the slow modes. When the time scale separation is allowed, the fast modes reach a stationary state (in the 'fast' time scale) during the slow evolution of modes with $s = 1, 2, \cdots, M$. That is, one takes $\mathrm{d}\xi_i/\mathrm{d}t = 0$ for $i = M + 1, \cdots, N$, so that the equation,

$$\xi_i = \gamma_i^{-1} h_i(\boldsymbol{\xi}), i = M + 1, \cdots, N, \tag{6.53}$$

is formally a simple algebraic equation. The solution of Eq.(6.53) is then plugged into Eq.(6.52), giving,

$$\frac{\mathrm{d}}{\mathrm{d}t} \xi_s + \gamma_s \xi_s = h_s(\xi_1, \cdots, \xi_M; \xi_{M+1}[\xi_s], \cdots, \xi_N[\xi_s]), s = 1, 2, \cdots, M, \tag{6.54}$$

so the projection of dynamics on to slow variables is completed. Equations (6.53) and (6.54) close the loop (Figure 6.9). This projection on slow dynamics can be made systematically.

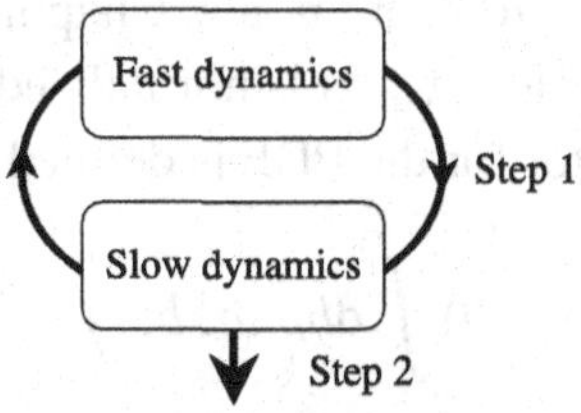

Fig. 6.9. Loop of slow variables and fast variables.

6.2.1.2 *Generalized Langevin equation for the probability density function*

Now the projection is defined for the evolution of the probability density function (pdf). (Here, the pdf is assumed to exist for this dynamical system.) A generalized master equation is introduced. Consider a set of variables (a, b), in which a is a relevant variable (i.e. of interest), with a slow time scale, and b has a fast time scale and will be eliminated. We consider a pdf $\rho(a, b; t)$. Note that the probability density function may not always be well defined for some dynamic problems. Here, we simply *assume* the existence of $\rho(a, b; t)$, for which the dynamical equation is written as,

$$\frac{\partial}{\partial t}\rho(a, b; t) = \mathcal{L}\rho(a, b; t), \tag{6.55}$$

where $\mathcal{L}$ represents a general evolution operator. In Liouvillian dynamics, $\mathcal{L} = \{\rho, H\}$, where H is the Hamiltonian and $\{\cdots\}$ is the Poisson bracket.

The projected pdf for the relevant parameter is introduced by integrating over the irrelevant variables as,

$$\mathcal{S}(a; t) \equiv \int db\rho(a, b; t). \tag{6.56}$$

The dynamical operator may be represented as,

$$\mathcal{L}(a, b; t) = \mathcal{L}_a + \mathcal{L}_b + \mathcal{L}_{\text{int}}, \tag{6.57}$$

where $\mathcal{L}_a$ describes a–a interaction (among slow variables), so this slow dynamics is resolved. The term $\mathcal{L}_b$ indicates the b–b interaction (among fast variables) but this time scale is not resolved (i.e. is ultimately statistically averaged over). The interaction term $\mathcal{L}_{\text{int}}$ the a–b interaction, i.e. denotes disparate-scale interaction. The treatment of the coupling term $\mathcal{L}_{\text{int}}$ is the key issue. We assume that, on the fast time scale, the pdf of the variable b reaches an equilibrium distribution $\rho_{\text{eq}}(b)$,

$$\mathcal{L}_b\rho_{\text{eq}}(b) = 0. \tag{6.58}$$

This first step is analogous to that in the Chapman–Enskog expansion. The equilibrium distribution function $\rho_{\text{eq}}(b)$ is normalized as $\int \mathrm{d}b\rho_{\text{eq}}(b) = 1$.

Now the projection operator for the PDF is defined as,

$$\mathcal{P}\rho\,(a,b;t) \equiv \rho_{\text{eq}}\,(b) \int \mathrm{d}b\rho\,(a,b;t) = \rho_{\text{eq}}\,(b)\,\mathcal{S}\,(a;t)\,, \tag{6.59}$$

i.e. the projection operator $\mathcal{P}$ projects out the slow variables. This operator $\mathcal{P}$ can be shown to be a true projection operator, i.e. to satisfy the idempotency requirement,

$$\mathcal{P}\mathcal{P} = \mathcal{P}.$$

This relation is confirmed by applying $\mathcal{P}$ to Eq.(6.59) as,

$$\mathcal{P}\mathcal{P}\rho\,(a,b;t) \equiv \rho_{\text{eq}}\,(b) \int \mathrm{d}b\rho_{\text{eq}}\,(b)\,\mathcal{S}\,(a;t) = \rho_{\text{eq}}\,(b)\,\mathcal{S}\,(a;t) = \mathcal{P}\rho\,(a,b;t)\,, \tag{6.60}$$

i.e. $\mathcal{P}\mathcal{P} = \mathcal{P}$. Another projection operator is $\mathcal{Q}$,

$$\mathcal{Q} \equiv 1 - \mathcal{P}, \tag{6.61}$$

which gives a projection on to the fast-scale dynamics, which is also introduced. Operating with $\mathcal{P}$ and $\mathcal{Q}$ on Eq.(6.55), respectively, and abbreviating $\rho_1 = \mathcal{P}\rho\,(a,b;t)$ and $\rho_2 = \mathcal{Q}\rho\,(a,b;t)$ $(\rho_1 + \rho_2 = \rho)$, one obtains,

$$\frac{\partial}{\partial t}\rho_1 - \mathcal{P}\mathcal{L}\rho_1 = \mathcal{P}\mathcal{L}\rho_2, \tag{6.62a}$$

$$\frac{\partial}{\partial t}\rho_2 - \mathcal{Q}\mathcal{L}\rho_2 = \mathcal{Q}\mathcal{L}\rho_1. \tag{6.62b}$$

Equation (6.62b) can be solved as,

$$\rho_2\,(t) = \exp\left(\int_0^t \mathrm{d}t\mathcal{Q}\mathcal{L}\right)\rho_2\,(0) + \int_0^t \mathrm{d}s \exp\left(\int_0^s \mathrm{d}t'\mathcal{Q}\mathcal{L}\right)\mathcal{Q}\mathcal{L}\rho_1\,(t-s)\,. \tag{6.63}$$

Substituting Eq.(6.63) into Eq.(6.62a), one has,

$$\frac{\partial}{\partial t}\rho_1 = \mathcal{P}\mathcal{L}\mathcal{P}\rho + \mathcal{P}\mathcal{L}\exp\left(\int_0^t \mathrm{d}t\mathcal{Q}\mathcal{L}\right)\mathcal{Q}\rho\,(0) + \int_0^t \mathrm{d}s\psi\,(s)\,\rho\,(t-s)\,, \tag{6.64a}$$

with the generalized memory kernel,

$$\psi\,(s) = \mathcal{P}\mathcal{L}\exp\left(\int_0^s \mathrm{d}t'\mathcal{Q}\mathcal{L}\right)\mathcal{Q}\mathcal{L}\mathcal{P}. \tag{6.64b}$$

Here the relations $\rho_2(0) = \mathcal{Q}\rho(0)$ and $\rho_1(t-s) = \mathcal{P}\rho(t-s)$ are used. Equation (6.64) is a generalized master equation extracting the evolution of the pdf for the relevant (i.e. slow) variables – the a's. It is manifestly non-Markovian, representing the projection onto the relevant variable. It is also useful to rewrite Eq.(6.14) for the evolution equation of ρ_1, i.e.

$$\frac{\partial}{\partial t}\rho_1 - \mathcal{P}\mathcal{L}\rho_1 - \int_0^t \mathrm{d}s\Gamma(s)\,\rho_1(t-s) = \mathcal{P}\mathcal{L}\exp\left(\int_0^t \mathrm{d}t\,\mathcal{Q}\mathcal{L}\right)\mathcal{Q}\rho(0), \quad (6.65a)$$

where the memory function is redefined as,

$$\Gamma(s) = \mathcal{P}\mathcal{L}\exp\left(\int_0^s \mathrm{d}t'\mathcal{Q}\mathcal{L}\right)\mathcal{Q}\mathcal{L}. \quad (6.65b)$$

Now, Eq.(6.65) has the structure of a generalized Langevin equation. The second term on the left-hand side accounts for the self-dynamics (on the slow scale) and the third term gives the memory function, which arises from coupling to the fast variables. The right-hand side of Eq.(6.65a) shows the impact of fluctuating terms which evolve from initial conditions (for all degrees of freedom, including fast variables). This works as a random kick term for the evolution of ρ_1.

The memory function and the kick (right-hand side) are not set arbitrarily, but given in a consistent manner, so that the magnitude of the memory function is self-consistently determined by the intensity of the random kick. A relation which corresponds to the fluctuation–dissipation theorem can be deduced. This should not come as a surprise to the reader. The whole aim of the M–Z elimination procedure is to represent irrelevant variables as noise and drag (or, more generally, memory scrambling) for the effective Brownian motion of relevant variables. Such an approach naturally builds upon the presumption of a fluctuation–dissipation theorem structure.

When the operator $\mathcal{L}$ does not include explicit time dependence (usually the case), the integral $\int_0^s \mathrm{d}t'\mathcal{Q}\mathcal{L}$ is $\mathcal{Q}\mathcal{L}s$. Then Eq.(6.65) takes the form,

$$\frac{\partial}{\partial t}\rho_1 - \mathcal{P}\mathcal{L}\rho_1 - \int_0^t \mathrm{d}s\Gamma(s)\,\rho_1(t-s) = \mathcal{P}\mathcal{L}e^{\mathcal{Q}\mathcal{L}t}\mathcal{Q}\rho(0), \quad (6.66)$$

and,

$$\Gamma(s) = \mathcal{P}\mathcal{L}e^{\mathcal{Q}\mathcal{L}s}\mathcal{Q}\mathcal{L}. \quad (6.67)$$

6.2.2 *Memory function and most probable path*

The method of the projection operator can also be applied to the dynamical equation, and a nonlinear force separated into a memory function and a fluctuating force. Mori's method of projection operator is applied to extract the memory function from the nonlinear force (Mori, 1965; Mori and Fujisaka, 2001; Mori *et al.*, 2003).

6.2.2.1 Formalism and fluctuation dissipation relation

A formal relation for the projection operator and the separation of memory function from the fluctuating force is explained. A set of macro variables (observables which are of interest), $A(t) = (A_i(t))$, $(i = 1, \cdots, n)$ in the turbulent state is considered. The variables that govern $A(t)$ in turn are denoted, $X(t) = (X_i(t))$, $(i = 1, \cdots, n)$, and the dynamical evolution equation is expressed as,

$$\frac{\mathrm{d}}{\mathrm{d}t} A(t) = \sum_{i=1}^{N} \dot{X}_i \frac{\partial}{\partial X_i} A(t) \equiv \Lambda A(t) . \tag{6.68}$$

Observables have fewer degrees of freedom than the governing dynamical variables, i.e. $n < N$. A formal solution of Eq.(6.68) is,

$$A(t) = \exp[\Lambda t] A(0) . \tag{6.69}$$

(For the case of Vlasov plasmas, the phase space variable X is the location and velocity of particles (x, v), and A is a distribution function. For fluid turbulence, X is composed of perturbations of density, velocity, pressure, etc., of various Fourier components, and A denotes the quantity of interest, e.g. the mean velocity, mean flux, test mode, etc.) The nonlinear force $\Lambda A(t)$ is expressed in terms of the sum of the memory function and fluctuating force. A formal explanation is given here.

A long time average,

$$\langle Z \rangle = \lim_{T \to \infty} \frac{1}{T} \int_0^T \mathrm{d}t Z(t) \tag{6.70}$$

is introduced. In this section, the symbol $\langle \cdots \rangle$ indicates the time average. The correlation between the observable A at two time points, $A(t')$ and $A(t' + t)$ (both of which are fluctuating) is defined by,

$$\left\langle A_i(t) A_{i'}^{\dagger}(0) \right\rangle = \lim_{T \to \infty} \frac{1}{T} \int_0^T \mathrm{d}t' A_i(t + t') A_{i'}^{\dagger}(t') , \tag{6.71}$$

where $A_{i'}^{\dagger}$ is the i'-element of the transposed row matrix and the mean is subtracted from A so that $\langle A(t) \rangle = 0$ holds. The long-time average $\left\langle A_i(t) A_{i'}^{\dagger}(0) \right\rangle$ is the (i, i')-element of the $n \times n$ matrix $\left\langle A(t) A^{\dagger}(0) \right\rangle$. In the approach of the Mori-formalism, *we assume that the long-time average in Eq.(6.71) exists and*

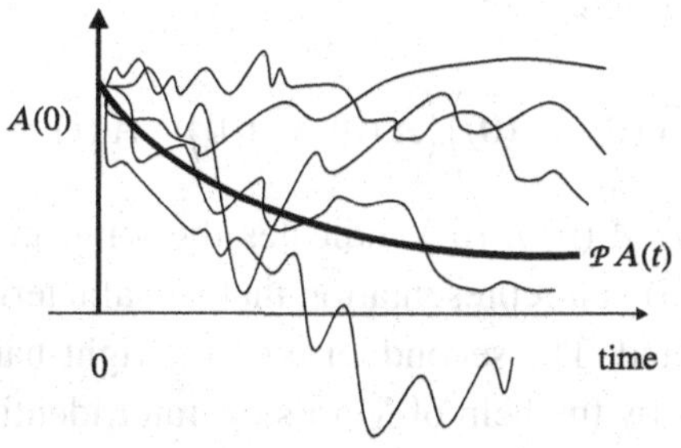

Fig. 6.10. Various realizations of time evolutions $A(t)$ (which are subject to chaotic variation) and the most probable path $\mathcal{P}A(t)$.

that the average does not depend on the choice of initial condition $X(0)$. With this assumption, the statistical average is constructed based on the long-time average.

A projection of a quantity Z on the observable $A(0)$ is defined as,

$$\mathcal{P}Z(t) \equiv \left\langle Z(t)\, A^{\dagger}(0)\right\rangle \left\langle A(0)\, A^{\dagger}(0)\right\rangle^{-1} A(0)\,. \tag{6.72}$$

It should be noted that the origin of time integral $t=0$ is chosen arbitrarily, and the following results do not depend on the choice of $t=0$. Nevertheless, the argument (0) is written here explicitly, in order to elucidate the roles of the two-time correlation and nonlinear decorrelation.

This operator (6.72) satisfies the relation $\mathcal{P}A(0)=A(0)$. Therefore, applying $\mathcal{P}$ once on the left and right-hand side of Eq.(6.72), one has $\mathcal{P}^2 Z(t) = \langle Z(t)\, A^{\dagger}(0)\rangle \langle A(0)\, A^{\dagger}(0)\rangle^{-1} \mathcal{P}A(0) = \mathcal{P}Z(t)$, i.e. the relation,

$$\mathcal{P}^2 = \mathcal{P} \tag{6.73a}$$

is deduced. For

$$\mathcal{Q} = 1 - \mathcal{P},$$

it follows that,

$$\mathcal{P}\mathcal{Q} = \mathcal{Q}\mathcal{P} = \mathcal{P} - \mathcal{P}^2 = 0. \tag{6.73b}$$

That establishes that this operator $\mathcal{P}$ is a projection operator (Figure 6.10).

By use of this projection operator, one can show that a fluctuating force and a memory function, which satisfy a fluctuation dissipation theorem relation, exist for Eq.(6.68). Rewriting Eq.(6.69) as $\mathrm{d}A(t)/\mathrm{d}t = \mathrm{e}^{t\Lambda}\dot{A}(0) = \mathrm{e}^{t\Lambda}(\mathcal{P}+\mathcal{Q})\,\dot{A}(0)$, Eq.(6.68) is decomposed as,

$$\frac{\mathrm{d}A(t)}{\mathrm{d}t} = \mathrm{e}^{t\Lambda}\mathcal{P}\dot{A}(0) + \mathrm{e}^{t\Lambda}\mathcal{Q}\dot{A}(0)\,. \tag{6.74}$$

The first term is expressed as,

$$\mathrm{e}^{t\Lambda}\mathcal{P}\dot{A}(0)=\left\langle \dot{A}(0)\,A^{\dagger}(0)\right\rangle \left\langle A(0)\,A^{\dagger}(0)\right\rangle^{-1}A(t)\equiv i\Omega A(t)\,. \tag{6.75}$$

Note that the quantity $\langle A(0)\,B(0)\rangle$ indicates the long-time average of the variable AB, thus Eq.(6.75) is a representation that a characteristic 'frequency' of the variable $A(t)$ is extracted. The second term on the right-hand side of Eq.(6.74) is divided into two terms by the help of Mori's operator identity,

$$\mathrm{e}^{t\Lambda}=\mathrm{e}^{t\mathcal{Q}\Lambda}+\int_0^t \mathrm{d}s\mathrm{e}^{(t-s)\Lambda}\mathcal{P}\Lambda\mathrm{e}^{s\mathcal{Q}\Lambda}, \tag{6.76}$$

so,

$$\mathrm{e}^{t\Lambda}\mathcal{Q}\dot{A}(0)=\mathrm{e}^{t\mathcal{Q}\Lambda}\mathcal{Q}\dot{A}(0)+\int_0^t \mathrm{d}s\mathrm{e}^{(t-s)\Lambda}\mathcal{P}\Lambda\mathrm{e}^{s\mathcal{Q}\Lambda}\mathcal{Q}\dot{A}(0)\,. \tag{6.77}$$

(The identity Eq.(6.76) is understood by noting that a functional $h=\mathrm{e}^{t\Lambda}$ is a solution of an equation $\partial h/\partial t=\Lambda h$, i.e. $\partial h/\partial t-\mathcal{Q}\Lambda h=\mathcal{P}\Lambda h$, with the initial condition $h(0)=1$.) The first term on the right-hand side of Eq.(6.76) is related to the fluctuating force. For a fluctuating quantity,

$$\tilde{R}(t)\equiv\mathrm{e}^{t\mathcal{Q}\Lambda}\mathcal{Q}\dot{A}(0)\,, \tag{6.78}$$

$\mathcal{P}\tilde{R}(t)=0$ holds due to the relation $\mathcal{P}\mathcal{Q}=0$; Therefore the relation,

$$\left\langle \tilde{R}(t)\,A^{\dagger}(0)\right\rangle=0, \tag{6.79}$$

holds by definition of operator $\mathcal{P}$, Eq.(6.72). Noting the fact that $\mathcal{P}A(t)$ is the projection to extract a component $A(0)$ as is shown in Eq.(6.72), one sees that $\left\langle \tilde{R}(t)\,\mathcal{P}A(t)\right\rangle$ is in proportion to $\left\langle \tilde{R}(t)\,A^{\dagger}(0)\right\rangle$. Therefore Eq.(6.79) leads to,

$$\left\langle \tilde{R}(t)\,\mathcal{P}A(t)\right\rangle=0, \tag{6.80}$$

which shows the orthogonality relation between the fluctuating force $\tilde{R}(t)$ and the projected trajectory $\mathcal{P}A(t)$.

The second term on the right-hand side of Eq.(6.77) constitutes a memory function. The integrand of this term is rewritten by use of the fluctuating force as $\mathrm{e}^{(t-s)\Lambda}\mathcal{P}\Lambda\tilde{R}(t)$ with the help of Eq.(6.78), which is then explicitly written as,

$$\begin{aligned}\mathrm{e}^{(t-s)\Lambda}\mathcal{P}\Lambda\tilde{R}(t)&=\mathrm{e}^{(t-s)\Lambda}\left\langle \Lambda\tilde{R}(t)\,A^{\dagger}(0)\right\rangle\left\langle A(0)\,A^{\dagger}(0)\right\rangle^{-1}A(0)\\&=-\left\langle \tilde{R}(t)\,\tilde{R}^{\dagger}(0)\right\rangle\left\langle A(0)\,A^{\dagger}(0)\right\rangle^{-1}A(t-s)\,.\end{aligned} \tag{6.81}$$

Here, we used the identity $\left\langle \left(\Lambda \tilde{R}(t) \right) A^{\dagger}(0) \right\rangle = -\left\langle \tilde{R}(t)\, \tilde{R}^{\dagger}(0) \right\rangle$, which follows from Eq.(6.79) and the definition $\tilde{R}(0) = \mathcal{Q}\Lambda A(0)$. Thus, the second term on the right-hand side of Eq.(6.77) is expressed in terms of the convolution with the memory function,

$$\int_0^t \mathrm{d}s \mathrm{e}^{(t-s)\Lambda} \mathcal{P} \Lambda \mathrm{e}^{s\mathcal{Q}\Lambda} \mathcal{Q} \dot{A}(0) = -\int_0^t \mathrm{d}s \Gamma(s)\, A(t-s), \tag{6.82}$$

where the memory function $\Gamma(t)$ is defined as

$$\Gamma(t) = -\left\langle \tilde{R}(t)\, \tilde{R}^{\dagger}(0) \right\rangle \left\langle A(0)\, A^{\dagger}(0) \right\rangle^{-1}. \tag{6.83}$$

Summarizing the above argument, the dynamical equation (6.74) can ultimately be rewritten as,

$$\frac{\mathrm{d}A(t)}{\mathrm{d}t} = i\Omega A(t) + \tilde{R}(t) - \int_0^t \mathrm{d}s \Gamma(s)\, A(t-s), \tag{6.84}$$

where the mean linear frequency Ω, the fluctuating force $\tilde{R}(t)$ and the memory function $\Gamma(t)$ are defined by Eqs.(6.75), (6.78) and (6.83), respectively. This has a non-Markovian form, owing to the fact that irrelevant (finer scale) variables are eliminated while extracting the mean damping of the observable $A(t)$.

The result Eq.(6.84) means that, under the assumption of the existence of the long-time average (which is independent of the initial conditions), the non-linear dynamical equation (6.74) can be decomposed as Eq.(6.84), in which (i) the fluctuating force is orthogonal to the most probable path $A(t)$, and (ii) the memory function is given by the correlation function of the fluctuating force. Equation (6.83) shows that the structure of the fluctuation–dissipation theorem can be extended to a nonlinear and non-equilibrium system.

6.2.2.2 *Memory function and nonlinear force*

Calculation of the memory function is an important goal of renormalized turbulence theory. Equation (6.84) is an existence theorem, and further theoretical study is necessary to obtain an explicit memory function for a physically interesting problem.

One of the outcomes of this decomposition theorem is an explicit procedure in the data analysis of nonlinear simulations. The need to respond to recent progress in direct nonlinear simulation compels us to develop a method to extract a relevant memory function from the particular realization of the turbulent state, which the computation gives us. Mori has noted that, although direct calculation of the quantity $\left\langle \tilde{R}(t)\,\tilde{R}^{\dagger}(0)\right\rangle$ is not possible in direct simulation because it contains the propagator $\exp(\mathcal{Q}\Lambda t)$, the calculation of the correlation of the nonlinear force itself is possible to evaluate. The nonlinear force,

$$F(t) \equiv \frac{\mathrm{d}A(t)}{\mathrm{d}t} - i\Omega A(t) = \tilde{R}(t) - \int_0^t \mathrm{d}s\Gamma(s)\,A(t-s) \tag{6.85}$$

is calculable, because both $\mathrm{d}A(t)/\mathrm{d}t$ and Ω are measurable in direct simulations. The correlation function,

$$\Phi(t) \equiv \left\langle F(t)\,F^{\dagger}(0)\right\rangle \left\langle A(0)\,A^{\dagger}(0)\right\rangle^{-1} \tag{6.86}$$

is introduced, and the substitution of the right-hand side of Eq.(6.85) into Eq.(6.86) provides,

$$\Phi(t) = \Gamma(t) - \int_0^t \mathrm{d}s\Gamma(t-s) \int_0^s \mathrm{d}s'\Xi\left(s-s'\right)\Gamma\left(s'\right), \tag{6.87}$$

where $\Xi(t)$ is defined by,

$$\mathcal{P}A(t) = \Xi(t)\,A(0),$$

i.e. showing the most probable path. The Laplace transform of Eq.(6.87) gives,

$$\Phi(z) = \Gamma(z) - \Gamma(z)\,\Xi(z)\,\Gamma(z), \tag{6.88}$$

where $\Phi(z)$, $\Gamma(z)$ and $\Xi(z)$ are Laplace transforms of $\Phi(t)$, $\Gamma(t)$ and $\Xi(t)$, respectively. The correlation function of the nonlinear force $\Phi(t)$ and the projected intensity $\Xi(t)$ are explicitly calculable in nonlinear simulations. Thus, the memory function $\Gamma(t)$ can be directly deduced from the simulation result. The memory function is quite likely useful in constructing reduced degree of freedom models which encapsulate the results of full simulations. Such reduced models are often of critical importance to practical applications.

In the case where the time scale separation holds, i.e. the decay of the memory function is much faster than the eddy damping time of the relevant variable, Γ_0^{-1}, which is given by

$$\Gamma_0 \equiv \int_0^{\infty} \mathrm{d}s\Gamma(s), \tag{6.89}$$

the Markovian approximation holds. One has,

$$\Gamma(z) \sim \Gamma_0, \text{ and } \Xi(z) \sim \frac{1}{(z+\Gamma_0)}. \tag{6.90}$$

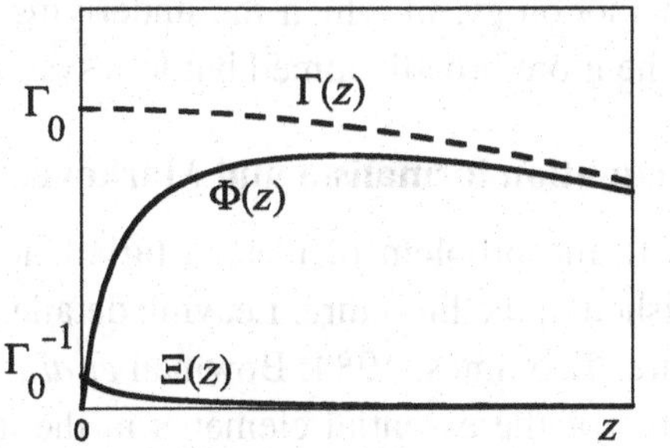

Fig. 6.11. Schematic drawing of the Laplace transforms $\Phi(z)$, $\Gamma(z)$ and $\Xi(z)$.

Figure 6.11 illustrates schematic behaviours of $\Phi(z)$, $\Gamma(z)$ and $\Xi(z)$. Representing a rapid disappearance of the memory function (due to a fast time variation of the fluctuating force), $\Gamma(z)$ extends to a larger value of z. In contrast, the autocorrelation $\Xi(t)$ decays with the transport time Γ_0, so that $\Xi(z)$ is localized in the region $|z| < \Gamma_0$. Because of this systematic motion, which decays in a time scale of Γ_0^{-1}, the correlation of the nonlinear force $\Phi(t)$ is characterized by two components, a fast component (fluctuations) and slow component (turbulent transport time scale).

A long time tail is also analyzed by use of Eq.(6.84). Correlation function $G(t) \equiv \langle A(t)\, A^\dagger(0)\rangle \langle A(0)\, A^\dagger(0)\rangle^{-1}$ and memory function $\Gamma(t)$ decay in time. Let us denote the characteristic times scales for them as τ_G and τ_Γ, respectively, i.e. $G(t) \to 0$ as $t > \tau_G$, and $\Gamma(t) \to 0$ as $t > \tau_\Gamma$. If $\tau_G \gg \tau_\Gamma$ is satisfied, the Markovian approximation holds for Eq.(6.84). In contrast, the long time tail appears for the case of $\tau_G \sim \tau_\Gamma$. From Eq.(6.84), the relation between $G(t)$ and $\Gamma(t)$ is just $\frac{\mathrm{d}}{\mathrm{d}t} G(t) = -\int_0^t \mathrm{d}s \Gamma(s)\, G(t-s)$ (for a case of scaler variable and $\Omega = 0$ in Eq.(6.84)). A long time tail is explicitly shown in Mori and Okamura (2007) and Mori (2008), by using an approximation that $G(t)$ and $\Gamma(t)$ have a same functional form, $G(t/\tau_G) = \Gamma(t/\tau_\Gamma)\,\Gamma(0)^{-1}$. In the case of $\tau_G = \tau_\Gamma$, $G(t)$ is given as $G(t) = \tau_G t^{-1} J_1(2t/\tau_G)$, where J_1 is the Bessel function. A formula similar to Eq.(6.50b) is deduced.

Put in perspective, Mori–Zwanzig theory represents both a useful tool for the elimination of irrelevant degrees of freedom by construction of an effective Langevin equation, and an important step in the development of problem reduction theory. Mori–Zwanzig theory is a bridge or intermediate step between zero-memory time Fokker–Planck models and full RG theories, in that M–Z theory allows finite or even 'long' memory (unlike Fokker–Planck models), but stops short of imposing or requiring full-scale invariance (i.e. it still requires a scale separation). Mori–Zwanzig theory is also important as an example of a rigorous, systematic closure methodology, in which the underlying assumptions are clear. It thus complements the more broadly aimed but less systematic DIA.

6.3 Langevin equation formalism and Markovian approximation

In applying the DIA to the problem of plasma turbulence, results from intense work have been published in the literature. Leaving detailed description of the formalism to the literature (Krommes, 1984; Bowman *et al.*, 1993; Krommes, 1996; Krommes, 1999), some of the essential elements in the application of DIA and Markovian approximation are explained here.

6.3.1 Langevin equation approximation

Deduction of the Langevin equation, by use of DIA, from the original nonlinear equation has been discussed. Elements in this calculation are explained here, based on the discussion in Ottaviani *et al.* (1991). In the fluid representation of plasma turbulence, where the dynamics of field quantities $\{\varphi^{\alpha}\}$ (φ^{α} represents perturbations of, e.g. density, pressure, electrostatic potential, etc.) is concluded, one encounters the equation for the $\boldsymbol{k}$-Fourier components $\varphi_{\boldsymbol{k}}^{\alpha}(t)$,

$$\frac{\partial}{\partial t}\varphi_{\boldsymbol{k}}^{\alpha}(t) + M_{\boldsymbol{k}\mu}^{\alpha}\varphi_{\boldsymbol{k}}^{\mu}(t) + \frac{1}{2}\sum_{\boldsymbol{p}+\boldsymbol{q}+\boldsymbol{k}=0} N_{\boldsymbol{kpq}}{}_{\mu\beta}^{\alpha}\varphi_{\boldsymbol{p}}^{*\mu}(t)\,\varphi_{\boldsymbol{q}}^{*\beta}(t) = 0, \tag{6.91}$$

where $M_{\boldsymbol{k}\mu}^{\alpha}$ indicates the linear matrix which dictates the linear dispersion relation (i.e. the linear eigenfrequency and linear damping rate), $N_{\boldsymbol{kpq}}{}_{\mu\beta}^{\alpha}$ denotes the nonlinear coupling coefficient for three-wave coupling, and the convention for the sum over repeated indices is employed. The correlation function,

$$C_{\boldsymbol{k}}{}^{\alpha,\beta}\left(t,t'\right) \equiv \left\langle \varphi_{\boldsymbol{k}}^{\alpha}(t)\,\varphi_{\boldsymbol{k}}^{\beta *}\left(t'\right)\right\rangle \tag{6.92}$$

is deduced from the dynamical equation (6.91). The objective of the closure modelling is to close the equation of the correlation function with appropriate renormalization of the decorrelation rate.

One of the successful applications of the DIA method to this dynamical equation is a reduction of the Langevin equation approximation. As is explained in the preceding subsection on the Mori formalism, the nonlinear interaction term (the third term in Eq.(6.91)) is decomposed into the memory function and the fluctuating force,

$$\frac{1}{2}\sum_{\boldsymbol{p}+\boldsymbol{q}=\boldsymbol{k}} N_{\boldsymbol{kpq}}{}_{\mu\beta}^{\alpha}\varphi_{\boldsymbol{p}}^{\mu}(t)\,\varphi_{\boldsymbol{q}}^{\beta}(t) = \int_0^t \mathrm{d}t'\Gamma_{\boldsymbol{k}\,\mu}^{\alpha}\left(t,t'\right)\varphi_{\boldsymbol{k}}^{\mu}\left(t'\right) - r_{\boldsymbol{k}}^{\alpha}(t), \tag{6.93}$$

where $\Gamma_{k\mu}^{\ \alpha}(t, t')$ is the memory function and $r_k^\alpha(t)$ is the random force acting on the component $\varphi_k^\alpha(t)$. With this in mind, the form of an approximate Langevin equation for Eq.(6.91) is postulated for the stochastic variable $\zeta_k^\alpha(t)$ as,

$$\frac{\partial}{\partial t}\zeta_k^\alpha(t) + M_{k\mu}^{\ \alpha}\zeta_k^\mu(t) + \int_0^t dt'\hat{\Gamma}_{k\mu}^{\alpha}(t, t')\,\zeta_k^\mu(t') = F_k^\alpha(t), \tag{6.94}$$

where $F_k^\alpha(t)$ is a fluctuating force due to nonlinearity. The correlation function of the stochastic variable $\zeta_k^\alpha(t)$ is introduced as,

$$\hat{C}_k^{\alpha,\beta}(t, t') \equiv \left\langle \zeta_k^\alpha(t)\,\zeta_k^{\beta *}(t')\right\rangle, \tag{6.95}$$

and modelling for $\hat{\Gamma}_{k\mu}^{\alpha}$ and F_k^α in Eq.(6.94) should be developed so that the relation,

$$\hat{C}_k^{\alpha,\beta}(t, t') \simeq C_k^{\ \alpha,\beta}(t, t') \tag{6.96}$$

holds to good approximation. An appropriate DIA expression for $\hat{\Gamma}_{k\mu}^{\alpha}$ is given as,

$$\hat{\Gamma}_{k\beta}^{\alpha}(t, t') = -\sum_{p,q} N_{kpq}{}_{\mu\nu}^{\ \alpha} N_{pqk}^{*}{}_{\nu'\beta}^{\ \mu'} R_{p}^{*}{}_{\mu'}^{\mu}\,\hat{C}_q^{\nu,\nu'}(t, t'), \tag{6.97}$$

and the correlation function for the fluctuating force,

$$\hat{S}_k^{\alpha,\beta}(t, t') \equiv \left\langle F_k^\alpha(t)\,F_k^{\beta *}(t')\right\rangle \tag{6.98}$$

is given by,

$$\hat{S}_k^{\alpha\beta}(t, t') = \frac{1}{2}\sum_{p,q} N_{kpq}{}_{\mu\nu}^{\ \alpha} N_{pqk}^{*}{}_{\nu'\beta}^{\ \mu'}\hat{C}_p^{*\mu,\mu'}(t, t')\,\hat{C}_q^{*\nu,\nu'}(t, t'). \tag{6.99}$$

In Eq.(6.97), the term $R_{p}{}_{\mu'}^{\mu}$ is the Green's function, which satisfies,

$$\frac{\partial}{\partial t}R_{k}{}_{\beta}^{\alpha}(t, t') + M_{k\mu}^{\ \alpha}R_{k}{}_{\beta}^{\mu}(t, t') + \int_0^t dt'\hat{\Gamma}_{k\mu}^{\alpha}(t, t')\,R_{k}{}_{\beta}^{\mu}(t, t') = \delta_\beta^\alpha\delta(t - t'), \tag{6.100}$$

where δ_β^α is the Kronecker's delta and $\delta(t - t')$ is a delta function. In the set of equations, Eqs.(6.94), (6.97), (6.99) and (6.100), the memory function and fluctuating force are given by the two-time correlation function $\hat{C}_k^{\alpha,\beta}(t, t')$, so that the model equation is *closed*, up to the second-order moments.

6.3.2 Markovian approximation

If one further assumes that the memory effect is negligible, i.e.

$$\hat{\Gamma}^{\alpha}_{k\beta}\left(t,t'\right) = \hat{\gamma}^{\alpha}_{k\beta}\left(t\right)\delta\left(t-t'\right), \tag{6.101a}$$

and the fluctuating force is a white noise,

$$\hat{S}^{\alpha,\beta}_{k}\left(t,t'\right) = \hat{s}^{\alpha,\beta}_{k}\left(t\right)\delta\left(t-t'\right), \tag{6.101b}$$

(where the time dependence in $\hat{\gamma}^{\alpha}_{k\beta}(t)$ or $\hat{s}^{\alpha,\beta}_{k}(t)$ denotes the change associated with the evolution of fluctuation intensity, which is assumed to be much slower than the correlation time of the fluctuating force), the Langevin equation and Green's function are simplified to,

$$\frac{\partial}{\partial t}\zeta^{\alpha}_{k}\left(t\right) + \left(M_{k}{}^{\alpha}_{\mu} + \hat{\gamma}^{\alpha}_{k\beta}\left(t\right)\right)\zeta^{\mu}_{k}\left(t\right) = F^{\alpha}_{k}\left(t\right), \tag{6.102a}$$

$$\frac{\partial}{\partial t}R_{k}{}^{\alpha}_{\beta}\left(t,t'\right) + \left(M_{k}{}^{\alpha}_{\mu} + \hat{\gamma}^{\alpha}_{k\beta}\left(t\right)\right)R_{k}{}^{\mu}_{\beta}\left(t,t'\right) = \delta^{\alpha}_{\beta}\delta\left(t-t'\right). \tag{6.102b}$$

The system is closed in terms of the one time correlation function,

$$I_{k}{}^{\alpha,\beta}\left(t\right) = \hat{C}^{\alpha,\beta}_{k}\left(t,t\right). \tag{6.103}$$

The magnitude of the memory function (nonlinear damping rate) and that of the fluctuating force are given by

$$\hat{\gamma}^{\alpha}_{k\beta}\left(t\right) = -\sum_{p,q} N_{kpq}{}^{\alpha}_{\mu\nu} N^{*}_{pqk}{}^{\mu'}_{\nu'\beta}\Theta_{kpq}\left(t\right) I^{*\nu,\nu'}_{q}\left(t\right), \tag{6.104a}$$

$$\hat{s}^{\alpha\beta}_{k}\left(t\right) = \frac{1}{2}\sum_{p,q} N_{kpq}{}^{\alpha}_{\mu\nu} N^{*}_{pqk}{}^{\mu'}_{\nu'\beta} I^{*\mu,\mu'}_{p}\left(t\right) I^{*\nu,\nu'}_{q}\left(t\right) \mathrm{Re}\Theta_{kpq}\left(t\right), \tag{6.104b}$$

where $\Theta_{kpq}(t)$ is a triad interaction time, i.e., the time during which three waves keep phase coherence. In a case of single evolving field (such as H-M equation), the triad interaction time is given as

$$\Theta_{kpq}\left(t\right) = \int_{0}^{t} dt' R_{k}\left(t,t'\right) R_{p}\left(t,t'\right) R_{q}\left(t,t'\right). \tag{6.105a}$$

In a stationary state, one has

$$\Theta_{kpq} = \frac{1}{M_{k} + M_{p} + M_{q} + \hat{\gamma}_{k} + \hat{\gamma}_{p} + \hat{\gamma}_{q}}. \tag{6.105b}$$

The set of dynamical equations is statistically tractable, in comparison with the original equation, because the fluctuating force is assumed to be Gaussian white noise, and the memory effects are neglected. The structure of this system closely resembles that of the renormalized Burgers equation, discussed in Section 1 of this chapter.

6.4 Closure model for drift waves

The formalisms discussed above are now applied to an example, so that some explicit relations are deduced. The nonlinear theory of drift waves is explained in Chapters 4 and 5, where special attention is paid to wave–particle interactions and wave–wave interactions. Here, we revisit the problem of drift wave turbulence clarifying the procedure and approximations which are made in closure modelling. Processes associated with the coherent damping on, as well as incoherent emission into, the test mode by background turbulence are illustrated, noting the conservation relation. Access to the nonlinear stationary state is also discussed.

For transparency of the argument, illustrations are made for the Hasegawa–Mima equation (H–M equation). This illuminates the essential elements in drift wave turbulence and their effect on closure. These include wave dispersion, wave-number space spectral evolution by nonlinearity and relation to the H-theorem.

6.4.1 Hasegawa–Mima equation

The simplest model equation which takes into account the nonlinearity of the $E \times B$ drift motion is the H–M equation (as explained in Appendix 1). We now repeat the discussion of Subection 5.3.4:

$$\frac{\partial}{\partial t}\left(\varphi - \nabla_\perp^2 \varphi\right) + \nabla_y \varphi - \nabla_\perp \varphi \; \times \hat{z} \cdot \nabla_\perp \nabla_\perp^2 \varphi = 0, \tag{6.106}$$

where normalizations are employed as,

$$\frac{\rho_s}{L_n}\omega_{\mathrm{ci}} t \to t, \; \left(\frac{x}{\rho_s}, \frac{y}{\rho_s}\right) \to (x, y), \; \frac{L_n}{\rho_s}\frac{e\tilde{\phi}}{T_e} \to \varphi, \tag{6.107}$$

$\rho_s = c_s/\omega_{\mathrm{ci}}$, ω_{ci}, is the ion cyclotron frequency, $\hat{z}$ is the direction of the mean magnetic field, and L_n is the gradient scale length of the mean plasma density, $L_n^{-1} = -\mathrm{d}\left(\ln n_0\right)/\mathrm{d}x$. The geometry of the inhomogeneous plasma and magnetic field is illustrated in Figure 6.12. The fluctuation is decomposed into Fourier components in space as:

$$\varphi\left(\boldsymbol{x}, t\right) = \sum_{\boldsymbol{k}} \varphi_{\boldsymbol{k}}\left(t\right) \exp\left(i\boldsymbol{k}\boldsymbol{x}\right), \tag{6.108}$$

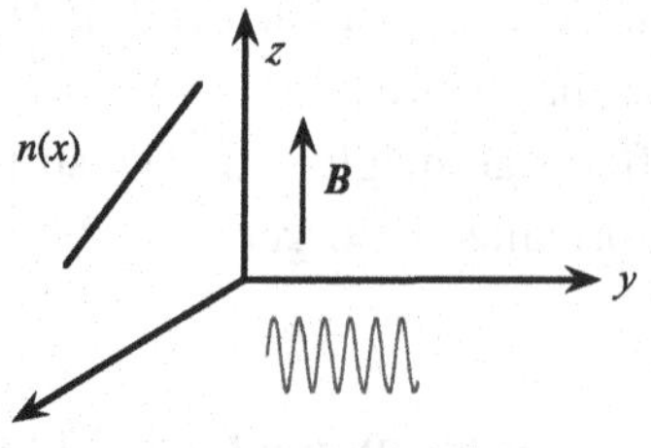

Fig. 6.12. Geometry of inhomogeneous plasma and magnetic field.

where the variable $\boldsymbol{x}$ is in two-dimensional space, so that $\boldsymbol{k}$ covers the two-dimensional Fourier space. (The suffix $\perp$, which denotes the direction perpendicular to the main magnetic field, is suppressed for simplicity of notation.) By this transformation, Eq.(6.106) becomes,

$$\frac{\partial}{\partial t}\varphi_{\boldsymbol{k}}(t) + i\omega_{\boldsymbol{k}}\varphi_{\boldsymbol{k}}(t) + \frac{1}{2}\sum_{\boldsymbol{k}=\boldsymbol{k}'+\boldsymbol{k}''} N_{\boldsymbol{k},\boldsymbol{k}',\boldsymbol{k}''}\varphi_{\boldsymbol{k}'}(t)\,\varphi_{\boldsymbol{k}''}(t) = 0, \tag{6.109}$$

where,

$$\omega_{\boldsymbol{k}} = \frac{k_y}{1+k^2} \tag{6.110}$$

is the linear wave dispersion relation and the nonlinear coupling coefficient is given as,

$$N_{\boldsymbol{k},\boldsymbol{k}',\boldsymbol{k}''} = \frac{-1}{1+k^2}\left(\boldsymbol{k}'\times\boldsymbol{k}''\cdot\hat{z}\right)\left(k''^2 - k'^2\right). \tag{6.111}$$

The reader may now recognize that the variables have changed as $\phi \rightarrow \varphi$, $V_{\boldsymbol{k},\boldsymbol{k}',\boldsymbol{k}''} \rightarrow -(1/2)N_{\boldsymbol{k},\boldsymbol{k}',\boldsymbol{k}''}$, and so on, compared with Eq.(5.56). In Eq.(6.110), the term $1+k^2$ on the right denotes the 'effective mass' (i.e. the first parenthesis) on the left of Eq.(6.106); the second term $\boldsymbol{k}'\times\boldsymbol{k}''\cdot\hat{z}$ comes from the operator $\nabla_\perp\varphi\ \times\hat{z}\ \cdot\nabla_\perp$, and the coefficient $k''^2 - k'^2$ represents the symmetrization. Equation (6.111) shows that the nonlinear coupling vanishes if $\boldsymbol{k}'$ and $\boldsymbol{k}''$ are parallel (or if $k''^2 = k'^2$ holds).

6.4.2 Application of closure modelling

Deduction of the spectral equation is explained here, step by step, illustrating the physics insights that motivate and support the approximations.

In constructing the spectral equation, iterative closure is employed, where eddy damping is assumed. The eddy damping is later determined by the self-consistency relation. Introduction of the eddy damping rate means the Markovian approximation is used, i.e. the duration of memory is assumed to be much shorter than the evolution of the spectrum. The time-scale separation is assumed in derivation of the spectral equation.

6.4.2.1 Hierarchy equations

Multiplying Eq.(6.109) by $\varphi_{\boldsymbol{k}}^{*}(t)$ and adding its complex conjugate, one obtains an equation,

$$\frac{\partial}{\partial t}|\varphi_{\boldsymbol{k}}(t)|^{2}+\mathrm{Re}\sum_{\boldsymbol{k}=\boldsymbol{k}'+\boldsymbol{k}''}N_{\boldsymbol{k},\boldsymbol{k}',\boldsymbol{k}''}\left\langle\varphi_{\boldsymbol{k}'}(t)\,\varphi_{\boldsymbol{k}''}(t)\,\varphi_{\boldsymbol{k}}^{*}(t)\right\rangle=0. \tag{6.112}$$

The essence of closure modelling is evaluation of the triad interaction term, $\left\langle\varphi_{\boldsymbol{k}'}(t)\,\varphi_{\boldsymbol{k}''}(t)\,\varphi_{\boldsymbol{k}}^{*}(t)\right\rangle$, in terms of the low-order correlations. As is shown by Eq.(6.109), the driving source for $\varphi_{\boldsymbol{k}}(t)$ contains a component which is proportional to $\varphi_{\boldsymbol{k}'}(t)\,\varphi_{\boldsymbol{k}''}(t)$. Therefore, $\varphi_{\boldsymbol{k}}(t)$ includes an element that is coherent to the *direct beat* $\varphi_{\boldsymbol{k}'}(t)\,\varphi_{\boldsymbol{k}''}(t)$, which we symbolically write as ${\varphi_{\boldsymbol{k}}}^{(c)}(t)$. By the same thinking, the element in $\varphi_{\boldsymbol{k}'}(t)$ that is coherent to $\varphi_{\boldsymbol{k}''}(t)\,\varphi_{\boldsymbol{k}}(t)$ is written as ${\varphi_{\boldsymbol{k}'}}^{(c)}(t)$, and the one in $\varphi_{\boldsymbol{k}''}(t)$ is denoted as ${\varphi_{\boldsymbol{k}''}}^{(c)}(t)$. With these notations, the correlation $\left\langle\varphi_{\boldsymbol{k}'}(t)\,\varphi_{\boldsymbol{k}''}(t)\,\varphi_{\boldsymbol{k}}^{*}(t)\right\rangle$ can be evaluated as,

$$\begin{aligned}\left\langle\varphi_{\boldsymbol{k}'}(t)\,\varphi_{\boldsymbol{k}''}(t)\,{\varphi_{\boldsymbol{k}}}^{*}(t)\right\rangle&=\left\langle\varphi_{\boldsymbol{k}'}(t)\,\varphi_{\boldsymbol{k}''}(t)\,\varphi_{\boldsymbol{k}}^{(c)*}(t)\right\rangle\\&+\left\langle\varphi_{\boldsymbol{k}'}^{(c)}(t)\,\varphi_{\boldsymbol{k}''}(t)\,\varphi_{\boldsymbol{k}}^{*}(t)\right\rangle+\left\langle\varphi_{\boldsymbol{k}'}(t)\,\varphi_{\boldsymbol{k}''}^{(c)}(t)\,\varphi_{\boldsymbol{k}}^{*}(t)\right\rangle.\end{aligned} \tag{6.113}$$

6.4.2.2 Response to a direct beat

Three terms on the right-hand side of Eq.(6.113) are evaluated as follows. Let us choose $\boldsymbol{k}$ for a label for a test mode. The mode $\boldsymbol{k}$ interacts with other modes through various combinations $(\boldsymbol{k}', \boldsymbol{k}'')$. Among possible combinations, let us take a particular set of $(\boldsymbol{k}', \boldsymbol{k}'')$, and Eq.(6.109) behaves thus,

$$\begin{aligned}&\frac{\partial}{\partial t}\varphi_{\boldsymbol{k}}(t)+i\omega_{\boldsymbol{k}}\varphi_{\boldsymbol{k}}(t)+(1/2)\sum_{\boldsymbol{k}=\boldsymbol{p}'+\boldsymbol{p}'',\boldsymbol{k}'\neq\boldsymbol{p}',\boldsymbol{p}''}N_{\boldsymbol{k},\boldsymbol{p}',\boldsymbol{p}''}\varphi_{\boldsymbol{p}'}(t)\,\varphi_{\boldsymbol{p}''}(t)\\&=-\frac{1}{2}N_{\boldsymbol{k},\boldsymbol{k}',\boldsymbol{k}''}\varphi_{\boldsymbol{k}'}(t)\,\varphi_{\boldsymbol{k}''}(t)-\frac{1}{2}N_{\boldsymbol{k},\boldsymbol{k}'',\boldsymbol{k}'}\varphi_{\boldsymbol{k}''}(t)\,\varphi_{\boldsymbol{k}'}(t),\end{aligned} \tag{6.114}$$

where $\boldsymbol{p}'$ and $\boldsymbol{p}''$ are not equal to $\boldsymbol{k}'$ but cover all other modes. Two terms on the right-hand side are identical (Figure 6.13). This equation illustrates the impact of the nonlinear source $N_{\boldsymbol{k},\boldsymbol{k}',\boldsymbol{k}''}\varphi_{\boldsymbol{k}'}(t)\,\varphi_{\boldsymbol{k}''}(t)$ on the test mode $\varphi_{\boldsymbol{k}}(t)$.

The nonlinear term $\sum_{\boldsymbol{k}=\boldsymbol{p}'+\boldsymbol{p}'',\boldsymbol{k}'\neq\boldsymbol{p}',\boldsymbol{p}''}N_{\boldsymbol{k},\boldsymbol{p}',\boldsymbol{p}''}\varphi_{\boldsymbol{p}'}(t)\,\varphi_{\boldsymbol{p}''}(t)$ can be related to the memory function on the test mode and the fluctuating force, as explained in

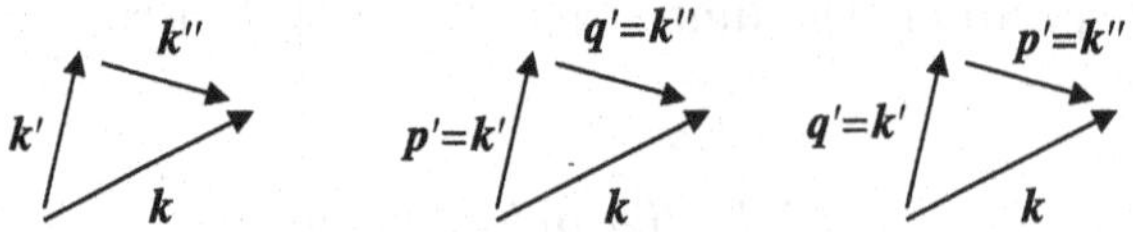

Fig. 6.13. Three-wave interaction (left). The source for the direct beat comes from the combinations centre and right.

Section 6.2. For transparency of the argument, the Markovian approximation is used for the memory function in the closure modelling described in this subsection. With this approximation, the memory function is replaced by the eddy-damping rate as,

$$(1/2) \sum_{k=p'+p'',k'\neq p',p''} N_{k,p',p''}\varphi_{p'}(t)\,\varphi_{p''}(t) = \hat{\hat{\gamma}}_k\varphi_k(t) - \hat{F}_k(t), \qquad (6.115)$$

where $\hat{\hat{\gamma}}$ is a slowly varying, eddy-damping rate and $\hat{F}_k(t)$ is a rapidly changing fluctuating force. This is analogous to separation of the total nonlinear term in Eq.(6.109) into the eddy damping rate and fluctuating force as,

$$\frac{1}{2}\sum_{k=k'+k''} N_{k,k',k''}\varphi_{k'}(t)\,\varphi_{k''}(t) = \hat{\gamma}_k\varphi_k(t) - F_k(t). \qquad (6.116)$$

Substituting Eq.(6.115) into Eq.(6.114), one has,

$$\frac{\partial}{\partial t}\varphi_k(t) + \left(i\omega_k + \hat{\hat{\gamma}}_k\right)\varphi_k(t) = \hat{F}_k(t) - N_{k,k',k''}\varphi_{k'}(t)\,\varphi_{k''}(t). \qquad (6.117)$$

The response of $\varphi_k(t)$, which is induced by imposition of the source terms $\hat{F}_k$ and $-N_{k,k',k''}\varphi_{k'}(t)\,\varphi_{k''}(t)$, is written as,

$$\varphi_k(t) = \int_{-\infty}^{t} \mathrm{d}t' \exp\left(\left(i\omega_k + \hat{\hat{\gamma}}_k\right)(t'-t)\right) \times \left(\hat{F}_k(t') - N_{k,k',k''}\varphi_{k'}(t')\,\varphi_{k''}(t')\right). \qquad (6.118)$$

6.4.2.3 *Steps in closure modelling*

The main physical considerations for closure modelling are as follows:

(i) The number of excited fluctuations should be so large that the memory function $\hat{\hat{\gamma}}_k$ is approximately equal to that for the total nonlinear terms, i.e. sufficient to justify the validity of the test wave hypothesis,

$$\hat{\hat{\gamma}}_k = \hat{\gamma}_k. \qquad (6.119a)$$

(ii) The excited modes are nearly independent, so that the fluctuating force $\hat{F}_k(t)$ in Eq.(6.106) is incoherent with $\varphi_{k'}(t)\,\varphi_{k''}(t)$, i.e.

$$\left\langle \hat{F}_k^*(t)\,\varphi_{k'}(t)\,\varphi_{k''}(t)\right\rangle = 0. \tag{6.119b}$$

(iii) The two-time correlation function is expressed in terms of one-time correlation functions as explained below.

With the ansatz (6.119b), the component ${\varphi_k}^{(c)}(t)$ is evaluated from Eq.(6.118) as,

$$\varphi_k^{(c)}(t) = -\int_{-\infty}^{t} \mathrm{d}t' \exp\left(\left(i\omega_k + \hat{\gamma}_k\right)\left(t'-t\right)\right) N_{k,k',k''}\varphi_{k'}\left(t'\right)\varphi_{k''}\left(t'\right). \tag{6.120}$$

With the help of Eq.(6.120), the first term on the right-hand side of Eq.(6.114) is evaluated as,

$$\left\langle \varphi_{k'}(t)\,\varphi_{k''}(t)\,\varphi_k^{(c)*}(t)\right\rangle = -\,N^*_{k,k',k''}\int_{-\infty}^{t} \mathrm{d}t' \exp\left(\left(-i\omega_k + \hat{\gamma}_k\right)\left(t'-t\right)\right)$$
$$\times\left\langle \varphi_{k'}(t)\,\varphi^*_{k'}\left(t'\right)\varphi_{k''}(t)\,\varphi^*_{k''}\left(t'\right)\right\rangle. \tag{6.121}$$

Within the assumption of almost-independent fluctuations (i.e. quasi-Gaussian statistics),

$$\left\langle \varphi_{k'}(t)\,\varphi^*_{k'}\left(t'\right)\varphi_{k''}(t)\,\varphi^*_{k''}\left(t'\right)\right\rangle = \left\langle \varphi_{k'}(t)\,\varphi^*_{k'}\left(t'\right)\right\rangle\left\langle \varphi_{k''}(t)\,\varphi^*_{k''}\left(t'\right)\right\rangle. \tag{6.122}$$

The next step is to express the two-time correlation function by use of the one-time correlation functions. It is noted that, in the limit of the Markovian approximation, the eddy-damping rate is equal to the decorrelation rate of the spectral function. This is confirmed as follows. Substituting Eq. (6.106) into Eq.(6.109), the solution is,

$$\varphi_k(t) = \int_{-\infty}^{t} \mathrm{d}t' \exp\left(\left(i\omega_k + \hat{\gamma}_k\right)\left(t'-t\right)\right) F_k\left(t'\right), \tag{6.123}$$

so that the correlation function is,

$$\left\langle \varphi_k^*(t)\,\varphi_k(t+\tau)\right\rangle = \exp\left(-i\omega_k\tau - \hat{\gamma}_k\left|\tau\right|\right)\left\langle \varphi_k^*(t)\,\varphi_k(t)\right\rangle. \tag{6.124}$$

Therefore, the decorrelation rate of the spectral function is equal to the eddy-damping rate. Similar relations hold for $\left\langle \varphi_{k'}(t)\,\varphi^*_{k'}\left(t'\right)\right\rangle$ and $\left\langle \varphi_{k''}(t)\,\varphi^*_{k''}\left(t'\right)\right\rangle$, so that combination of Eqs.(6.121), (6.122) and (6.124) provides a closed relation for the first term on the right-hand side of Eq.(6.114) as,

$$\left\langle \varphi_{k'}(t)\varphi_{k''}(t)\,\varphi_k^{(c)*}(t)\right\rangle = -N^*_{k,k',k''}\left\langle \varphi_{k'}(t)\,\varphi^*_{k'}(t)\right\rangle\left\langle \varphi_{k''}(t)\,\varphi^*_{k''}(t)\right\rangle$$
$$\times \int_{-\infty}^{t} \mathrm{d}t' \exp\left(i\left(-\omega_k+\omega_{k'}+\omega_{k''}\right)\left(t'-t\right)+\left(\hat{\gamma}_k+\hat{\gamma}_{k'}+\hat{\gamma}_{k''}\right)\left(t'-t\right)\right). \tag{6.125}$$

That is,

$$\left\langle \varphi_{k'}(t)\varphi_{k''}(t)\varphi_k^{(c)*}(t)\right\rangle = -N^*_{k,k',k''}\Theta_{k,k',k''}\left\langle \varphi_{k'}(t)\varphi^*_{k'}(t)\right\rangle\left\langle \varphi_{k''}(t)\varphi^*_{k''}(t)\right\rangle, \tag{6.126a}$$

and,

$$\Theta_{k,k',k''} = \frac{1}{i\left(-\omega_k+\omega_{k'}+\omega_{k''}\right)+\left(\hat{\gamma}_k+\hat{\gamma}_{k'}+\hat{\gamma}_{k''}\right)}, \tag{6.126b}$$

where $\Theta_{k,k',k''}$ is a triad interaction time. Repeating the same procedure for the second and third terms on the right-hand side of Eq.(6.114), one obtains,

$$\left\langle \varphi_{k'}^{(c)}(t)\,\varphi_{k''}(t)\,\varphi_k^*(t)\right\rangle$$
$$= -N^*_{k',-k'',k}\Theta_{k,k',k''}\left\langle \varphi_{k''}(t)\,\varphi^*_{k''}(t)\right\rangle\left\langle \varphi_k(t)\,\varphi^*_k(t)\right\rangle, \tag{6.127a}$$

$$\left\langle \varphi_{k'}(t)\,\varphi_{k''}^{(c)}(t)\,\varphi_k^*(t)\right\rangle$$
$$= -N^*_{k'',-k',k}\Theta_{k,k',k''}\left\langle \varphi_{k'}(t)\,\varphi^*_{k'}(t)\right\rangle\left\langle \varphi_k(t)\,\varphi^*_k(t)\right\rangle, \tag{6.127b}$$

where the relation $\Theta_{k',-k'',k} = \Theta_{k'',-k',k} = \Theta_{k,k',k''}$ is used. For abbreviation of expression, a one-time spectral function is written as,

$$I_k = \left\langle \varphi_k(t)\,\varphi_k^*(t)\right\rangle. \tag{6.128}$$

Combination of Eqs.(6.114), (6.126) and (6.127) gives,

$$\sum_{k=k'+k''} N_{k,k',k''}\left\langle \varphi_{k'}(t)\,\varphi_{k''}(t)\,\varphi_k^*(t)\right\rangle$$
$$= -\sum_{k=k'+k''}\left|N_{k,k',k''}\right|^2\Theta_{k,k',k''}I_{k'}I_{k''}$$
$$-\sum_{k=k'+k''} N_{k,k',k''}N^*_{k',-k'',k}\Theta_{k,k',k''}I_{k'}I_k$$
$$-\sum_{k=k'+k''} N_{k,k',k''}N^*_{k'',-k',k}\Theta_{k,k',k''}I_{k''}I_k. \tag{6.129}$$

The first term on the right-hand side of Eq.(6.129) does not include I_k explicitly, so that it acts as a source for the excitation of I_k. The second and third terms are in proportion to I_k, i.e. they behave as the damping (or growth, in particular circumstances) terms for the test mode. The dynamic equation (6.112) takes the form,

$$\frac{\partial}{\partial t} I_k + \hat{\Gamma}_k I_k = \sum_{k=k'+k''} \left| N_{k,k',k''} \right|^2 \mathrm{Re}\, \Theta_{k,k',k''} I_{k'} I_{k''}, \tag{6.130a}$$

with

$$\hat{\Gamma}_k = - \sum_{k=k'+k''} \mathrm{Re}\, \Theta_{k,k',k''} N_{k,k',k''} \left(N^*_{k'',-k',k} I_{k''} + N^*_{k',-k'',k} I_{k'} \right), \tag{6.130b}$$

where two terms on the right-hand side give identical contributions. The consistency condition, $\hat{\Gamma}_k = 2\hat{\gamma}_k$, i.e.

$$\hat{\gamma}_k = - \sum_{k=k'+k''} \mathrm{Re}\, \Theta_{k,k',k''} N_{k,k',k''} N^*_{k',-k'',k} I_{k'} \tag{6.131}$$

closes the loop. Equations (6.102), (6.131) and (6.126b) form a set of equations, which together describe the evolution of the spectrum taking into account the nonlinear coupling between turbulent components.

6.4.3 On triad interaction time

The set of equations (6.130), (6.131) and (6.126b) is a closed system for the spectral functions I_k, which is slowly evolving in time, and does not include the higher order correlations. In this sense, this is a set of 'closed' equations up to the second-order moments. In these coupled equations, the triad interaction time $\mathrm{Re}\, \Theta_{k,k',k''}$ plays the key role in determining the nonlinear coupling coefficients. It takes the form,

$$\mathrm{Re}\, \Theta_{k,k',k''} = \frac{\hat{\gamma}_k + \hat{\gamma}_{k'} + \hat{\gamma}_{k''}}{\left(\omega_{k'} + \omega_{k''} - \omega_k\right)^2 + \left(\hat{\gamma}_k + \hat{\gamma}_{k'} + \hat{\gamma}_{k''}\right)^2}, \tag{6.132}$$

and denotes the time that the interaction of three modes persists. Several limiting forms for the turbulent decorrelation rate are implied.

The properties of the coherent damping rate, $\hat{\gamma}_k$ can be illuminated by considering various limits. The consistency relation for the nonlinear damping rate (6.131) provides,

$$\hat{\gamma}_k = \sum_{k=k'+k''} N_{k,k'',k'} N^*_{-k',k'',-k} I_{k'} \frac{\hat{\gamma}_k + \hat{\gamma}_{k'} + \hat{\gamma}_{k''}}{\left(\omega_{k'} + \omega_{k''} - \omega_k\right)^2 + \left(\hat{\gamma}_k + \hat{\gamma}_{k'} + \hat{\gamma}_{k''}\right)^2}.$$

In the weak turbulence limit, $\hat{\gamma}_k \rightarrow 0$, the nonlinear damping rate is given as,

$$\hat{\gamma}_k = \sum_{k=k'+k''} N_{k,k'',k'} N^*_{-k',k'',-k} I_{k'} \pi \delta \left(\omega_{k'} + \omega_{k''} - \omega_k\right). \tag{6.133}$$

This is the result which is given in the quasi-linear limit, which is discussed in Chapter 5, see Eq.(5.63). On resonance, $\omega_{k'} + \omega_{k''} = \omega_k$, the scrambling is finite for arbitrarily small amplitude. Thus, the fundamental origin of the irreversibility is three-wave interaction resonance. In an opposite limit, $\hat{\gamma}_k >> \left|\omega_{k'} + \omega_{k''} - \omega_k\right|$, one has a form,

$$\hat{\gamma}_k = \sum_{k=k'+k''} N_{k,k'',k'} N^*_{-k',k'',-k} I_{k'} \frac{1}{\hat{\gamma}_k + \hat{\gamma}_{k'} + \hat{\gamma}_{k''}}. \tag{6.134}$$

This provides an order of magnitude estimate,

$$\hat{\gamma}_k \simeq \left(\sum_{k=k'+k''} N^2 I_{k'} \right)^{1/2} \sim (N\varphi)_{\text{rms}}, \tag{6.135a}$$

with

$$(N\varphi)_{\text{rms}} \simeq \left(\frac{k^3}{1+k^2} \frac{\tilde{E}_y}{B} \right)_{\text{rms}} \sim \left(k\tilde{V}\right)_{\text{rms}}, \tag{6.135b}$$

where $\tilde{V}$ is the $E \times B$ velocity associated with the fluctuating field. This result shows that, in the strong turbulence limit (where the nonlinear damping rate is faster than the triad dispersion time), the nonlinear damping rate is of the order of the eddy circulation time $\left(k\tilde{V}\right)_{\text{rms}}$. It is the root-mean square time for an $E \times B$ motion to move a distance of one wavelength. Considering that the nonlinear damping rate equals the decorrelation rate of the auto-correlation function, Eq.(6.124), Eq.(6.135) is equivalent to the Kubo number,

$$\mathcal{K} = \frac{\text{auto-correlation time}}{\text{eddy turn-over time}} = \frac{\left(k\tilde{V}\right)_{\text{rms}}}{\hat{\gamma}_k}, \tag{6.136}$$

being close to unity, i.e.

$$\mathcal{K} \simeq 1, \tag{6.137}$$

in the states of stationary turbulence that the H–M equation describes.

The condition that the nonlinear damping rate is in the range of the wave frequency, $\hat{\gamma}_k \sim \omega_*$ can be rewritten in a dimensional form as,

$$\frac{\tilde{n}}{n} \simeq \frac{e\tilde{\phi}}{T_e} \simeq \frac{1}{k_x L_n}, \tag{6.138}$$

which is often referred to as 'mixing length estimate'.

The formula for the triad interaction time Eq.(6.132) shows that whether the fluctuations are in the strong turbulence regime or not depends on the dispersion relation of the waves. For instance, the sound waves, for which the relation $\omega_k = c_s k$ holds, are non-dispersive. The frequency mismatch vanishes if the wave number matching holds. Therefore, non-dispersive waves are *always* in the "strong turbulence regime". In contrast, for dispersive waves, the condition $\hat{\gamma}_k \sim \omega_k$ sets a boundary in the wave amplitude for the strong turbulence regime. For drift waves, since the dispersion relation is,

$$\omega_k = \frac{V_{\mathrm{de}} k_y}{1 + k^2 \rho_s^2},$$

in the long wavelength limit $(1 >> k^2\rho_s^2)$ the drift waves are non-dispersive. In contrast, they are dispersive in the short wavelength limit, $1 << k^2\rho_s^2$. Therefore, long wavelength drift wave turbulence is *always* in a "strong turbulence regime".

6.4.4 Spectrum

In a stationary state,

$$2\hat{\gamma}_k I_k = \sum_{k=k'+k''} \left| N_{k,k',k''} \right|^2 \mathrm{Re}\, \Theta_{k,k',k''} I_{k'} I_{k-k'}, \tag{6.139}$$

that is, the damping by the nonlinear interaction (left-hand side) balances with the incoherent emission (right-hand side). In other words,

$$I_k = \frac{1}{2\hat{\gamma}_k} \sum_{k=k'+k''} \left| N_{k,k',k-k'} \right|^2 \mathrm{Re}\, \Theta_{k,k',k-k'} I_{k'} I_{k-k'}. \tag{6.140}$$

In order to highlight the non-local transfer of energy in the wave number space, it might be useful to write,

$$I_{k-k'} = I_k - \boldsymbol{k}' \cdot \frac{\partial}{\partial \boldsymbol{k}} I_k + \frac{1}{2}\left(\boldsymbol{k}' \cdot \frac{\partial}{\partial \boldsymbol{k}}\right)^2 I_k + \cdots. \tag{6.141}$$

Substitution of Eq.(6.141) into Eq.(6.140) provides,

$$\begin{aligned}
&\frac{1}{2\hat{\gamma}_k} \sum_{k=k'+k''} \left|N_{k,k',k-k'}\right|^2 \mathrm{Re}\ \Theta_{k,k',k-k'} I_{k'} \left(k' \cdot \frac{\partial}{\partial k}\right)^2 I_k \\
&- \frac{1}{\hat{\gamma}_k} \sum_{k=k'+k''} \left|N_{k,k',k-k'}\right|^2 \mathrm{Re}\ \Theta_{k,k',k-k'} I_{k'} k' \cdot \frac{\partial}{\partial k} I_k \\
&+ \left(\frac{1}{\hat{\gamma}_k} \sum_{k=k'+k''} \left|N_{k,k',k-k'}\right|^2 \mathrm{Re}\ \Theta_{k,k',k-k'} I_{k'} - 2 \right) I_k \\
&= 0,
\end{aligned} \qquad (6.142)$$

where the first term on the left-hand side shows the k-space diffusion, and the second term is the k-space flow.

6.4.5 Example of dynamical evolution – access to statistical equilibrium and H-theorem

6.4.5.1 Statistical equilibrium in nonlinear dynamics

Evolution of the spectrum follows the nonlinear interaction, which is formulated in Eqs.(6.130), (6.131) and (6.126b), as well as the source and sink in the wave number space. The competition between the (linear) growth, nonlinear energy exchange, and dissipation by the molecular viscosity determines the final stationary spectrum. One characteristic example of nonlinear dynamics is the problem of access to the statistical equilibrium, in the absence of the source and sink (Gang *et al.*, 1991). In this problem, the closure theory also provides a powerful method of analysis.

Let us consider the modes that obey the H–M equation. The range of wave numbers is specified by $k_{\min} < k < k_{\max}$. It is known that the H–M equation has the following conserved quantities, namely energy E and potential enstrophy Ω, i.e., $E = \iint \mathrm{d}^2x \left(\varphi^2 + (\nabla\varphi)^2\right)$, and $\Omega = \iint \mathrm{d}^2x \left(\varphi - \nabla^2\varphi\right)^2$, respectively. (See Subsection 5.3.4 and Appendix 1.) The difference between the H–M equation and the corresponding expressions for ordinary 2D fluids is that the density perturbation appears in the plasma dynamics, $\tilde{n}/n \simeq e\tilde{\phi}/T_e$. For example, the density perturbation (i.e. the pressure perturbation) appears as the first term in the parenthesis of E.

In a Fourier representation, one has two conserved quantities,

$$E = \sum_k \left(1 + k^2\right) I_k, \text{ and } \quad \Omega = \sum_k \left(1 + k^2\right)^2 I_k. \qquad (6.143)$$

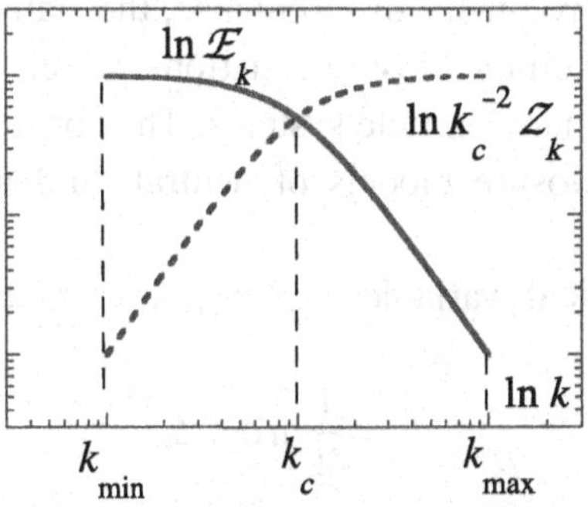

Fig. 6.14. Spectral distribution of the statistical equilibrium which is realized by the nonlinear interaction of the H–M equation.

These "dual conservation" relations require that, if the potential enstrophy spectrum spreads to the higher ks, the energy is preferentially transported to the lower ks, as is illustrated in Figure 2.15.

The energy density and the enstrophy density for the k-mode are written as,

$$E_k = \left(1 + k^2\right) I_k, \quad Z_k = k^2 E_k = k^2 \left(1 + k^2\right) I_k, \tag{6.144}$$

where $\sum_k E_k$ and $\sum_k Z_k$ are conserved during the nonlinear interaction of excited modes. Therefore, in the statistical equilibrium, partition is expected for the variable,

$$\alpha E_k + \beta Z_k,$$

where the coefficients α and β are determined by the condition that $\sum_k E_k$ and $\sum_k Z_k$ are identical to the initial conditions. The equi-partition of the quantity $\alpha E_k + \beta Z_k$ means that this quantity is independent of the wave number k, i.e. the spectrum of the energy density E_k takes a form $E_k \propto \left(\alpha + \beta k^2\right)^{-1}$, that is,

$$E_k = \frac{E_{k \to 0}}{1 + k^2 k_c^{-2}}, \tag{6.145}$$

where the ratio α / β is rewritten as k_c^2. This statistical equilibrium distribution indicates the approximate equi-partition of energy in the long wavelength regime, $k^2 < k_c^2$, while the approximate equi-partition of the enstrophy density Z_k holds in the short wavelength limit, $k^2 > k_c^2$. The spectral distribution function is illustrated in Figure 6.14.

6.4.5.2 H-theorem

Access of the spectrum to the statistical equilibrium one (in the absence of the source and sink) is governed by the H-theorem, and closure modelling is useful for explicitly demonstrating access to the statistical equilibrium.

In the statistical closure theory of turbulence, the statistical evolution of turbulence is described based on the closure equations, which play a similar role to the Boltzmann equation in many particle systems. The construction of the H-theorem has been extended to closure models of neutral fluid turbulence by Carnevale (1982).

For a set of dynamical variables $(z_1, z_2, z_3, \cdots)$, an entropy functional is introduced as,

$$S = \frac{1}{2} \ln \det \mathbf{Z}, \tag{6.146}$$

where $\mathbf{Z}$ is a matrix, the (i, j)-th element of which is given by $Z_{ij} = \langle z_i z_j \rangle$. The statistical information for this dynamical variable is specified by the set of two-body correlations $\langle z_i z_j \rangle$, and is given by $-S$. Therefore, reduction of the information corresponds to an increment in S. Choosing I_k as for the dynamical variable of interest z_i, the entropy function for the H–M equation is constructed. In the modelling of the closure, the correlation between two different components is taken to be near zero, so the off-diagonal elements are much smaller than the diagonal elements. Then, one has the estimate,

$$\det Z = \prod_k I_k,$$

so that the entropy functional is introduced as,

$$S = \frac{1}{2} \sum_k \ln I_k. \tag{6.147}$$

The evolution of the entropy functional is given by,

$$\frac{\partial}{\partial t} S = \frac{1}{2} \sum_k \frac{1}{I_k} \frac{\partial}{\partial t} I_k. \tag{6.148}$$

The closure modelling provides explicit expression for the evolution of the entropy functional. Equation (6.130a) is rewritten, upon rearranging the wave number index, as:

$$\frac{\partial}{\partial t} I_k = \sum_{\mathbf{k}=\mathbf{k}'+\mathbf{k}''} \left| \mathbf{k}' \times \mathbf{k}'' \right|^2 \frac{k''^2 - k'^2}{1+k^2} \mathrm{Re}\, \Theta_{\mathbf{k},\mathbf{k}',\mathbf{k}''}$$
$$\times \left(\frac{k''^2 - k'^2}{1+k^2} I_{k'} I_{k''} + \frac{k'^2 - k^2}{1+k''^2} I_{k'} I_k + \frac{k^2 - k''^2}{1+k'^2} I_{k''} I_k \right), \tag{6.149}$$

where the first term in the parenthesis on the right-hand side shows the excitation of the element $I_{\boldsymbol{k}}$ by background turbulence, and the second and third terms denote the turbulent damping. One has,

$$\frac{1}{I_{\boldsymbol{k}}}\frac{\partial}{\partial t}I_{\boldsymbol{k}} = \sum_{\boldsymbol{k}=\boldsymbol{k}'+\boldsymbol{k}''} \left|\boldsymbol{k}' \times \boldsymbol{k}''\right|^2 \mathrm{Re}\,\Theta_{\boldsymbol{k},\boldsymbol{k}',\boldsymbol{k}''} I_{\boldsymbol{k}} I_{\boldsymbol{k}'} I_{\boldsymbol{k}''}$$
$$\times \frac{k''^2 - k'^2}{(1+k^2)\,I_{\boldsymbol{k}}}\left(\frac{k''^2 - k'^2}{\left(1+k^2\right)I_{\boldsymbol{k}}} + \frac{k'^2 - k^2}{\left(1+k''^2\right)I_{\boldsymbol{k}''}} + \frac{k^2 - k''^2}{\left(1+k'^2\right)I_{\boldsymbol{k}'}}\right), \tag{6.150}$$

It is useful to rewrite Eq.(6.148) as,

$$\frac{\partial}{\partial t}S = \frac{1}{6}\left(\sum_{\boldsymbol{k}}\frac{1}{I_{\boldsymbol{k}}}\frac{\partial}{\partial t}I_{\boldsymbol{k}} + \sum_{\boldsymbol{k}'}\frac{1}{I_{\boldsymbol{k}'}}\frac{\partial}{\partial t}I_{\boldsymbol{k}'} + \sum_{\boldsymbol{k}''}\frac{1}{I_{\boldsymbol{k}''}}\frac{\partial}{\partial t}I_{\boldsymbol{k}''}\right), \tag{6.151}$$

in order to take advantage of the symmetry property. Rotating symbols $\boldsymbol{k}$, $\boldsymbol{k}'$ and $\boldsymbol{k}''$ in Eq.(6.150) and substituting them into Eq.(6.151), one has,

$$\frac{\partial}{\partial t}S = \frac{1}{6}\sum_{\boldsymbol{k}}\sum_{\boldsymbol{k}=\boldsymbol{k}'+\boldsymbol{k}''} \left|\boldsymbol{k}' \times \boldsymbol{k}''\right|^2 \mathrm{Re}\Theta_{\boldsymbol{k},\boldsymbol{k}',\boldsymbol{k}''} I_{\boldsymbol{k}} I_{\boldsymbol{k}'} I_{\boldsymbol{k}''}$$
$$\times \left(\frac{k^2 - k''^2}{\left(1+k'^2\right)I_{\boldsymbol{k}'}} + \frac{k'^2 - k^2}{\left(1+k''^2\right)I_{\boldsymbol{k}''}} + \frac{k''^2 - k'^2}{\left(1+k^2\right)I_{\boldsymbol{k}}}\right)^2, \tag{6.152}$$

where the relation $I_{-\boldsymbol{k}} = I_{\boldsymbol{k}}$ is used. All terms on the right-hand side of Eq. (6.152) are positive definite, thus Eq.(6.152) shows that,

$$\frac{\partial}{\partial t}S \geq 0. \tag{6.153}$$

The entropy functional is an increasing function of time. One can also show that the spectral function, Eq.(6.145), satisfies the relation,

$$\frac{\left(k^2 - k''^2\right)}{\left(1+k'^2\right)I_{\boldsymbol{k}'}} \propto \left(k^2 - k''^2\right)\left(1 + k'^2 k_c^{-2}\right).$$

Thus, one has the relation,

$$\frac{k^2 - k''^2}{\left(1+k'^2\right)I_{\boldsymbol{k}'}} + \frac{k'^2 - k^2}{\left(1+k''^2\right)I_{\boldsymbol{k}''}} + \frac{k''^2 - k'^2}{\left(1+k^2\right)I_{\boldsymbol{k}}} = 0, \tag{6.154}$$

i.e. $\partial S/\partial t = 0$ holds for the spectral function (6.145).

It is also possible to show that the class of spectrum that is given by Eq.(6.145) is the only solution, for which Eq.(6.153) vanishes. This is shown explicitly. Assume $\left(1+k^2\right) I_k$ includes an additional term in the denominator such that,

$$\frac{1}{\left(1+k^2\right) I_k} \propto \left(1+k^2 k_c^{-2}+\alpha k^{2m}\right), \; m>1. \tag{6.155}$$

Then one has,

$$\begin{aligned} &\frac{k^2-k''^2}{\left(1+k'^2\right) I_{k'}}+\frac{k'^2-k^2}{\left(1+k''^2\right) I_{k''}}+\frac{k''^2-k'^2}{\left(1+k^2\right) I_k} \\ &=\alpha\left(k^2-k''^2\right) k'^{2m}+\alpha\left(k'^2-k^2\right) k''^{2m} \\ &+\alpha\left(k''^2-k'^2\right) k^{2m}. \end{aligned} \tag{6.156}$$

This gives a positive contribution to $\partial S/\partial t$, so the entropy functional is the maximal for the spectral function (6.145). In other words, the spectrum evolves to the state Eq.(6.145), due to the nonlinear interaction.

In this state, the exchange of the quantity,

$$Z_k=\left(1+k^2 k_c^{-2}\right)\left(1+k^2\right) I_k$$

among different Fourier components leads to its equipartition. The problem of the statistical equilibrium in the absence of the source and sink is different from the case where the stationary turbulence is realized by the balance between the source and sink, and coupling between them is carried out by nonlinear interaction. In the case of stationary driven turbulence (with sink), the state is characterized by the flow of energy in the mode number space, and the flow is a parameter that characterizes the non-equilibrium state.

6.5 Closure of kinetic equation

From the viewpoint of reducing degrees of freedom in plasma dynamics, it is relevant to present a short discussion on the fluid moment equation for collisionless plasmas. Fluid moment equations are much simpler than the original kinetic equation, in which the evolution of particle velocity is kept. Because of their simplicity, focused studies have been developed analyzing nonlinear interactions, as explained in Chapter 5.

As is discussed in, e.g. Eq.(4.53), the equation of the i-th moment includes the $(i+1)$-th moment. In the case of Eq.(4.53c), the moment $\overline{v_z v^2}$ appears, where the over-bar is an average over the distribution function $f(\boldsymbol{v})$. This third-order moment needs to be expressed in terms of the density, velocity or pressure, in order to 'close' the set of equations (4.53). In deducing a closed set of equations for moments, higher-order moments must be modelled by lower-order moments. Thus, 'closure' theory has also been developed in plasma physics.

Systematic deduction of the fluid moment equations has been developed in, e.g. Braginskii (1965), where the expansion parameter is the mean free time $1/\nu$ (interval between collisions) relative to the characteristic dynamical time (frequency, particle transit time, etc.) However, as the collision frequency of plasma particles becomes lower, the $1/\nu$-expansion fails. The transit time of particles governs the response as well as the closure of fluid moments.

There are a couple of closure models for kinetic plasmas where Landau damping is taken into account, as is explained below. The modelling is illustrated by the equations,

$$\frac{\partial}{\partial t} n + \nabla \cdot n\boldsymbol{V} = 0, \tag{6.158a}$$

$$mn\left(\frac{\partial}{\partial t}\boldsymbol{V} + \boldsymbol{V}\cdot\nabla\boldsymbol{V}\right) = en\left(\boldsymbol{E} + \frac{1}{c}\boldsymbol{V}\times\boldsymbol{B}\right) - \nabla p - \nabla^2\Pi, \tag{6.158b}$$

$$\frac{3}{2}n\left(\frac{\partial}{\partial t}T + \boldsymbol{V}\cdot\nabla T\right) = -p\nabla\cdot\boldsymbol{V} - \Pi : \nabla\boldsymbol{V} - \nabla\cdot\boldsymbol{q} + Q, \tag{6.158c}$$

where n is the density, $\boldsymbol{V}$ is the velocity, T is the temperature and p is the pressure. The stress tensor Π and the heat flux $\boldsymbol{q}$ represent the flux of momentum and energy, and need to be modelled in terms of lower-order variables. The linear response of Π and $\boldsymbol{q}$ to perturbations has been explored so as to have forms (e.g., for ions),

$$\boldsymbol{b}\cdot\nabla\cdot\Pi_{\parallel} = -n_0 m_i \mu_{\parallel} \nabla_{\parallel}^2 \tilde{V}_{\parallel} - \lambda_{1i} n_0 \nabla_{\parallel}\tilde{T}, \tag{6.159a}$$

$$q_{\parallel} = -\lambda_{2i} p_0 \tilde{V}_{\parallel} - n_0 \chi_{\parallel} \nabla_{\parallel}\tilde{T}, \tag{6.159b}$$

where $\boldsymbol{b}$ is the unit vector in the direction of the main magnetic field, n_0 and p_0 are the mean density and pressure, respectively, and the tilde ˜ indicates the perturbations.

One of the first efforts in parallel direction was given in Lee and Diamond (1986), where the effective viscosity was estimated as,

$$\mu_{\parallel} \simeq \min\left(\frac{v_{\mathrm{Ti}}^2}{\nu_{\mathrm{ii}}}, \frac{v_{\mathrm{Ti}}^2}{\omega}\right), \tag{6.160a}$$

where ν_{ii} is the ion–ion collision frequency and v_{Ti} is the thermal velocity of ions. This study has pointed out two essential features. First, the ion velocity inhomogeneity $\nabla_{\parallel}^2 \tilde{V}_{\parallel}$ decays with the coefficient v_{Ti}^2/ω in the collisionless limit. This is the result of ion Landau damping, where waves are subject to decay owing to coupling with ions. The other element is that the damping with coefficient v_{Ti}^2/ω has the non-Markovian form. The coefficient is dependent on the frequency, so the damping term in real coordinates is expressed in a form of memory function. The appearance of the non-Markovian form is also explained in the preceding sections of this chapter.

The interaction between waves and plasma particles also affects the evolution of energy (Waltz, 1988; Hamaguchi and Horton, 1990). Thus, the effective thermal conductivity, which represents the wave–particle interaction, was introduced in the limit of $\nu \to 0$ as, e.g. (Waltz, 1988),

$$\mu_{\parallel} = \chi_{\parallel} \simeq \min\left(\frac{v_{Ti}}{k_{\parallel}}, \frac{v_{Ti}^2}{\omega}\right), \tag{6.160b}$$

where the limiting time scale by the transit frequency $v_{Ti}k_{\parallel}$ is introduced in order to cover the low-frequency perturbations, $v_{Ti}k_{\parallel} > \omega$, as well. Equations (6.160a) and (6.160b) illustrate the essence that the key elements, i.e. (1) the kinetic interactions are modelled in terms of the effective diffusion operator, (2) the step size v_{Ti}, and limiting time scales of $v_{Ti}k_{\parallel}$ and ω (in stead of ν), and (3) the non-Markovian nature of the interaction. For instance, the model heat flux (Hammett and Perkins, 1990),

$$\tilde{q}_k = -n_0 \chi_{\parallel} \nabla \tilde{T} = -i \frac{n_0 v_{Ti}}{|k_{\parallel}|} k_{\parallel} \tilde{T} \tag{6.161a}$$

is rewritten in real space in a form of space integral. Fourier composition gives

$$\tilde{q}(z) = -n_0 v_{Ti} \int_0^{v_{Ti}/\omega} \mathrm{d}l \frac{\tilde{T}(z+l) - \tilde{T}(z-l)}{l}, \tag{6.161b}$$

i.e. the influence from very long distance is screened by a finite frequency. (In Eq.(6.161b), the integral over l is cut-off at $l = v_{Ti}/\omega$ according to Eq.(6.160b).) This integral form is natural for kinetic interaction, in which particles with long mean-free-path keep the memory along the trajectory. The response of particles at the location $z = z$ is influenced by the perturbations at $z = z \pm l$, $|l| < v_{Ti}/\omega$.

The numerical coefficient of the order of unity in Eqs.(6.160a) and (6.160b) may be determined so as to reproduce the known linear dispersion of waves accurately. In the adiabatic limit, $v_{Ti}k_{\parallel} >> \omega$, Hammett and Perkins (1990) propose a form,

$$\mu_{\parallel} = 0 \text{ and } \chi_{\parallel} \simeq \frac{2}{\sqrt{\pi}} \frac{v_{Ti}}{k_{\parallel}} \text{ with } \gamma_T = 3, \tag{6.162}$$

where γ_T is the specific heat ratio. By this choice of numerical coefficients, the linear dispersion is better reproduced. The forms (6.160) and (6.162) may easily be implemented in simulations. If one takes the linear modes with Fourier space representation, more accurate forms can be deduced. For instance, Chang and Callen (1992) propose the forms for $\mu_\parallel$, λ_{1i}, λ_{2i} and $\chi_\parallel$ in Eq.(6.159) as,

$$\mu_\parallel = -\frac{i2v_{\mathrm{Ti}}}{5\pi k_\parallel} Z(\xi_1),$$

$$\lambda_{1i} = -\frac{1}{5} Z'(\xi_2) \simeq \lambda_{3i},$$

$$\chi_\parallel = -\frac{i9v_{\mathrm{Ti}}}{5\pi k_\parallel} Z(\xi_3), \tag{6.163}$$

where Z is the plasma dispersion function, $\xi_1 = (3/5)\left(\omega/k_\parallel v_{\mathrm{Ti}}\right)$, $\xi_2 = \sqrt{3/10}\,(\omega/k_\parallel v_{\mathrm{Ti}})$ and $\xi_3 = (36/25)\left(\omega/k_\parallel v_{\mathrm{Ti}}\right)$.

One issue which must be noted is the intrinsic dissipation (time irreversibility) caused by closure modelling, like Eq.(6.163). Landau resonance causes damping of the wave through phase mixing, but the original equation itself has a property of time reversibility. The time-reversal property was improved in the modelling of Mattor and Parker (1997) and Sugama *et al.* (2001). Leaving details to the literature, we note here that modelling reproduces the entropy production through stable modes, and that it does not introduce the artificial production of entropy that occurred in simpler models, such as (6.162).

From these lessons, we see that the large degrees of freedom of particle motion can be modelled in a form of fluid moment, by a closure theory. So far, success has been made in depicting the coherent part in the distribution functions. As is discussed in previous sections of this chapter, the incoherent part can be as important as the coherent part. The importance of the latter is discussed in Chapter 8.

6.6 Short note on prospects for closure theory

The advancement of direct nonlinear simulation stimulates modern motivation of the closure theory. Figures 6.15 and 6.16 illustrate some of the prototypical examples in achievement of simulations. Plasma turbulence is characterized by fluctuations of multiple scales. Fluctuations in the range of ion gyroradius develop in toroidal plasmas so as to form a complex pattern of fluctuations (Figure 6.15a). If one focuses on nonlinear dynamics at much shorter scale lengths (electron gyroradius), other complex dynamics develop in finer scales (Figure 6.15b). More and more detailed direct simulations are being executed, at each distinct scale length. These fluctuations coexist in two different scales, in reality, and mutual interaction

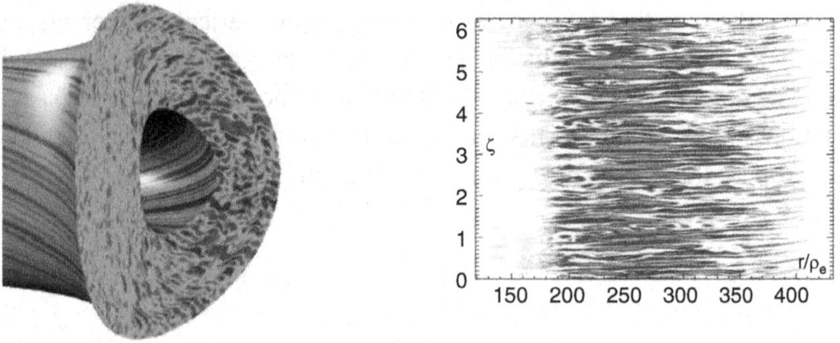

Fig. 6.15. Advance of direct nonlinear simulation in toroidal plasmas. Left: ion-temperature-gradient-driven turbulence, and fluctuation on the poloidal cross-section of tokamak is illustrated (quoted from Candy and Waltz). Right: electron-temperature-gradient-driven turbulence, shown on the plane of plasma radius and toroidal angle (quoted from Lin *et al.* (2007)).

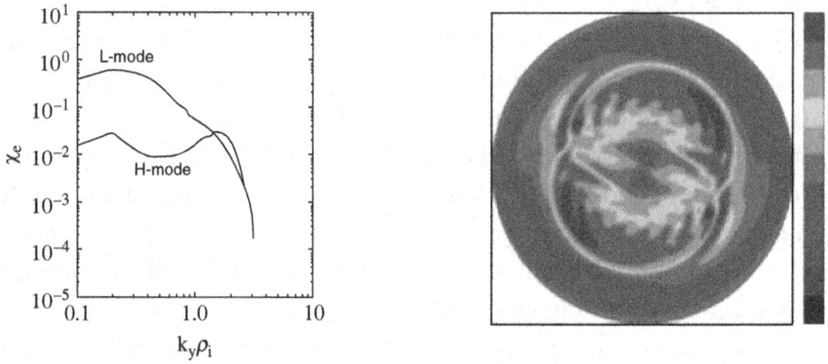

Fig. 6.16. Coexistence of fluctuations at different scale lengths. Left: fine and hyper-fine scale fluctuations coexist, and the contribution of each Fourier component to turbulent transport is illustrated (quoted from Jenko (2005)). Right: microscopic fluctuation and global instability coexist, and the contour of perturbation is shown on the poloidal cross-section (quoted from Yagi *et al.* (2005)).

among them also plays an important role in the evolution of turbulence (see Itoh and Itoh (2000) and Itoh and Itoh (2001) for explanations of bifurcation induced by multi-scale coupling). Direct nonlinear simulation, which takes into account mutual interaction, is shown in Figure 6.16a. The fluctuation spectrum has peaks at the scale lengths of ion and electron gyroradii. The other type of nonlinear interaction between fluctuations in different scales is also demonstrated in Figure 6.16b. In the latter example, a large-scale perturbation is induced (which has the pitch of poloidal mode number $m = 2$), and fine-scale rippling coexists. In addition, a large-scale transport code has been developed.

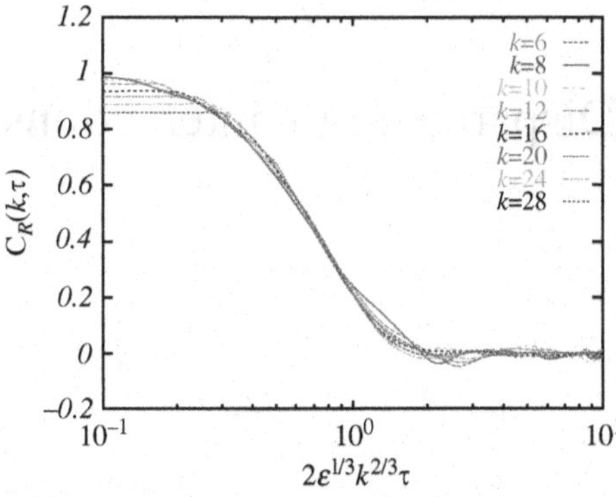

Fig. 6.17. Auto-correlation function of the fluctuating force for various choices of test wave number. Units of time are normalized to $\epsilon^{-1/3}k^{-2/3}$ (Gotoh *et al.*, 2002; Gotoh, 2006).

These evolutions of research demand correct modelling of fluctuations in the unresolved scales. The impact of the fluctuations in unresolved scales can be modelled through closures. In particular, the influence of the fluctuating force (the incoherent residual part) needs particular care, so as to satisfy the conservation relation. Thus there arises a task to test closure modelling in idealized and controlled circumstances. One example is briefly quoted here for illustration. The fluctuating force is often approximated by Gaussian white noise (an example is shown in Section 6.3). This is an idealized approximation, because the correlation time of the incoherent part of nonlinear interactions is small but finite. It is necessary to observe (or confirm, if possible) in what manner this approximation holds. Such an investigation of a statistical model has been performed in the field of neutral fluid turbulence (Gotoh, 2006). A large-scale direct nonlinear simulation for isotropic and homogeneous turbulence is performed. In this simulation, a scale of observation, k_c, is chosen. The fluctuating force, which is acting on the Fourier component at this test scale driven by fluctuation of finer scales, is measured. The auto-correlation function $C(\tau)$ of this fluctuating force is evaluated. Figure 6.17 illustrates how $C(\tau)$ behaves for different choices of test wave numbers. It was shown that the auto-correlation function $C(\tau)$ becomes more and more peaked at $\tau = 0$ as k_c increases. In the range of performed numerical simulations, the work reports a hypothesis that the half-width at half-maximum decreases as $k_c^{-2/3}$. This result shows that the approximation of the fluctuating force as Gaussian white noise may hold in an asymptotic limit, $k_c \rightarrow \infty$, but the convergence is very slow.

Based on these observations, we see that direct simulations and nonlinear theory stand side-by-side in the progress of the physics of turbulence.

7

Disparate scale interactions

Long and short delimit each other.

(Lao Tzu, "Tao Te Ching")

7.1 Short overview

In this chapter, we describe one generic class of nonlinear interaction in plasmas, called disparate scale interaction. One of the characteristic features of plasma turbulence is that there can be several explicit distinct scale lengths in the dynamics. For example, the ion gyroradius and electron gyroradius define intrinsic lengths in magnetized plasmas. The Debye length $\lambda_{\mathrm{De}} = v_{\mathrm{T},e}/\omega_{\mathrm{p}e}$ gives the boundary for collective oscillation, and the collisionless skin depth, c/ω_{pe}, is the scale of magnetic perturbation screening. These characteristic scale lengths define related modes in plasmas.

One evident reason why these scales are disparate is that the electron mass and ion mass differ substantially. Thus the fluctuations at different scale lengths can have different properties (in the dispersion, eigenvectors, etc.). Such a separation is not limited to *linear* dispersion, but occurs even in *nonlinear* dynamics. This is because the unstable modes are coupled with stable modes within the same group of fluctuations (those with a common scale length). The plasma response often leads to the result that instability is possible for a particular class of wave numbers. For instance, the drift waves in magnetized plasma (which preferentially propagate in the direction of diamagnetic drift velocity) can be unstable only if the wave number in the direction of the magnetic field $k_{\parallel}$ is much smaller than that in the direction of propagation. Therefore, nonlinear interaction within a *like-scale*, which increases $k_{\parallel}$ significantly, allows a transfer of the fluctuation energy to strongly damped modes.

This is in contrast with the familiar case of the Kolmogorov cascade in neutral fluids. In this consideration, the kinetic energy, which is contained in an observable

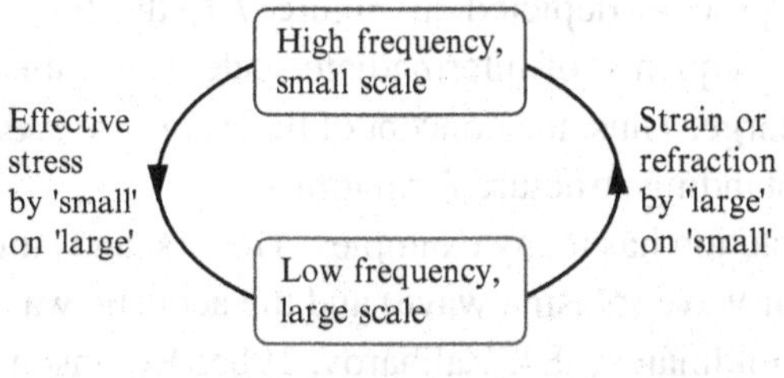

Fig. 7.1. Interaction of small-scale fluctuations and large-scale fluctuations.

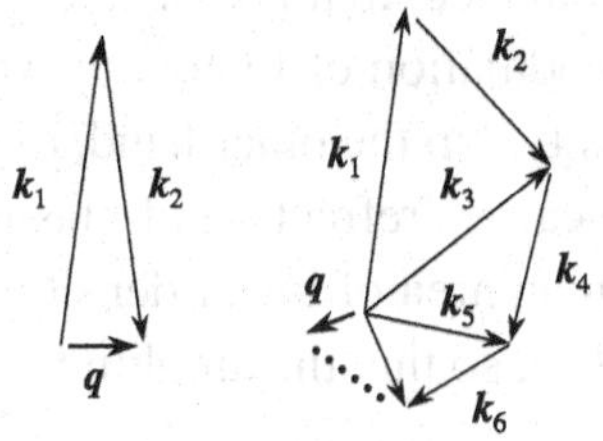

Fig. 7.2. Comparison of disparate scale interaction (left) and inverse cascade (right). The long-wavelength perturbation (characterized by q) is directly generated by, e.g., parametric instability in the disparate-scale interactions.

scale L, is transferred to, and dissipated (by molecular viscosity) at, the microscale l_{d}. The ratio L/l_{d} is evaluated as $L/l_{\mathrm{d}} \simeq R_e^{3/4}$, where R_e is the Reynolds number. Between these two scales, L and l_{d}, there is no preferred scale.

The presence of multiple-scale dynamics in plasmas allows a new class of multi-scale nonlinear interaction, i.e., *disparate scale interaction*. Consider the situation that two kinds of fluctuations (with different scales) coexist, as is illustrated in Figure 7.1. The fluctuations with high frequency and small spatial scale (referred to as 'small scale') can provide a 'source' for the fluctuations with low frequency and large spatial scale (referred to as 'large scale'), by inducing a stress which acts upon them, as fluxes in their density, momentum and energy. (This mechanism is explained in Chapter 3.) On the other hand, the large-scale perturbation acts as a 'strain field' in the presence of which the small-scale fluctuation evolves.

Disparate scale interaction is a prototypical process for the structure in turbulent plasma (Diamond *et al.*, 2005b). Large-scale structure (flows, density modulation, etc.) are generated as a result of the evolution of perturbations which break the symmetry of the turbulence. This mechanism is explained in detail in this chapter. The process of 'formation of large scale by turbulence' has some similarity to the 'inverse cascade' in fluid dynamics. Of course both of them share common physics, but at least noticeable difference between them exists. In the process of the inverse-cascade, the energy transfer from the small scales to the large scales occurs by interaction between two excitations of comparable scale (Figure 7.2).

In contrast, in the process depicted in Figure 7.1, the transfer of energy does not occur through a sequence of intermediate scales, but rather proceeds directly between small and large. Thus, the concept of the *disparate-scale interaction* plays a key role in understanding structure formation.

This type of interaction has many examples. The first and simplest is the interaction of the Langmuir wave (plasma wave) and the acoustic wave (ion sound wave) (Zakharov, 1984; Goldman, 1984; Zakharov, 1985; Robinson, 1997). This is the *classic example* of disparate scale interaction, and has significantly impacted our understanding of plasma turbulence. In this case, the ponderomotive pressure associated with the amplitude modulation of Langmuir waves (at large scale) induces the depletion of plasma density. On the other hand, the density modulation associated with the acoustic wave causes refraction of plasma waves, so that the plasma waves tend to accumulate in an area of lower density. These two effects close the interaction loop (in Figure 7.1), so that the amplitude modulation of plasma waves grows in time.

The second example is the system of drift waves (DW, small scale) and zonal flows (ZF, large scale), which is important to our understanding of toroidal plasmas (Sagdeev *et al.*, 1978; Hasegawa *et al.*, 1979; Diamond *et al.*, 1998; Hinton and Rosenbluth, 1999; Smolyakov *et al.*, 1999; Champeaux and Diamond, 2001; Jenko *et al.*, 2001; Manfredi *et al.*, 2001; Li and Kishimoto, 2002; Diamond *et al.*, 2005b; Itoh *et al.*, 2006). Zonal flows are $\boldsymbol{E} \times \boldsymbol{B}$ drift flows on magnetic surfaces, with electric perturbation constant on the magnetic surface but changing rapidly across it. In the system of DW–ZF, the small-scale drift waves induce the transport of momentum (Reynolds stress). The divergence of the (off-diagonal) stress component amplifies the zonal flow shear. On the other hand, zonal shears stretch the drift wave packet. The coupling between them leads to the growth of zonal flows from drift wave turbulence. In this process, the energy of the drift waves is transferred to zonal flows, so that the small-scale fluctuation level and the associated transport is reduced. This is an important nonlinear process for the self-organization and confinement of toroidal magnetized plasmas, and will be described in detail in Volume 2 of this series of books. Other examples of disparate-scale interaction are also listed in Table 7.1.

In the following, we explain the disparate-scale nonlinear interaction by carefully considering the example of the system of the Langmuir wave and an acoustic wave. We examine two simplified limits. In Section 7.2, we consider the case where a (nearly) monochromatic Langmuir wave is subject to disparate-scale interaction. Through this example, the positive feedback loop between modulation of the envelope of small-scale oscillation and excitation of the acoustic wave is revealed. This loop crystallizes the elementary process that makes this particular coupling interesting and effective. The general methodology of envelope modulation is introduced, and illustrated using the particular example of the

Table 7.1. *Examples of disparate-scale interactions. Symbols* (s) *and* (l) *identify the small scale and large scale, respectively*

Example of system	Small scale	Large scale
(s) plasma wave and (l) acoustic wave	plasmon presssure (ponderomotive force)	refractions of plasmon ray by density perturbation
(s) drift wave (DW) and (l) zonal flow (ZF)	Reynolds stress on ZF	stretching and tilting of DW by shear
(s) MHD turbulence and (l) mean B field	mean induction of B (L) $\rightarrow$ dynamo	bending of mean field (l) by fluctuations (s)
(s) acoustic wave and (l) vortex	acoustic ponderomotive force	refraction of ray by vortex
(s) internal wave and (l) current	wave Reynolds stress	induced k-diffusions by random refraction of wave packets

Zakharov equations. By considering nonlinear evolution, the appearance of a singularity in a finite time (collapse) in the model is discussed (Section 7.4). The collapse of plasma wave packets is an alternate route to dissipation of the wave energy at a very small scale.

Study of the plain Langmuir wave is illustrative (for the understanding of self-focusing), but its applicability is limited. This is because, in reality, plasma waves are not necessarily (quasi-)plane waves, but more often appear in the form of packets or turbulence. Acoustic fluctuations are not a single coherent wave. Thus, theoretical methods to analyze the disparate-scale interaction in wave turbulence are necessary. In Section 7.3, the case of Langmuir wave turbulence is discussed. A theoretical approach, based on the quasi-particle picture, is explained. This method is a basic tool for studying plasma turbulence which exhibits disparate-scale interaction.

7.2 Langmuir waves and self-focusing

7.2.1 Zakharov equations

In this section, we illustrate the basic physics of the interaction between Langmuir waves (plasma waves, plasmons) and the ion sound wave (ion-acoustic wave) which induces self-focusing of the Langmuir waves. Plasma waves have wavelengths such that $k\lambda_{\mathrm{De}} < 1$. The Debye length λ_{De} is shorter than the characteristic wavelength of an ion sound wave. The scale lengths of these two kinds of waves are separated, but their nonlinear interaction constitutes one prototypical element of plasma dynamics. The basic physics that plays a role is:

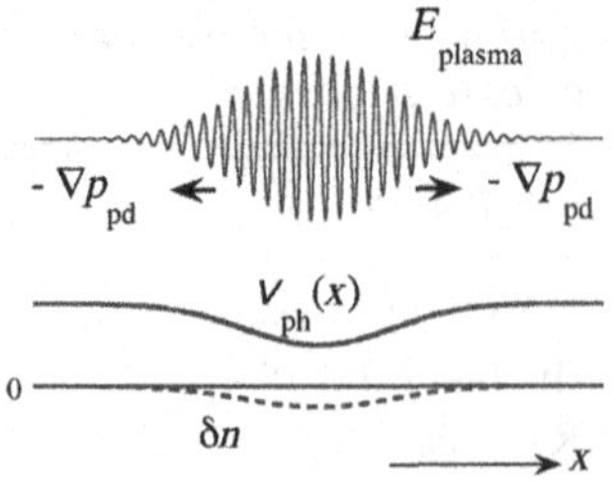

Fig. 7.3. Inhomogeneous plasma waves generate the pressure field for the mean electron dynamics. Density perturbation of the acoustic wave δn_e (dotted line) gives rise to the modulation of the refractive index of plasma waves (thick solid line).

Plasma wave → forms pressure field for acoustic waves.
Acoustic wave → density perturbation refracts plasma waves.

In the presence of plasma waves, electrons oscillate at the plasma wave frequency. The rapid electron motion associated with this wave produces the pressure field via ponderomotive force, discussed below. In other words, if the amplitude of the plasma wave is inhomogeneous, as is illustrated in Figure 7.3, then ambient electrons are repelled from the region of large amplitude. (The effective force, which arises from inhomogeneous and rapid oscillations, is known as 'ponderomotive force'.) Thus inhomogeneity of plasma waves causes pressure perturbations that couple to the acoustic wave. For the validity of this argument, the scale length of inhomogeneity of plasma waves must be longer than the wavelength of the plasma waves. The reciprocal influence of acoustic waves on plasma waves is caused by the fact that the dielectric function for plasma waves depends on the plasma density. Thus, the density perturbation associated with the ion sound wave causes modulation in the refractive index of plasma waves. The change of refractive index leads to modification of the plasma wave ray and intensity. These two processes constitute the loop that induces nonlinear instability of the plasma wave–acoustic wave system (Goldman, 1984; Zakharov, 1985; Robinson, 1997).

A heuristic description can be developed by application of the envelope formalism to plasma wave–acoustic wave interaction. Let us write the inhomogeneous plasma waves as,

$$\tilde{E} = E(\boldsymbol{x}, t)\, \boldsymbol{e}_0 \exp(i\boldsymbol{k} \cdot \boldsymbol{x} - \omega t), \tag{7.1}$$

where $E(\boldsymbol{x}, t)$ indicates the slowly-varying (in space and time) envelope, $\boldsymbol{e}_0$ denotes the polarization of the wave field, and $\exp(i\boldsymbol{k} \cdot \boldsymbol{x} - \omega t)$ is the rapidly oscillating plasma wave carrier. For plasma waves, the dispersion relation,

$$\omega^2 = \omega_{\text{pe}}^2 + \gamma_T k^2 v_{\text{T},e}^2 \tag{7.2}$$

holds, where ω_{pe} is the plasma frequency,

$$\omega_{\text{pe}}^2 = \frac{4\pi n_e e^2}{m_e},$$

γ_T is the specific heat ratio, and $v_{\text{T},e}$ is the electron thermal velocity. In the dynamic expression, the dispersion relation (7.1) takes the form,

$$-\frac{\partial^2}{\partial t^2}\tilde{\boldsymbol{E}} = \omega_{\text{pe}}^2 \tilde{\boldsymbol{E}} - \gamma_T v_{\text{T},e} \nabla^2 \tilde{\boldsymbol{E}}. \tag{7.3}$$

We now consider the two elements of the feedback loop in sequence.

(a) Influence of acoustic waves on plasma waves

When acoustic waves are present, the density perturbation is denoted by δn_e, so the plasma frequency becomes,

$$\omega_{\text{pe}}^2 = \omega_{\text{p0}}^2 (1 + \delta n) \quad \text{with} \quad \delta n \equiv \frac{\delta n_e}{n_{e,0}}. \tag{7.4}$$

Here, ω_{p0}^2 is defined at the unperturbed density $n_{e,0}$ as $\omega_{\text{p0}}^2 = 4\pi n_{e,0} e^2 / m_e$. Thus, in this case, the dynamical equation of plasma waves Eq.(7.3) takes the form,

$$-\frac{\partial^2}{\partial t^2}\tilde{\boldsymbol{E}} = \omega_{\text{p0}}^2 (1 + \delta n)\, \tilde{\boldsymbol{E}} - \gamma_T v_{\text{T},e}^2 \nabla^2 \tilde{\boldsymbol{E}}. \tag{7.5}$$

If one substitutes Eq.(7.1) into Eq.(7.5), the rapidly varying terms balance, and the slowly varying envelope equation emerges as,

$$\frac{i}{\omega_{\text{p0}}}\frac{\partial}{\partial t} E + \lambda_{\text{De}}^2 \nabla^2 E = \delta n E, \tag{7.6}$$

where we use the relation $\lambda_{\text{De}}^2 = \gamma_T v_{\text{T},e}^2 \omega_{\text{p0}}^{-2}$. In Eq.(7.6), the second term on the left-hand side is the diffraction term, which is caused by the dispersion of plasma waves. The modification of refraction, owing to the acoustic wave, appears on the right. Equation (7.6) describes how the envelope of the plasma waves is affected by the acoustic waves.

(b) Influence of plasma waves on acoustic waves

The influence of plasma waves on the acoustic wave is modelled by considering the contribution to the electron pressure from rapid oscillation by the plasma waves. In response to the plasma waves, electrons execute rapid oscillatory motion, but the kinetic energy of ions associated with this rapid oscillation is $m_e Z^2/m_i$-times smaller than that of electrons, owing to their heavier mass. (Here, Z is the charge per ion divided by the unit charge e.) In the slow-time scale, which is relevant to acoustic waves, the rapid electron oscillation by the plasma waves induces an effective wave or radiation pressure,

$$p_{\text{pw}} \equiv \left\{ \frac{\partial}{\partial \omega} (\omega \epsilon) \bigg|_{\omega_{\text{p0}}} \right\} \frac{|E|^2}{8\pi},$$

where ϵ is a dielectric function, which becomes unity in vacuum. The contribution from the response of ions to the rapid oscillation can be neglected. The energy density p_{pw} is that of the plasma waves, so electrons are repelled, on average (which is taken in a time scale longer than that of the plasma waves' frequency), from the region where p_{pw} takes a large value. This ponderomotive force (associated with the gradient of p_{pw}) induces ion motion on a slow timescale. In addition to the thermal pressure, p, p_{pw} also appears in the ion equation of motion, so,

$$m_i n_{i0} \frac{\partial}{\partial t} \boldsymbol{V} = -\nabla \left(p + p_{\text{pw}} \right).$$

The dynamic equation for the acoustic wave is then given as,

$$\left(\frac{\partial^2}{\partial t^2} - c_{\text{s}}^2 \nabla^2 \right) \delta n = \frac{\nabla^2 |E|^2}{4\pi n_0 m_i}. \tag{7.7}$$

In deriving Eq.(7.7), the relation $\nabla^2 \left| \tilde{E} \right|^2 = \nabla^2 |E|^2$, and the approximation $\partial(\omega\epsilon)/\partial\omega|_{\omega_{\text{p0}}} \simeq 1$ is applied to the plasma waves, and the relation $p_{\text{T}} = p_0(1 + \delta n)$ is used for (isothermal) acoustic waves. Equation (7.7) illustrates that the inhomogeneity of the envelope of plasma waves can excite the acoustic waves.

Equations (7.6) and (7.7) form a set of equations that describes the interaction of acoustic waves and the envelope of plasma waves. It is convenient to introduce dimensionless variables as,

$$\omega_{\text{p0}} t \to t, \quad \lambda_{\text{De}}^{-1} x \to x, \quad \frac{E}{\sqrt{4\pi n_0 T_e}} \to E. \tag{7.8}$$

(The new variable E is the oscillation velocity of an electron at the plasma wave frequency, normalized to the electron thermal velocity.) In these rescaled variables, the coupled equations (7.6) and (7.7) take the form,

$$\left(i\frac{\partial}{\partial t}+\nabla^2\right)E-\delta nE=0 \tag{7.9a}$$

$$\left(\frac{\partial^2}{\partial t^2}-\frac{m_e}{m_i}\nabla^2\right)\delta n-\frac{m_e}{m_i}\nabla^2|E|^2=0. \tag{7.9b}$$

This set of equations is knows as the dimensionless 'Zakharov equations', and they are coupled envelope equations for the:

(a) plasma wave amplitude $E(\boldsymbol{x},t)$;
(b) density perturbation δn.

In the absence of nonlinear coupling, Eq.(7.9a) (i.e., with $\delta n\to 0$) becomes the Schrödinger equation for a free particle, and Eq.(7.9b) (with $|E|\to 0$) reduces to the acoustic wave equation. We again emphasize that in deriving the nonlinear coupling between classes of waves, the space-time scale separation is crucial.

7.2.2 Subsonic and supersonic limits

Depending on the velocity of envelope propagation, relative to the ion sound velocity, the nonlinearly coupled equations show different characters. Let us take the characteristic time scale τ and the characteristic scale length L for the envelope of plasma waves. The time derivative term in Eq.(7.9b) has the order of magnitude estimate $\partial^2/\partial t^2\sim\tau^{-2}$, while the spatial derivative terms have $m_e m_i^{-1}\nabla^2\sim m_e m_i^{-1}L^{-2}$. When evolution of the envelope is slow,

$$\tau^{-2}\ll m_e m_i^{-1}L^{-2}\text{ (i.e., rate of change }\ll c_s/L), \tag{7.10a}$$

the envelope modulation propagates much slower than the acoustic wave. This is the *subsonic* (*adiabatic*) limit. In contrast, if inequality

$$\tau^{-2}\gg m_e m_i^{-1}L^{-2}\text{ (i.e., rate of change }\gg c_s/L) \tag{7.10b}$$

holds, the modulation propagates much faster than the acoustic wave. This is the *supersonic* (*non-adiabatic*) limit.

Note that the plasma wave envelope propagates (in the linear response regime) at the group velocity, $v_g=\partial\omega_k/\partial k$. The dispersion relation $\omega_k=\omega_{pe}\sqrt{1+k^2\lambda_{De}^2}$ yields,

$$v_g=\frac{k\lambda_{De}v_{T,e}}{\sqrt{1+k^2\lambda_{De}^2}},$$

which is of the order of $k\lambda_{\mathrm{De}} v_{\mathrm{T},e}$ for the long wavelength limit $k\lambda_{\mathrm{De}} \ll 1$. Thus, the subsonic limit or supersonic limit depends on the wave number, in part. However, the propagation velocity also depends on the amplitude of plasma waves in a nonlinear regime.

7.2.3 Subsonic limit

In the subsonic limit, Eq.(7.10a), the inertia of ion motion (at slow time scale) is unimportant, and Eq.(7.7) is given by the balance between the ponderomotive force and the gradient in kinetic pressure. Thus, we have,

$$m_e m_i^{-1} \nabla^2 \left(\delta n + |E|^2 \right) = 0,$$

so

$$\delta n \cong - |E|^2 \tag{7.11}$$

is satisfied. Substituting Eq.(7.11) into Eq.(7.9a), the coupled Zakharov equations reduce to one combined equation,

$$\left(i \frac{\partial}{\partial t} + \nabla^2 \right) E + |E|^2 E = 0. \tag{7.12}$$

This is the so-called nonlinear Schrödinger equation (NLS equation), and is the adiabatic Zakharov equation. Note that density perturbations here are local depletions (i.e., $\delta n < 0$), and so are called cavitons.

7.2.4 Illustration of self-focusing

At this point, it is useful to recall the optical self-focusing problem, in order to understand the common physics content of focusing and the adiabatic Zakharov equation. Let us consider a light beam propagating in a nonlinear medium, in which the refractive index ($n^2 = c^2/v_{\mathrm{ph}}^2$) varies with the intensity of the light $|E|^2$, so,

$$n^2 = 1 + \Delta n \frac{|E|^2}{E_c^2},$$

where E_c^2 indicates the critical intensity above which the modification of the refractive index becomes apparent. The equation of light propagation is given as,

$$\nabla^2 \tilde{E} + \frac{\omega^2}{c^2} n^2 \tilde{E} = 0,$$

where ω is the frequency of the light and c is the speed of light in a vacuum. Substituting the wave and envelope modulation $\tilde{\boldsymbol{E}} = E(\boldsymbol{x}, t)\, \boldsymbol{e}_0 \exp(ik \cdot z - \omega t)$ into this wave propagation equation, we have,

$$\left(2ik\frac{\partial}{\partial z} + \nabla^2\right) E + \Delta n k^2 |E|^2 E = 0, \tag{7.13}$$

where z is taken in the direction of the propagation and $k^2 = \omega^2 c^{-2}$. Terms of $O\left(k^{-2}\, \partial^2 E/\partial z \partial r\right)$ have been neglected. This is the NLS equation, describing the self-focusing of light in a nonlinear medium. The physics of self-focusing is the change of the phase speed owing to the intensity of the light,

$$v_{\text{ph}}^2 = \frac{c^2}{n^2} = \frac{c^2}{1 + \Delta n |E|^2}.$$

Thus, when $\Delta n > 0$ holds, the phase velocity becomes *lower* in regions of *high intensity*. This is illustrated by considering propagation of phase fronts (i.e., contoured surface of constant phase) in a beam of finite width. (Rays are perpendicular to the phase front.) As is illustrated in Figure 7.4, the phase front (iso-phase surface) is deformed in the region of high intensity. When $\Delta n > 0$ holds, the phase front lags behind in the region of high intensity, so that the phase front becomes concave. The local propagation direction of the wave is perpendicular to the phase front. When the phase front becomes concave, the local propagation directions are no longer parallel, but instead tend to focus. As a result, the light beam focuses itself. The peak intensity of light becomes higher and higher as the light propagates. It suggests formation of a singularity in the light intensity field, unless an

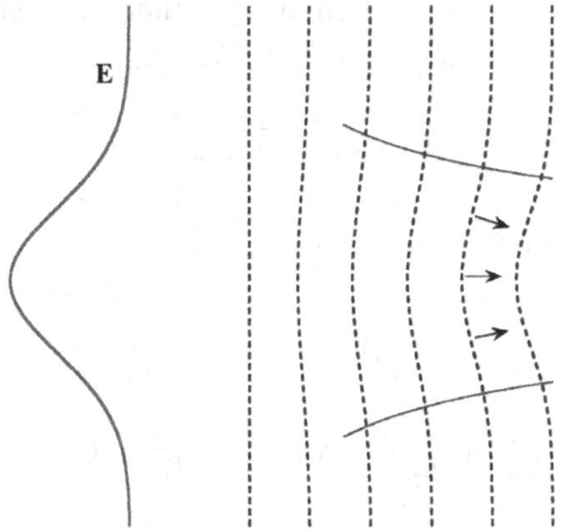

Fig. 7.4. Propagation of a light beam in a nonlinear medium with $\Delta n > 0$. Phase fronts are shown by dotted lines, and a local direction of propagation is denoted by arrows. Thin lines illustrate knees of the beam intensity profile.

additional mechanism alters this self-focusing process. The problem of singularity formation will be discussed later in this chapter.

7.2.5 Linear theory of self-focusing

The dynamics of self-focusing in the NLS equation can be analyzed using a linear analysis. The NLS equation (7.12) is a complex equation, so that the envelope functions in terms of two fields, i.e., the amplitude and the phase,

$$E = A \exp(i\varphi) . \tag{7.14}$$

Both the amplitude A and the phase φ are (slow) functions of space and time. Substituting the complex form of E into Eq.(7.12), the real and imaginary parts reduce to a set of coupled equations for the (real) fields A, φ,

$$\frac{1}{A}\frac{\partial}{\partial t}A + \nabla^2\varphi + \frac{2\nabla\varphi \cdot \nabla A}{A} = 0, \tag{7.15a}$$

$$\frac{\partial}{\partial t}\varphi + (\nabla\varphi)^2 - \left(\frac{\nabla^2 A}{A} + |A|^2\right) = 0. \tag{7.15b}$$

The second term on the left-hand side of Eq.(7.15a) (the $\nabla^2\varphi$ term) shows that the amplitude increases if the phase front is concave ($\nabla^2\varphi < 0$) as is illustrated in Figure 7.4. In Eq.(7.15b), the fourth term on the left-hand side (the $|A|^2$ term) indicates that the larger amplitude causes variation in the phase. Thus the spatial variations in the intensity field induce bending of the phase front in Figure 7.4. These features induce self-focusing of the plasma waves.

A systematic analysis of the growth of amplitude modulation for plasma waves can be performed by linearizing Eqs.(7.15). We put,

$$A = A_0 + \tilde{A}, \quad \varphi = \tilde{\varphi}$$

in Eq.(7.15), and retain the linear terms in $\tilde{A}$, $\tilde{\varphi}$,

$$\frac{\partial}{\partial t}\tilde{A} + A_0\nabla^2\varphi = 0 \tag{7.16a}$$

$$A_0\frac{\partial}{\partial t}\tilde{\varphi} - \nabla^2\tilde{A} - A_0^2\tilde{A} = 0. \tag{7.16b}$$

Putting the perturbation in the form,

$$\left(\tilde{A}, \tilde{\varphi}\right) \propto \exp(i\boldsymbol{q} \cdot \boldsymbol{x} - i\Omega t),$$

where q and Ω denote slow spatio-temporal variation of envelope, we find the dispersion relation,

$$\Omega^2 = -q^2\left(A_0^2 - q^2\right). \tag{7.17}$$

That is, the perturbation grows if $A_0^2 > q^2$. The term $q^2A_0^2$ on the right of Eq.(7.17) is the destabilizing term for the self-focusing, and the q^4 term denotes the effect of diffraction, which spreads or blurs modulation. Thus, the NLS equation describes the amplification of the peak plasma wave intensity $|E|^2$ and the perturbed density of long wavelength, if the amplitude of plasma waves exceeds the threshold, $A_0 > q$, i.e., in a dimensional form,

$$\frac{|E|^2}{4\pi n_0 T_e} > q^2\lambda_{\mathrm{De}}^2. \tag{7.18}$$

(The condition is interpreted that the electron oscillation velocity induced by the plasma wave is larger than $q\lambda_{\mathrm{De}}v_{\mathrm{T},e}$. It is more easily satisfied if q is smaller.) This criterion shows a competition between the self-attraction of plasma waves (through depleting background plasma density on a large scale) and the spreading of plasma waves associated with their diffraction.

Equation (7.18) indicates that the plane plasma waves are, at any amplitude, unstable against amplitude modulation by choosing the long wavelength of modulation. In reality, there is a lower bound of the light intensity for instability. For instance, system size limits the lowest allowable value for q; in addition, small but finite dissipation (which is ignored for transparency of the argument) can suppress the instability, because the maximum growth rate in Eq.(7.17), $A_0^2/2$, scales as $|E|^2$.

When linear theory predicts the instability to occur, the central problem is how the focusing evolves. Such nonlinear evolution is discussed at the end of Section 7.4, which deals with Langmuir collapse.

7.3 Langmuir wave turbulence

The Zakharov equations in Section 7.2 illustrate the essence of self-focusing of plasma waves by disparate scale interactions in plasmas. This set of equations was derived for a plane plasma wave interacting with narrow-band ion acoustic waves. In more general circumstances, plasma waves are excited as turbulence, and the acoustic spectrum might also be composed of broad band wavelengths. The spatio-temporal evolution of the envelope of Langmuir turbulence may or may not be slower than the acoustic speed.

In order to analyze the evolution of plasma wave turbulence by disparate scale interactions, it is illuminating to use the quasi-particle description. The scale separation between the carrier plasma waves and the modulator (ion sound waves) is exploited. As is explained in Chapter 5, the Manley–Rowe relation holds for three-wave interactions. When a scale separation in frequency and wave number applies for the plasma waves and acoustic waves, the action density of plasma waves is invariant during the interaction with an acoustic wave. Of course, the action density of plasma waves changes in time on account of nonlinear interaction among plasma waves, which occurs on the range of their common scales. These effects can be incorporated into theory as a collision operator on the right-hand side of the quasi-particle kinetic equation.

7.3.1 Action density

The energy density and action density (in the phase space) of plasma waves are introduced as,

$$E_k \equiv \frac{\partial}{\partial\omega}(\omega\epsilon)\bigg|_{\omega_k} \frac{\left|\tilde{\boldsymbol{E}}_k\right|^2}{8\pi} \quad \text{and} \quad N = \frac{E_k}{\omega_k}, \tag{7.19}$$

where $\tilde{\boldsymbol{E}}_k$ is the electric field of the plasma wave at the wave number k, and the dispersion relation of plasma waves is given by $\omega_k^2 = \omega_{\mathrm{pe}}^2 + \gamma_T k^2 v_{\mathrm{T},e}^2$. Now, $N(\boldsymbol{k}, \boldsymbol{x}, t)$ is a population density of waves (excitons). Regarding the variables $(\boldsymbol{k}, \boldsymbol{x}, t)$ of $N(\boldsymbol{k}, \boldsymbol{x}, t)$, $\boldsymbol{k}$ stands for the wave vector of the carrier wave (that is the short scale, corresponding to the plasma waves), and $\boldsymbol{x}$ and t denote a slow scale which appears in the envelope.

7.3.2 Disparate scale interaction between Langmuir turbulence and acoustic turbulence

The coupling between Langmuir turbulence and acoustic turbulence is studied employing the quasi-particle approach. In derivation of the model, we take the limit where the time and spatial scales of these two kinds of turbulence are well separated. As is illustrated in Eq.(7.9), an expansion parameter that leads to scale separation is, in this case, essentially the ratio of electron mass to ion, $\sqrt{m_\mathrm{e}/m_\mathrm{i}} \ll 1$. The perturbations of the Langmuir turbulence field (i.e., radiation pressure, etc.) are expressed in terms of the action density $N = \omega_k^{-1} E_k$ defined in Eq.(7.19), where $\tilde{\boldsymbol{E}}_k$ has slow spatio-temporal variations. The acoustic waves are characterized by the density and velocity perturbations $\tilde{n}$ and $\tilde{\boldsymbol{V}}$. The quantities $\tilde{n}$ and $\tilde{\boldsymbol{V}}$ have spatio-temporal dependencies which are slow compared to $\boldsymbol{k}$ and

ω_k of Langmuir wave turbulence. Under the circumstance of scale separation, the quantity $N(\boldsymbol{k}, \boldsymbol{x}, t)$, the number distribution of waves, is conserved along the trajectory. That is, the number of quanta of the $(\boldsymbol{k}, \omega_k)$ component of the Langmuir wave moves in the $(\boldsymbol{k}, \boldsymbol{x})$-space as a '*particle*' interacting with the field of acoustic waves. The conservation relation,

$$\frac{\mathrm{d}N(\boldsymbol{k}, \boldsymbol{x}, t)}{\mathrm{d}t} = 0,$$

in the presence of the acoustic waves is rewritten as,

$$\frac{\mathrm{d}N}{\mathrm{d}t} = \frac{\partial N}{\partial t} + \left(\boldsymbol{v}_{\mathrm{g}} + \tilde{\boldsymbol{V}}\right) \cdot \frac{\partial N}{\partial \boldsymbol{x}} - \frac{\partial}{\partial \boldsymbol{x}} \left(\omega_k + \boldsymbol{k} \cdot \tilde{\boldsymbol{V}}\right) \cdot \frac{\partial N}{\partial \boldsymbol{k}} = 0. \tag{7.20}$$

Note that, in addition to interaction with ion acoustic waves, the excitation (by an instability or external supply), damping (by such as the Landau damping for $\omega \simeq k v_{\mathrm{T},e}$) and energy transfer among Langmuir waves through self-nonlinear interactions can take place. These processes, which occur on the scale of Langmuir waves, lead to the rapid, non-adiabatic evolution of Langmuir wave action density. The rate of change in $N(\boldsymbol{k}, \boldsymbol{x}, \boldsymbol{t})$ through these mechanisms is schematically written $\Gamma(k, \omega; x, t)$. (As for the wave field, the slow spatio-temporal dependence is expressed in terms of $\boldsymbol{x}, t$.) Thus, the evolution equation of Langmuir wave action under the influence of the acoustic waves is written as,

$$\frac{\partial N}{\partial t} + \left(\boldsymbol{v}_{\mathrm{g}} + \tilde{\boldsymbol{V}}\right) \cdot \frac{\partial N}{\partial \boldsymbol{x}} - \frac{\partial}{\partial \boldsymbol{x}} \left(\omega_k + \boldsymbol{k} \cdot \tilde{\boldsymbol{V}}\right) \cdot \frac{\partial N}{\partial \boldsymbol{k}} = \Gamma(k, \omega; x, t; N)\, N. \tag{7.21}$$

This equation states that the wave quanta N in the phase space $N(\boldsymbol{x}, \boldsymbol{k})$, having a finite life time $-\Gamma^{-1}$, follow trajectories determined by the eikonal equation,

$$\frac{\mathrm{d}\boldsymbol{x}}{\mathrm{d}t} = \frac{\partial \omega_k}{\partial \boldsymbol{k}} + \tilde{\boldsymbol{V}} \quad \text{and} \quad \frac{\mathrm{d}\boldsymbol{k}}{\mathrm{d}t} = -\frac{\partial}{\partial \boldsymbol{x}} \left(\omega_k + \boldsymbol{k} \cdot \tilde{\boldsymbol{V}}\right).$$

The dynamical equation (7.21) is simplified in order to study the response against the acoustic waves. In the presence of long-scale perturbations, the wave frequency is modified such that,

$$\omega_k = \omega_{k0} + \tilde{\omega}_k,$$

(ω_{k0} is given in the absence of acoustic waves), and the *unperturbed orbit of quasi-particles* may be defined as,

$$\frac{\mathrm{d}\boldsymbol{x}}{\mathrm{d}t} = \frac{\partial \omega_{k0}}{\partial_k} = \boldsymbol{v}_{\mathrm{g}}, \qquad \frac{\mathrm{d}\boldsymbol{k}}{\mathrm{d}t} = -\frac{\partial}{\partial \boldsymbol{x}} \omega_{k0}. \tag{7.22}$$

The term $\Gamma(k, \omega; x, t; N)\, N$ includes linear terms of N (e.g., due to the linear damping/growth or collisional damping, etc.) and nonlinear terms of N

(e.g., owing to the nonlinear interactions in small scales). Therefore, as in the Chapman–Enskog approach, the mean distribution $\langle N \rangle$ is determined by the relation,

$$\boldsymbol{v}_{\mathrm{g}} \cdot \frac{\partial \langle N \rangle}{\partial \boldsymbol{x}} - \frac{\partial \omega_k}{\partial \boldsymbol{x}} \cdot \frac{\partial \langle N \rangle}{\partial \boldsymbol{k}} = \Gamma\left(k, \omega; x, t; \langle N \rangle\right) \langle N \rangle .$$

The deviation from the mean, $\tilde{N}$, is induced by coupling with the acoustic wave. The distribution N and the term $\Gamma(k, \omega; x, t; N)\, N$ are rewritten as,

$$N = \langle N \rangle + \tilde{N}, \tag{7.23a}$$

$$\Gamma\left(k, \omega; x, t; N\right) N = \Gamma\left(k, \omega; x, t; \langle N \rangle\right) \langle N \rangle - \hat{\Gamma} \tilde{N} + \cdots, \tag{7.23b}$$

where $\hat{\Gamma}\tilde{N}$ is the first-order correction (so that $\hat{\Gamma}$ is independent of $\tilde{N}$). In the usual circumstances, where the linear growth (growth rate: γ_{L}) is balanced by the quadratic nonlinearity between different plasma waves (self-nonlinearity), the estimate,

$$\hat{\Gamma} \simeq \gamma_{\mathrm{L}}$$

holds. Keeping the first-order terms with respect to $\tilde{N}$, Eq.(7.21) becomes,

$$\begin{aligned} &\frac{\partial \tilde{N}}{\partial t} + \boldsymbol{v}_{\mathrm{g}} \cdot \frac{\partial \tilde{N}}{\partial \boldsymbol{x}} - \frac{\partial}{\partial \boldsymbol{x}} \omega_{k0} \cdot \frac{\partial \tilde{N}}{\partial \boldsymbol{k}} + \hat{\Gamma} \tilde{N} \\ &\quad = -\left(\frac{\partial \tilde{\omega}_k}{\partial \boldsymbol{k}} + \tilde{\boldsymbol{V}} \right) \cdot \frac{\partial \langle N \rangle}{\partial \boldsymbol{x}} + \frac{\partial}{\partial \boldsymbol{x}} \left(\tilde{\omega}_k + \boldsymbol{k} \cdot \tilde{\boldsymbol{V}} \right) \cdot \frac{\partial \langle N \rangle}{\partial \boldsymbol{k}}. \end{aligned} \tag{7.24}$$

The analogy between this equation and the Boltzmann equation is evident in light of the Chapman–Enskog expansion. The plasma waves, the amplitude of which is modulated by acoustic waves, is illustrated in Figure 7.5(a). In this circumstance,

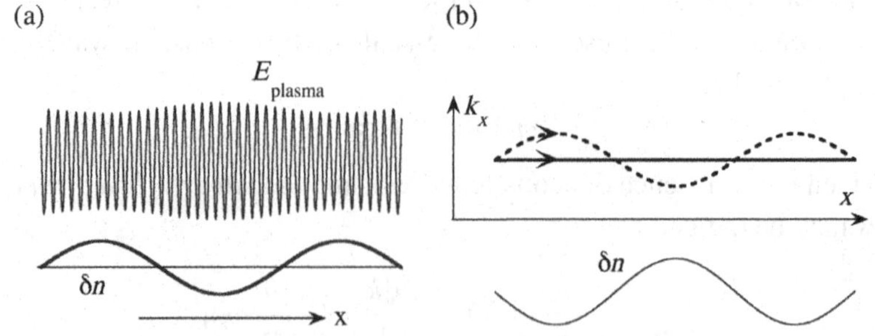

Fig. 7.5. Plasma waves coexist with the acoustic wave (a). The trajectory of quasi-particles of plasma waves in the presence of the acoustic wave (dashed line) and the unperturbed trajectory (solid line), (b). (Schematic drawing.)

the trajectory of the quasi-particle (plasma waves) is shown by the dashed line in Figure 7.5(b), and the unperturbed orbit (7.22) is illustrated by the solid line. Equation (7.20) indicates that the action is unchanged by acoustic waves along the perturbed orbit. As was the case for solving the Boltzmann and Vlasov equations, the contribution from the acoustic waves is separated in the right-hand side of Eq.(7.24). Treating this as a source, the action of plasma wave quasi-particles is calculated by integrating along the unperturbed trajectory of quasi-particles. The reaction of the Langmuir wave turbulence on the acoustic wave takes place by the pressure perturbation on electrons i.e. $-\nabla \int \mathrm{d}k\, \omega_k N$. Adding this term to the pressure perturbation associated with the sound waves, one has,

$$\frac{\partial^2}{\partial t^2}\tilde{n} - \nabla^2 \tilde{n} = -\nabla^2 \frac{2}{m_i} \int \mathrm{d}k\, \omega_k N. \tag{7.25}$$

The set of equations (7.24) and (7.25) describes the evolution of the coupled Langmuir–acoustic turbulence through disparate scale interactions. From the viewpoint of investigating mean scales (the scale of acoustic waves), the evolution equation for the Langmuir waves (7.21) is effectively an energy equation for *sub-grid scales*. In this analogy, the effect of plasma waves (*unresolved* on the scale of Eq.(7.25)) on *resolved* acoustic waves occurs through the radiation stress term on the right-hand side of Eq.(7.25).

7.3.3 Evolution of the Langmuir wave action density

In order to study the mutual interaction between plasma waves and acoustic waves, we consider that the modulation of density in the acoustic wave fluctuations, $\delta n = \tilde{n}/n_0$, is the first-order term, and put the action density of plasmons, Eq.(7.23a), $N = \langle N \rangle + \tilde{N}$, where $\tilde{N}$ is of the order of $\delta n = \tilde{n}/n_0$.

Consideration of energy

The average action density $\langle N \rangle$ evolves much slower than the acoustic wave fluctuations. In the absence of acoustic waves, the stationary solution, which is determined by the term $\Gamma\,(k, \omega; x, t; N)$, gives $\langle N \rangle$ as is explained in (7.23b). By coupling with acoustic waves, N evolves slowly in time. In interacting with acoustic waves, the action N is conserved. Thus the change of energy density of plasma waves E_k follows,

$$\frac{\mathrm{d}}{\mathrm{d}t} E_k = N \frac{\mathrm{d}}{\mathrm{d}t} \omega_k.$$

Noting the relations,

$$\frac{\mathrm{d}\omega_k}{\mathrm{d}t} = \frac{\partial \omega_k}{\partial \boldsymbol{k}} \cdot \frac{\mathrm{d}\boldsymbol{k}}{\mathrm{d}t}, \qquad \frac{\partial \omega_k}{\partial \boldsymbol{k}} = \boldsymbol{v}_{\mathrm{g}}$$

and the dynamics of the refractive index by density perturbation $\mathrm{d}\boldsymbol{k}/\mathrm{d}t = -\,\partial/\partial\boldsymbol{x}\left(\omega_{\mathrm{p}0}\delta n\right)$, one has,

$$\frac{\mathrm{d}}{\mathrm{d}t}E_k = -N\omega_{\mathrm{p}0}\boldsymbol{v}_{\mathrm{gr}}\frac{\partial}{\partial\boldsymbol{x}}\delta n.$$

Putting $N = \langle N\rangle + \tilde{N}$ into this relation, and noting that the correlation $\langle N\rangle\,\delta n$ vanishes in a long-time average but that of $\tilde{N}\delta n$ may survive, we have,

$$\frac{\mathrm{d}}{\mathrm{d}t}\langle E_k\rangle = -\omega_{\mathrm{p}0}\boldsymbol{v}_{\mathrm{gr}}\cdot\left\langle\tilde{N}\frac{\partial}{\partial\boldsymbol{x}}\delta n\right\rangle. \tag{7.26}$$

This relation illustrates that the change of energy of plasma waves, which is transferred to acoustic waves, is given by the correlation $\left\langle\tilde{N}\,\partial\delta n/\partial\boldsymbol{x}\right\rangle$.

Wave kinetic equation of action density

For this purpose, we analyze the case that the plasma wave turbulence is homogeneous (in the unperturbed state), and the Doppler shift by the ion fluid motion $\boldsymbol{k}\cdot\tilde{\boldsymbol{V}}$ is smaller than the effect of the modulation of the refractive index, $\tilde{\omega}_k$. The relation,

$$\frac{\partial\tilde{\omega}_k}{\partial\boldsymbol{x}} = \nabla\tilde{n}\left(\frac{\partial\tilde{\omega}_k}{\partial n_e}\right)$$

is employed. Under this circumstance, putting Eq.(7.26) into Eq.(7.21), together with Eq.(7.24), yields the responses of $\tilde{N}$ and $\langle N\rangle$ to the acoustic waves as,

$$\frac{\partial\tilde{N}}{\partial t} + \boldsymbol{v}_{\mathrm{gr}}\cdot\frac{\partial\tilde{N}}{\partial\boldsymbol{x}} - \frac{\partial}{\partial\boldsymbol{x}}\omega_{k0}\cdot\frac{\partial\tilde{N}}{\partial\boldsymbol{k}} + \hat{\Gamma}\tilde{N} = \frac{\partial\omega_k}{\partial n_e}\nabla\tilde{n}\cdot\frac{\partial\langle N\rangle}{\partial\boldsymbol{k}}, \tag{7.27a}$$

$$\frac{\partial\langle N\rangle}{\partial t} + \langle\Gamma N\rangle = \frac{\partial}{\partial\boldsymbol{k}}\cdot\left\langle\frac{\partial\omega_k}{\partial n_e}\nabla\tilde{n}\tilde{N}\right\rangle. \tag{7.27b}$$

A similar way of thinking leads the back interaction of plasma quasi-particles on the acoustic wave, Eq.(7.25) into the form,

$$\frac{\partial^2}{\partial t^2}\tilde{n} - \nabla^2\tilde{n} = -\nabla^2\frac{2}{m_i}\int\mathrm{d}k\,\omega_{k0}\tilde{N}. \tag{7.28}$$

The coupled system of Eqs.(7.27) and (7.28) has a clear similarity to the Vlasov–Maxwell equations which describe the evolutions of particle distribution function and fields. The correspondence between the disparate scale interaction and the Vlasov–Maxwell system is summarized in Table 7.2.

Table 7.2. *Analogy and correspondence between the Vlasov–Maxwell system and the disparate scale interaction in Langmuir turbulence are summarized*

	Vlasov plasma	Quasi-particles and disparate scale interaction
Particle	electron, ion	plasma wave (plasmon)
Velocity	particle velocity	group velocity of plasma wave, or $\boldsymbol{k}$
Distribution	$f(\boldsymbol{x}, \boldsymbol{v})$	$N(\boldsymbol{x}, \boldsymbol{v}_{\rm g})$, or $N(\boldsymbol{x}, \boldsymbol{k})$
Field	electro(magnetic) field	ion acoustic wave
Change *of velocity*	acceleration by fields	modification of $\boldsymbol{v}_{\rm g}$ or $\boldsymbol{k}$ by refraction
Dynamical equation for particles	Vlasov equation	wave kinetic equation for N
Equation for field	Poisson equation	equation for acoustic wave

7.3.4 Response of distribution of quasi-particles

It is convenient to use the normalized density perturbation δn, so we take the spatio temporal structure of acoustic wave fluctuations to be,

$$\delta n = \frac{\tilde{n}}{n} = \sum_{q,\Omega} \delta n_{q,\Omega} \exp\left(i\boldsymbol{q} \cdot \boldsymbol{x} - i\Omega t\right), \quad \tilde{N} = \sum_{q,\Omega} \tilde{N}_{q,\Omega} \exp\left(i\boldsymbol{q} \cdot \boldsymbol{x} - i\Omega t\right). \tag{7.29}$$

As is the case in subsection 7.2.5, $\boldsymbol{q}$ and Ω stand for the slow spatio-temporal variation associated with acoustic waves. For transparency of argument, we take,

$$\frac{\partial \omega_k}{\partial n_e} = \frac{1}{2}\frac{\omega_{\rm p0}}{n_0},$$

where n_0 and $\omega_{\rm p0}$ are the density and plasma frequency at unperturbed state. Equations (7.27) and (7.29) immediately give the response,

$$\tilde{N}_{q,\Omega} = -\frac{\delta\tilde{n}\omega_{\rm p0}}{\Omega - \boldsymbol{q} \cdot \boldsymbol{v}_{\rm g} + i\hat{\Gamma}} \boldsymbol{q} \cdot \frac{\partial \langle N \rangle}{\partial \boldsymbol{k}}. \tag{7.30}$$

Applying quasi-linear theory to calculating the mean evolution, as explained in Chapter 3, then yields the change of plasmon energy as,

$$\frac{\mathrm{d}}{\mathrm{d}t}\langle E_k \rangle = -\omega_{\rm p0}^2 \sum_{q,\Omega} \boldsymbol{v}_{\rm gr} \cdot \boldsymbol{q} \frac{i\left|\delta n_{q,\Omega}\right|^2}{\Omega - \boldsymbol{q} \cdot \boldsymbol{v}_{\rm gr} + i\hat{\Gamma}} \boldsymbol{q} \cdot \frac{\partial \langle N \rangle}{\partial \boldsymbol{k}}. \tag{7.31}$$

When the self-interaction process of plasma waves is weaker than the decorrelation due to the dispersion of waves, (i.e., $\tau_{ac} < \tau_c, \tau_T$) one may take,

$$\frac{i}{\Omega - \boldsymbol{q} \cdot \boldsymbol{v}_g + i\hat{\Gamma}} \simeq \pi\delta\left(\Omega - \boldsymbol{q} \cdot \boldsymbol{v}_g\right)$$

where $\delta\left(\Omega - \boldsymbol{q} \cdot \boldsymbol{v}_g\right)$ is Dirac's delta function. Then,

$$\frac{\mathrm{d}}{\mathrm{d}t}\langle E_k\rangle = -\omega_{p0}\sum_{q,\Omega}\pi\delta\left(\Omega - \boldsymbol{q}\cdot\boldsymbol{v}_g\right)\left|\delta n_{q,\Omega}\right|^2 \boldsymbol{v}_g \cdot \boldsymbol{q}\boldsymbol{q}\cdot\frac{\partial\langle N\rangle}{\partial\boldsymbol{k}}. \tag{7.32}$$

This describes the relation between the population density $\langle N\rangle$ and the direction of energy transfer between plasma waves and acoustic waves. Since the group velocity $\boldsymbol{v}_g = \partial\omega_k/\partial\boldsymbol{k} = \gamma_T\omega_k^{-1}v_{T,e}^2\boldsymbol{k}$ is positive for plasma waves, the energy evolution rate satisfies the condition,

$$\frac{\mathrm{d}}{\mathrm{d}t}\langle E_k\rangle < 0 \quad \text{if} \quad \frac{\partial\langle N\rangle}{\partial k} > 0. \tag{7.33}$$

The energy is transformed from plasma wave quasi-particles to acoustic waves in the case of a population inversion $\partial\langle N\rangle/\partial k > 0$. That is, the acoustic waves grow in time at the expense of plasma wave quasi-particles. This resembles the inverse cascade, in that the long-wavelength modes accumulate energy from short-wavelength perturbations. It is important to stress that the difference is that the energy transfer takes place through the disparate scale interactions, not by local couplings. The energy from plasma waves is *directly* transferred to acoustic waves, without exciting intermediate scale fluctuations.

The transfer of energy from plasma waves to acoustic waves occurs through diffusion of the plasma wave quasi-particles. Substituting Eq.(7.30) into Eq.(7.27b), one obtains the evolution of the mean action density as,

$$\frac{\partial\langle N\rangle}{\partial t} + \langle\Gamma N\rangle = \frac{\partial}{\partial\boldsymbol{k}}\cdot\sum_{q,\Omega}\boldsymbol{q}\frac{i\omega_{p0}^2(\delta\tilde{n})^2}{\Omega - \boldsymbol{q}\cdot\boldsymbol{v}_g + i\hat{\Gamma}}\boldsymbol{q}\cdot\frac{\partial\langle N\rangle}{\partial\boldsymbol{k}}. \tag{7.34}$$

This is a diffusion equation for $\langle N\rangle$, i.e.,

$$\frac{\partial\langle N\rangle}{\partial t} + \langle\Gamma N\rangle = \frac{\partial}{\partial\boldsymbol{k}}\cdot\boldsymbol{D}\cdot\frac{\partial\langle N\rangle}{\partial\boldsymbol{k}} \tag{7.35a}$$

with,

$$\boldsymbol{D} = \sum_{q,\Omega}\frac{i\omega_{p0}^2(\delta\tilde{n})^2}{\Omega - \boldsymbol{q}\cdot\boldsymbol{v}_g + i\hat{\Gamma}}\boldsymbol{q}\boldsymbol{q}. \tag{7.35b}$$

It is evident that the total action density $\int \mathrm{d}k \, \langle N \rangle$ is conserved in the interaction with acoustic waves, because operating with $\int \mathrm{d}k$ to the right-hand side of Eq.(7.35a) vanishes. The total action $\int \mathrm{d}k \, \langle N \rangle$ is determined by the balance between the source and sink, i.e., $\int \mathrm{d}k \, \langle \Gamma N \rangle$. This redistribution of action density leads to the energy exchange between plasma waves and acoustic waves. Operating with $\int \mathrm{d}k \, \omega_k$ on Eq.(7.35a), the energy transfer rate, $(\partial/\partial t) \int \mathrm{d}k \, \omega_k \langle N \rangle$, induced by acoustic waves is seen to be equal to,

$$-\int \mathrm{d}k \, \frac{\partial \omega_k}{\partial \boldsymbol{k}} \cdot \boldsymbol{D} \cdot \frac{\partial \langle N \rangle}{\partial \boldsymbol{k}},$$

which is equivalent to Eq.(7.32). Thus, energy relaxation occurs if $\boldsymbol{v}_{\mathrm{g}} \cdot \boldsymbol{D} \cdot \partial \langle N \rangle / \partial \boldsymbol{k} > 0$ holds.

The diffusion of $\langle N \rangle$ and the energy transfer are explained in Figure 7.6, using an example of a one-dimensional problem. (The vector $\boldsymbol{q}$ is in one dimension.) The density $\langle N \rangle$ is subject to flattening and turns into the one shown by dashed lines. The areas 'A' and 'B' in this figure are equal. However, the frequency is higher in the domain 'A' than that in 'B', so that the energy content in 'A' is larger than that in 'B'. The difference between energies in 'A' and 'B' is converted into acoustic waves.

The $\boldsymbol{k}$-space diffusion coefficient for plasma wave quasi-particles is evaluated as,

$$D \simeq \left\langle q^2 \omega_{\mathrm{p}}^2 \delta n^2 \right\rangle \tau_{\mathrm{ac}}, \quad \tau_{\mathrm{ac}} = \left| \Delta \left(\Omega - \boldsymbol{q} \cdot \boldsymbol{v}_{\mathrm{g}} + i \hat{\Gamma} \right) \right|^{-1}, \tag{7.36}$$

where $\left| \Delta \left(\Omega - \boldsymbol{q} \cdot \boldsymbol{v}_{\mathrm{g}} + i \hat{\Gamma} \right) \right|$ is the width of the resonance of quasi-particles with the acoustic wave field. Equations (7.32) or (7.35) indicate that relaxation

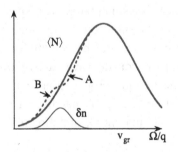

Fig. 7.6. Relaxation of the quasi-particle action density owing to interaction with acoustic waves (in a one-dimensional problem). The density $\langle N \rangle$ is subject to flattening and turns into the one shown by a dashed line. The areas 'A' and 'B' in this figure are equal. However, the frequency of plasma waves is higher in domain 'A' than that in 'B', so that the energy content in 'A' is larger than that in 'B'. The difference between energies in 'A' and 'B' is converted into acoustic waves.

of the plasma wave action occurs if the velocity of a quasi-particle coincides with the phase velocity of acoustic waves, $\Omega \simeq \boldsymbol{q} \cdot \boldsymbol{v}_g$. The group velocity of plasma waves $\boldsymbol{v}_g = \gamma_T \omega_k^{-1} v_{T,e}^2 \boldsymbol{k}$ and the dispersion of the acoustic wave $\Omega = c_s q / \sqrt{1 + q^2 \lambda_{De}^2} \simeq c_s q$ lead to the resonance condition,

$$\frac{\boldsymbol{q} \cdot \boldsymbol{k}}{qk} \simeq \frac{c_s}{\gamma_T v_{T,e}} \frac{1}{k \lambda_{De}}. \tag{7.37}$$

In a one-dimensional situation $\boldsymbol{q} \cdot \boldsymbol{k} \simeq qk$, this condition is satisfied for the wavelength $k\lambda_{De} \simeq O\left(\sqrt{m_e/m_i}\right)$. Oblique propagation of acoustic wave $\boldsymbol{q} \cdot \boldsymbol{k} \ll qk$ allows substantial resonance for the cases of $k\lambda_{De} > \sqrt{m_e/m_i}$. The subsonic limit, Eq.(7.10a), holds for the case $\Omega \ll \boldsymbol{q} \cdot \boldsymbol{v}_g$, i.e., $k\lambda_{De} \ll \sqrt{m_e/m_i}$, or very oblique propagation of modulation. (In the configuration of oblique propagation of modulation by the acoustic wave, the growth rate of modulation is smaller.)

Evolution of the mean action density is given by the induced diffusion equation. Resonant diffusion with the ion acoustic wave is irreversible. The origin of the irreversibility is discussed here. In the limit of small self-nonlinear interaction, $\Gamma \to 0$, the evolution equation of quasi-particles, Eq.(7.20), preserves time-reversal symmetry, as in the Vlasov equation. As is explained in Chapter 3, the irreversibility of resonant quasi-linear diffusion is ultimately rooted in orbit chaos. In the case of quasi-particle interaction with acoustic wave fields, *ray chaos* occurs by the overlap of multiple resonances at $\Omega \simeq \boldsymbol{q} \cdot \boldsymbol{v}_g$ for various values of the velocity of quasi-particles $\boldsymbol{v}_g$. The dispersion of the acoustic wave is weak, $\Omega = c_s q / \sqrt{1 + q^2 \lambda_{De}^2}$, and $\Omega \simeq q c_s$ for a wide range of Ω. The overlap of wave–quasi-particle resonance can happen as follows. Although the dispersion of acoustic waves is weak, there is a small but finite dispersion for acoustic waves. In addition, acoustic waves may be subject to a damping process owing to the kinetic interaction of ions. This provides a finite bandwidth for acoustic wave fluctuations, and thus induces the overlap of acoustic wave–quasi-particle resonances. The other route to overlapping is the oblique propagation of multiple acoustic waves, by which the resonance condition $\Omega \simeq \boldsymbol{q} \cdot \boldsymbol{v}_g$ can be satisfied for quasi-particles with a wide range of group velocity magnitude and direction.

7.3.5 Growth rate of modulation of plasma waves

The growth rate of the envelope of self-focusing plasma waves is derived for the subsonic limit in the preceding subsections. More general cases can be investigated by use of the quasi-particle approach. Substituting the response of quasi-particles (7.30) into Eq.(7.28), the dispersion relation of the acoustic wave field is given as,

$$-\Omega^2 + q^2 c_s^2 = \frac{q^2}{m_i}\omega_{p0}^2 \int \mathrm{d}k \, \frac{1}{\Omega - \boldsymbol{q}\cdot\boldsymbol{v}_g + i\hat{\Gamma}} \boldsymbol{q}\cdot\frac{\partial \langle N \rangle}{\partial \boldsymbol{k}}. \quad (7.38)$$

In the perturbation analysis where the right-hand side of Eq.(7.38) is smaller than $q^2 c_s^2$, one has,

$$\Omega = q c_s + i\frac{\pi q^2}{2m_i}\frac{\omega_{p0}^2}{q c_s}\int \mathrm{d}k\, \delta\left(\Omega - \boldsymbol{q}\cdot\boldsymbol{v}_g\right)\boldsymbol{q}\cdot\frac{\partial \langle N \rangle}{\partial \boldsymbol{k}}, \quad (7.39)$$

(for the case of a small self-nonlinearity term $\hat{\Gamma} \to 0$). Instability is possible if,

$$\frac{\partial \langle N \rangle}{\partial \boldsymbol{k}} > 0 \quad \text{at} \quad \Omega \simeq \boldsymbol{q}\cdot\boldsymbol{v}_g.$$

7.3.6 *Trapping of quasi-particles*

In order to proceed to understand nonlinear evolution of the modulation of plasma wave turbulence, the perturbation of the quasi-particle trajectory is illustrated here. Let us consider the motion of a quasi-particle in a single and coherent acoustic wave perturbation. The case where the lifetime of the quasi-particle (through a self-nonlinear mechanism) is small, $\Gamma \to 0$, is explained. This situation corresponds to the 'collisionless limit', i.e., the Vlasov plasma.

When the quasi-particle is subject to a longer-wavelength perturbation (ion sound wave in this case), the particle trajectory is deformed and trapping of the orbit in the eikonal phase space takes place. For transparency of argument, we consider again the simple limit in Section 7.3.3 (which lead to Eq.(7.26)). That is, we consider the case where the plasma wave turbulence is homogeneous (in the unperturbed state), and the Doppler shift by the ion fluid motion $\boldsymbol{k}\cdot\tilde{\boldsymbol{V}}$ is smaller than the effect of the modulation of the refractive index, $\tilde{\omega}_k$. The dynamic equations of quasi-particles in the presence of large-scale fields, $\mathrm{d}\boldsymbol{x}/\mathrm{d}t = \partial\omega_k/\partial\boldsymbol{k} + \tilde{\boldsymbol{V}}$ and $\mathrm{d}\boldsymbol{k}/\mathrm{d}t = -\partial\left(\tilde{\omega}_k + \boldsymbol{k}\cdot\tilde{\boldsymbol{V}}\right)/\partial\boldsymbol{x}$, are simplified to $\mathrm{d}\boldsymbol{x}/\mathrm{d}t = \partial\omega_k/\partial\boldsymbol{k}$ and $\mathrm{d}\boldsymbol{k}/\mathrm{d}t = -\partial\tilde{\omega}_k/\partial\boldsymbol{x}$. In the presence of the long-scale perturbations, $n = n_0 + \tilde{n}$, the relation,

$$\partial\tilde{\omega}_k/\partial\boldsymbol{x} = \nabla\tilde{n}\left(\partial\omega_k/\partial n_e\right)$$

is employed. Noting the relation $\partial\omega_k/\partial n_e = (1/2)n_0^{-1}\omega_{p0}$ for the plasma wave, one has,

$$\frac{\mathrm{d}\boldsymbol{x}}{\mathrm{d}t} = \boldsymbol{v}_\mathrm{g} \simeq \gamma_T \omega_{\mathrm{p}0}^{-1} v_{\mathrm{T},e}^2 \left(\boldsymbol{k}_0 + \tilde{\boldsymbol{k}}\right) \tag{7.40a}$$

$$\frac{\mathrm{d}\tilde{\boldsymbol{k}}}{\mathrm{d}t} = -\frac{\nabla \tilde{n}}{n_0}\omega_{\mathrm{p}0}, \tag{7.40b}$$

where k_0 denotes the wave number that satisfies the resonance condition $\boldsymbol{v}_\mathrm{g} = \Omega/q$ and $\tilde{k}$ is the modulation of the wave number by the acoustic mode. It is evident from this relation that the quasi-particles undergo a bounce motion (in the frame which is moving together with the large-scale wave). If one explicitly puts,

$$\frac{\tilde{n}}{n} = -\delta n \cos\left(qx - \Omega t\right),$$

(i.e., the origin is taken at the trough of the density perturbation of the acoustic wave), the relative displacement of the quasi-particle in the frame moving with the acoustic mode, $\xi = x - \Omega t/q$, obeys the equation,

$$\frac{\mathrm{d}^2\xi}{\mathrm{d}t^2} = -\delta n \gamma_T v_{\mathrm{T},e}^2 q \sin\left(q\xi\right). \tag{7.41}$$

That is, the orbit Eq.(7.40) is given by the elliptic function. Figure 7.7 illustrates the trajectories of quasi-particles. Trapped orbits and transiting orbits are separated

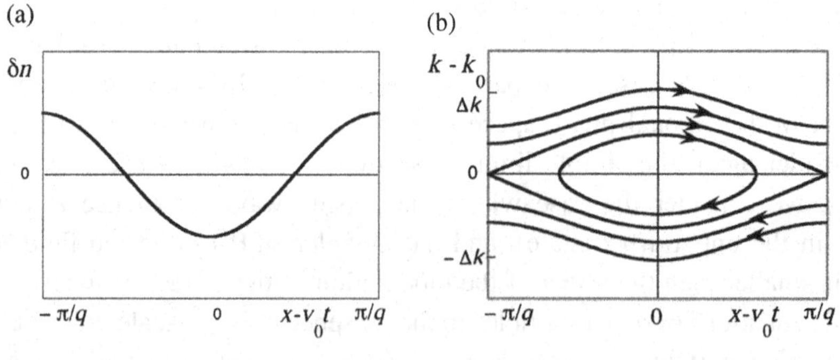

Fig. 7.7. Trapping of a quasi-particle in the trough of a long-wavelength perturbation. Here $v_0 = \Omega/q$ indicates the phase velocity of the long-wavelength mode and k_0 denotes the wave number that satisfies the resonance condition $v_\mathrm{g} = \Omega/q$. (a) shows the density perturbation of the acoustic mode, and (b) illustrates the trajectory of the quasi-particle (plasmon) in the phase space.

by a separatrix. Near the trough of the large-scale density perturbation, the bounce frequency is given as,

$$\omega_b = \sqrt{\gamma_T \delta n}\, v_{T,e} q. \tag{7.42}$$

(Bounce frequency is the inverse of rotation time of a trapped quasi-particle.) The bounce frequency is in proportion to the square-root of the density perturbation of the acoustic waves. Multiplying $d\xi/dt$ to the right and left-hand sides of Eq.(7.41), and integrating once in time, one has an integral of motion,

$$\frac{1}{2}\left(\frac{d\xi}{dt}\right)^2 = \delta n \gamma_T v_{T,e}^2 \cos(q\xi) + C, \tag{7.43}$$

where C is an integration constant. The trajectory forms a separatrix for the quasi-particle orbit with $C - \delta n \gamma_T v_{T,e}^2$. The largest modulation of the velocity of the quasi-particles is given as,

$$\Delta v_g = 2\sqrt{\delta n \gamma_T}\, v_{T,e}, \tag{7.44}$$

or, in terms of the wave number,

$$\Delta k = 2\sqrt{\frac{\delta n}{\gamma_T}\frac{\omega_{p0}}{v_{T,e}}}.$$

7.3.7 Saturation of modulational instability

The modulational instability of plasma wave turbulence is saturated when the perturbation amplitude becomes high. The analogy between the systems of:

(i) quasi-particle (plasma wave) and perturbed field (sound wave);
(ii) particle and perturbed electric field in Vlasov plasma

immediately tells that the modulational instability can be saturated through:

(a) flattening of the distribution function through phase space diffusion;
(b) nonlinear Landau damping by coupling to stable sound waves;
(c) trapping of quasi-particles in the trough of the field.

To identify the dominant process, the key parameters are the bounce frequency of the quasi-particle in the trough of the large-scale field, ω_b, and the dispersion in the wave number of the quasi-particle due to the large-scale fields, Δk. A typical example of Δk is the width of island in the phase space for the quasi-particle orbit.

When multiple acoustic waves exist so that resonance can occur at various values of velocity Ω/q, the separation of two neighbouring phase velocities $\Delta(\Omega/q)$

is introduced. If the separation between two phase velocities $\Delta(\Omega/q)$ is larger than the variation of the velocity of the quasi-particle in a trough of a large-scale field, $(\partial v_g/\partial k)\,\Delta k$, i.e.,

$$\Delta(\Omega/q) > (\partial v_g/\partial k)\,\Delta k, \tag{7.45a}$$

the quasi-particle orbit is trapped (if the lifetime is long enough) in a trough of one wave. In contrast, if the separation is smaller than the change in quasi-particle velocity,

$$\Delta(\Omega/q) < (\partial v_g/\partial k)\,\Delta k, \tag{7.45b}$$

so that 'island-overlapping' of resonances occurs. We introduce a 'Chirikov parameter' for a quasi-particle dynamics in long-wavelength perturbation fields as,

$$\mathcal{S} = \frac{(\partial v_g/\partial k)\,\Delta k}{\Delta(\Omega/q)}. \tag{7.46}$$

Thus, the island overlapping condition is given by $\mathcal{S} > 1$.

The other key parameter is the Kubo number of quasi-particles. Here, Kubo number $\mathcal{K}$ is the ratio between the lifetime of the quasi-particle (which is limited by the self-nonlinear effects of short-wavelength perturbations) and the bounce time of the quasi-particles in the trough of longer wavelength perturbations,

$$\mathcal{K} = \frac{\omega_b}{\Gamma}. \tag{7.47}$$

When $\mathcal{K}$ is much smaller than unity, a quasi-particle loses its memory before completing the circumnavigation in a trough of large-scale waves. Thus trapping does not occur. In contrast, if $\mathcal{K}$ is much larger than unity, the trapping of quasi-particles plays the dominant role in determining the nonlinear evolution. Note that $\mathcal{K}$ is closely related to the Strouhal number, familiar from fluid turbulence. The Strouhal number $\mathcal{S}_t$ is given by,

$$\mathcal{S}_t = \frac{\tilde{V}\tau_c}{l_c}.$$

If we take $\tilde{V}/l_c \sim 1/\tau_b$, i.e., to identify the 'bounce time' with an 'eddy circulation time', $\mathcal{S}_t$ may be re-written,

$$\mathcal{S}_t = \frac{\tau_c}{\tau_b},$$

which is essentially the Kubo number.

In the $(\mathcal{K}, \mathcal{S})$ diagram, important nonlinear processes are summarized (Diamond *et al.*, 2005b; Balescu, 2005). First, when the modulation amplitude is small and only one wave is considered $(\mathcal{K}, \mathcal{S} \to 0)$, the method of modulational parametric instability applies, and the instability criterion of Section 7.3.5 is deduced.

When $\mathcal{K}$ is small but $\mathcal{S}$ is larger than unity, ray chaos occurs. That is, the motion of the quasi-particle becomes stochastic (without being trapped in one particular trough), and diffuse in phase space. The quasi-linear diffusion approach (Chapter 3) applies. The evolution equation for the action density is given as,

$$\frac{\partial}{\partial t} \langle N \rangle - \frac{\partial}{\partial k} D_k \frac{\partial \langle N \rangle}{\partial k} = -\Gamma \langle N \rangle , \tag{7.48}$$

where D_k is the diffusion coefficient of the quasi-particle in the field of acoustic wave turbulence (Eq.(7.36)).

In the other extreme limit, i.e., $\mathcal{K} \to \infty$ but $\mathcal{S}$ is small, a quasi-particle moves (without decorrelation) along its perturbed orbit. Thus, ultimately a BGK state (Bernstein *et al.*, 1957; Kaw *et al.*, 1975, 2002) is approached. The integral of motion, C in Eq.(7.43), is given as,

$$C(x_0, k_0) = \frac{1}{2} \left(\gamma_T \frac{v_{\mathrm{T},e}^2}{\omega_{\mathrm{p}0}} k_0 - \frac{\Omega}{q} \right)^2 - \delta n \gamma_T v_{\mathrm{T},e}^2 \cos(q x_0) ,$$

where (x_0, k_0) is the position of the quasi-particle in the phase space at $t = 0$. Each trajectory in Figure 7.7(b) is characterized by the constant of motion $C(x_0, k_0)$. The distribution of the action density $N(x, k)$ is constant along this trajectory in a stationary state. By use of these integrals of motion, an exact solution for the distribution function is given in the form,

$$N(x, k_x) = N\left(C\left(x_0, k_{x0}\right)\right) . \tag{7.49}$$

When $\mathcal{S}$ is small, the beat wave excitation by modulated quasi-particles couples to the (more stable) acoustic waves. Then, nonlinear Landau damping of quasi-particles is the means for saturation of the perturbations. Coherent structure may be sustained by the balance between the (linear) modulational instability and nonlinear Landau damping of quasi-particles.

In the intermediate regime of $\mathcal{K}, \mathcal{S} \sim 1$, turbulent trapping of quasi-particles occurs. The various theoretical approaches are summarized in Figure 7.8.

7.4 Collapse of Langmuir turbulence

7.4.1 Problem definition

The nonlinear mechanism through disparate scale interaction is shown to generate large-scale perturbations. These might be observed as "mesoscale structures"

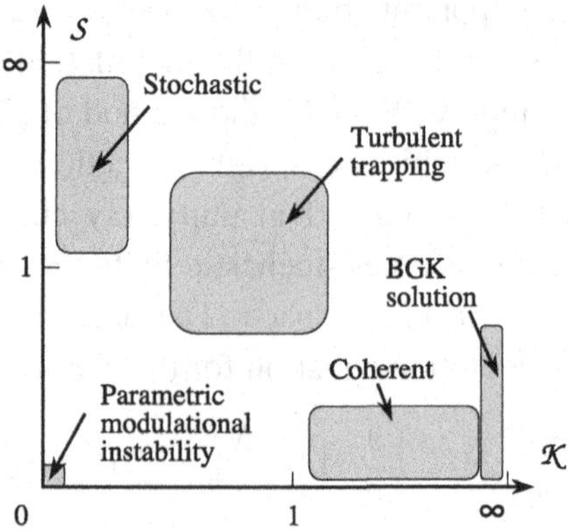

Fig. 7.8. Parameter domains for various theoretical approaches.

when global quantities are observed. This mechanism is, at the same time, the origin of a new route to enhanced energy dissipation at small scales. The standard route of energy dissipation occurs through the cascade into finer-scale perturbations, for which the understanding from Kolmogorov's analysis is very powerful. The evolution of self-focusing (which is induced by disparate scale interactions) can generate the compression of plasma wave energy into a small area. This phenomenon is known as the 'collapse' of plasma waves (Armstrong *et al.*, 1962; Bespalov and Talanov, 1966).

Recall, the problem of singularity in fluid turbulence. The dissipation rate per unit volume ϵ satisfies the relation $\epsilon = \left\langle \nu \, (\nabla V)^2 \right\rangle$, where ν is the (molecular) viscosity. If the energy is continuously injected into the system and is dissipated at microscale, it is plausible (but unproven) that the dissipation rate ϵ remains independent of ν, stays in the vicinity of the mean value (possibly with fluctuations in time) and does not vanish. Thus, the relation $(\nabla V)^2 = \epsilon/\nu$ implies that $|\nabla V| \to \infty$ in the invicid limit $\nu \to 0$. Thus, singularity formation is expected to occur. This hypothesis of singularity formation in Navier–Stokes turbulence has not yet been proven vigorously, but lies at the heart of turbulence physics. ("Singularity" does not, of course, indicate divergence of the velocity, but rather the formation of steep velocity gradients. In reality, the model of continuum for the fluid breaks down; new dynamics for energy dissipation occurs at the scale where the fluid description no longer holds.)

The collapse of the plasma wave is explained here, in order to illuminate an alternative (in comparison with the Kolmogorov cascade process) self-similar route to dissipation.

7.4.2 Adiabatic Zakharov equation

In order to illuminate the physics of collapse through disparate-scale interaction, we employ the adiabatic Zakharov equation, Eq.(7.12),

$$\left(i\frac{\partial}{\partial t}+\nabla^2\right)E+|E|^2E=0.$$

(The envelope is denoted by E, and the normalization is explained in Eq.(7.8).) The Zakharov equation indicates that coherent plasma waves can be subject self-focusing if the criterion Eq.(7.18) is satisfied. Once the focusing starts, the local enhancement of wave intensity will further accelerate the formation of localized structures, so that a singularity may form.

The nonlinear state of the focusing depends critically on the dimensionality of the system. In fact, the NLS equation in one-dimensional space is known to be integrable: the nonlinear stationary solutions (soliton solutions) were found to be,

$$E\left(x,t\right)=E_0\mathrm{sech}\left(\sqrt{\frac{1}{2}}E_0x\right)\exp\left(\frac{i}{2}E_0^2t\right). \tag{7.50}$$

(Noting the Galilean invariance of the adiabatic Zakharov equation (7.12), one may say that solutions with finite propagation velocity are easily constructed.)

In a three-dimensional system, focusing continues and collapse occurs. This process is explained in this subsection, but a qualitative explanation is given before going into details. Let us consider the situation where the wave field of high intensity $|E|^2$ is localized in a region with a size l. The conservation of the total plasma wave energy, which is in proportion to $|E|^2\,l^d$, requires that $|E|^2\,l^d$ remain constant (here, d is the number of space dimensions). When the size of the bunch changes, $l\to l'$, the squared intensity changes as $|E|^2\to\left(l/l'\right)^d$. The diffraction term increases as $\left(l/l'\right)^2$. Thus the nonlinear attraction term works more strongly than the diffraction term in the three-dimensional case $d=3$, so that contraction of the system size is not restricted. In the one-dimensional case, the increment of diffraction is larger upon contraction, so that the self-focusing stops at finite amplitude. In that case, diffraction balances self-focusing, thus producing a soliton structure.

7.4.3 Collapse of plasma waves with spherical symmetry

The envelope with spherical symmetry, $E\left(r,t\right)$, in the adiabatic Zakharov equation satisfies,

$$i\frac{\partial}{\partial t}E+\frac{1}{r^2}\frac{\partial}{\partial r}r^2\frac{\partial}{\partial r}E+|E|^2E=0. \tag{7.51}$$

With the help of an analogy with quantum physics, we can identify integrals of this equation. First is the conservation of the total plasmon number. If one introduces the 'flux',

$$F = i\left(E\,\partial E^*/\partial r - E^*\,\partial E/\partial r\right),$$

Eq.(7.51) immediately gives the conservation relation for intensity distribution,

$$\partial\,|E|^2/\partial t + r^{-2}\,\partial\left(r^2 F\right)/\partial r = 0.$$

Thus one finds that the total intensity of the wave (i.e., the total number of plasmons, I_1) is conserved,

$$I_1 = \int_0^\infty \mathrm{d}r\, r^2\,|E|^2\,. \tag{7.52a}$$

The total Hamiltonian I_2 is also conserved in time, i.e.,

$$I_2 = \int_0^\infty \mathrm{d}r\, r^2 \left(\left|\frac{\partial E}{\partial r}\right|^2 - \frac{1}{2}\,|E|^4\right) \tag{7.52b}$$

is a constant of motion. According to the analogy of quantum physics, the first term and the second term in the integral are the kinetic energy and (attractive) potential energy, respectively. The second term on the right-hand side of Eq.(7.52b) is the 'attractive potential', which increases when the amplitude of the wave becomes larger. Thus, the 'total energy' I_2 (for which $|\partial E/\partial r|^2$ is the 'kinetic energy density' and $|E|^4$ is the 'potential energy') is positive for a small amplitude wave, and I_2 is negative for large amplitude waves. In other words, the small amplitude wave is considered to be a 'positive energy state' and the large amplitude wave is considered to be a 'negative energy state'.

By use of these two constants of motion, the evolution of the mean radius,

$$\left\langle r^2\right\rangle = (I_1)^{-1}\int_0^\infty \mathrm{d}r\, r^4\,|E|^2\,,$$

can be studied. Noting the conserved form for the intensity,

$$\partial\,|E|^2/\partial t + r^{-2}\,\partial\left(r^2 F\right)/\partial r = 0,$$

one has the relation,

$$\partial^2\left(r^2\,|E|^2\right)/\partial t^2 + r^{-2}\,\partial\left(r^4\,\partial F/\partial t\right)/\partial r - 2r\,\partial F/\partial t = 0.$$

Therefore, the mean squared radius obeys the evolution equation,

$$I_1 \frac{\partial^2}{\partial t^2}\left\langle r^2\right\rangle = 2\int_0^\infty \mathrm{d}r\, r^3 \frac{\partial}{\partial t} F.$$

The integral of the right-hand side of this relation is calculated by use of the NLS equation. Thus, we have the identity,

$$\frac{\partial^2}{\partial t^2}\left\langle r^2\right\rangle = 8\frac{I_2}{I_1} - \frac{2}{I_1}\int_0^\infty \mathrm{d}r\, r^2 |E|^4, \tag{7.53a}$$

that is,

$$\frac{\partial^2}{\partial t^2}\left\langle r^2\right\rangle < 8\frac{I_2}{I_1}. \tag{7.53b}$$

The mean squared radius $\left\langle r^2\right\rangle$ must satisfy the condition,

$$\left\langle r^2\right\rangle < 8\frac{I_2}{I_1}t^2 + \frac{\partial}{\partial t}\left\langle r^2\right\rangle\bigg|_0 t + \left\langle r^2\right\rangle\bigg|_0 . \tag{7.54}$$

The integral I_1 is positive definite. Therefore, if the initial value of I_2 is negative, the mean squared radius approaches zero after a finite time. The mean-squared radius $\left\langle r^2\right\rangle$ is positive definite, by definition, so that the solution must encounter a singularity in a finite time. The time for the collapse, τ_{collapse}, where $\left\langle r^2\right\rangle = 0$ holds, can be calculated from Eq.(7.54). For instance, starting from a stationary initial condition, $\frac{\partial}{\partial t}\left\langle r^2\right\rangle\big|_0 = 0$, one has the bound,

$$\tau_{\text{collapse}} < \sqrt{\frac{-I_1}{8I_2}\left\langle r^2\right\rangle\bigg|_0}. \tag{7.55}$$

In the large amplitude limit, where the second term is much larger than the first term in the integrand of Eq.(7.52b), the ratio $-I_1/I_2$ is of the order of $|E|^{-2}$. The normalization is shown in Eq.(7.8), i.e., $\omega_{\mathrm{p}0}t \to t$, $\lambda_{\mathrm{De}}^{-1}x \to x$, and E is the oscillation velocity of an electron at the plasma wave frequency ($\tilde{v}_{\mathrm{pw}}$), normalized to the electron thermal velocity. Then, the estimate $\tau_{\text{collapse}} \sim \sqrt{|E|^{-2}\left\langle r^2\right\rangle\big|_0}$ (apart from a factor of order unity) can be interpreted as the time for the onset of collapse as,

$$\tau_{\text{collapse}} \sim \frac{l}{\tilde{v}_{\mathrm{pw}}}, \tag{7.56}$$

where l is the initial size of the hump of plasma waves. (An example from numerical calculation is given in Figure 7.9.)

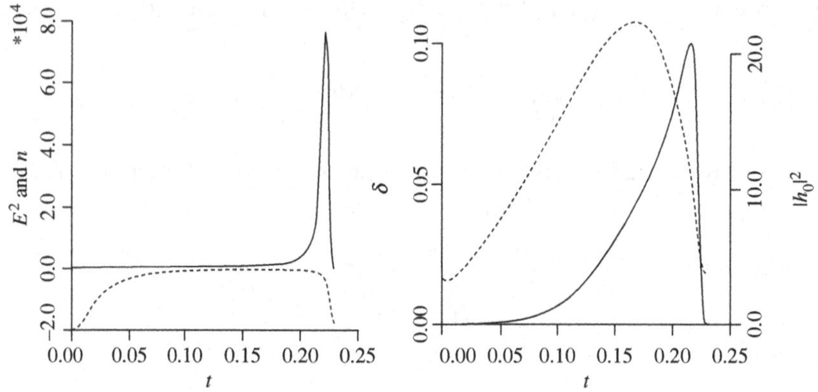

Fig. 7.9. Example of spherical collapse of Langmuir waves. Intensity of waves $|E|^2$ and perturbation density (dashed line) at the centre $r=0$ are given in (a). (Parameters are: pump wave intensity $|E_0|=5$, $m_\mathrm{i}/m_\mathrm{e}=2000$. Normalization units for time, length, electric field and perturbed number density are $(3m_\mathrm{i}/2m_\mathrm{e})\,\omega_\mathrm{pe}^{-1}$, $(3/2)\sqrt{m_\mathrm{i}/m_\mathrm{e}}\lambda_\mathrm{De}$, $8\sqrt{\pi n_0 T_\mathrm{e} m_\mathrm{e}/3m_\mathrm{i}}$ and $n_0 4m_\mathrm{e}/3m_\mathrm{i}$, respectively.) In the early phase of the collapse, the eigenmode with lowest eigenvalue $\mathbf{e}_0\,(\mathbf{x}, t)$, the amplitude of which is $h_0\,(t)$, grows dominantly. Amplitude $|h_0\,(t)|^2$ and radius of collapsing state $\delta\,(t)$ are shown in (b); $\delta(t)$ is defined by the radius that maximizes $r^2|\mathbf{e}_0\,(\mathbf{r}, t)|^2$. [quoted from (Dubois *et al.*, 1988)].

Note that the relation between $\partial^2\left\langle r^2\right\rangle/\partial t^2$ and I_2/I_1, like Eq.(7.52), is derived for the two-dimensional ($d=2$) and the one-dimensional ($d=1$) cases. One obtains after some manipulation,

$$\frac{\partial^2}{\partial t^2}\left\langle r^2\right\rangle = 8\frac{I_2}{I_1}, \quad (d=2) \tag{7.57a}$$

$$\frac{\partial^2}{\partial t^2}\left\langle r^2\right\rangle = 8\frac{I_2}{I_1} + \frac{2}{I_1}\int_0^\infty \mathrm{d}r\,|E|^4, \quad (d=1) \tag{7.57b}$$

(where the weight of volume element r^2 in the integrand of Eq.(7.52) is appropriately adjusted). We see that the proportionality between $\partial^2\left\langle r^2\right\rangle/\partial t^2$ and I_2/I_1 holds for the 2D case. Collapse occurs in this case, if the *initial total energy* is negative $I_2 < 0$. In contrast, for the one-dimensional problem, $d = 1$, collapse is prohibited by the second term on the right-hand side of Eq.(7.57b), even if the condition $I_2 < 0$ is satisfied. For an arbitrary negative value of I_2, the second term on the right-hand side eventually overcomes the first term if focusing continues. Thus, contraction must stop at a finite size. Comparison between these three cases is illustrated schematically in Figure 7.10. (Note that the physics of singularity formation works in other turbulence, too. For instance, extension to singularity formation

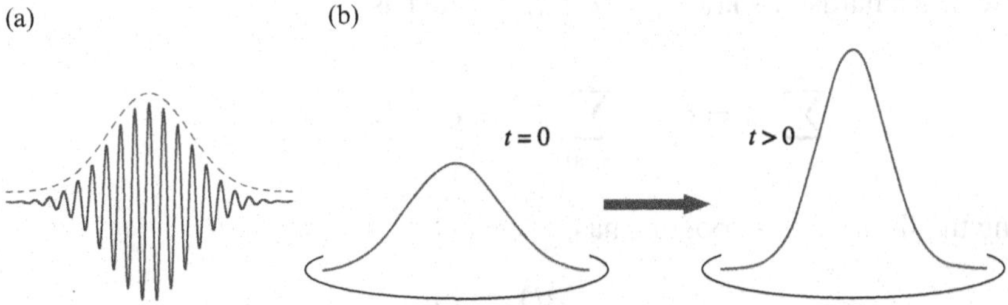

Fig. 7.10. Schematic description of the cases of 1D (a), 2D and 3D (b).

in electron-temperature-gradient-driven turbulence was discussed in (Gurcan and Diamond, 2004).)

7.4.4 Note on 'cascade versus collapse'

The collapse via disparate scale nonlinear interaction is a new route to the dissipation of wave energy. This process is compared to the Kolmogorov cascade process. Equation (7.56) shows that the time of collapse is shorter if the intensity of the Langmuir wave is higher, and is longer if the size of an original spot is larger. Such a dimensional argument is also available for the Kolmogorov cascade.

We revisit the relation (7.56) for the case of cascade. The scale length of the largest eddy is evaluated by the energy input scale $l_{\mathrm{E}} = k_{\mathrm{E}}^{-1}$, where k_{E} is defined by the condition that most of energy resides near k_{E}, i.e.,

$$K \cong K_0 \epsilon^{2/3} \int_{k_{\mathrm{E}}}^{\infty} l^{-5/3} \mathrm{d}k,$$

where ϵ is the dissipation rate of the energy per unit volume and K is the kinetic energy density. From this consideration, we have $k_{\mathrm{E}} = \epsilon K^{-3/2}$, that is,

$$l_{\mathrm{E}} = \epsilon^{-1} K^{3/2}.$$

The characteristic time for energy dissipation is evaluated by considering the sequence of cascades which make an eddy into smaller eddies. Let us consider a process that the n-th eddy (size l_n, velocity v_n) is broken into the $(n+1)$-th eddy (Figure 2.12(b)),

$$l_{n+1} = \alpha l_n,$$

(where l_n is given by l_{E} at $n=0$). The time for the cascade from the n-th eddy to the $(n+1)$-th eddy, induced by the nonlinearity $\boldsymbol{v} \cdot \nabla$, is given by $\tau_n = l_n / v_n$, thus,

$$\textit{time for the cascade to the n-th eddy} = \sum_{n=0}^{n} \tau_n = \sum_{n=0}^{n} \frac{l_n}{v_n}.$$

By use of Richardson's law, $v_n = \epsilon^{1/3} l_n^{1/3}$, one has

$$\sum_{n=0}^{n} \tau_n = \epsilon^{-1/3} \sum_{n=0}^{n} l_n^{2/3} = \epsilon^{-1/3} l_0^{2/3} \sum_{n=0}^{n} \alpha^{2n/3}.$$

Taking the limit of $n \to \infty$, one has,

$$\tau_{\text{cascade}} = \sum_{n=0}^{\infty} \tau_n = \frac{l_{\text{E}}^{2/3}}{\epsilon^{-1/3}} \frac{1}{1-\alpha^{2/3}} \sim \frac{1}{1-\alpha^{2/3}} \frac{l_{\text{E}}}{\tilde{v}}, \tag{7.58a}$$

where $\tilde{v} = \epsilon^{-1/3} l_{\text{E}}^{-1/3}$ is the turbulent velocity. The time for the energy cascade is approximately given by the eddy turn-over time, τ_{eddy},

$$\tau_{\text{eddy}} = \frac{l_{\text{E}}}{\sqrt{K}} = \frac{l_{\text{E}}}{\tilde{v}}. \tag{7.58b}$$

Comparing Eqs.(7.56) and (7.58), one observes a similarity which appears in the dependence as,

$$\tau \sim l/\tilde{v}.$$

The difference is, however, noticeable: the time given by (7.57) scales (up to the microscopic scale where dissipation works). In contrast, the equation (7.56) indicates that the energy is concentrated in the area of focus, which has a small radius. The cascade is considered as a stationary state, while the alternative path such as collapse is an explosive and dynamical phenomenon, which resembles steepening en route to shock formation.

8

Cascades, structures and transport in phase space turbulence

It is that science does not try to explain, nor searches for interpretations but primarily constructs models. A model is a mathematical construction, which supplemented with some verbal explanation, describes the observed phenomena. Such mathematical construction is proved if and only if it works, that it describes precisely a wide range of phenomena. Furthermore, it has to satisfy certain aesthetic criteria, i.e., it has to be more or less simple compared to the described phenomena.

(J. Von Neumann)

8.1 Motivation: basic concepts of phase space turbulence

8.1.1 Issues in phase space turbulence

Up to now, our discussion of plasma turbulence has developed by following the two parallel roads shown in Figure 8.1. Following the first, well trodden, path, we have developed the theory of nonlinear mode interaction and turbulence as applied to fundamentally *fluid* dynamical systems, such as the Navier–Stokes (NS) equation or the quasi-geostrophic (QG) Hasegawa–Mima equation. Along the way, we have developed basic models such as the scaling theory of eddy cascades as in the Kolmogorov theory, the theory of coherent and stochastic wave interactions, renormalized theories of fluid and wave turbulence, the Mori–Zwanzig memory function formalism for elimination of irrelevant variables, and the theory of structure formation in Langmuir turbulence by disparate scale interaction. Following the second, less familiar trail, we have described the theory of kinetic Vlasov turbulence – i.e. turbulence where the fundamental dynamical field is the phase space density $f(x, v, t)$ and the basic equation is the Vlasov equation or one of its gyrokinetic variants. Along the way, we have discussed quasi-linear theory of mean field ($\langle f \rangle$) evolution and the theory of resonance broadening and nonlinear response $\delta f / \delta E$, and of nonlinear wave–particle interaction. We note, though, that our discussion of Vlasov turbulence has been cast *entirely* within the framework

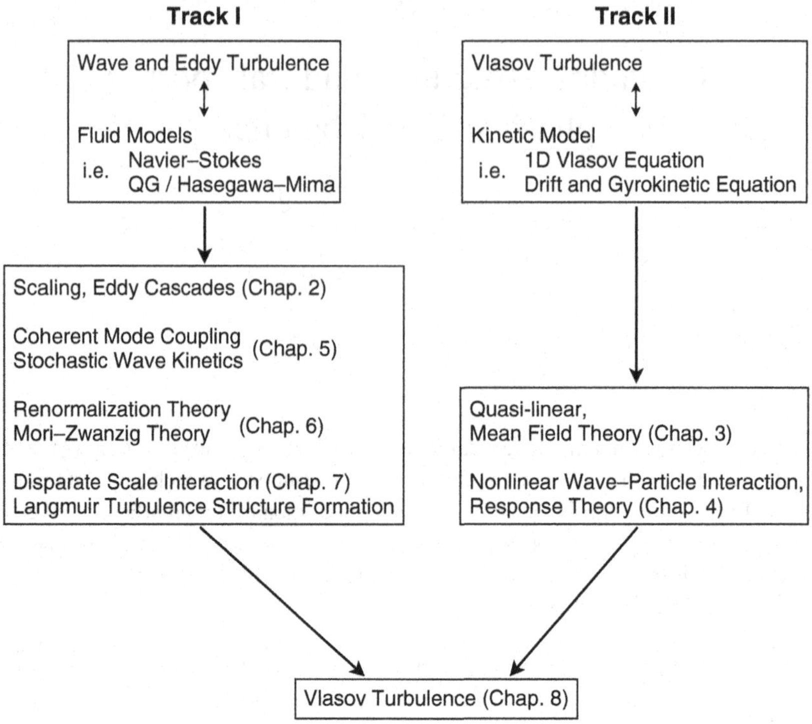

Fig. 8.1. Two paths in the discussion of plasma turbulence.

of linear and nonlinear response theory – i.e. we have focused exclusively upon the impact of Vlasov nonlinearity on the distribution function response δf to a fluctuating electric field δE, and on the consequent construction of macroscopic evolution equations. As a result of this focus, the approaches developed so far, such as the (coarse grained) quasi-linear equation or the wave kinetic equation for $\langle E^2 \rangle_{\mathbf{k}}$, have *not* really described the dynamics of Vlasov turbulence at the level of its governing nonlinear phase space equation, which is the Vlasov equation. In particular, we have not adapted familiar turbulence concepts such as eddies, coherent vortices, cascades, mixing, etc. to the description of Vlasov turbulence. Furthermore, we have so far largely separated the phenomena of wave–particle resonance – which we have treated primarily using quasi-linear theory, from the process of mode–mode coupling, which we have treated using fluid equations or modal amplitude equations.

8.1.1.1 Vlasov–Poisson system

In this chapter, we present an extensive discussion of phase space turbulence, as governed by the Vlasov–Poisson system. Recall the Vlasov equation is simply the Boltzmann equation,

$$\frac{\partial f}{\partial t} + v\frac{\partial f}{\partial x} + \frac{qE}{m}\frac{\partial f}{\partial v} = C(f) \tag{8.1a}$$

for vanishing $C(f)$. As with the well-known distinction between the Navier–Stokes and Euler equations, one must take care to distinguish the limit of $C(f)$ small but finite, from the case where $C = 0$. The Poisson equation,

$$\frac{\partial E}{\partial x} = 4\pi n_0 q \int \mathrm{d}v\, f + 4\pi q_{\mathrm{os}} \hat{n}_{\mathrm{os}} \tag{8.1b}$$

allows a *linear* feedback channel between the electric field which evolves f and the charge density which f itself produces. For completeness, we include the possibility of additional charge perturbations induced by other species (denoted here by $q_{\mathrm{os}}\hat{n}_{\mathrm{os}}$), which couple to f evolution via Poisson's equation. The Vlasov equation is a statement of the local conservation (up to collisional coarse-graining) of a (scalar) phase space density $f(x, v, t)$ along the particle trajectories set by its Hamiltonian characteristic equations,

$$\frac{\mathrm{d}x}{\mathrm{d}t} = v, \qquad \frac{\mathrm{d}v}{\mathrm{d}t} = \frac{q}{m}E(x, t). \tag{8.2}$$

Thus, Vlasov–Poisson turbulence may be thought of as "active scalar" turbulence, as is 2D fluid or quasi-geostrophic (QG) turbulence.

8.1.1.2 Analogy between Vlasov system and quasi-geostrophic system

Indeed, there is a close and instructive analogy between the quasi-geostrophic or Hasegawa–Mima system and the Vlasov–Poisson system. In the QG system, potential vorticity (PV) $Q(\mathbf{x}, t)$ is conservatively advected (modulo a small viscous cut-off) along the streamlines of incompressible flow $\mathbf{v}$, itself determined by the PV field $Q(\mathbf{x}, t)$ (Vallis, 2006), so the QG equation is just,

$$\partial_t Q + \{Q, \phi\} - \nu\nabla^2 Q = 0, \tag{8.2a}$$

where the velocity is expressed in terms of a stream function,

$$\mathbf{v} = \nabla\phi \times \hat{z}, \tag{8.2b}$$

ν is the viscosity, and the advected PV is related to the stream functions ϕ via,

$$Q = Q_0 + \beta x + \nabla^2\phi. \tag{8.2c}$$

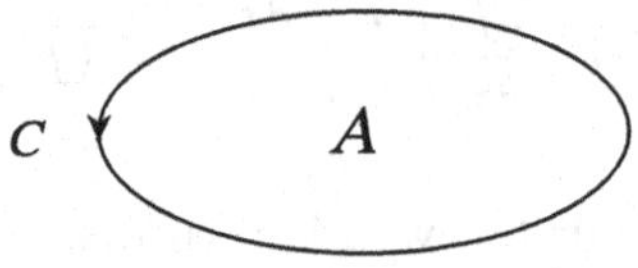

Fig. 8.2. Contour and area of integration.

(See the description of coordinates in Appendix 1.)[1] Note that the stream function is related to potential vorticity via a Poisson equation. In turn, the Vlasov–Poisson system may be rewritten in the form,

$$\partial_t f + \{f, H\} = C(f) \tag{8.3a}$$

$$f = \langle f(v, t)\rangle + \delta f(\mathbf{x}, \mathbf{v}, t) \tag{8.3b}$$

$$H = \frac{p^2}{2m} + q\phi(\mathbf{x}), \tag{8.3c}$$

where q is a charge, ϕ is an electrostatic potential, p is a momentum, and f and ϕ are related through Poisson's equation, given by Eq. (8.1b).

8.1.1.3 Circulation in QG system revisited

Circulation and its conservation are central to PV dynamics and, as we shall see, to Vlasov plasma dynamics. It is well known that an inviscid, barotropic fluid obeys Kelvin's circulation theorem,

$$\frac{\mathrm{d}\Gamma}{\mathrm{d}t} = 0, \tag{8.4a}$$

where,

$$\Gamma = \oint_C \mathbf{v} \cdot \mathbf{dl} = \int_A \boldsymbol{\omega} \cdot \mathbf{da}. \tag{8.4b}$$

Here C is a closed contour which bounds the area (Fig. 8.2). The element of thus fluid may be thought of as carrying a circulation $\Gamma \sim Vl \sim \omega A$. Conservation of circulation is fundamental to vortex dynamics and enstrophy prediction, discussed in Chapter 2, and to the very notion of an "eddy", which is little more than a conceptual cartoon of an element of circulation on a scale l. *So, Kelvin's theorem is indeed a central element of turbulence theory* (P. A. Davidson, 2004). For the QG system, a new twist is that the conserved potential vorticity is the sum of

[1] We note here that the x-direction is taken in the direction of inhomogeneity of the mean profile and y- is an ignorable coordinate after the convention of plasma physics. In geofluid dynamics, the x-direction is in the longitudinal direction (ignorable coordinate) and y- is in the latitudinal direction (in the direction of mean inhomogeneity). Thus, our notation is **not** the standard one used in geofluid dynamics.

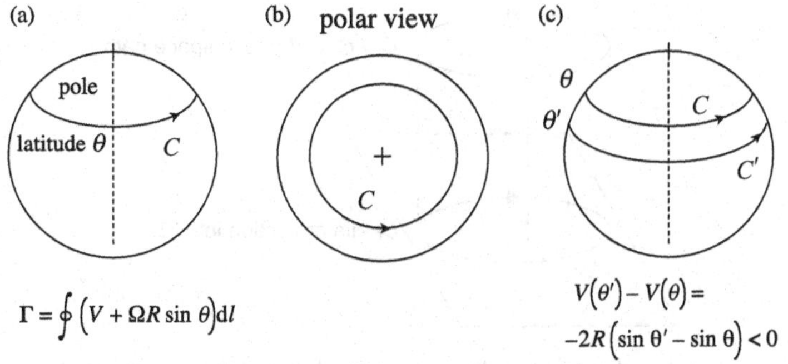

Fig. 8.3. Contour C at latitude θ and the circulation Γ (a). Polar view is illustrated in (b). When the contour is moved to lower latitude, $\theta \rightarrow \theta'$, conservation of circulation induces a westward circulation (c).

planetary (i.e. mean, i.e. $Q = Q_0 + \beta x$) and relative (i.e. fluctuating ω) pieces, so the integrated PV is,

$$\int_A Q \,\mathrm{d}a = \int_A (\omega + 2\Omega \sin\theta)\, da, \tag{8.5a}$$

and that the conserved circulation of, say, an eddy encircling the pole of latitude θ_0 (on a rotating sphere) is then,

$$\Gamma = \oint_C (V + 2\Omega R \sin\theta)\, \mathrm{d}l, \tag{8.5b}$$

(where Ω and R are rotation frequency and radius of the sphere, respectively), as shown in Figure 8.3. The novel implication here is that since total Γ is conserved, there must be trade-offs between planetary and relative contributions when C moves. This follows because advecting a PV patch or element to higher latitude necessarily implies an increase in planetary vorticity $Q_{\mathrm{ot}} = \beta x$, so conservation of *total* PV and circulation consequently imply that *relative* vorticity $\omega = \nabla^2 \phi$ and relative circulation $\int \omega \,\mathrm{d}a = \oint \mathbf{V} \cdot \mathrm{d}\mathbf{l}$ must *decrease*, thus generating a westward flow, i.e. see Figure 8.3. Thus, in QG fluids, Kelvin's theorem links eddy dynamics to the effective mean vorticity profile, and so governs how localized eddies interact with the mean PV gradient.

8.1.1.4 Circulation theorem for Vlasov system and granulations

Perhaps not surprisingly, then, the Vlasov equation also satisfies a circulation theorem, which we now present for the general case of electrostatic dynamics in 3D. Consider C_Γ, a closed phase space trajectory in the 6-dimensional phase space, and $C_{\mathbf{r}}$, its projection into 3-dimensional configuration space, i.e. see Figure 8.4.

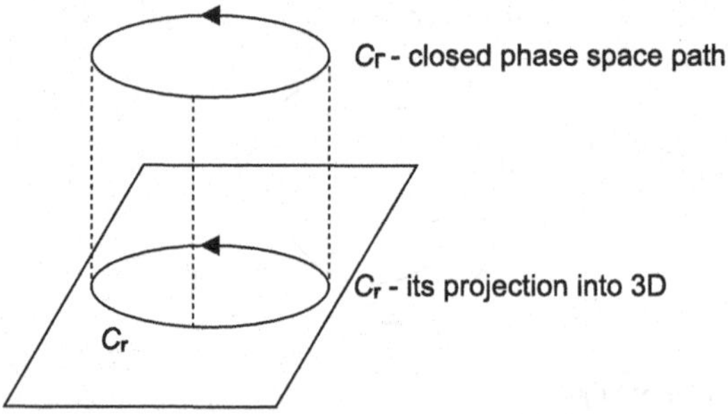

Fig. 8.4. Closed path in a phase space and its projection into real space.

Let s specify the location along the path C_Γ, so that $\mathbf{r}(s)$ corresponds to the trajectory $C_\mathbf{r}$ and $(\mathbf{r}(s), \mathbf{v}(s))$ corresponds to the trajectory C_Γ. Then the circulation is simply,

$$\Gamma = \oint \mathbf{v}(s) \cdot \mathbf{dr} \tag{8.6a}$$

and,

$$\frac{\mathrm{d}\Gamma}{\mathrm{d}t} = \oint \left[\frac{\mathrm{d}\mathbf{v}(s)}{\mathrm{d}t} \cdot \mathbf{dr} + \mathbf{v}(s) \cdot \frac{\mathrm{d}}{\mathrm{d}t}\mathbf{dr}(s) \right]. \tag{8.6b}$$

Since

$$\frac{\mathrm{d}\mathbf{v}(s)}{\mathrm{d}t} = -\frac{q}{m}\nabla\phi, \tag{8.6c}$$

and

$$\mathbf{v}(s)\frac{\mathrm{d}}{\mathrm{d}t}\mathbf{dr}(s) = \frac{\mathrm{d}}{\mathrm{d}t}\left(\frac{1}{2}v^2(s)\right), \tag{8.6d}$$

we easily see that,

$$\mathrm{d}\Gamma/\mathrm{d}t = 0, \tag{8.6e}$$

so the *collisionless electrostatic Vlasov–Poisson system indeed conserves phase space circulation*. The existence of a Kelvin's theorem for the Vlasov–Poisson system then suggests that an "eddy" is a viable and useful concept for phase space turbulence, as well as for fluids. For the 2D phase space of (x, v), the circulation is simply the familiar integral $\Gamma = \oint v\,\mathrm{d}x$, so a Vlasov eddy or *granulation* may be thought of as a chunk of phase space fluid with effective circulation equal to its

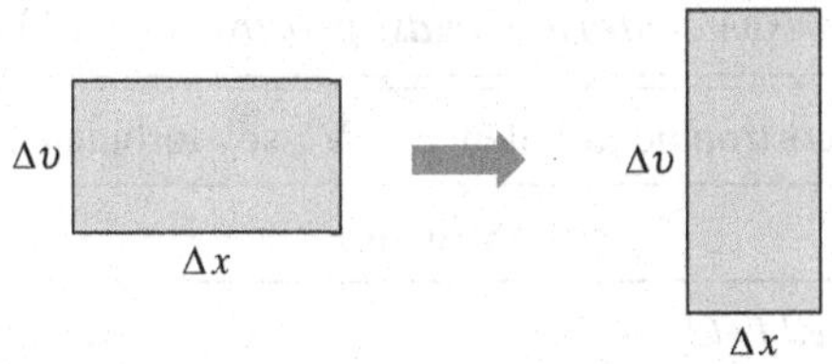

Fig. 8.5. Conservation of a volume of an element in the phase space.

conserved phase volume $\Delta x \Delta v$, a circulation time equal to the orbit bounce time and an associated conserved phase space density,

$$\left[f(x, v, t)\right]_{\Delta x, \Delta v} = \int_{x-\Delta x/2}^{x+\Delta x/2} \frac{\mathrm{d}x'}{\Delta x} \int_{v-\Delta v/2}^{v+\Delta v/2} \frac{\mathrm{d}v'}{\Delta v} f\left(x', v', t\right). \quad (8.7)$$

Usually, the velocity scale of a phase space eddy will correspond to its trapping width Δv_{T}, so the eddy circulation time scale $\tau = \Delta x / \Delta v$ will correspond to the wave–particle correlation time τ_{c}. Just as $(\Delta v_{\mathrm{T}}) \tau_{\mathrm{c}} = k^{-1}$, $(\Delta v) \tau = \Delta x$. Note that Kelvin's theorem requires that if, say Δv increases on account of acceleration, Δx must then decrease concomitantly, to conserve circulation. This process of stretching while conserving phase volume is depicted in Figure 8.5. It suggests that a turbulent Vlasov fluid may be thought of as a tangle of thin stretched strands, each with its own value of locally conserved f, and that fine scale structure and sharp phase space gradients must develop and ultimately be limited by collisions. The correspondence between the conservative QG system and the Vlasov–Poisson system is summarized in Table 8.1.

8.1.2 Granulation – what and why

8.1.2.1 Dynamics of granulations

Like a fluid eddy, a granulation need not be related to a wave or collective mode perturbation, for which the dielectric function,

$$\epsilon(k, \omega) \rightarrow 0,$$

but may be nonlinearly driven instead, as in a fluid turbulence cascade. To see this, consider the behaviour of the two-point correlation function, discussed in detail later in this chapter. If one takes the phase space density fluctuation $\delta f = f^{\mathrm{c}}$, i.e. just the coherent response – as in the case of a wave, one has,

Table 8.1. *Comparison/contrast of quasi-geostrophic and Vlasov turbulence*

Quasi-geostrophic turbulence	Vlasov turbulence
Structure	
Conserved field	
potential vorticity – Q	phase space density – f
Evolution	
$\{Q, \phi\}$	$\{f, H\}$
Evolver	
stream function – ϕ	electrostatic potential – ϕ, $H = p^2/2m + \phi$
Dissipation	
$-\nu\nabla^2$	$C(f)$
Feedback	
$Q = Q_0 + \beta x + \nabla^2\phi$	Poisson equation
Circulation	
$\oint_C (V + 2\Omega R \sin\theta_0)\, \mathrm{d}l$	$\oint \mathbf{v}(s) \cdot \mathbf{dr}$
Planetary + Relative vorticity	$f = \langle f \rangle + \delta f$
Element	
Vortex Patch	Granulation
$\int Q\, da$ conserved	Phase volume conserved

$$\langle \delta f(1)\, \delta f(2) \rangle^{\mathrm{c}} = \left\langle f^{\mathrm{c}}(1)\, f^{\mathrm{c}}(2) \right\rangle$$
$$= \mathrm{Re} \sum_{\mathbf{k},\omega} \frac{q^2}{m^2} \frac{\left|E_{k,\omega}\right|^2 (\partial \langle f \rangle / \partial v)^2 \exp\{ik(x_2 - x_1)\}}{(\omega - kv_1 + i/\tau_{ck})^* (\omega - kv_2 + i/\tau_{ck})} \tag{8.8a}$$

so,

$$\lim_{1\to 2} \left\langle f^{\mathrm{c}}(1)\, f^{\mathrm{c}}(2) \right\rangle = 2\tau_{\mathrm{c}} D\, (\partial \langle f \rangle / \partial v)^2\,, \tag{8.8b}$$

where D is just the familiar quasi-linear diffusion coefficient. Note that $\lim_{1\to 2} \langle f^{\mathrm{c}}(1)\, f^{\mathrm{c}}(2) \rangle$ is manifestly *finite* at small separations. In contrast, by using phase space density conservation to write the *exact* evolution equation for $\langle \delta f(1)\, \delta f(2) \rangle$ in the relative coordinates x_-, v_-, we see that,

$$\frac{\partial}{\partial t} \langle \delta f(1)\, \delta f(2) \rangle + v_- \frac{\partial}{\partial x_-} \langle \delta f(1)\, \delta f(2) \rangle$$
$$+ \frac{q}{m} \left\langle (E(1) - E(2)) \frac{\partial}{\partial v_-} \delta f(1)\, \delta f(2) \right\rangle + (\delta f(2)\, C(\delta f(1)) + (1 \leftrightarrow 2))$$
$$= -\frac{q}{m} \langle E(1)\, \delta f(2) \rangle \frac{\partial \langle f(1) \rangle}{\partial v_1} + (1 \leftrightarrow 2)\,. \tag{8.9a}$$

This may be condensed to the schematic form,

$$\left(\partial_t + T_{1,2}\left(x_-, v_-\right) + \nu\right)\left\langle \delta f^2 \right\rangle = P, \tag{8.9b}$$

so we see that correlation evolves via a balance between *production* by mean distribution function relaxation (i.e., $P\,(1,2)$, where $\lim_{x_-,v_-\to 0} P\,(1,2)$ is finite) and relative dispersion (i.e., $T_{1,2}\,(x_-, v_-)$, where $\lim_{x_-,v_-\to 0} T_{1,2}\,(x_-, v_-) \to 0$), cut off at small scale by collisions only. Observe that the stationary $\left\langle \delta f^2 \right\rangle$ tends to diverge as $1 \to 2$, so finiteness requires a collisional cut-off. *The small-scale divergence of* $\langle \delta f\,(1)\,\delta f\,(2)\rangle$, *as contrasted to the finiteness of* $\langle \delta f\,(1)\,\delta f\,(2)\rangle^{\mathrm{c}}$, *establishes that there must indeed be a constituent or element of the total phase space density fluctuation in addition to the familiar coherent response,* f^{c} (i.e. the piece linearly proportional to E). That piece is the *granulation* or *incoherent* fluctuation $\tilde{f}$, which is produced by nonlinear phase space mixing and which is associated with the element of conserved circulation $\Delta x\,\Delta v$ (Lynden-Bell, 1967; Kadomtsev and Pogutse, 1970; Dupree, 1970, 1972; Boutros-Ghali and Dupree, 1981; Diamond *et al.*, 1982; Balescu and Misguish, 1984; Suzuki, 1984; Terry and Diamond, 1984; McComb, 1990; Berk *et al.*, 1999). A large portion of this chapter is concerned with determining the macroscopic consequences of granulations.

8.1.2.2 Evolution correlation in QG turbulence

The analogy between 2D, quasi-geostrophic and Vlasov turbulence may be extended further, in order to illustrate the fundamental cascade and balance relations in Vlasov turbulence. Both the QG and Vlasov equations describe the conservative advection of a field by an incompressible flow. Thus, the key balance in QG turbulence is focused on that of the correlation of fluctuating potential enstrophy $\left\langle \delta Q^2 \right\rangle$, as a direct consequence of PV evolution,

$$\frac{\mathrm{d}Q}{\mathrm{d}t} - \nu\nabla^2 Q = +\tilde{f}. \tag{8.10a}$$

This may be condensed to the schematic form,

$$\left(\partial_t + \mathbf{V}\cdot\nabla\right)\delta Q = -\tilde{V}_x\frac{\partial\,\langle Q\rangle}{\partial x} + \tilde{f} + \nu\nabla^2 Q, \tag{8.10b}$$

so,

$$\begin{aligned}
&\left(\partial_t + \mathbf{V}\,(1)\cdot\nabla_1 + \mathbf{V}\,(2)\cdot\nabla_2\right)\langle \delta Q\,(1)\,\delta Q\,(2)\rangle \\
&= -\left\langle \tilde{V}_x\,(1)\,\delta Q\,(2)\right\rangle\frac{\partial\,\langle Q\,(1)\rangle}{\partial x} + \left\langle \tilde{f}\,(1)\,\delta Q\,(2)\right\rangle \\
&\quad - \nu\,\langle \nabla_1\delta Q\,(1)\cdot\nabla_2\delta Q\,(2)\rangle + (1 \leftrightarrow 2)\,.
\end{aligned} \tag{8.10c}$$

Here the evolution of the correlation function $\langle \delta Q\,(1)\,\delta Q\,(2)\rangle$ – which necessarily determines the potential enstrophy spectrum, etc. – results from:

(i) production by forcing $\tilde{f}$ and interaction with the mean PV gradient $\partial \langle Q \rangle / \partial x$;
(ii) nonlinear transfer due to relative advection;
(iii) viscous damping at small scales.

8.1.2.3 'Phasestrophy' in Vlasov turbulence

The nonlinear transfer mechanism in Eq.(8.10) is just that of the forward enstrophy cascade in 2D turbulence, namely, the scale-invariant self-similarity of $\langle \delta f^2 \rangle$. Then, it is apparent that for Vlasov turbulence, the quadratic quantity of interest, must be the "phasestrophy", i.e. the mean-square phase space density $\langle \delta f^2 \rangle$, the spectrum of which is set by the two-point correlation function $\langle \delta f\,(1)\,\delta f\,(2)\rangle$. When $\delta f \ll \langle f \rangle$, the integrated phasestrophy equals the fluctuation entropy. Note that while all powers of f are conserved in the absence of collisions, we especially focus on the quadratic quantity phasestrophy, since,

(i) it alone is conserved on a finite mesh or interval in k,
(ii) it is directly related to field energy, etc., via the linear Poisson's equation.

Also note that the Vlasov equation states that total f, i.e. $\langle f \rangle + \delta f$, is conserved, so that scattering in v necessarily entails trade-offs between the amplitude of the mean $\langle f \rangle$ and the fluctuations δf. As in the case of enstrophy, phasestrophy evolution is determined by the balance between production by mean relaxation $\sim -\,(q/m)\,\langle E \delta f\rangle\,\partial \langle f \rangle / \partial v$ and collisional dissipation $\sim \nu \langle \delta f^2 \rangle$, mediated by stretching and dissipation of phase space fluid elements. As we shall see, the phasestrophy cascade closely resembles the forward cascade of enstrophy in 2D fluid turbulence, since both correspond to the increase of mean-square gradients, in space and phase space, due to stretching of iso-vorticity and iso-phase space density contours, respectively. Recall that in the enstrophy cascade, a scale-independent transfer of fluctuation enstrophy to small scale occurs, with,

$$\left\langle \tilde{\omega}^2\,(k) \right\rangle / \tau\,(k) \sim \eta.$$

Since $\left\langle \tilde{\omega}^2\,(k) \right\rangle \sim k^3 \left\langle \tilde{V}^2\,(k) \right\rangle$ and $1/\tau\,(k) \sim k \left[k \left\langle \tilde{V}^2\,(k) \right\rangle \right]^{1/2}$, we have

$$\left\langle \tilde{\omega}^2\,(k) \right\rangle \sim \eta^{2/3}/k,$$

and $l_{\mathrm{d}} \sim \left(\nu^2/\eta\right)^{1/4}$. For a driven, self-similar phasestrophy cascade, we thus expect,

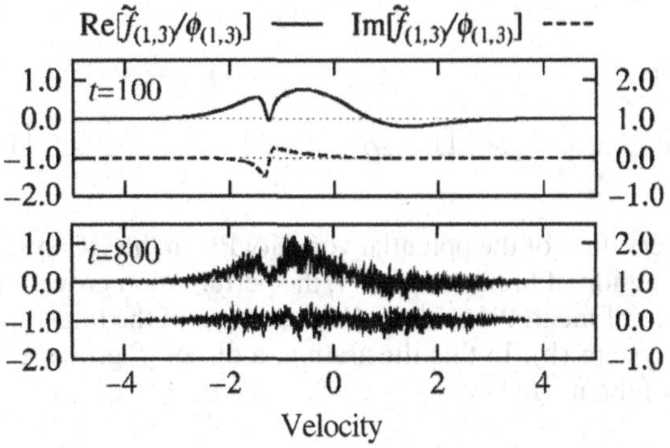

Fig. 8.6. Evolution of the distribution function from the linear phase (top) to the turbulent phase (lower) in direct nonlinear simulation. In the turbulent state, small-scale and sharp corrugations are driven until smeared by (small-but-finite) collisions. [Quoted from (Watanabe *et al.*, 2002), which explains details.]

$$\frac{\left\langle \delta f^2(l) \right\rangle}{\tau(l)} \sim \alpha, \tag{8.11a}$$

where α is the phasestrophy flux in scale l and $\tau(l)$ is the lifetime of a phase space eddy. From dimensional considerations and Poisson's equation, we can *estimate* the lifetime $\tau(l)$ to be,

$$\frac{1}{\tau(l)} \sim \frac{q}{m} E \frac{\partial}{\partial v} \sim l\omega_p^2 \frac{\delta f \Delta v}{\Delta v} \sim l\omega_p^2 \delta f. \tag{8.11b}$$

Thus, the cascade balance for phasestrophy is,

$$\left(l\omega_p^2\right) \delta f^3 \sim \alpha, \tag{8.11c}$$

so $\delta f(l) \sim \left(\alpha / l\omega_p^2\right)^{1/3}$. Straightforward manipulation then gives $\left\langle \delta f^2 \right\rangle_k \sim \alpha^{2/3} k^{-1/3}$, so we see that considerable fine-scale structure is generated by stochastic acceleration in phase space. The phasestrophy cascade is terminated when the phase element decay rate $\sim l\omega_p^2 \delta f$ becomes comparable to the collisional diffusion rate $D_{v,\mathrm{col}}/\Delta v^2$. This gives $l_\mathrm{d} \sim \left(D_{v,\mathrm{col}}/\omega_p^2 \Delta v^2\right)$ as the spatial dissipation scale. Equivalently, the velocity coarse-graining scale Δv_c corresponding to a spatial scale l is $\Delta v_\mathrm{c} \sim \left(D_{v,\mathrm{col}}/l\omega_p^2\right)^{1/2}$, so that scales with $\Delta v < \Delta v_\mathrm{c}$ are smoothed out by collisions. The function Δv_c thus sets a lower bound on the thickness of phase space elements produced by stretching.

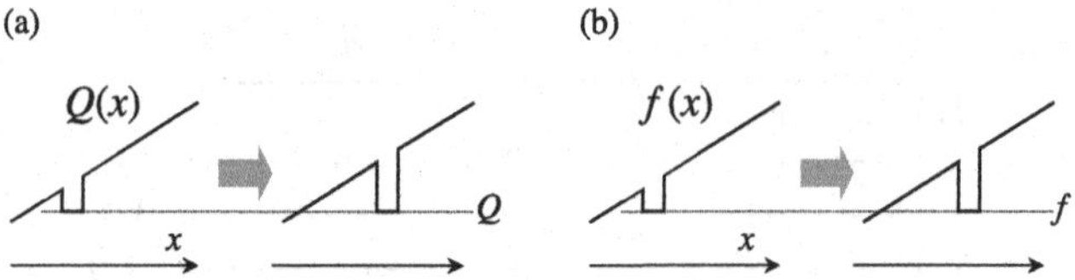

Fig. 8.7. Conservation of the potential vorticity PV in the QG system (a). When a dip moves to a region of higher mean PV, the perturbation grows. Associated with this, the gradient of mean PV relaxes. Conservation of the total number density f in the Vlasov system (b). In this illustration, a dip of f grows at the expense of the relaxation of the mean.

8.1.2.4 Generation of eddy in QG system

An essential and instructive element of the theory of phase space structures is the link via dynamics which it establishes between production and the dynamics of a localized structure, as opposed to a wave or eigenmode. Here, "production" refers to growth of fluctuation energy via its extraction from mean gradients. A prototypical example of this type of reasoning, originally due to G.I. Taylor, appears in the theory of mean flow–fluctuation interaction in 2D QG fluids, and deals with shear perturbation growth (Taylor, 1915). Periodicity of zonal flows and the relation between the mean Magnus force and PV flux tell us at the outset that the zonal flow $\langle V_y \rangle$ satisfies,

$$\frac{\mathrm{d}\langle V_y \rangle}{\mathrm{d}t} = \left\langle \tilde{V}_x \tilde{\omega} \right\rangle, \tag{8.12a}$$

while conservation of PV implies that the sum of planetary and relative vorticity is conserved, since $Q = \langle Q \rangle + \delta Q = Q_0 + \beta x + \tilde{\omega}$. Then displacing a vortex patch from x_0 to x means that its relative vorticity must change, since necessarily Q is constant at $Q = \langle Q(x_0) \rangle$, yet $\langle Q(x) \rangle \simeq \langle Q(x_0) \rangle + (x - x_0)\, \mathrm{d}\langle Q \rangle / \mathrm{d}x$. Thus the fluctuation δQ must compensate for the change in mean PV, i.e.

$$\delta Q = -(x - x_0) \frac{\mathrm{d}\langle Q \rangle}{\mathrm{d}x}, \tag{8.12b}$$

to conserve total PV, as shown in Figure 8.7. Since $\delta Q = \tilde{\omega}$, we have,

$$\begin{aligned} \frac{\mathrm{d}\langle V_y \rangle}{\mathrm{d}t} &= -\frac{\mathrm{d}\langle Q \rangle}{\mathrm{d}x} \frac{\mathrm{d}}{\mathrm{d}t} \frac{\langle \delta x^2 \rangle}{2} \\ &= -\frac{\mathrm{d}\langle Q \rangle}{\mathrm{d}x} \frac{\mathrm{d}}{\mathrm{d}t} \frac{\langle \xi^2 \rangle}{2}, \end{aligned} \tag{8.12c}$$

where ξ is the x-displacement of a fluid element. Note that $\mathrm{d}\left\langle\xi^2/2\right\rangle/\mathrm{d}t$ is just the Lagrangian fluid element diffusivity D_Q, so,

$$\frac{\mathrm{d}\left\langle V_y\right\rangle}{\mathrm{d}t} = -D_Q\frac{\mathrm{d}\left\langle Q\right\rangle}{\mathrm{d}x}. \tag{8.12d}$$

Equation (8.12d) tells us that:

(i) diffusive relaxation of mean PV gradients will drive mean zonal flows;
(ii) for the geophysically relevant case where the latitudinal derivative of $\langle Q\rangle$ ($\cong\beta$) is positive, any scattering process which increases the latitudinal variance of fluid particles must necessarily produce a net westward zonal flow.
(iii) since, in general, shear flow instability requires that $\left\langle\xi^2\right\rangle$ increases everywhere, while *total* flow x-momentum must be conserved (i.e. shear instabilities displace stream lines in both directions) so that $\int \mathrm{d}x\left\langle V_y\right\rangle$ is constant, $\mathrm{d}\langle Q\rangle/\mathrm{d}x$ *must change sign for some* x, in order for perturbations to grow.

The result that $\mathrm{d}\langle Q\rangle/\mathrm{d}x \to 0$ at some x is a necessary condition for instability is equivalent to Rayleigh's famous inflection point theorem. It is derived here in a short, physically transparent way, without the cumbersome methods of eigenmode theory. This, in turn, suggests that the inflection point condition is more fundamental than linear theory, as was suggested by C. C. Lin, on the basis of considering the restoring force for vortex interchange, and subsequently proven via *nonlinear* stability arguments by V. I. Arnold. We note in passing here that the result of Eq. (8.12c) is related to the well-known Charney–Drazin theorem, which constrains zonal flow momentum. The Charney–Drazin theorem will be discussed extensively in Volume 2 of this series.

8.1.2.5 Growth of granulations

A similar approach can be used to determine the condition for growth of *localized* phase space density perturbations. We can expect such an approach to bear fruit here, since the Vlasov system has the property that total f is conserved, so that δf must adjust when a localized fluctuation is scattered up or down the profile of $\langle f\rangle$, as indicated in Figure 8.7

To see this clearly, we note that since f and $f^2 = (\langle f\rangle + \delta f)^2$ are conserved along phase space trajectories, we can write directly,

$$\int \mathrm{d}\Gamma\frac{\mathrm{d}}{\mathrm{d}t}\left(\langle f\rangle + \delta f\right)^2 = 0, \tag{8.13a}$$

so,

$$\frac{\partial}{\partial t}\int \mathrm{d}\Gamma\left\langle\delta f^2\right\rangle = -\frac{\partial}{\partial t}\int \mathrm{d}\Gamma\left(2\delta f\langle f\rangle + \langle f\rangle^2\right). \tag{8.13b}$$

Here $\int d\Gamma$ is an integral over the fluctuation's phase volume $dxdv$. As we are interested in localized fluctuations δf – i.e. 'blobs' or holes in phase space – we can expand $\langle f \rangle$ near the fluctuation centroid v_0 and thereafter treat it as static, giving,

$$\begin{aligned}\frac{\partial}{\partial t}\int d\Gamma \left\langle \delta f^2 \right\rangle &\cong -2\frac{\partial}{\partial t}\int d\Gamma \ (v - v_0)\,\delta f \left.\frac{\partial \langle f \rangle}{\partial v}\right|_{v_0} \\ &= -2\left.\frac{\partial \langle f \rangle}{\partial v}\right|_{v_0} \frac{1}{m}\frac{dp_f}{dt}.\end{aligned} \tag{8.13c}$$

Here $p_f = m\int d\Gamma \ (v - v_0)\,\delta f$ is the net momentum associated with the phase space density fluctuation δf. For a single species plasma, momentum conservation requires that $dp/dt = 0$, so fluctuations cannot grow by accelerating up the mean phase space gradient since they have no place to deposit their momentum, and $\partial_t \left\langle \delta f^2 \right\rangle = 0$ is forced. However, for a two-species, electron–ion plasma, the relevant momentum conservation constraint becomes,

$$\frac{d}{dt}\left(p_{f_e} + p_{f_i}\right) = 0, \tag{8.13d}$$

enabling momentum exchange between species. Using Eq. (8.13c) to re-express the relation between momentum evolution and fluctuation growth then gives,

$$\frac{m_e}{\partial \langle f_e \rangle / \partial v|_{v_0}}\partial_t \int d\Gamma \left\langle \delta f_e^2 \right\rangle = -\frac{m_i}{\partial \langle f_i \rangle / \partial v|_{v_0}}\partial_t \int d\Gamma \left\langle \delta f_i^2 \right\rangle. \tag{8.13e}$$

Thus, we learn growth of a localized structure at v_0 is possible if $(\partial \langle f_e \rangle / \partial v) \times (\partial \langle f_i \rangle / \partial v)|_{v_0} < 0$ – i.e. the slopes of the electron and ion distribution functions are opposite at v_0. This condition is usually encountered in situations where the electrons carry a net current, so that $\langle f_e \rangle$ is shifted relative to $\langle f_i \rangle$, and is shown in Figure 3.2(b). While this is superficially reminiscent of the familiar textbook example of current-driven ion-acoustic instability (CDIA), it is important to keep in mind that:

(a) *linear instability – mediated by waves* – requires minimal overlap of $\langle f_e \rangle$ and $\langle f_i \rangle$, so that electron growth (due to inverse dissipation) exceeds ion Landau damping;
(b) *granulation growth, which is nonlinear, is larger* for significant overlap of the electron and ion distribution functions, and even requires significant $\partial \langle f_i \rangle / \partial v$, in order to optimize the collisionless exchange of momentum between species, as indicated by Eq. (8.13e). In this limit, the linear CDIA is strongly stabilized.

This contrast is readily apparent from consideration of the evolution of $\left\langle \delta f^2 \right\rangle$ for localized fluctuations. For an ion granulation at v_0, phase space density conservation for homogeneous turbulence implies,

$$\frac{\partial}{\partial t}\left\langle \delta f_i^2 \right\rangle = -\frac{q}{m}\left\langle E\delta f_i \right\rangle \frac{\partial \left\langle f_i \right\rangle}{\partial v}, \tag{8.14a}$$

so, since δf is localized in velocity,

$$\frac{1}{\partial \left\langle f_i \right\rangle / \partial v|_{v_0}} \partial_t \int \mathrm{d}v \left\langle \delta f_i^2 \right\rangle = -\frac{q}{m_i} \int \left\langle E\delta f_i \right\rangle = -\frac{q}{m_i} \left\langle E\tilde{n}_i \right\rangle. \tag{8.14b}$$

However, momentum balance requires that,

$$\frac{\mathrm{d}\left\langle p_{f_i} \right\rangle}{\mathrm{d}t} = +q \left\langle E\tilde{n}_i \right\rangle, \tag{8.14c}$$

and,

$$\frac{\mathrm{d}\left\langle p_{f_i} \right\rangle}{\mathrm{d}t} + \frac{\mathrm{d}\left\langle p_{f_e} \right\rangle}{\mathrm{d}t} = 0, \tag{8.14d}$$

(in the stationary state) giving,

$$\frac{m_i}{\partial \left\langle f_i \right\rangle / \partial v|_{v_0}} \partial_t \int \mathrm{d}v \left\langle \delta f_i^2 \right\rangle = +\frac{\mathrm{d}\left\langle p_{f_e} \right\rangle}{\mathrm{d}t}. \tag{8.14e}$$

Now, since electrons are not trapped, and so are weakly scattered, a simple quasi-linear estimate of $\mathrm{d}\left\langle p_{f_e} \right\rangle / \mathrm{d}t$ gives,

$$\frac{\mathrm{d}\left\langle p_{f_e} \right\rangle}{\mathrm{d}t} = -m_e \int \mathrm{d}v \, D \frac{\partial \left\langle f_e \right\rangle}{\partial v}. \tag{8.14f}$$

Here D is the velocity space quasi-linear diffusion coefficient, which is a function of velocity. Since fluctuations are localized in phase space, we can assume $D(v)$ is peaked at v_0. Thus, combining Eqs. (8.14e) and (8.14f) finally gives,

$$\partial_t \int \mathrm{d}v_i \left\langle \delta f_i^2 \right\rangle = -\frac{m_e}{m_i} \frac{\partial \left\langle f_i \right\rangle}{\partial v}\bigg|_{v_0} \left(\int \mathrm{d}v_e \, D(v) \frac{\partial \left\langle f_e \right\rangle}{\partial v} \right). \tag{8.14g}$$

Since D is maximal for $v \leq v_0$, we see again that,

$$\frac{\partial \left\langle f_i \right\rangle}{\partial v} \frac{\partial \left\langle f_e \right\rangle}{\partial v}\bigg|_{v_0} < 0$$

is necessary for growth of ion granulations. Also, it is clear that growth is nonlinear (i.e. amplitude dependent).

8.2 Statistical theory of phase space turbulence

8.2.1 Structure of the theory

We now present the *statistical* theory of Vlasov turbulence. We construct the theory with the aim of calculating the structure and evolution of the two-point phase space density correlation function $\langle \delta f(1)\, \delta f(2) \rangle$, from which we may extract all other quantities, spectra, fluxes, etc. (Here the argument (1) indicates the position and velocity of the particle 1.) After a discussion of the general structure of the theory, we proceed to in-depth studies of production, relative dispersion and the various nonlinear states which may be realized.

The basic equation is the Vlasov equation, retaining a weak residual level of collisionality. As we are interested in the fluctuation phasestrophy for the fluctuation in the distribution function, δf, we write,

$$\begin{aligned}
&\frac{\partial}{\partial t}\delta f(1) + v_1 \frac{\partial}{\partial x_1}\delta f(1) + \frac{q}{m}E(1)\frac{\partial}{\partial v_1}\delta f(1) \\
&\quad = -\frac{q}{m}E(1)\frac{\partial}{\partial v_1}\langle f(1)\rangle + C(\delta f(1)) . \qquad (8.15)
\end{aligned}$$

The equation for two-point phase space density correlation is then obtained by multiplying Eq.(8.15) by $\delta f(2)$ and symmetrizing, which gives,

$$\begin{aligned}
&\frac{\partial}{\partial t}\langle \delta f(1)\, \delta f(2)\rangle + \left(v_1 \frac{\partial}{\partial x_1} + v_2 \frac{\partial}{\partial x_2} \right) \langle \delta f(1)\, \delta f(2)\rangle \\
&\quad + \frac{q}{m}\frac{\partial}{\partial v_1}\langle E(1)\, \delta f(1)\, \delta f(2)\rangle + \frac{q}{m}\frac{\partial}{\partial v_2}\langle E(2)\, \delta f(1)\, \delta f(2)\rangle \\
&\quad = -\frac{q}{m}\langle E(1)\, \delta f(2)\rangle \frac{\partial}{\partial v_1}\langle f\rangle - \frac{q}{m}\langle E(2)\, \delta f(1)\rangle \frac{\partial}{\partial v_2}\langle f\rangle \\
&\qquad + \langle \delta f(2)\, C(\delta f(1))\rangle + \langle \delta f(1)\, C(\delta f(2))\rangle . \qquad (8.16)
\end{aligned}$$

Equation (8.16) tells us that $\langle \delta f(1) \delta f(2) \rangle$ evolves via:

(i) linear dispersion, due to relative particle streaming (the 2nd term on the left-hand side);
(ii) mode–mode coupling, via triplets associated with particle scattering by fluctuating electric fields (the 3rd and 4th terms on the left-hand side);
(iii) production, due to the relaxation of $\langle f \rangle$ (the 1st and 2nd terms on the right-hand side);
(iv) collisional dissipation (the 3rd and 4th terms on the right-hand side). Hereafter, we take C to be a Krook operator, unless otherwise noted.

8.2.1.1 Relative evolution operator $T_{1,2}$

It is instructive to group the linear dispersion, mode–mode coupling and collision terms together into a relative evolution operator $T_{1,2}$, which is defined as,

$$\begin{aligned} T_{1,2}\left\langle \delta f\left(1\right)\delta f\left(2\right)\right\rangle &= \left(v_1\frac{\partial}{\partial x_1}+v_2\frac{\partial}{\partial x_2}\right)\left\langle \delta f\left(1\right)\delta f\left(2\right)\right\rangle \\ &+\frac{q}{m}\frac{\partial}{\partial v_1}\left\langle E\left(1\right)\delta f\left(1\right)\delta f\left(2\right)\right\rangle \\ &+\frac{q}{m}\frac{\partial}{\partial v_2}\left\langle E\left(2\right)\delta f\left(1\right)\delta f\left(2\right)\right\rangle+\nu\left\langle \delta f^2\right\rangle, \end{aligned} \tag{8.17a}$$

where a simple Krook collision model is introduced as,

$$\left\langle \delta f\left(2\right)C\left(\delta f\left(1\right)\right)\right\rangle+\left\langle \delta f\left(1\right)C\left(\delta f\left(2\right)\right)\right\rangle=\nu\left\langle \delta f^2\right\rangle. \tag{8.17b}$$

As mentioned in Section 8.1, its also useful to replace $x_{1,2}$ and $v_{1,2}$ with the centroid and relative coordinates,

$$x_\pm=\frac{1}{2}\left(x_1\pm x_2\right), \tag{8.18a}$$

$$v_\pm=\frac{1}{2}\left(v_1\pm v_2\right). \tag{8.18b}$$

The correlation function $\langle\delta f\left(1\right)\delta f\left(2\right)\rangle$ is far more sensitive to the relative coordinate dependencies, because the variable v_- describes the scale (in velocity space) of particle resonance or trapping, while the variable v_+ denotes the variation of the order of thermal velocity v_{Th}, so the ordering $|\partial/\partial v_-|\gg|\partial/\partial v_+|$ holds. Spatial homogeneity ensures that $|\partial/\partial x_-|\gg|\partial/\partial x_+|$. Thus, we can discard the centroid (x_+, v_+) dependency of $T_{1,2}$ to obtain,

$$\begin{aligned} T_{1,2}\left\langle \delta f^2\left(x_-,v_-\right)\right\rangle &= v_-\frac{\partial}{\partial x_-}\left\langle \delta f\left(1\right)\delta f\left(2\right)\right\rangle \\ +\frac{q}{m}&\left\langle\left(E\left(1\right)-E\left(2\right)\right)\frac{\partial}{\partial v_-}\delta f\left(1\right)\delta f\left(2\right)\right\rangle+\nu\left\langle \delta f^2\left(x_-,v_-\right)\right\rangle, \end{aligned} \tag{8.19}$$

where the centroid dependency of correlation is also neglected, so $\langle\delta f\left(1\right)\delta f\left(2\right)\rangle \rightarrow \left\langle\delta f^2\left(x_-,v_-\right)\right\rangle$.

8.2.1.2 Limiting behaviours and necessity of granulations

Interesting limiting behaviours may be observed from Eq.(8.19). From this we see, in the limit of $1\rightarrow 2$,

$$\lim_{1\rightarrow 2}T_{1,2}\left\langle \delta f^2\left(x_-,v_-\right)\right\rangle=\nu\left\langle \delta f^2\left(x_-,v_-\right)\right\rangle, \tag{8.20}$$

so relative evolution vanishes, apart from collisions (i.e., $T_{1,2}$ becomes very small in the limit of $1 \to 2$). In contrast, the production term (the 1st and 2nd terms on the right-hand side of Eq.(8.16)),

$$P(1,2) = -\frac{q}{m}\langle E(1)\,\delta f(2)\rangle \frac{\partial}{\partial v_1}\langle f\rangle - \frac{q}{m}\langle E(2)\,\delta f(1)\rangle \frac{\partial}{\partial v_2}\langle f\rangle \tag{8.21}$$

is well behaved as $1 \to 2$, so we hereafter neglect its dependence on relative separation (x_-, v_-). Thus, $\langle \delta f(1)\,\delta f(2)\rangle$ is sharply peaked at small separation, and the singularity of $\langle \delta f(1)\,\delta f(2)\rangle$ as $x_-, v- \to 0$ is regulated *only* by collisional dissipation. This behaviour is precisely analogous to the behaviour of the two-point velocity fluctuation correlation $\left\langle \tilde{V}(1)\,\tilde{V}(2)\right\rangle$ in Navier–Stokes turbulence. We remark in passing that collisional dissipation is well known to be necessary in order to regulate small-scale entropy fluctuations in turbulent Vlasov systems.

The physical origin of $P(1,2)$ is relaxation of the mean distribution $\langle f\rangle$, as may be seen from the following argument, which neglects collisions. Consideration of phase space density ensures that,

$$\frac{\mathrm{d}}{\mathrm{d}t}f^2 = \frac{\mathrm{d}}{\mathrm{d}t}(\langle f\rangle + \delta f)^2 = 0, \tag{8.22a}$$

so its average over phase space follows,

$$\int \mathrm{d}\Gamma \frac{\mathrm{d}}{\mathrm{d}t}(\delta f)^2 = -\int \mathrm{d}\Gamma \frac{\partial}{\partial t}\langle f\rangle^2, \tag{8.22b}$$

where the surface term on the right-hand side vanishes. Since evolution of the mean $\langle f\rangle$ obeys the relation,

$$\frac{\partial}{\partial t}\langle f\rangle = -\frac{q}{m}\frac{\partial}{\partial v}\langle E\delta f\rangle, \tag{8.22c}$$

an integration by parts on the right-hand side of Eq.(8.22b) gives,

$$\int \mathrm{d}\Gamma \frac{\partial}{\partial t}(\delta f)^2 = -2\frac{q}{m}\int \mathrm{d}\Gamma\,\langle E\delta f\rangle \frac{\partial}{\partial v}\langle f\rangle, \tag{8.22d}$$

which is equivalent to Eq.(8.21). Equation (8.22c) manifestly links production of perturbation phasestrophy to relaxation of $\langle f\rangle$, consistent with intuition.

We see that, absent collisions, $\left\langle \delta f^2\right\rangle$ diverges as $1 \to 2$, while $P(1,2)$ remains finite. We now address, the origin of the divergence in $\left\langle \delta f^2\right\rangle$ in the limit of small collisionality. The fluctuation δf has, in general, two components, the coherent and the incoherent part. The coherent correlation function $\langle f^c(1) f^c(2)\rangle$ satisfies the relation (as is explained in Chapter 3),

$$\langle f^c(1) f^c(2)\rangle = 2\tau_{\text{cor}} D_{\text{QL}} \left(\frac{\partial}{\partial v}\langle f\rangle\right)^2, \tag{8.23}$$

where τ_{cor} is the wave–particle correlation time $\sim (k^2 D_{\text{QL}})^{-1/3}$ and D_{QL} is the quasi-linear diffusion coefficient. The right-hand side is finite as $1 \to 2$, even in the absence of collisions. Thus, the coherent correlation function does not contribute to the divergent behaviour of the correlation function $\langle \delta f(1)\, \delta f(2)\rangle$. Hence, we confirm that δf must contain an additional constituent beyond f^c, so,

$$\delta f = f^c + \tilde{f}. \tag{8.24}$$

This incoherent fluctuation $\tilde{f}$ is the 'granulation' or 'phase space eddy' piece. We are shown later in this chapter that it drives the dynamical friction contribution to the evolution of $\langle f\rangle$, which enters in addition to the diffusive relaxation (driven by the coherent part of δf).

8.2.1.3 Impact of granulations on the evolution of the mean

To illustrate dynamical friction self-consistently, we employ the approach of the Lenard–Balescu theory (discussed in Chapter 2) to construct an evolution equation for $\langle f\rangle$ which incorporates the effect of phase space density granulations. The novel contribution from granulations is a drag term, which enters in addition to the usual quasi-linear diffusion (presented in Chapter 3).

The physics of the granulation drag is momentum loss via radiation of waves (ultimately damped), much like the way a ship loses momentum by propagation of a wave wake. Recall that emission of waves by discrete particles is explained in Chapter 2. There, fluctuations associated with the discreteness of particles are retained, in parallel to the (smooth) fluctuations due to collective modes that satisfy the dispersion relation. By analogy with this, if there are 'granulations' in the phase space, in addition to the (smooth) fluctuations that are coherent with eigenmodes, these granulations can emit waves because of their effective discreteness. We explained in the previous subsections of this chapter that 'granulations' must exist in phase space turbulence, and they are 'produced' in conjunction with the relaxation of the mean distribution function. Thus, naturally we are motivated to study the influence of granulations. (A noticeable difference between the argument here and that in Chapter 2 is that, while the 'discreteness' of particles is prescribed for thermal fluctuations, the magnitude and distribution of granulations must be determined self-consistently here, via a turbulence theory.)

Damping of the wave wake emitted by granulations opens a channel for collisionless momentum exchange between species, either with or without linear instability. Indeed, such momentum change processes induce the novel, nonlinear instability mechanisms mentioned earlier in this chapter. As in the case of forward enstrophy cascade in 2D turbulence, conservation of total phase space density links phasestrophy production to the relaxation of the mean $\langle f \rangle$, including both mean diffusion and drag. Since stationarity requires a balance (akin to the spectral balance discussed in Chapters 4 and 5) between phasestrophy production P and phasestrophy transfer, we have from Eq.(8.17),

$$\frac{\partial}{\partial t}\left\langle \delta f^2 \right\rangle + \frac{1}{\tau\left(\Delta x, \Delta v\right)}\left\langle \delta f^2 \right\rangle = P, \tag{8.25a}$$

where nonlinear interactions in the phase space are physically written by use of $\tau\left(\Delta x, \Delta v\right)$. Of course, the phase space element lifetime $\tau\left(\Delta x, \Delta v\right)$ is directly analogous to $\tau\left(l\right)$, the eddy lifetime for scale l. This relation gives,

$$\left\langle \delta f^2 \right\rangle = \tau\left(\Delta x, \Delta v\right) P, \tag{8.25b}$$

in a stationary state. The similarity to the dynamics of production in the Prandtl mixing theory discussed in Chapter 2 should be obvious. Hence, we see that the phase space eddy lifetime $\tau\left(\Delta x, \Delta v\right)$, along with production, sets $\left\langle \delta f^2 \right\rangle$. As in the case of two-particle dispersion in the 2D enstrophy-cascade range (as in Richardson, see Chapter 2), $\tau\left(\Delta x, \Delta v\right)$ can be related to the exponentiation time for relative separation of orbits stochasticized by the turbulence. Since production depends on the fluctuations,

$$P = P\left[\left\langle \delta f^2 \right\rangle\right],$$

we can thus 'close the loop' of theoretical construction and obtain a phasestrophy balance condition. The loop between phasestrophy, radiated electric field fluctuations, production and lifetime is illustrated in Figure 8.8. In the two following subsections, 8.2.2 and 8.2.3, the production P and the phase space eddy lifetime $\tau\left(\Delta x, \Delta v\right)$ are analyzed.

This development of the theory of Vlasov turbulence is illustrated in Table 8.3. The detailed comparison and contrast of quasi-geostrophic Hasegawa–Mima (QG H–M) turbulence and Vlasov turbulence is summarized in Table 8.4.

8.2.2 Physics of production and relaxation

We now turn to the detailed physics of the phasestrophy production term $P(1, 2)$. Since incoherent fluctuations (i.e., granulations) are present, $\delta f = f^c + \tilde{f}$, we have (absorbing the factor of 2),

Table 8.2. *Theoretical development*

Basic concepts	Vlasov turbulence
Eddy	Phase space density granulation
scale: l	scale: Δx, Δv
intensity: enstrophy $\langle \lvert\nabla \times v\rvert^2 \rangle$	intensity: phasestrophy $\langle \delta f^2 \rangle$
Enstrophy cascade	Phasestrophy cascade
Lenard–Balescu operator	Lenard–Balescu operator
$\to$ drag due to discreteness	$\to$ drag due to granulations
Production versus cascade	Production versus straining

Table 8.3. *Comparison and contrast of quasi-geostrophic Hasegawa–Mima (QG–HM) turbulence and Vlasov turbulence*

Notion	QG–HM turbulence	Vlasov turbulence
Correlation	potential enstrophy $\langle \delta Q(1) \delta Q(2) \rangle$	fluctuation phasestrophy $\langle \delta f(1) \delta f(2) \rangle$
Production	$-\langle V_x \delta Q \rangle \frac{\partial}{\partial x} \langle Q \rangle$	$-\frac{q}{m} \langle E \delta f \rangle \frac{\partial}{\partial v} \langle f \rangle$
Transfer/ dispersion	$\delta \mathbf{V} \cdot \nabla_-$ $\tau^{-1} \sim \alpha^{1/3}$	$T_{1,2}$ $\tau \sim \tau_c \sim (k^2 D)^{-1/3}$
Cascade	enstrophy $\langle \delta Q^2 \rangle \sim \alpha^{2/3} k^{-1}$	phasestrophy $\langle \delta f^2 \rangle \sim \alpha^{2/3} k^{-1/3}$
Dissipation scale	l_d	l_d, $\Delta v(l_d)$

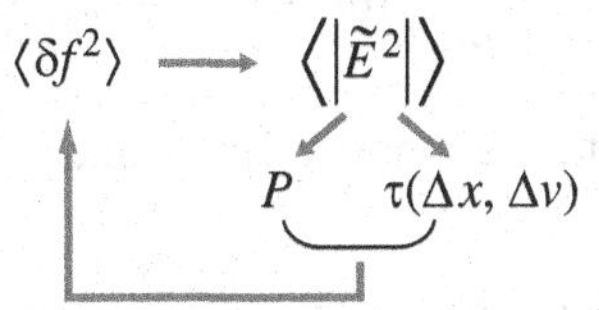

Fig. 8.8. A self-consistent loop between the phasestrophy, radiated electric field fluctuations, production and lifetime of granulations.

$$
\begin{aligned}
P(1,2) &= -\frac{q}{m} \langle E \delta f \rangle \frac{\partial}{\partial v} \langle f \rangle \\
&= -\frac{q}{m} \left\langle E f^c \right\rangle \frac{\partial}{\partial v} \langle f \rangle - \frac{q}{m} \left\langle E \tilde{f} \right\rangle \frac{\partial}{\partial v} \langle f \rangle . \qquad (8.26)
\end{aligned}
$$

Here, $\langle E f^c \rangle$ yields the familiar coherent relaxation term, related to quasi-linear diffusion discussed in Chapter 3, while $\left\langle E \tilde{f} \right\rangle$ gives the new dynamical friction term (see also Adam *et al.* (1979) and Laval and Pesme (1983)). This is closely

Table 8.4. *Comparison and contrast of test particle model and theory of phase space granulations*

Notion	Test particle model	Phase space granulation
Regime	Near equilibrium Thermal fluctuations Linearly stable modes	Non-equilibrium Turbulent fluctuations, granulations mode–mode coupling Linearly stable *or* nonlinearly saturated modes
Content	Emission balances absorption Incoherent $\tilde{f} \leftrightarrow$ discreteness Dressed-test particle	Phasestrophy cascade driven production Incoherent $\tilde{f} \leftrightarrow$ granulations Clump – phase space eddy
Structure	$\left\langle \tilde{f} \right\rangle = \frac{1}{n} \langle f \rangle \, \delta(x_-)\delta(v_-)$ $J(v) \to$ diffusion + drag $D \leftrightarrow$ stochastic acceleration drag from discreteness	$\left\langle \tilde{f}^2 \right\rangle$ from closure theory $J(v) \to$ diffusion + drag $D \leftrightarrow$ stochastic acceleration drag from granulations

analogous to the outcome of the Lenard–Balescu theory of Chapter 2, where $\tilde{f}$ is due to discreteness and f^c is a linear coherent response.

8.2.2.1 *Property of coherent terms*

For a (k, ω)–Fourier component, we write the coherent part f^c as,

$$f^c_{k,\omega} = -\frac{q}{m} R\,(\omega - kv)\, E_{k,\omega} \frac{\partial}{\partial v} \langle f \rangle \,, \tag{8.27a}$$

where $R(\omega - kv)$ is the particle response function,

$$R\,(\omega - kv) = \frac{i}{\omega - kv + i\tau^{-1}_{c,k,\omega}}, \tag{8.27b}$$

and $\tau_{c,k,\omega}$ is the wave–particle coherence time for the (k, ω)–Fourier component. The coherent production term P_c is then just,

$$P_c = D_{\mathrm{QL}} \left(\frac{\partial}{\partial v} \langle f \rangle \right)^2 . \tag{8.27c}$$

Note that P_c has the classic form of production as given by mixing length theory (see the discussion of pipe flow in Chapter 2), in that it says in essence that the rearrangement of $\langle f \rangle$ produces secular growth of $\left\langle \delta f^2 \right\rangle$. This process is highlighted as follows. The perturbation due to rearrangement of $\langle f \rangle$ takes the form,

$$\delta f = \langle f(v) \rangle - \langle f(v - \Delta v) \rangle \sim \Delta v \frac{\partial}{\partial v} \langle f \rangle , \tag{8.28a}$$

that is, the mean of the statistically averaged square δf^2 is,

$$\left\langle \delta f^2 \right\rangle \sim (\Delta v)^2 \left(\frac{\partial}{\partial v} \langle f \rangle \right)^2 . \tag{8.28b}$$

In a diffusion process, the mean deviation evolves as,

$$(\Delta v)^2 \sim D_{\mathrm{QL}} t, \tag{8.28c}$$

so the time derivative of Eq.(8.28b) is given by,

$$\frac{\partial}{\partial t} \left\langle \delta f^2 \right\rangle \sim D_{\mathrm{QL}} \left(\frac{\partial}{\partial v} \langle f \rangle \right)^2 . \tag{8.28d}$$

Note that Eq.(8.28d) states that fluctuation phasestrophy must grow secularly on a transport time-scale, given the presence of a turbulence spectrum $\left\langle E^2 \right\rangle_{k,\omega}$, and phase space gradients $\partial \langle f \rangle / \partial v$. It is useful to remark here that the growth on transport time-scales discussed here is also the origin of the 'growing weight' problem in long time runs of δf PIC (particle in cell) simulation codes (Nevins *et al.*, 2005). Here the term 'weight' refers to a parameter associated with a particle that tracks its effective δf. On long, transport timescales, this unavoidable growth of δf without concomitant evolution of $\langle f \rangle$ and without dissipation of δf fluctuations via collisions, will lead to unphysical weight growth and thus to unacceptably high noise levels in the simulation.

8.2.2.2 A note on productions

The dynamical friction term has the appearance of a Fokker–Planck drag, since a Fokker–Planck equation for $\langle f \rangle$ has the generic structure,

$$\frac{\partial}{\partial t} \langle f \rangle = -\frac{\partial}{\partial v} J(v) , \tag{8.29a}$$

where the flux in the phase space takes the form,

$$J(v) = -D \frac{\partial}{\partial v} \langle f \rangle + F \langle f \rangle . \tag{8.29b}$$

Field emitted by granulations

To actually calculate the dynamical friction term, we must relate $\tilde{f}$ (and f^c) to E via Poisson's equation. The explicit relations are discussed below. In this explanation, we address the phase space dynamics of *ions*, and introduce the electron response in terms of a "response function" $\hat{\chi}_e$ as,

$$\nabla^2\phi = -4\pi n_0 q \int \mathrm{d}v \delta f - 4\pi\hat{\chi}_e \frac{n_0 q^2 \phi}{T_e}, \tag{8.30a}$$

where ϕ is the (fluctuating) electrostatic potential, n_0 is the mean number density, q is a unit charge, T_e is electron temperature, and $\hat{\chi}_e$ is a linear susceptibility (response function) of electrons. For simplicity, here we consider ion phase space turbulence. Taking $\delta f = f^c + \tilde{f}$, we rewrite Eq.(8.30a) as,

$$\epsilon(k,\omega)\,\phi_{k,\omega} = -\frac{4\pi n_0 q}{k^2}\int \mathrm{d}v\,\tilde{f}_{k,\omega}, \tag{8.30b}$$

where the contributions of f^c and $\hat{\chi}_e$ are included in the dielectric function $\epsilon(k,\omega)$, which is,

$$\epsilon(k,\omega) = 1 - \frac{\omega_{p,i}^2}{k}\int \mathrm{d}v \frac{1}{\omega - kv}\frac{\partial}{\partial v}\langle f\rangle - \frac{\hat{\chi}_e}{k^2\lambda_{\mathrm{De}}^2}. \tag{8.30c}$$

Here, λ_{De} is the Debye length and $\omega_{p,i}$ is the ion plasma oscillation frequency. Note that, just as in the test particle model in Chapter 2, the incoherent part plays the role of a source in Eq.(8.30b), and the coherent fluctuation f^c forms part of the screening response to the incoherent fluctuation $\tilde{f}$, (due to granulations). For stable or over-saturated modes (i.e., waves which are nonlinearly over-damped, beyond marginal saturation), we can then write the potential in terms of the screened incoherent fluctuation,

$$\phi_{k,\omega} = -\frac{4\pi n_0 q}{\epsilon(k,\omega)\,k^2}\int \mathrm{d}v\,\tilde{f}_{k,\omega}. \tag{8.31a}$$

Note that Eq.(8.31a) is rigorously valid *only* in the long-time asymptotic limit, where "long" is set by the time required for the damped or over-saturated collective mode response to decay. This is a consequence of the structure of the full solution to Eq.(8.30b), which is,

$$\phi_{k,\omega} = -\frac{4\pi n_0 q}{\epsilon(k,\omega)\,k^2}\int \mathrm{d}v\,\tilde{f}_{k,\omega} + \sum_j \phi_{j,0}\exp i\left(k_j x - \omega_j t\right)$$
$$\rightarrow -\frac{4\pi n_0 q}{\epsilon(k,\omega)\,k^2}\int \mathrm{d}v\,\tilde{f}_{k,\omega}, \tag{8.31b}$$

for Im $\omega_j < 0$. Obviously, then, the notion of a "screened granulation" requires reconsideration as we approach marginally from below or in transient states with growing modes.

Further progress follows by relating both $\langle E f^c \rangle$ and $\left\langle E \tilde{f} \right\rangle$ to the incoherent, or granulation, correlation function $\left\langle \tilde{f} \tilde{f} \right\rangle$, which is the effective source in the theory, again analogous to the discreteness correlation in the test particle model in Chapter 2. Taking the spectrum to be sufficiently broad so the auto-correlation time is short and renormalization unnecessary, we have,

$$\left|\phi_{k,\omega}\right|^2 = \left(\frac{4\pi n_0 q}{k^2}\right)^2 \iint \mathrm{d}v_1 \mathrm{d}v_2 \frac{\left\langle \tilde{f}(v_1)\, \tilde{f}(v_2) \right\rangle_{k,\omega}}{\left|\epsilon(k,\omega)\right|^2}, \tag{8.32}$$

so that the production term (8.26),

$$P = -\frac{q}{m}\left\langle E f^c \right\rangle \frac{\partial}{\partial v}\langle f \rangle - \frac{q}{m}\left\langle E \tilde{f} \right\rangle \frac{\partial}{\partial v}\langle f \rangle \equiv P_c + P_G, \tag{8.33a}$$

is given by the use of Eq.(8.32). Substituting Eq.(8.31b) into Eq.(8.27a), the coherent piece P_c is calculated as,

$$\frac{q}{m}\left\langle E f^c \right\rangle = -\frac{\partial}{\partial v}\langle f \rangle \sum_{k,\omega} \frac{q^2 k^2 \pi}{m^2} \delta(\omega - kv) \left(\frac{4\pi n_0 q}{k^2}\right)^2 \iint \mathrm{d}v_1 \mathrm{d}v_2 \frac{\left\langle \tilde{f} \tilde{f} \right\rangle_{k,\omega}}{\left|\epsilon(k,\omega)\right|^2}. \tag{8.33b}$$

The incoherent or granulation-induced correlation contribution to P_G (dynamical friction term) is given by,

$$\frac{q}{m}\left\langle E \tilde{f} \right\rangle = \frac{\partial}{\partial v}\langle f \rangle \sum_{k,\omega} \frac{qk}{m} \frac{4\pi n_0 q}{k^2} \iint \mathrm{d}v_1 \mathrm{d}v_2 \frac{\mathrm{Im}\,\epsilon(k,\omega)}{\left|\epsilon(k,\omega)\right|^2} \left\langle \tilde{f} \tilde{f} \right\rangle_{k,\omega}. \tag{8.33c}$$

Note that, in contrast to the usual practice in quasi-linear theory, here in Eq.(8.33), both k and ω are summed over, since the latter is not tied or restricted to wave resonances (i.e., $\omega \neq \omega_k$), and frequency broadening occurs. Indeed, we shall see that ballistic mode Doppler emission is a significant constituent in the spectrum, along with collective mode lines.

8.2.2.3 Introduction of modeling for the structure of granulations

To progress from here, we must simplify the correlation function of granulations $\left\langle \tilde{f}(v_1)\, \tilde{f}(v_2) \right\rangle$. To do so, keep in mind that:

(i) The correlation $\left\langle \tilde{f}(v_1)\tilde{f}(v_2)\right\rangle$ is sharply peaked at small relative velocity, i.e.,

$$\left\langle \tilde{f}\tilde{f}\right\rangle \sim F\left(\frac{v_-}{\Delta v}\right),$$

where Δv belongs to the class of fine-scale widths (of the order of wave–particle resonance or trapping width, etc.). Thus the dependence on the centroid x_+, v_+, is neglected.

(ii) For weak turbulence, we can derive $\left\langle \tilde{f}\tilde{f}\right\rangle_{k,\omega}$ from $\left\langle \tilde{f}\tilde{f}\right\rangle_k$ via the linear particle propagator, as,

$$\begin{aligned}\left\langle \tilde{f}(v_1)\tilde{f}(v_2)\right\rangle_{k,\omega} &= \mathrm{Re}\left\{\int_0^{\infty} \mathrm{d}\tau e^{i(\omega-kv)\tau} + \int_{-\infty}^{0} \mathrm{d}\tau e^{-i(\omega-kv)\tau}\right\} \\ &\quad \times \left\langle \tilde{f}(v_1)\tilde{f}(v_2)\right\rangle_k \\ &\cong 2\pi\delta(\omega-kv)\left\langle \tilde{f}(v_1)\tilde{f}(v_2)\right\rangle_k,\end{aligned} \tag{8.34}$$

so that,

$$\int \mathrm{d}v \left\langle \tilde{f}(v_1)\tilde{f}(v_2)\right\rangle_{k,\omega} = \frac{2\pi}{|k|}\left\langle \tilde{f}(u)\tilde{f}(v)\right\rangle_k, \tag{8.35}$$

where $u = \omega/k$ is the fluctuation phase velocity.

Equation (8.34) should be thought of as the resonant particle limit of the more general result,

$$\left\langle \tilde{f}(v_1)\tilde{f}(v_2)\right\rangle_{k,\omega} \cong \frac{2\tau_{c,k}^{-1}}{(\omega-kv)^2+\tau_{c,k}^{-2}}\left\langle \tilde{f}(v_1)\tilde{f}(v_2)\right\rangle_k,$$

where $\tau_{c,k}$ is a correlation time. *A distinguishing property of resonant particle, phase space turbulence is the linear proportionality of* $\left\langle \tilde{f}\tilde{f}\right\rangle_{k,\omega}$ *to* $\left\langle \tilde{f}\tilde{f}\right\rangle_k$ *via the amplitude-independent, resonant particle propagator* $\pi\delta(\omega-kv)$. This is the case of interest to our discussion of phase space density granulations. In the non-resonant limit,

$$\left\langle \tilde{f}\tilde{f}\right\rangle_{k,\omega} \cong \frac{2\tau_{c,k}^{-1}}{(\omega-kv)^2+\tau_{c,k}^{-2}}\left\langle \tilde{f}\tilde{f}\right\rangle_k \cong \frac{2\tau_{c,k}^{-1}}{\omega^2}\left\langle \tilde{f}\tilde{f}\right\rangle_k,$$

as is usually encountered. Finally, then,

$$\iint \mathrm{d}v_1\mathrm{d}v_2\left\langle \tilde{f}(v_1)\tilde{f}(v_2)\right\rangle_{k,\omega} = \frac{2\pi}{|k|}\left\langle \tilde{n}\tilde{f}(u)\right\rangle_k, \tag{8.36}$$

where $\tilde{n}$ is the density fluctuation associated with the granulations. Equation (8.36) is a particularly simple and attractive result, tying $\langle \tilde{n}\tilde{n}\rangle_{k,\omega}$ to $2\pi|k|^{-1}\left\langle \tilde{n}\tilde{f}(u)\right\rangle_k$, namely the phase space density correlation function on resonance (i.e., at $v = u$), with characteristic frequency $\sim |k|\, v_{\mathrm{T}}$.

We can simplify the production correlations, i.e. Eqs.(8.33b) and (8.33c), by use of Eq.(8.36), to obtain,

$$\frac{q}{m}\left\langle Ef^{c}\right\rangle = -\frac{\partial}{\partial v}\langle f\rangle \sum_{k,\omega} \frac{q^2k^2\pi}{m^2}\delta(\omega - kv)\left(\frac{4\pi n_0 q}{k^2}\right)^2 \frac{2\pi}{|k|}\frac{\left\langle \tilde{n}\tilde{f}(u)\right\rangle_k}{|\epsilon(k,\omega)|^2}, \tag{8.37a}$$

and,

$$\frac{q}{m}\left\langle E\tilde{f}\right\rangle = -\sum_{k,\omega}\frac{qk}{m}\frac{4\pi n_0 q}{k^2}\frac{\operatorname{Im}\varepsilon(k,\omega)}{|\epsilon(k,\omega)|^2}\frac{2\pi}{|k|}\left\langle \tilde{f}(v)\tilde{f}(u)\right\rangle_k. \tag{8.37b}$$

Equations (8.37a) (8.37b) may then be combined to yield the net velocity current $J(v)$,

$$J(v) = \frac{q}{m}\langle E\delta f\rangle = \sum_{k,\omega}\frac{\omega_{\mathrm{pi}}^2}{k}\frac{2\pi^2}{|k|^2}\frac{G}{|\epsilon(k,\omega)|^2}, \tag{8.38a}$$

where,

$$\begin{aligned} G(v) = {} & \frac{\omega_{\mathrm{pi}}^2}{k}\delta(v-u)\left\{\left\langle \tilde{n}\tilde{f}(u)\right\rangle_k \frac{\partial}{\partial v}\langle f\rangle - \left\langle \tilde{f}(v)\tilde{f}(u)\right\rangle_k \frac{\partial}{\partial v}\langle f\rangle\bigg|_u\right\} \\ & - \left\langle \tilde{f}(v)\tilde{f}(u)\right\rangle_k \operatorname{Im}\epsilon_e(k,\omega), \end{aligned} \tag{8.38b}$$

and ϵ_e is the electrons contribution to the dielectric E.

8.2.2.4 Like-particle and inter-particle interactions

Equation (8.38) merits some detailed discussion. First, note that Eq.(8.38b) implies that ion production P_i, Eq.(8.38a), may be decomposed into like-particle (i.e., ion–ion) and interspecies (i.e., ion–electron) contributions, i.e.,

$$P_i = P_{i,i} + P_{i,e}, \tag{8.39a}$$

where,

$$P_{i,i} = -\frac{\partial}{\partial v}\langle f_i\rangle \sum_{k,\omega}\frac{\omega_{\mathrm{pi}}^2}{k}\frac{2\pi^2}{|k|^2}\frac{G_{i,i}}{|\epsilon(k,\omega)|^2}, \tag{8.39b}$$

and,

$$P_{i,e} = \frac{\partial}{\partial v} \langle f_i \rangle \sum_{k,\omega} \frac{\omega_{\text{pi}}^2}{k} \frac{2\pi^2}{|k|^2} \frac{G_{i,e}}{|\epsilon(k,\omega)|^2}, \tag{8.39c}$$

with,

$$G_{i,i} = \frac{\omega_{\text{pi}}^2}{k} \delta(v-u) \left\{ \left\langle \tilde{n} \tilde{f}(u) \right\rangle_k \frac{\partial}{\partial v} \langle f_i \rangle - \left\langle \tilde{f}(v) \tilde{f}(u) \right\rangle_k \frac{\partial}{\partial v} \langle f_i \rangle \bigg|_u \right\} \tag{8.39d}$$

$$G_{i,e} = \left\langle \tilde{f}(v) \tilde{f}(u) \right\rangle_k \operatorname{Im}\epsilon_e(k,\omega). \tag{8.39e}$$

Second, since the granulation correlation function is sharply localized,

$$\left\langle \tilde{f}(u) \tilde{f}(v) \right\rangle_k \simeq \delta(u-v) \left\langle \tilde{n} \tilde{f}(u) \right\rangle_k, \tag{8.40}$$

so the second term on the right-hand side of Eq.(8.39d) vanishes, to good approximation, as,

$$G_{i,i} \cong \frac{\omega_{\text{pi}}^2}{k} \delta(v-u) \left\{ \left\langle \tilde{n} \tilde{f}(u) \right\rangle_k \frac{\partial}{\partial v} \langle f_i \rangle - \left\langle \tilde{n} \tilde{f}(u) \right\rangle_k \frac{\partial}{\partial v} \langle f_i \rangle \bigg|_u \right\}$$
$$\to 0. \tag{8.41}$$

From this consideration we see that the like-particle contribution to relaxation and production *vanishes*! This is precisely analogous to the vanishing of like-particle contributions to relaxation in one-dimension for the Lenard–Balescu theory, discussed in Chapter 2. The underlying physics is the same as well – in 1D, interactions that conserve energy and momentum leave the final state identical to the initial state, so no relaxation can occur. Here, rather than a physical "collision", the interaction in question is the scattering of a particle with velocity v by a fluctuation with phase velocity u.

In the event that $P_{i,i} \to 0$, we have,

$$P_{i,e} = \frac{\partial}{\partial v} \langle f_i \rangle \sum_{k,\omega} \frac{\omega_{\text{pi}}^2}{k} \frac{2\pi^2}{|k|^2} \frac{\operatorname{Im}\varepsilon_e(k,\omega)}{|\epsilon(k,\omega)|^2} \left\langle \tilde{f}(v) \tilde{f}(u) \right\rangle_k. \tag{8.42}$$

Once again, we see that,

$$\frac{\partial \langle f_i \rangle}{\partial v} \frac{\partial \langle f_e \rangle}{\partial v} < 0$$

is required for $P_{i,e} > 0$, and net production.

8.2.2.5 Momentum transfer channel

Two other features of these results merit special discussion. First, proximity to collective resonance (i.e., small $\epsilon(k, ku)$) can strongly enhance relaxation and transport, since in such cases, the granulations will radiate rather weakly damped waves, thus leaving a significant "wake". Second, $P_{i,e} \neq 0$ presents an interesting alternative to the quasi-linear momentum transfer "channel", discussed in Chapter 3, and so may have interesting implications for anomalous resistivity.

Recall from Chapter 3 that in quasi-linear theory (for *electrons* in 1D), resonant particle momentum is exchanged with wave momentum, while conserving the sum, so that,

$$\frac{\partial}{\partial t}\left\{\langle P_{\text{res}}\rangle + \sum_k kN_k\right\} = 0. \tag{8.43a}$$

Here,

$$\frac{\partial}{\partial t}\langle P_{\text{res}}\rangle = q\int \mathrm{d}v \left\langle Ef^c\right\rangle, \tag{8.43b}$$

with f^c given by the resonant, linear response, as in Eq.(8.27), and $\sum_k kN_k$ is the total momentum of waves. As a consequence, the evolution of resonant particle momentum is tied directly (and exclusively) to the wave growth, so it is difficult to simultaneously reconcile stationary turbulence with exchange of momentum by resonant particles. Indeed in 1D, Eq.(8.43b) has only the trivial solution of local plateau formulation (i.e., $\partial\langle f\rangle/\partial v \to 0$) for a stationary state, $\partial\left(\sum kN_k\right)/\partial t = 0$. In contrast, proper accounting for electron granulations opens a new channel for collisionless electron–ion momentum exchange, which does *not* rely on the presence of growing collective modes. To see this, note that for electrons,

$$\begin{aligned}\frac{\partial}{\partial t}\langle P_{\text{res}}\rangle_e &= -|q|\int \mathrm{d}v\left\langle E\delta f_{e,\text{res}}\right\rangle \\ &= -|q|\int \mathrm{d}v\left[\left\langle Ef_e^c\right\rangle + \left\langle E\tilde{f}_e\right\rangle\right]_{\text{res}}.\end{aligned} \tag{8.44a}$$

Since by analogy with Eq.(8.39a), electron phasestrophy production $P_e = P_{e,i} + P_{e,e}$, and since $P_e = (\partial\langle f_e\rangle/\partial v)\, J_e(v)$ (here $J_e(v)$ is the electron velocity space current), then in the absence of growing waves we have,

$$\begin{aligned}\frac{\partial}{\partial t}\langle P_{\text{res}}\rangle_e &= m_e\int \mathrm{d}v\left[J_{e,e}(v) + J_{e,i}(v)\right] \\ &= m_e\int \mathrm{d}v\, J_{e,i}(v),\end{aligned} \tag{8.44b}$$

because $J_{e,e}(v) \to 0$ in 1D. Thus, the evolution of resonant electron momentum is finally just,

$$\frac{\partial}{\partial t}\langle P_{\text{res}}\rangle_e = m_e \sum_{k,\omega} \frac{\omega_{\text{pi}}^2}{k} \frac{2\pi^2}{|k|^2} \frac{\text{Im}\epsilon_i(k,\omega)}{|\epsilon(k,\omega)|^2} \left\langle \tilde{n}\tilde{f}(u)\right\rangle_k. \tag{8.44c}$$

An interesting feature of Eq.(8.44c) is that it reconciles conceptually finite momentum loss by resonant electrons with stationary turbulence by replacing the dependence on wave growth in the (non-stationary) quasi-linear theory by proportionality to collisionless ion dissipation, $\text{Im}\varepsilon_i(k,\omega)$.

It should be noted that there is no "automatic" momentum transfer because an interesting value of the product $\text{Im}\epsilon_i(k,\omega)\left\langle \tilde{n}\tilde{f}(u)\right\rangle_k$ requires that:

(i) electron granulations be excited, so $\partial\langle f_e\rangle/\partial v|_u > 0$, assuming $\partial\langle f_i\rangle/\partial v|_u < 0$;
(ii) electron granulations resonate with ions – i.e., $u \sim v_{Th,i}$, for $\text{Im}\epsilon_i(k,\omega) \neq 0$.

These two conditions require significant overlap of the $\partial\langle f_e\rangle/\partial v|_u > 0$ and $\partial\langle f_i\rangle/\partial v|_u < 0$ regions. Finally, we note that the possibility of collisionless inter-species momentum exchange in the absence of unstable waves also offers a novel, alternative mechanism for anomalous resistivity, which is related, but also complementary, to the classical paradigm involving current-driven ion-acoustic instability.

We conclude this section with Table 8.4, which compares and contrasts the physics and treatment of production and transport in the test particle model (TPM) and Lenard–Balescu theory with their counterparts in the theory of phase space density granulation.

We emphasize that the 1D cancellation is rather special. Thus, should additional degrees of freedom be present in the resonance dynamics, like-particle interchanges, which conserve total particle Doppler frequency, will become possible. In this case, the cancellation no longer need occur. This is seen in the following example of drift wave turbulence. For particles in drift wave turbulence in the presence of a sheared mean flow in the $\hat{y}$-direction, $\mathbf{V}_E = V_E' x\hat{y}$ (where $Vn_0(x)$ is the mean density gradient, and the magnetic field is in the $\hat{z}$-direction), the effective Doppler frequency becomes,

$$\omega_{\text{Doppler}} = k_\parallel v_\parallel + k_y V_E' x,$$

so for that case, a class of scatterings or interchanges of $v_\parallel$ and x exists which leaves total ω_{Doppler} invariant. That is, the net transport in velocity and radius can occur via like-particle interactions which scatter both $v_\parallel$ and x but leave ω_{Doppler}

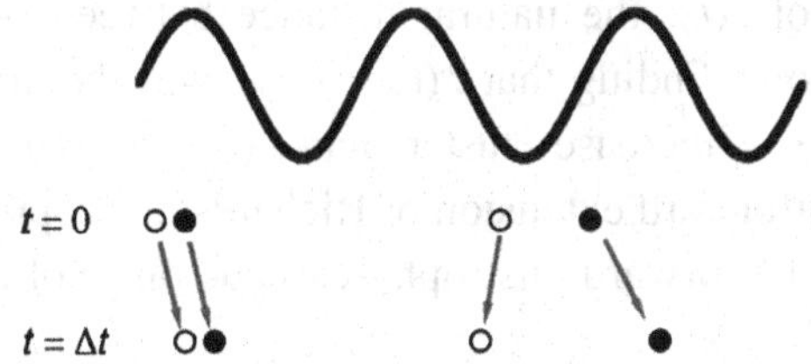

Fig. 8.9. Relative motion of two particles, being scattered by fluctuations. The growth of separation depends on the initial separation.

unchanged. Nevertheless, since a strong magnetic field always severely constrains possible wave–particle resonance, the 1D problem is an important and instructive limit which must always be kept in mind.

8.2.3 Physics of relative dispersion in Vlasov turbulence

Equation (8.16) neatly states the fundamental balance that governs Vlasov turbulence, namely that the structure of two-point correlation in phase space is set by a competition between production (discussed in detail in the previous section) and *relative dispersion*. By 'relative dispersion' we mean the tendency for two trajectories or particles to separate from one another, on account of relative streaming or relative scattering by fluctuating electric fields (see Figure 8.9.) In effect, the rate of relative dispersion assigns a characteristic lifetime $\tau_c\,(\Delta x, \Delta v)$ to a phase space element of scale, Δx, Δv, so that the stationary two-point correlation is simply $\langle \delta f\,(1)\,\delta f\,(2)\rangle \sim \tau_c\,(\Delta x, \Delta v)\,P\,(1, 2)$. As discussed in the previous section, this picture is essentially a generalization of the Prandtl mixing length theory for turbulent pipe flow (see Chapter 2) to the case of phase space. We should keep in mind that:

(i) Production is not simply diffusive mixing by an effective eddy viscosity, but rather due to a relaxation process involving both diffusion and dynamical friction.
(ii) The calculation of dispersion must account for the structure of the governing Vlasov equation, the statistical property of phase space orbits in Vlasov turbulence, and the effect of collisions at small scale.

In this section, we turn to the calculation of dispersion and the effective phase space element lifetime, $\tau_{\mathrm{sep}}\,(\Delta x, \Delta v)$. Along the way, we will further elucidate the relationship between the phasestrophy cascade and the evolution of $\langle \delta f(1)\delta f(2)\rangle$.

8.2.3.1 Richardson's theory revisited

The concept of relative dispersion has a long history in turbulence theory, starting with the seminal ideas of L. F. Richardson (discussed in Chapter 2) who considered the growth in time of $l(t)$, the natural distance between two particles in K41 turbulence. Richardson's finding that $l\,(t)^2 \sim \epsilon t^3$ was the first instance of super-diffusive kinetics (i.e., l increases faster than $l\,(t)^2 \sim D_0 t$, as for diffusion) in turbulence. A straightforward extension of Richardson's approach to dispersion in scales falling within the forward-enstrophy-cascade range of 2D turbulence gives,

$$\frac{\partial}{\partial t} l\,(t) \sim \eta^{1/3} l\,(t)\,,$$

indicating *exponential* growth set by the enstrophy dissipation rate $\eta^{1/3}$. We shall again encounter exponentially increasing relative separation in our study of Vlasov turbulence.

8.2.3.2 *Case of Vlasov turbulence*

In considering dispersion in Vlasov turbulence, two comments are necessary at the outset. First, here we aim to develop a *statistical* weak turbulence theory for the correlation $\langle \delta f^2 \rangle$. Possible local trapping could manifest itself by a net skewness $\langle \delta f^3 \rangle$, indicative of a preferred sign in the phase space fluctuation density, and/or by violation of the weak turbulence ansatz that the spectral auto-correlation time τ_{ac} be short in comparison with the local bounce time, i.e., $\tau_{ac} < \tau_b$. Note that in a dielectric medium, trapping is related to the sign of δf, since only one sign of δf (i.e., $\delta f > 0$ for the BGK mode, or $\delta f < 0$ for a phase space density hole, but not both) is consistent with the existence of a self-trapped, stable state on a scale k^{-1} and phase velocity u_0. The sign for δf which is selected is determined by the sign of the dielectric constant $\epsilon(k, ku_0)$. We will discuss the physics of self-trapping, hole formation, etc. in Volume 2. Obviously, local trapping can surely punctuate, restrict or eliminate relative particle dispersion in a globally fluctuating plasma. However, in this section, which deals exclusively with statistical theory, we hereafter ignore trapping. Thus, the sign of δf is not determined or addressed.

Second, absent trapping in localized structures, we can expect particle orbits to be stochastic, since phase space islands will surely overlap for a broad, multi-mode spectrum. Rigorously speaking, a state of stochasticity implies at least one positive Lyapunov exponent, so neighbouring (test) particle trajectories *must* separate, with divergence increasing exponentially in time. Similarly, then, we can expect particle dispersion to grow exponentially, with the dispersion rate related to the dynamics of the underlying phase space chaos.

It is useful now to recall explicitly that,

$$\frac{\partial}{\partial t} \langle \delta f\,(1)\, \delta f\,(2) \rangle + T_{1,2} \left[\langle \delta f\,(1)\, \delta f\,(2) \rangle \right] = P\,(1, 2)\,, \tag{8.45a}$$

where the two-point evolution operator $T_{1,2}$ is,

$$T_{1,2}\left[\langle \delta f\,(1)\,\delta f\,(2)\rangle\right] = \left(v_1\frac{\partial}{\partial x_1} + v_2\frac{\partial}{\partial x_2}\right)\langle \delta f\,(1)\,\delta f\,(2)\rangle$$
$$+\frac{q}{m}\frac{\partial}{\partial v_1}\langle E\,(1)\,\delta f\,(1)\,\delta f\,(2)\rangle$$
$$+\frac{q}{m}\frac{\partial}{\partial v_2}\langle E\,(2)\,\delta f\,(1)\,\delta f\,(2)\rangle$$
$$+\,\nu\,\langle \delta f\,(1)\,\delta f\,(2)\rangle\,. \tag{8.45b}$$

Hence, we see immediately that any calculation of dispersion in phase space requires some closure or renormalization of the triplet terms $\sim \langle E(1)\delta f(1)\delta f(2)\rangle$ in Eq.(8.45b). Perhaps the simplest, most direct and most transparent closure is via a mean field or quasi-linear approach, as with the closure for $\langle f\rangle$ evolution. Note that here, as is quasi-linear theory, we are concerned primarily with *resonant* scattering processes.

8.2.3.3 Closure modeling for triplet correlations

To construct a quasi-linear theory for the triplet correlation $\langle E(1)\delta f(1)\delta f(2)\rangle$, we proceed by:

(i) first, calculating an effective two-point coherent response, $\langle \delta f(1)\delta f(2)\rangle^c_{k,\omega}$, which is *phase coherent* with the electric field component $E_{k,\omega}$ at *both* phase space points (1) and (2). This is simply the two-point analogue of the familiar one-point coherent response (8.27a), $f^c_{k,\omega} = -\,(q/m)\times R\,(\omega - kv)\,E_{k,\omega}\partial\,\langle f\rangle\,/\partial v$;

(ii) then iterating to derive a closed equation for $\langle \delta f(1)\delta f(2)\rangle$ evolution in terms of the field spectrum $\langle E^2\rangle_{k,\omega}$. This equation has the form of a bivariate diffusion equation in velocity space.

Let us progress along these lines. To obtain the coherent response $\langle \delta f\,(1)\,\delta f\,(2)\rangle^c_{k,\omega}$, we simply linearize Eq.(8.45a) in $E_{k,\omega}$, neglecting collisions (i.e., take $\tau_{\mathrm{ac}} \ll \tau_{\mathrm{coll}}$). This gives,

$$\langle \delta f\,(1)\,\delta f\,(2)\rangle^c_{k,\omega} = \mathrm{Re}\frac{q}{m}E_{k,\omega}$$
$$\times\left\{\mathrm{e}^{ikx_1}R\,(\omega - kv_1)\,\frac{\partial}{\partial v_1} + \mathrm{e}^{ikx_2}R\,(\omega - kv_2)\,\frac{\partial}{\partial v_2}\right\}\langle \delta f\,(1)\,\delta f\,(2)\rangle\,, \tag{8.46}$$

where $R(\omega - kv_1)$ is the wave–particle resonance function, discussed in Eq.(8.27). In practice, we may take $R(\omega - kv) \simeq \pi\delta(\omega - kv)$. Note that $\langle \delta f(1)\delta f(2)\rangle^c_{k,\omega}$

is simply the *sum* of the independent responses of particle 1 at x_1, v_1 plus that for particle 2 at x_2, v_2. These responses are dynamically independent, (i.e., correlated *only* via the driving mean correlator $\langle \delta f(1)\delta(2)\rangle$), as required by the factorizability of the Vlasov hierarchy. (In deriving the Vlasov equation, the joint probability $f(1,2)$ is approximated by multiplication of one-particle distribution functions.) Note also that $(\delta f(1)\delta f(2))^c_{k,\omega}$ is phase coherent with $\exp\left(i\theta_{k,\omega}\right)$, where $\theta_{k,\omega}$ is the phase of the k, ω field, $E_{k,\omega}$, i.e., $E_{k,\omega} = A_{k,\omega}\exp\left(i\theta_{k,\omega}\right)$. In this approach, phase coherency is fundamental, since it links $f(1)f(2)$ to $E_{k,\omega}$ at *both* points 1 and 2.

Then the quasi-linear equation for $\langle \delta f(1)\delta(2)\rangle$ follows from simply substituting Eq.(8.46) into the triplet terms of Eq.(8.45b), e.g.,

$$\frac{\partial}{\partial v_1}\langle E(1)\,\delta f(1)\,\delta f(2)\rangle \simeq \frac{\partial}{\partial v_1}\left\langle E(1)\,(\delta f(1)\,\delta f(2))^c\right\rangle.$$

With this approximation, Eq.(8.45) becomes,

$$\begin{aligned}
&\left(\frac{\partial}{\partial t} + v_1\frac{\partial}{\partial x_1} + v_2\frac{\partial}{\partial x_2} + \nu\right)\langle \delta f(1)\,\delta f(2)\rangle \\
&-\left(\frac{\partial}{\partial v_1}D_{11}\frac{\partial}{\partial v_1} + \frac{\partial}{\partial v_2}D_{22}\frac{\partial}{\partial v_2}\right)\langle \delta f(1)\,\delta f(2)\rangle \\
&-\left(\frac{\partial}{\partial v_2}D_{21}\frac{\partial}{\partial v_1} + \frac{\partial}{\partial v_1}D_{12}\frac{\partial}{\partial v_2}\right)\langle \delta f(1)\,\delta f(2)\rangle = P(1,2),
\end{aligned} \tag{8.47a}$$

where,

$$D_{jj} = \sum_{k,\omega}\frac{q^2}{m^2}\left|E_{k,\omega}\right|^2 R\left(\omega - kv_j\right) \quad (j = 1 \text{ or } 2), \tag{8.47b}$$

and,

$$D_{12} = \sum_{k,\omega}\frac{q^2}{m^2}\mathrm{e}^{ik(x_1-x_2)}\left|E_{k,\omega}\right|^2 R\left(\omega - kv_2\right), \tag{8.47c}$$

with $1 \leftrightarrow 2$ for $D_{1,2}$. Equation (8.47a) is a bivariate diffusion equation for $\langle \delta f(1)\delta(2)\rangle$. Observe that D_{11} and D_{22} are usual quasi-linear diffusion coefficients, while D_{12} and D_{21} represent *correlated* diffusion, in that they approach unity for $|kx_-| < 1$, and tend to oscillate and so cancel for $|kx_-| > 1$. Note also that correlated scattering will occur only if v_1 and v_2 resonate with the same portion of the electric field spectrum, so $|v_-| < \Delta v_{\mathrm{T}}$ is required as well.

8.2.3.4 Alternative derivation

It is worthwhile to elaborate on the derivation of Eq.(8.47), prior to embarking on a discussion of its physics. There are at least three ways to derive Eq.(8.47). These are:

(i) the two-point quasi-linear approach, as implemented above;
(ii) a bi-variate Fokker–Planck calculation. In this approach, the validity of which is rooted in particle stochasticity, $\langle \delta f(1)\delta f(2)\rangle$ is assumed to evolve via independent random walks of particle 1 and particle 2;
(iii) a DIA-type closure of the two-point correlation equation, as presented in (Boutros-Ghali and Dupree, 1981).

It is instructive to discuss the bivariate Fokker–Planck calculation in some detail. The essence of Fokker–Planck theory is evolution via small, uncorrelated random scattering events, which add in coherently to produce a diffusive evolution. Thus $T_{1,2}\left\langle \delta f^2\right\rangle$ becomes,

$$T_{1,2}\left\langle \delta f\left(1\right)\delta f\left(2\right)\right\rangle = \left(v_1\frac{\partial}{\partial x_1}+v_2\frac{\partial}{\partial x_2}\right)\left\langle \delta f\left(1\right)\delta f\left(2\right)\right\rangle + \frac{\Delta\left\langle \delta f\left(1\right)\delta f\left(2\right)\right\rangle}{\Delta t}. \tag{8.48}$$

Here $\Delta\left\langle \delta f(1)\delta f(2)\right\rangle/\Delta t$ symbolically represents non-deterministic evolution of $\langle \delta f(1)\delta f(2)\rangle$ due to stochastic scattering events Δv_1 and Δv_2 in a scattering time step of Δt. (We expect $\Delta t\sim\tau_{\mathrm{ac}}$.) Thus, $\Delta\left\langle \delta f(1)\delta f(2)\right\rangle/\Delta t$ is represented in terms of the transition probability T as,

$$\begin{aligned}\frac{\Delta\left\langle \delta f\left(1\right)\delta f\left(2\right)\right\rangle}{\Delta t} &= \left[\iint \mathrm{d}\left(\Delta v_1\right)\mathrm{d}\left(\Delta v_2\right)T\left(\Delta v_1, v_1;\, \Delta v_2, v_2;\, \Delta t\right)\right.\\ &\quad\left.\times\left\langle \delta f\left(v_1-\Delta v_1\right)\delta f\left(v_2-\Delta v_2\right)\right\rangle - \left\langle \delta f\left(1\right)\delta f\left(2\right)\right\rangle\right]\Big/\Delta t.\end{aligned} \tag{8.49}$$

Here $T\left(\Delta v_1, v_1;\, \Delta v_2, v_2;\, \Delta t\right)$ is the two-particle step probability distribution function or transition probability. Since particle scattering is statistically uncorrelated, the transition probability may be factorized, so,

$$T\left(\Delta v_1, v_1;\, \Delta v_2, v_2;\, \Delta t\right) = T\left(\Delta v_1, v_1;\, \Delta t\right)T\left(\Delta v_2, v_2;\, \Delta t\right). \tag{8.50}$$

With this factorization, the usual Fokker–Planck expansion of the right-hand side of Eq.(8.49) in small step size gives,

$$\frac{\Delta\langle\delta f(1)\,\delta f(2)\rangle}{\Delta t} =$$
$$-\frac{\partial}{\partial v_1}\left[\left(\frac{\langle\Delta v_1\rangle}{\Delta t}+\frac{\langle\Delta v_2\rangle}{\Delta t}\right)\langle\delta f(1)\,\delta f(2)\rangle\right.$$
$$\left.-\frac{\partial}{\partial v_1}\left\{\left(\frac{\langle\Delta v_1\Delta v_1\rangle}{2\Delta t}+\frac{\langle\Delta v_1\Delta v_2\rangle}{2\Delta t}\right)\langle\delta f(1)\,\delta f(2)\rangle\right\}\right]$$
$$-\frac{\partial}{\partial v_2}\left[\left(\frac{\langle\Delta v_2\rangle}{\Delta t}+\frac{\langle\Delta v_1\rangle}{\Delta t}\right)\langle\delta f(1)\,\delta f(2)\rangle\right.$$
$$\left.-\frac{\partial}{\partial v_2}\left\{\left(\frac{\langle\Delta v_2\Delta v_2\rangle}{2\Delta t}+\frac{\langle\Delta v_2\Delta v_1\rangle}{2\Delta t}\right)\langle\delta f(1)\,\delta f(2)\rangle\right\}\right], \quad (8.51a)$$

which has a form,

$$\frac{\Delta\langle\delta f(1)\,\delta f(2)\rangle}{\Delta t} =$$
$$-\frac{\partial}{\partial v_1}\left[(F_{11}+F_{12})\langle\delta f(1)\,\delta f(2)\rangle-\frac{\partial}{\partial v_1}\{(D_{11}+D_{12})\langle\delta f(1)\,\delta f(2)\rangle\}\right]$$
$$-\frac{\partial}{\partial v_2}\left[(F_{22}+F_{21})\langle\delta f(1)\,\delta f(2)\rangle-\frac{\partial}{\partial v_2}\{(D_{22}+D_{21})\langle\delta f(1)\,\delta f(2)\rangle\}\right]. \quad (8.51b)$$

Now clearly D_{11} and D_{12} (along with their counterparts with $1 \leftrightarrow 2$) correspond to single particle and correlated or cross-diffusions, respectively. Likewise, F_{11} and F_{12} (and their counterparts with $1 \leftrightarrow 2$) correspond to drag and cross-drag. It is well known that for a 1D Hamiltonian system, Liouville's theorem requires that,

$$\frac{\partial}{\partial v}\frac{\langle\Delta v\Delta v\rangle}{2\Delta t}=\frac{\langle\Delta v\rangle}{\Delta t}, \quad (8.52)$$

so *dynamical friction cancels the gradient of diffusion*. Similar cancellations occur between the cross-diffusions and cross-drags. Thus, Eq.(8.51) reduces to,

$$T_{1,2}\langle\delta f(1)\,\delta f(2)\rangle=\left(v_1\frac{\partial}{\partial x_1}+v_2\frac{\partial}{\partial x_2}\right)\langle\delta f(1)\,\delta f(2)\rangle$$
$$-\left(\frac{\partial}{\partial v_1}D_{11}\frac{\partial}{\partial v_1}+\frac{\partial}{\partial v_1}D_{12}\frac{\partial}{\partial v_2}+\frac{\partial}{\partial v_2}D_{22}\frac{\partial}{\partial v_2}+\frac{\partial}{\partial v_2}D_{21}\frac{\partial}{\partial v_1}\right)\langle\delta f(1)\,\delta f(2)\rangle, \quad (8.53)$$

where $D_{ij}=\left\langle\Delta v_i\Delta v_j\right\rangle/2\Delta t$ (i, j are 1 or 2) and is identical to Eq.(8.47). This rather formal discussion is useful since it establishes that the fundamental physics in Eq.(8.47) is just resonant diffusive scattering dynamics, which results from particle stochasticity. As we shall see, when coupled to free streaming, this results in exponential divergence of orbits, which then determines the particle dispersion rate.

We remark in passing here that the full closure theory for $\langle \delta f(1)\delta f(2)\rangle$ (Krommes, 1984) offers little of immediate utility beyond what is presented here. It is extremely tedious, conceptually unclear (at this moment) and calculationally intractable. Thus we do not discuss it in detail here.

8.2.3.5 Physics of two-particle dispersion

We now turn to a discussion of the physics of the two-particle dispersion process, as given by Eq.(8.47). First, it is clear that the essential physics is resonant scattering in velocity space, due to random acceleration by the electric field spectrum. This scattering can be uncorrelated (giving D_{11}, D_{22}) or correlated (giving D_{12}, D_{21}).

Second, given that the aim here is to calculate a two-point correlation function, and since the two-point correlation function is simply the Fourier transform of the associated field, it is useful to relate Eq.(8.53) to our earlier discussion of spectral evolution, in Chapters 5 and 6. To this end, note that Fourier transformation of Eq.(8.53) (which is in real space) gives,

$$\begin{aligned}
&\frac{\partial}{\partial t}\langle \delta f(1)\,\delta f(2)\rangle_k + ik\,(v_1 - v_2)\,\langle \delta f(1)\,\delta f(2)\rangle_k \\
&\quad - \frac{\partial}{\partial v_1} D_{11}\frac{\partial}{\partial v_1}\langle \delta f(1)\,\delta f(2)\rangle_k - \frac{\partial}{\partial v_2} D_{22}\frac{\partial}{\partial v_2}\langle \delta f(1)\,\delta f(2)\rangle_k \\
&\quad - \sum_{\substack{\pm p,q \\ p\pm q=k}} \left(\frac{\partial}{\partial v_1}\left\{\left\langle E^2\right\rangle_p R\,(\omega - pv)\,\frac{\partial}{\partial v_2}\langle \delta f(1)\,\delta f(2)\rangle_q\right\}\right. \\
&\qquad\qquad \left. + \frac{\partial}{\partial v_2}\left\{\left\langle E^2\right\rangle_p R\,(\omega + pv)\,\frac{\partial}{\partial v_1}\langle \delta f(1)\,\delta f(2)\rangle_q\right\}\right) \\
&\quad = P_k\,(1,2)\,.
\end{aligned} \tag{8.54}$$

Thus, we see immediately that:

- single-particle scattering corresponds to coherent mode coupling and a Markovian 'eddy viscosity' in velocity space;
- correlated diffusion corresponds to incoherent mode coupling and thus to nonlinear noise.

Stated equivalently, the *interplay* of single particle and correlated diffusion corresponds to 'cascading', i.e., the process whereby small scales are generated from larger ones. Note here that since free streaming couples to velocity scattering and diffusion, that too enters the rate at which small scales are generated.

Third, the net stationary value of $\langle \delta f(1)\delta f(2)\rangle$ is set by the *balance* of production with the lifetime of the scale of size x_- and v_-.

8.2.3.6 Calculation of dispersion time

The actual calculation of the dispersion time $\tau_c(x_-, v_-)$ is most expeditiously pursued by working in relative coordinates. To this end, as for Eq.(8.18), we write,

$$x_\pm = \frac{1}{2}(x_1 \pm x_2),$$

$$v_\pm = \frac{1}{2}(v_1 \pm v_2),$$

and discard the slow x_+, v_+ dependence in the T_{12} operator. Thus, Equation (8.47) can be simplified to the form,

$$\left(\frac{\partial}{\partial t} + v_- \frac{\partial}{\partial x_-} - \frac{\partial}{\partial v_-} D_{rel} \frac{\partial}{\partial v_-}\right) \langle \delta f(1)\, \delta f(2)\rangle = P(1,2), \tag{8.55}$$

where,

$$\begin{aligned} D_{\mathrm{rel}} &= D_{11} + D_{22} - D_{12} - D_{21} \\ &= \sum_{k,\,\omega} \frac{q^2}{m^2} (1 - \cos(kx_-)) \left\langle E^2 \right\rangle_{k,\omega} R(\omega - kv). \end{aligned} \tag{8.56a}$$

Here, $D_{\mathrm{rel}}(x_-)$ is the relative diffusion function, which gives a measure of how rapidly particles (separated by x_- in phase space) diffuse apart. Note that for $\left(k^2 x_-^2\right) > 1$ (Balescu, 2005),

$$D_{rel} \simeq D_{11} + D_{22}, \tag{8.56b}$$

so that diffusion then asymptotes to the value for two uncorrelated particles. For $\left\langle k^2 x_-{}^2 \right\rangle < 1$,

$$D_{\mathrm{rel}} \simeq \frac{k_0^2 x_-^2}{2} D, \tag{8.56c}$$

so $D_{\mathrm{rel}} \to 0$ as $x_-^2 \to 0$. Here k_0^2 is a spectral average, i.e., $k_0^2 = \langle k^2 \rangle$ and $D = (D_{11} + D_{22})/2$.

While exact calculation of the evolution of relative dispersion is lengthy and intricate, the essential behaviour can be determined by working in the $\left(k^2x_-^2\right) < 1$ limit, which captures key features, like the peaking of $\langle\delta f(1)\delta f(2)\rangle$ on small scales. Indeed, for $\left(k^2x_-^2\right) \ll 1$, $\langle\delta f(1)\delta f(2)\rangle \simeq \left\langle\tilde{f}(1)\tilde{f}(2)\right\rangle$, so no detailed calculations are needed. Of course, as we shall see, the existence of an individual small-scale peak in $\langle\delta f(1)\delta f(2)\rangle$ *requires* collisionality to be weak, i.e., $1/\tau_c \gg \nu$, to ensure that the small-scale structure of the correlation function is not smeared out. Note that the condition $1/\tau_c \gg \nu$ (which defines an effective 'Reynolds number' $Re_{\text{eff}} \sim 1/\tau_c\nu$) is equivalent to $\Delta v_T > 1/kv$, i.e., the requirement that the turbulently broadened resonance width exceeds the width set by collisional broadening. In this case we can rewrite Eq.(8.55) as an evolution equation for F, the probability density function (pdf) of relative separations x_-, v_-. As here we are concerned only with relative dispersion, and have tacitly assumed that $1/\tau_c \gg \nu$, we can now drop ν and $P(1,2)$. Thus, F satisfies the simple kinetic equation,

$$\frac{\partial F}{\partial t} + v_-\frac{\partial F}{\partial x_-} - \frac{\partial}{\partial v_-}Dk_0^2x_-^2\frac{\partial F}{\partial v_-} = 0. \tag{8.57}$$

It is now straightforward to derive a coupled set of moment equations from Eq.(8.57). Defining the moments by,

$$\langle A\left(x_-, v_-\right)\rangle \equiv \frac{\iint dx_- dv_- A\left(x_-, v_-\right) F\left(x_-, v_-; t\right)}{\iint dx_- dv_- F\left(x_-, v_-; t\right)}, \tag{8.58a}$$

we have:

$$\frac{\partial}{\partial t}\left\langle x_-^2\right\rangle = 2\left\langle x_- v_-\right\rangle, \tag{8.58b}$$

$$\frac{\partial}{\partial t}\left\langle x_- v_-\right\rangle = \left\langle v_-^2\right\rangle, \tag{8.58c}$$

$$\frac{\partial}{\partial t}\left\langle v_-^2\right\rangle = 2\left\langle D_{\text{rel}}\right\rangle. \tag{8.58d}$$

Equations (8.58b–d) combine to give,

$$\frac{\partial^3}{\partial t^3}\left\langle x_-^2\right\rangle = 4\left\langle D_{\text{rel}}\right\rangle = 4Dk_0^2\left\langle x_-^2\right\rangle, \tag{8.58e}$$

which tells us that mean square trajectory separations increase exponentially in time, at the rate $\left(Dk_0^2\right)^{1/3}$, as long as the condition $\left(k^2x_-^2\right) < 1$ is satisfied. More precisely, Eq.(8.58e), when solved for the initial conditions $\partial\langle x_-\rangle/\partial t = \langle v_-\rangle$ and $\partial\langle v_-\rangle/\partial t = 0$, gives,

$$\left\langle x_-(t)^2\right\rangle = \frac{1}{3}\left(x_-^2 + 2x_-v_-\tau_c + 2v_-^2\tau_c^2\right)\exp\left(\frac{t}{\tau_c}\right), \tag{8.59a}$$

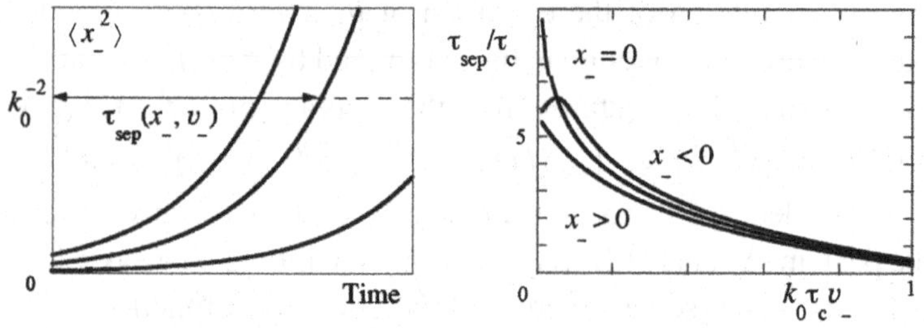

Fig. 8.10. Evolution of the statistical average of separation for various initial separations (left). The separation time τ_{sep} is defined for each initial condition. The separation time is shown as a function of initial separation in velocity space v_-.

where,

$$\tau_c^{-1} = \left(4Dk_0^2\right)^{1/3}. \tag{8.59b}$$

This gives the characteristic decorrelation rate for granulations in phase space turbulence. Equation (8.58e) has three eigensolutions, two of which are damped. To obtain the result of Eq.(8.59a), we neglect the damped solutions, which are time asymptotically subdominant.

Now finally, as we are most interested in the relative dispersion time as a function of given (initial) phase space separation x_-, v_- (i.e., corresponding to a given scale), it is appropriate to define a scale-dependent separation time $\tau_{\text{sep}}(x_-, v_-)$ by the condition,

$$k_0^2 \left\langle x_- \left(\tau_{\text{sep}}\right)^2 \right\rangle = 1, \tag{8.59c}$$

i.e., as the time needed for the pair to disperse k_0^{-1} (Fig. 8.10(a)). Hence,

$$\tau_{\text{sep}}\left(x_-, v_-\right) = \tau_c \ln\left\{3k_0^{-2}\left(x_-^2 + 2x_- v_- \tau_c + 2v_-^2 \tau_c^2\right)^{-1}\right\}. \tag{8.59d}$$

We note this expression applies only when the argument of the logarithmic function is positive. An example is illustrated in Figure 8.10(b).

The expressions for τ_{sep} and τ_c in Eqs.(8.59d) and (8.59b) are the principal results of this section, and so merit some further discussion. First, we note that the basic time scale for relative dispersion is τ_c (Eq.(8.59b)), the wave–particle turbulent decorrelation time, which is also the Lyapunov time for separation of stochastic particle orbits. Thus, the calculated dispersion time is consistent with expectations from dynamical systems theory.

Second, note that $\tau_{\text{sep}} > \tau_c$ for small separations (where $k_0^2 x_-^2 \ll 1$) and/or $k_0 v_- \tau_c = v_-/\Delta v_{\text{T}} < 1$. Thus τ_{sep} is sharply peaked on scales small compared

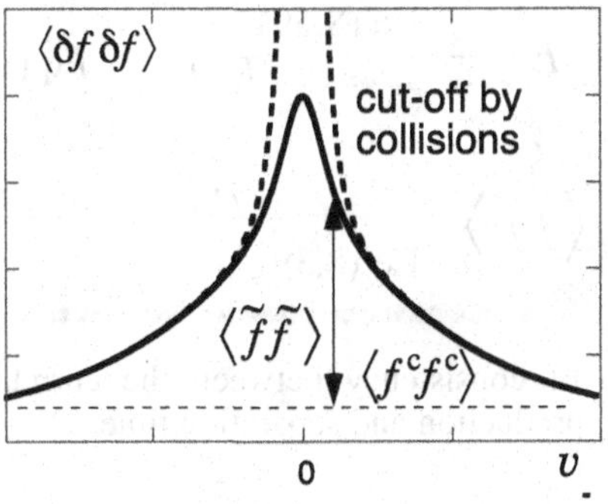

Fig. 8.11. Correlation function $\langle \delta f(1)\delta f(2)\rangle$ as a function of the separation of particles. The collisionless limit (dotted line) and the case where cut-off by collision works at $v_- = 0$. Contributions from the coherent component and granulations are also noted.

with the basic turbulence correlation scales. This peaking is consistent with the collisionless singularity of $\langle \delta f(1)\delta f(2)\rangle$, which is required for the limit $1 \to 2$, as discussed in Eq.(8.25b). Figure 8.11 illustrates schematically the correlation function $\langle \delta f(1)\delta f(2)\rangle$ and contributions from the coherent component and granulations. Retaining finite collisionality removes the singularity, and truncates the peaking of $\langle \delta f(1)\delta f(2)\rangle$ on scales $v_- < \nu/k_0$. As noted above, for this truncation to be observable, $\nu/k_0 < \Delta v_{\mathrm{T}}$ is necessary. As we shall see, even in the absence of collisions, $\int \mathrm{d}v_- \tau_{\mathrm{sep}}(x_-, v_-)$ is finite, so all physical observables are well behaved.

Third, this entire calculation is predicted on the existence of k_0^2, i.e., we assume that,

$$k_0^2 = D^{-1} \sum_{k,\omega} k^2 \frac{q^2}{m^2} \left\langle E^2 \right\rangle_{k,\omega} R(\omega - kv) \tag{8.60c}$$

is finite. The requisite spectral convergence must indeed be demonstrated a posteriori. Absence of such convergence necessitates a different approach to the calculation of the phase space density correlation function.

Fourth, we remark that τ_c is also the effective 'turn-over time' or scale lifetime which determines the phasestrophy cascade in a turbulent Vlasov plasma. Here, small scales are generated by the coupled processes of relative streaming and relative scattering, rather than by eddy shearing, as in a turbulent fluid.

The structure of the theory is summarized in Figure 8.12, where the explanations in this chapter are revisited. The granulations radiate the (non-modal) electric field. The intensity and spectrum of the electric field, which is excited by granulations, is given in Eq.(8.32). By use of the excited electric field, the rate of production is evaluated in Eq.(8.39), as a functional of the granulation correlation function.

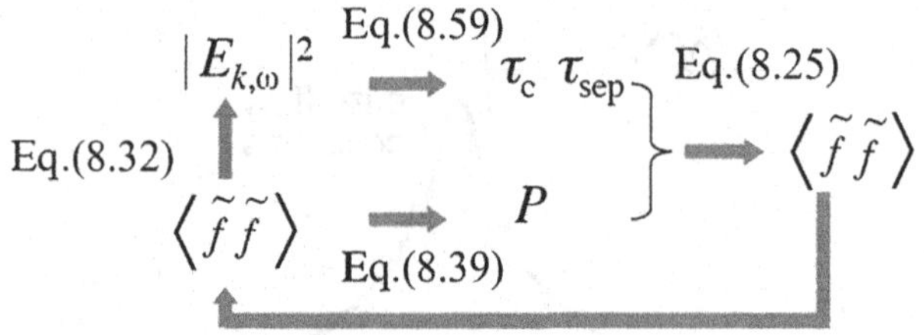

Fig. 8.12. The loop of consistency between the correlation of granulation, radiated electric field, production and separation time.

The triplet correlation is estimated by using the electric field spectrum, so that the separation time is estimated by Eq.(8.59). The self-consistency loop closes at Eq.(8.25), in which the granulation intensity is determined by the balance between the production and lifetime. This set of equations constitutes the theory that determines phase space density granulations.

8.3 Physics of relaxation and turbulent states with granulation

In the preceding sections of Chapter 8, we have discussed the physics of phase space density granulation at length, and have derived the equation of evolution for the correlation function $\langle \delta f\,(1)\,\delta f\,(2) \rangle$. We now turn to the 'bottom line' – we face the question of *what actually happens when granulations are present.* In particular, we focus on two issues, which are:

(1) what types of saturated states are possible, and what is the role of granulations in the dynamics of these states?
(2) what types of novel, nonlinear instability mechanisms may occur via granulations?

The calculations required to answer these questions quantitatively are extremely lengthy and detailed. Hence, in this section we take a 'back-of-an-envelope' approach, and only sketch the essence and key elements of the calculations, along with their physical motivations and implications. Our aim is to help the reader understand the landscape of this 'terra nova'. Once motivated and oriented, a serious reader can then consult the original research literature for details.

As we have seen, the equation of evolution for the two-point correlation function takes the generic form,

$$\left(\frac{\partial}{\partial t} + \frac{1}{\tau_{c_1}}\right) \langle \delta f\,(1)\,\delta f\,(2) \rangle = P\,(1,2)\,, \tag{8.61}$$

so that, as in a turbulent shear flow, the correlation function is set by a balance between drive by gradient relaxation (in this case, the gradients of $\langle f \rangle$) and relative shearing and dispersion, as parametrized by $\tau_{c_1}\,(x_-, v_-)$. In a stationary state, we then have simply,

$$\langle \delta f\,(1)\,\delta f\,(1)\rangle = \tau_{c_1} P\,(1,2)\,. \tag{8.62}$$

The production term $P\,(1,2)$, as given by Eq. (8.42), is a function of the integrated granulation correlation function $\left\langle \tilde{n}\tilde{f}\,(u)\right\rangle_k$ and is, in essence, set by the:

(i) free energy stored in driving gradients, which makes $P\,(1,2) > 0$;
(ii) dielectric screening – which is particularly important near resonance between collective modes with $\omega = \omega_k$ and ballistic modes (i.e. granulations here), for which $\omega = ku$;
(iii) the spectral profile of Doppler emission by test granulations.

It is useful, then, to convert Eq. (8.62) to an integral equation for $\left\langle \tilde{n}\tilde{f}\,(u)\right\rangle_k$ by subtracting the coherent correlation function and $\left\langle \tilde{f} f^{c}\right\rangle$ cross-term from $\langle \delta f\,(1)\,\delta f\,(2)\rangle$, and then integrating over the relative velocity v_-. Now, since,

$$\delta f = f^{c} + \tilde{f}, \tag{8.63a}$$

we have,

$$\left\langle \tilde{f}\tilde{f}\right\rangle = \langle \delta f\,\delta f\rangle - \langle f^{c} f^{c}\rangle - 2\left\langle f^{c}\tilde{f}\right\rangle. \tag{8.63b}$$

The first subtraction ($\langle f^{c} f^{c}\rangle$) is proportional to the diffusive mixing term in $P\,(1,2)$, while the second ($2\left\langle f^{c}\tilde{f}\right\rangle$) is proportional to the drag term. Hence, we ultimately have just,

$$\left\langle \tilde{f}\,(1)\,\tilde{f}\,(2)\right\rangle = \left(\tau_{\mathrm{cl}}\,(x_-, v_-) - \tau_{\mathrm{c}}\right) P\,(1,2)\,, \tag{8.63c}$$

which may be re-written in terms of a $\tau_{\mathrm{cl,eff}}$, peaked sharply as $x_-, v_- \to 0$. Hereafter, we assume this sample subtraction to be in force, and do not distinguish between τ_{cl} and $\tau_{\mathrm{cl,eff}}$. Observe that $1/\tau_{\mathrm{c}} \gg \nu$ is necessary for $\tau_{\mathrm{cl}} - \tau_{\mathrm{c}}$ to exhibit a non-trivial range of scales between the integral scale and the collisional cut-off.

Since generically, $\tau_{\mathrm{cl}} = \tau_{\mathrm{cl}}\,(v_-/\Delta v_{\mathrm{T}}, k_0 x_-)$, (recall Δv_{T} is a trapping width and k_0 is an integral scale) we have,

$$\begin{aligned}\left\langle \tilde{n}\tilde{f}\right\rangle &= \int \mathrm{d}v_-\,\tau_{\mathrm{cl}} P\,(1,2)\\ &\cong \Delta v_{Tr}\tau_{\mathrm{cl}} P\,(1,2)\,.\end{aligned} \tag{8.64a}$$

Recall, though, that,

$$\tau_{\mathrm{cl}} = \tau_{\mathrm{c}} F\,(v_-/\Delta v_{\mathrm{T}}, k_0 x_-)\,, \tag{8.64b}$$

so the integration over v_- necessarily yields the product $\tau_c \Delta v_T$. Since the resonance width Δv_{T_r} and correlation time τ_c are related by the definition $\Delta v_T = 1/k\tau_c$, we then have,

$$\left\langle \tilde{n} \tilde{f} \right\rangle \cong (1/k_0)\ G\left(k_0 x_-\right) P\left(1, 2\right). \tag{8.64c}$$

Here $G\left(k_0 x_-\right)$ is the spatial structure function of the granulation defined by the shape of τ_{cl}. The key point here is that due to the reciprocal relation between τ_c and Δv_T, the integral equation for $\left\langle \tilde{n} \tilde{f} \right\rangle$ defined by Eq.(8.64c) is at least *formally homogeneous*. This homogeneity is ultimately a consequence of the *resonant*, linear propagator appearing in the relation $\langle \tilde{g}\tilde{g} \rangle_{k,\omega} = 2\pi\delta\left(\omega - kv\right) \langle \tilde{g}\tilde{g} \rangle_k$. Fourier transformation in space then gives,

$$\left\langle \tilde{n} \tilde{f} \right\rangle_k = A\left(k, k_0\right) P\left(1, 2\right), \tag{8.64d}$$

where $A\left(k, k_0\right)$ is the spatial form factor for the granulation correlation.

Recall from Eq.(8.42), that for our generic example case of ion granulations in CDIA (current driven ion acoustic) turbulence, production is given by,

$$P\left(1, 2\right) = \frac{\partial \left\langle f_i \right\rangle}{\partial v} \sum_{k,\omega} \frac{\omega_{pi}^2}{k} \frac{2\pi^2}{|k|^2} \frac{\operatorname{Im} \epsilon_e\left(k, \omega\right) \left\langle \tilde{f}\left(v\right) \tilde{f}\left(u\right) \right\rangle_k}{|\epsilon\left(k, \omega\right)|^2}. \tag{8.65}$$

Observe that only two elements in Eq.(8.59) can adjust to yield a stationary balance. These are:

- the electron distribution function, which defines $\operatorname{Im} \epsilon_e\left(k, \omega\right)$;
- the net damping of the wave resonance (i.e. mode) at $\omega = \omega_k$, which is set by $\operatorname{Im} \epsilon\left(k, \omega_k\right)$.

Thus, if we interpret Eq.(8.64d) as a stationarity condition for $\left\langle \tilde{n} \tilde{f} \right\rangle$, we see that a steady state, where production balances dissipation, requires either relaxation of $\langle f \rangle$ *or* a particular value of wave damping i.e. $\operatorname{Im} \epsilon\left(k, \omega_k\right)$. Relaxation of $\langle f \rangle$ (i.e. electron slowing down) may be calculated using the granulation-driven Lenard–Balescu equation. To determine the wave damping, note that the familiar pole approximation,

$$\begin{aligned} \frac{1}{|\epsilon\left(k, \omega\right)|^2} &\cong \frac{1}{\left(\omega - \omega_k\right)^2 \left(\partial\epsilon/\partial\omega|_{\omega_{\mathbf{k}}}\right)^2 + |\operatorname{Im}\epsilon|^2} \\ &\cong \left\{ \frac{1}{\left|\partial\epsilon/\partial\omega|_{\omega_{\mathbf{k}}}\right| \left|\operatorname{Im}\epsilon\left(k, \omega_{\mathbf{k}}\right)\right|} \right\} \delta\left(\omega - \omega_{\mathbf{k}}\right), \end{aligned} \tag{8.66}$$

allows us to perform the frequency summation in Eq.(8.65), and so to obtain,

$$P(1,2) \cong \frac{\partial \langle f_i \rangle}{\partial v} \sum_k \frac{\omega_{\mathrm{pi}}^2}{k} \frac{2\pi}{|k|^2} \frac{\operatorname{Im} \epsilon_e (k, \omega_k) \left\langle \tilde{f}(v) \tilde{f}(u_k) \right\rangle_k}{|\operatorname{Im} \epsilon (k, \omega_k)| \left| \partial \epsilon / \partial \omega|_{\omega_{\mathbf{k}}} \right|}, \tag{8.67}$$

where $u_k = \omega_k / k$. Hence, Eq.(8.65) then gives,

$$\left\langle \tilde{n} \tilde{f} \right\rangle_k = A(k, k_0) \sum_k \frac{\omega_{\mathrm{pi}}^2}{k} \frac{2\pi}{|k|^2} \frac{\operatorname{Im} \epsilon_e (k, \omega_k) \int \mathrm{d}v \, (\partial \langle f_i \rangle / \partial v) \left\langle \tilde{f}(v) \tilde{f}(u_k) \right\rangle_k}{|\operatorname{Im} \epsilon (k, \omega_k)| \left| \partial \epsilon / \partial \omega|_{\omega_{\mathbf{k}}} \right|}. \tag{8.68}$$

Though Eq.(8.68) is an integral relation, it transparently reveals the basic scaling of Imϵ $(k, \omega_{\mathbf{k}})$ enforced by the stationarity condition of Eq.(8.58), which is,

$$|\operatorname{Im} \epsilon (k, \omega_k)| \sim \frac{-A(k, k_c) \operatorname{Im} \epsilon_e (k, \omega_k)}{\left| \partial \epsilon / \partial \omega|_{\omega_k} \right|}. \tag{8.69a}$$

We see that Eq.(8.69), which is a sort of "eigenvalue condition", links mode dissipation to:

(i) the granulation structure form factor $A(k, k_0)$, which is a measure of the strength and scale of granulation emissivity;
(ii) $\operatorname{Im} \epsilon_e (k, \omega_k)$, which is a measure of net free energy (i.e. electron current) available to drive relaxation. Of course, $\operatorname{Im} \epsilon_e (k, \omega_k) > 0$ is required.

Note that schematically,

$$|\operatorname{Im} \epsilon (k, \omega_k)| \sim -\operatorname{Im} \epsilon_i (k, \omega_k) \operatorname{Im} \epsilon_e (k, \omega_k) A(k, k_c). \tag{8.69b}$$

Stationarity in the presence of noise emission requires that the modes be *over-saturated*, so as to ensure a fluctuation–dissipation type balance. Then,

$$\operatorname{Im} \epsilon (k, \omega_k) \sim \operatorname{Im} \epsilon_i (k, \omega_k) \operatorname{Im} \epsilon_e (k, \omega_k) A(k, k_0), \tag{8.69c}$$

so $\operatorname{Im} \epsilon < 0$, since $A > 0$ and $\operatorname{Im}\epsilon_i \operatorname{Im}\epsilon_e < 0$.

Equation (8.69c) is, in some sense, "the answer" for the granulation problem, since several key results follow directly from it. First, since the line width (at fixed k) for mode k is just $\Delta\omega_k = |\operatorname{Im} \epsilon (k, \omega_k)| / (\partial \epsilon / \partial \omega_k)$, we see that the frequency line width at fixed k for ion acoustic modes scales as,

$$\Delta\omega_k \sim |\operatorname{Im} \epsilon_i (k, \omega_k)| \, |\operatorname{Im} \epsilon_e (k, \omega_k)| \, A(k, k_0) / \left(\partial \epsilon / \partial \omega|_{\omega_k} \right). \tag{8.70a}$$

Now, since $\text{Im}\,\epsilon\,(k,\omega_k) = \text{Im}\,\epsilon_e\,(k,\omega_k) + \text{Im}\,\epsilon_i\,(k,\omega_k)$ and $\text{Im}\,\epsilon_e \text{Im}\,\epsilon_i < 0$ here, the effective growth or drive in the stationary state is,

$$\gamma_k^{\text{eff}} = \left(\text{Im}\,\epsilon_e \,/[1 - A\,(k,k_0)\,\text{Im}\,\epsilon_e\,(k,\omega_{\mathbf{k}})]\right) / \left.\frac{\partial \epsilon}{\partial \omega}\right|_{\omega_k}. \tag{8.70b}$$

This result may be interpreted as an enhancement of effective wave growth due to the presence of granulation noise, and so is a step beyond the quasi-linear theory of Chapter 3. The result of Eq.(8.70b) is, of course, directly related to Eq.(8.70a). The mechanism of growth enhancement is granulation noise emission. Here we also add the cautionary comment that the pole approximation fails for $A\,(k,k_0)\,\text{Im}\,\epsilon_e\,(k,\omega_k) \rightarrow 1$. To go further, note that Eq.(8.68) is, in principle, an integral equation for both the k and ω spectra, but the pole approximation and Eq.(8.70a) determine only the frequency spectrum. The saturated model k-spectrum can be determined by solving the balance condition,

$$\text{Im}\,\epsilon_i = \gamma_k^{\text{eff}} (\partial\epsilon/\partial\omega|_{\omega_k}), \tag{8.70c}$$

or equivalently,

$$\text{Im}\,\epsilon_{i,\text{NL}}\,(k,\omega_k) = \gamma_k^{\text{eff}} \left(\left.\frac{\partial \epsilon}{\partial \omega}\right|_{\omega_k} \right) - \text{Im}\,\epsilon_{i,\text{L}}\,(k,\omega_k)\,. \tag{8.70d}$$

Here $\text{Im}\,\epsilon_{i,\text{L}}$ and $\text{Im}\,\epsilon_{i,\text{NL}}$ are the linear (i.e. Landau damping) and nonlinear (i.e. nonlinear wave–wave and wave–particle scattering) pieces of the ion susceptibility. Equation (8.70d) has the now familiar structure of "nonlinear damping = (granulation enhanced) growth − linear damping". To actually calculate the saturated k-spectrum requires consideration of nonlinear wave–particle and wave–wave interaction processes, as discussed in Chapters 4 and 5. Finally, it is interesting to also notice that Eq.(8.70) defines a new dynamical stability condition, due to the effects of granulation enhancement. To see this, recall that the purely *linear* instability criterion is just,

$$\text{Im}\,\epsilon\,(k,\omega_k) = \text{Im}\,\epsilon_e\,(k,\omega_k) + \text{Im}\,\epsilon_i\,(k,\omega_k) > 0. \tag{8.71a}$$

However, upon including -f granulations, the condition for non-trivial saturation at a finite amplitude becomes,

$$\gamma_k^{\text{eff}} \left(\left.\frac{\partial \epsilon}{\partial \omega}\right|_{\omega_k} \right) + \text{Im}\,\epsilon_{i,\text{L}}\,(k,\omega_k) > 0. \tag{8.71b}$$

Here ϵ_{L} indicates the *linear* dielectric. Eq.(8.71b) states that

$$\frac{\operatorname{Im} \epsilon_e\,(k, \omega_k)}{1 - A\,(k, k_0) \operatorname{Im} \epsilon_e\,(k, \omega_k)} + \operatorname{Im} \epsilon_{i\,\mathrm{L}(k,\omega_k)} > 0 \tag{8.71c}$$

is required for net relaxation and a non-trivial saturated state. We see that the effect of granulations is to augment or boost the drive for relaxation of the free energy source – in this case the current. Equation (8.71c) suggests that the formation of granulations induces an element of subcriticality or hysteresis into the instability process. To see this, recall that linear instability requires a current sufficient to make $\operatorname{Im} \epsilon_{\mathrm{L}} > 0$. However, for $A\,(k, k_0) \operatorname{Im} \epsilon_e\,(k, \omega_k) > 0$, the current to satisfy Eq.(8.71c) is surely *smaller* than that required to satisfy Eq.(8.71a)! Thus, the theory suggests that a viable scenario in which:

(i) the driving current is induced, so linear instability is initiated, according to Eq.(8.71a);
(ii) particle orbits go stochastic, resulting in phase space turbulence;
(iii) granulations form;
(iv) the driving current is then lowered, so Eq.(8.71c) is still satisfied.

Such a scenario predicts *sustained, sub-critical turbulence* and so may be considered as a model of nonlinear instability or relaxation due to phase space density granulations.

The possibility of a self-sustaining state evolving out of free energy levels (i.e. currents) *below* what is required for linear instability naturally motivates us to consider the broader possibility of nonlinear growth of phase space density granulations. By "granulation growth", we mean an increase in fluctuation phasestrophy $\left\langle \delta f^2 \right\rangle$ at the expense of available free energy stored in $\langle f \rangle$. Note that here "fluctuation" is not limited to eigenmodes (i.e. waves that obey a dispersion relation $\omega = \omega\,(k)$ with line width $\Delta\omega_k < \omega_k$), but also includes structures localized in phase space, which are akin to an eddy in a turbulent fluid. Thus, from a statistical perspective, the structure growth problem must be formulated at the level of the two-point correlation equation for $\left\langle \delta f^2 \right\rangle$, and "growth" should thus be interpreted as the amplification of correlation on a certain scale, i.e. $\gamma = (1/\langle \delta f\,(1)\, \delta f\,(2)\rangle)\,(\partial \langle \delta f\,(1)\, \delta f\,(2)\rangle / \partial t)$, rather than as eigenmode growth.

Consideration of the physics of nonlinear granulation growth again takes us to the $\langle \delta f\,(1)\, \delta f\,(2)\rangle$ equation,

$$\frac{\partial}{\partial t} \langle \delta f\,(1)\, \delta f\,(2)\rangle + \frac{1}{\tau_{c_1}} \langle \delta f\,(1)\, \delta f\,(2)\rangle = P\,(1, 2)\,. \tag{8.72}$$

Multiplying through by τ_{c_1} and integrating over relative velocity (v_-) then gives,

$$(\gamma\tau_c + 1)\left\langle \tilde{n}\tilde{f}\right\rangle \cong \tau_c \Delta v_T P(1,2) \cong \frac{1}{k_0} P(1,2). \tag{8.73a}$$

Equation (8.73a) can be recognized as the balance condition of Eq. (8.65), now generalized to the case of a non-stationary state. Using the structure of $P(1,2)$, we can then write,

$$\begin{aligned}(\gamma\tau_c + 1)\left\langle \tilde{n}\tilde{f}\right\rangle_k &\cong A(k,k_0)\sum_{k,\omega} \frac{\omega_{\mathrm{Pi}}^2}{k}\frac{2\pi^2}{|k|^2}\frac{\mathrm{Im}\,\epsilon_e(k,\omega)}{|\epsilon(k,\omega)|^2} \\ &\times \left(\frac{\partial\langle f\rangle}{\partial v}\left\langle \tilde{f}(v)\tilde{f}(u)\right\rangle_k\right).\end{aligned} \tag{8.73b}$$

Equation (8.73b) constitutes a (non-stationary) spectral balance equation, formulated at the level of the Vlasov equation. Once again, using the pole approximation gives,

$$\begin{aligned}\left(\gamma_{g,k}\tau_c + 1\right)\left\langle \tilde{n}\tilde{f}\right\rangle_k &= A(k,k_0)\sum_k \frac{\omega_{\mathrm{Pi}}^2}{k}\frac{2\pi}{|k|^2} \\ &\times \frac{\mathrm{Im}\,\epsilon_e(k,\omega_k)\int \mathrm{d}v\,(\partial\langle f\rangle/\partial v)\left\langle \tilde{f}(v)\tilde{f}(u_k)\right\rangle_k}{|\mathrm{Im}\,\epsilon(k_0,\omega_k)|\,\left|\partial\epsilon/\partial\omega|_{\omega_{\mathbf{k}}}\right|},\end{aligned} \tag{8.73c}$$

which is effectively, the 'nonlinear eigenvalue' equation for growth of fluctuations on scale k. Proceeding more schematically, Eq. (8.73c) may re-written as,

$$\left(\gamma_{g,k}\tau_c + 1\right)\left\langle \tilde{n}\tilde{f}\right\rangle_k \sim \frac{A(k,k_0)\left(-\mathrm{Im}\,\epsilon_{\mathrm{i}}(k,\omega_k)\right)\left(\mathrm{Im}\,\epsilon_e(k,\omega_k)\right)\left\langle \tilde{n}\tilde{f}\right\rangle}{|\mathrm{Im}\,\epsilon(k,\omega_k)|\,\left|\partial\epsilon/\partial\omega|_{\omega_k}\right|}, \tag{8.73d}$$

so finally we see that the growth rate is just,

$$\gamma_{g,k} = \frac{1}{\tau_c}\left[\frac{A(k,k_0)\left(-\mathrm{Im}\,\epsilon_i(k,\omega_k)\right)\left(\mathrm{Im}\,\epsilon_e(k,\omega_k)\right)}{|\mathrm{Im}\,\epsilon(k,\omega_k)|\,\left|\partial\epsilon/\partial\omega|_{\omega_k}\right|} - 1\right]. \tag{8.73e}$$

At long last, Eq. (8.73e) gives the nonlinear granulation growth rate! $\gamma_{g,k} = 0$ gives the marginality condition. Several features of the granulation growth rate $\gamma_{g,k}$ are apparent from inspection of Eq. (8.73e). First, we note that growth is *nonlinear* (i.e. amplitude dependent), with basic scaling $\gamma_g \sim 1/\tau_c$. Thus, fluctuation growth and granulation instability are fundamentally *explosive*. Second, we see that the stationarity condition given by Eq. (8.68) sets the effective marginality condition for instability, since putting the quantity in brackets equal to zero recovers the stationarity condition of Eq. (8.68). Third, we note that exceeding marginality or criticality requires $(-\text{Im}\,\epsilon_i\,(k, \omega_k))\,(\text{Im}\,\epsilon_e\,(k, \omega_k)) > 0$, so free energy must be stored in the electrons – i.e. a net current must be carried. We also see that in effect, marginality requires electron free energy (i.e. current) to exceed a critical level, which is set by the condition that the right-hand side of Eq. (8.73e) > 0. This condition is different from that required for *linear* marginality. For linearly subcritical instability, $\text{Im}\epsilon_e$ and $\text{Im}\epsilon_i$ should be evaluated using linear susceptibilities. Speaking pragmatically, subcritical instability requires both a sufficiently large current and also that ion Landau damping be not too strong – i.e. $\text{Im}\epsilon < 0$, but not too strongly negative. In principle, subcritical nonlinear granulation growth *can* occur in linearly stable plasmas. Such an instability mechanism has already been observed in numerical simulations. Marginality can also be achieved by a state of over-saturated modes, where $\text{Im}\,\epsilon$ is negative but amplitude dependent. In this case, marginality is assumed when Eq. (8.69b) is satisfied. This can occur either via $\langle f_e \rangle$ profile adjustment or by an increase in the magnitude of collective mode dissipation (i.e. adjustment of $\text{Im}\,\epsilon\,(k, \omega_k)$). The physics mechanism allowing this subcritical instability mechanism is inter-species momentum transfer mediated by electron scattering of ion granulations. In this mechanism, the waves that support the 'wakes' of the granulation are damped, so they do not carry significant momentum nor do they play a significant role in the scattering process.

8.4 Phase space structures – a look ahead

Although lengthy, this chapter has only scratched the surface of the fascinating subject of phase space turbulence and phase space structures. We are especially cognizant of our omission of any discussion of intrinsic, dynamic phase space structures, such as solitons, collisionless shocks, BGK modes, double layers, phase space holes, etc. Indeed, our discussion here in Chapter 8, although lengthy is limited only to the *statistical* theory of phase space turbulence, as a logical extension of our treatment of quasi-linear theory (Chapter 3), nonlinear wave–particle scattering (Chapter 4) and nonlinear wave–wave interaction (Chapters 5, 6, 7). We defer discussion of dynamical phase space structures to Volume 2. We defer detailed discussion of the applications of phase space structures to Volume 2, as well. In particular, the subjects of anomalous resistivity and particle acceleration by shocks will be addressed there.

9
MHD turbulence

Yet not every solution of the equation of motion, even if it is exact, can actually occur in Nature.

(L. D. Landau and E. M. Lifshitz, Fluid Mechanics)

9.1 Introduction to MHD turbulence

In this chapter, we explain some of the basic ideas in MHD turbulence and turbulent transport, with special attention to incompressible and weakly compressible dynamics with a mean magnetic field. We emphasize intuition, ideas and basic notions rather than detailed results. Throughout Chapters 2 to 8, the examples were drawn primarily from cases involving electrostatic perturbations. In homogeneous plasmas, the fundamental excitations include electron plasma waves, ion sound waves and Alfvén waves. The electron plasma wave is the best venue to understand microscopic excitations by particle discreteness and particle resonance with collective motions. The ion sound wave is the other fundamental electrostatic mode, and is the simplest that mediates inter-species momentum exchange. In particular, adding inhomogeneity to magnetized plasmas, this wave connects to the drift wave, and the nonlinear evolution of drift waves is the focus of turbulence theory for confined plasmas. In addition, the combination of the plasma wave and ion sound wave defines the problem of disparate-scale interaction, which is a prototype mechanism for structure formation in plasmas. However, Alfvén waves are central to the dynamics of magnetized plasmas in nearly all circumstances. In particular, plasmas in nature are very often threaded by a magnetic field, and the dynamics of plasma evolution is thus coupled to the dynamics of the magnetic field. Therefore, MHD plasma turbulence, in which plasma motion and dynamics of magnetic field interact and are equally important, merits special focus in turbulence theory (Moffat, 1978; Parker, 1979; Montgomery *et al.*, 1979; Krause and Radler, 1980; Taylor, 1986; Biskamp, 1993; Yoshizawa, 1998; Roberts, 2000).

Motivated by these observations, in this chapter, we discuss the plasma turbulence of MHD waves. We place primary emphasis on understanding the case of *magnetized* MHD turbulence, in which a strong, externally fixed, large-scale magnetic field breaks symmetry, produces anisotropy and restricts nonlinear interactions. This limit to be contrasted to MHD turbulence in weakly magnetized (i.e. with only a disordered or small-scale field) or unmagnetized systems.

This chapter is organized into three sections, each of which presents an essential paradigm in MHD turbulence theory. First, the cascade theory of turbulence is discussed in the context of MHD turbulence. The interaction between Alfvén waves and vortical motion is explained by analogy with the theory of nonlinear wave–particle interaction in Vlasov plasmas, discussed in Chapter 4. Several new characteristic spectra in MHD turbulence are explained, and this discussion addresses the fundamental mechanisms for generation of small scales via cascades. Such mechanisms have applications to physical phenomena such as solar wind turbulence, ISM turbulence, and the solar dynamo, etc. Then we visit the physics of disparate-scale interaction in conjunction with the steepening of propagating Alfvén waves. This discussion presents the physics of an alternative route to small scales, namely via coherent phase-front steepening, and so is a natural complement to the discussion of cascades. It is relevant to understanding quasi-parallel Alfvénic shocks, which have been observed in the solar wind and which are also highly relevant to the dynamics of cosmic ray acceleration. In the last section, we encounter the problem of transport of flux and magnetic field in MHD turbulence. This issue appears in the dynamo problem, which has been pursued in order to understand plasma in space and astrophysical objects. The problem of turbulent diffusion of magnetic fields is central to dynamo and reconnection physics, as well as constituting an important application of the theory of mean field electrodynamics. At a fundamental level, the problem of magnetic field diffusion is one of transport in a system with a strong memory. These three sections should be thought of as three related but distinct ways of highlighting key issues in MHD turbulence. These three topics do have a common theme, namely the constraints which Alfvénically induced 'memory' exerts on turbulence and transport. Also, as stressed throughout this book, excitation by nonlinear noise is simultaneously included with coherent drag by nonlinear interaction in calculating transport by MHD turbulence. We show here that simultaneous consideration of these two effects, both of which are necessary to preserve conservation properties, is essential to determining the relevant relation between the magnetic perturbation and kinematic perturbation. The concept of quenching of mean field evolution, especifically the quench of turbulent transport coefficient D by the mean field, is illustrated. Thus, we show that the nonlinear theory and statistical physics considerations discussed throughout this book are relevant to MHD turbulence, as well.

9.2 Towards a scaling theory of incompressible MHD turbulence

In this section, the focus will be exclusively on *incompressible MHD*, which, for a uniform mean magnetic field,

$$\mathbf{B}_0 = B_0\hat{z},$$

is described by the well-known equations for coupled fluid $\mathbf{v}$ and magnetic field $\mathbf{B}$, namely,

$$\frac{\partial \mathbf{v}}{\partial t} + \mathbf{v}\cdot\nabla\,\mathbf{v} = \frac{-\nabla p}{\rho_0} + \frac{B_0}{4\pi\rho_0}\frac{\partial}{\partial z}\mathbf{B} + \frac{\mathbf{B}\cdot\nabla\mathbf{B}}{4\pi\rho_0} + \nu\nabla^2\mathbf{v} + \mathbf{F}_v{}^{\mathrm{ext}}, \tag{9.1a}$$

$$\frac{\partial \mathbf{B}}{\partial t} + \mathbf{v}\cdot\nabla\mathbf{B} = B_0\frac{\partial}{\partial z}\mathbf{v} + \mathbf{B}\cdot\nabla\mathbf{v} + \eta\nabla^2\mathbf{B} + \mathbf{F}_m{}^{\mathrm{ext}}. \tag{9.1b}$$

Here the mass density ρ_0 is constant, and magnetic pressure has been absorbed into p (Biskamp, 1993). Equations (9.1a), (9.1b) describe the evolution of two inter-penetrating fluids ($\mathbf{v}$ and $\mathbf{B}$), which are strongly coupled for large magnetic Reynolds number $R_m \sim v_0\ell_0/\eta$. Equivalently put, $\mathbf{B}$ is 'frozen into' the fluid, up to the resistive dissipation. The system can have two external stochastic forcings $\mathbf{F}_v^{\mathrm{ext}}$ and $\mathbf{F}_m^{\mathrm{ext}}$, though we take $\mathbf{F}_m^{\mathrm{ext}} \to 0$ here. There are two control parameters, R_e and R_m, or equivalently R_m and magnetic Prandtl number $P_m = \nu/\eta$.

9.2.1 Basic elements: waves and eddies in MHD turbulence

For a strongly magnetized system, we are concerned with small-scale turbulence consisting of small amplitude fluctuations with ($|\delta\mathbf{B}| < B_0$) and which are isotropic in the plane perpendicular to $\mathbf{B}_0$. Forcing is taken to be restricted to large scales, and *assumed* to result in a mean dissipation rate ϵ. Note that in contrast to the corresponding hydrodynamic system, MHD turbulence has two components or constituents, namely:

(i) shear Alfvén waves, with frequency $\omega_{\mathbf{k}} = k_{\|}v_{\mathrm{A}}$, where $v_{\mathrm{A}}^2 = B_0^2/4\pi\rho_0$. Note that a single shear Alfvén wave is an *exact* solution of the incompressible MHD equations. In the absence of dissipation or non-Alfvénic perturbations, then, an Alfvén wave will simply persist *ad-infinitum*;

(ii) 'eddies', namely zero frequency hyrdodynamic cells, which do *not* bend magnetic field lines (i.e. have $\mathbf{k}\cdot\mathbf{B}_0 = 0$). Eddies are characterized by a finite self-correlation time or lifetime $\tau_{\mathbf{k}}$. For strong B_0, $k_{\|}v_{\mathrm{A}} > 1/\tau_{\mathbf{k}}$, which is equivalent to $|\delta\mathbf{B}| \ll B_0$.

Note that in MHD, the waves are high frequency with respect to fluid eddies. Thus, as first recognized by Kraichnan and Iroshnikov, *two Alfvén waves must beat together and produce a low frequency virtual mode, in order to interact with fluid eddy turbulence* (Iroshnikov, 1964; Kraichnan, 1965). Such interaction is necessary for *any* cascade to small scale dissipation. Indeed, the generation of such non-Alfvénic perturbations is a key to the dynamics of MHD turbulence!

9.2.2 Cross-helicity and Alfvén wave interaction

It is instructive to discuss an analogy between magnetized MHD turbulence and Vlasov turbulence, as we note that the latter system is a popular paradigm, universally familiar to plasma physicists. Vlasov turbulence consists of two constituents, namely collective modes or 'waves', and 'particles'. For example, ion acoustic turbulence consists of ion acoustic waves, ions and Boltzmann electrons. The analogue in MHD of the 'collective mode' is the Alfvén wave, while the analogue of the 'particle' is the eddy. In both cases, the dispersive character of the collective modes (Alfvén waves are effectively dispersive via anisotropy and because of the existence of counter-streaming populations, since $k_{||} = \pm\mathbf{k}\cdot\mathbf{B}_0/|B_0|$; most plasma waves of interest are also dispersive) implies that strong nonlinear interaction occurs when two waves interact to generate a low frequency 'beat' or virtual mode. In the case of Vlasov turbulence, such a low frequency beat wave may resonate and exchange energy with the particles, even if the primary waves are non-resonant (i.e. have $\omega \gg kv$). This occurs via the familiar process of *nonlinear Landau damping*, which happens when,

$$\omega_k - \omega'_{k'} = (k - k')v. \tag{9.2}$$

In the case of MHD, the frequency and wave number matching conditions for Alfvén wave interaction require that,

$$\mathbf{k}_1 + \mathbf{k}_2 = \mathbf{k}_3, \tag{9.3a}$$

$$k_{||_1} v_A + k_{||_2} v_A = k_{||_3} v_A. \tag{9.3b}$$

Therefore, the only way to generate higher $|\mathbf{k}_\perp|$, and thus smaller scales, through coupling with vortical motion at $\omega \sim 0$ (i.e., small $|k_{||_3}|$) as in a cascade, is to have $k_{||_1} k_{||_2} < 0$, which means that the *two primary waves must be counter-propagating*! Note that counter-propagating waves necessarily generate low-frequency modes, which resemble the quasi-2D eddies or cells referred to earlier. Indeed, for $k_{||_3} v_A \lesssim 1/\tau_{k_3}$, the distinction between these two classes of fluctuations is lost. Hence, in strongly magnetized MHD turbulence, interaction between counter-propagating populations generates smaller *perpendicular* scales, thus initiating a cascade. Note that unidirectional propagating packets *cannot* interact, in incompressible MHD, as each Alfvén wave moves at the same speed and

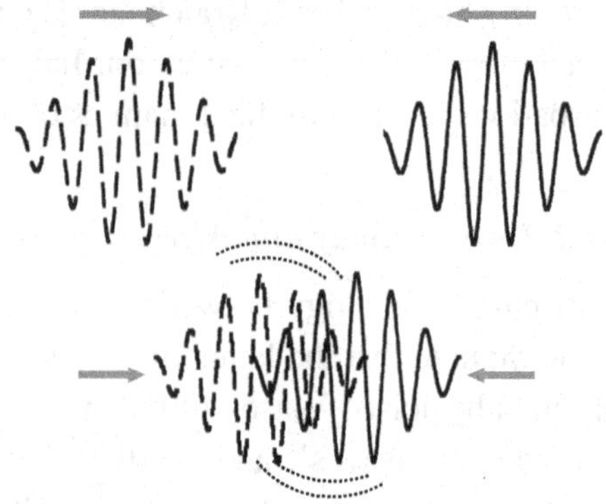

Fig. 9.1. Counter-propagating Alfvén wave streams interact.

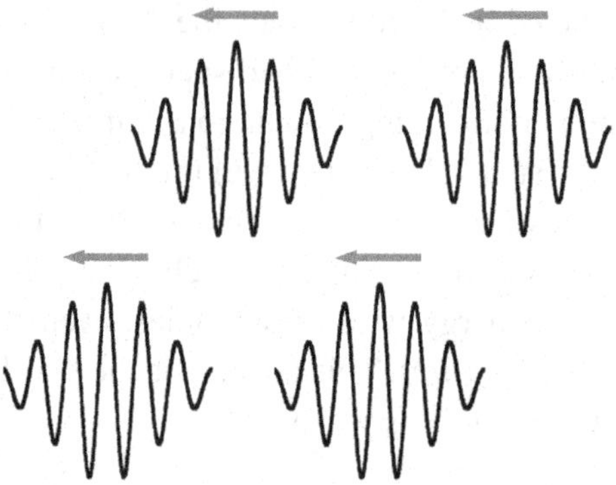

Fig. 9.2. Parallel propagating wave streams do not interact.

is, in fact, an exact solution of the incompressible MHD equations. Instead, two counter-streaming Alfvén populations must pass through one another, in order for "cascading" to occur (see Figures 9.1 and 9.2). This seminal insight is due to Kraichnan and Iroshnikov.

We note here that the requirement of counter-propagating populations constrains the cross-helicity of the system. The Elsasser variables $Z_\pm$, where,

$$Z_\pm = \mathbf{v} \pm \mathbf{B}, \tag{9.4a}$$

each correspond to one of the two Elsasser populations. Note that we use here the Alfvén unit, i.e., the velocity normalized to Alfvén velocity. Then $\mathbf{v}$ and $\mathbf{B}$ appear in a symmetric form. The net imbalance in the two population densities is thus,

$$N_+ - N_- = \mathbf{Z}_+ \cdot \mathbf{Z}_+ - \mathbf{Z}_- \cdot \mathbf{Z}_- = 4\mathbf{v} \cdot \mathbf{B}, \tag{9.4b}$$

where the total cross-helicity is just,

$$H_c = \int \mathrm{d}^3 x \mathbf{v} \cdot \mathbf{B}. \tag{9.4c}$$

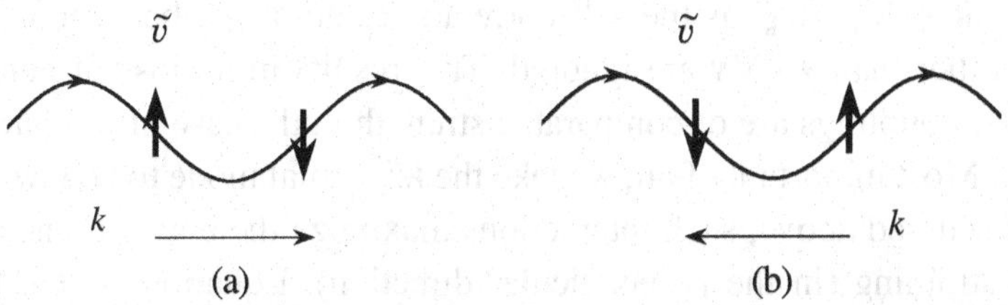

Fig. 9.3. Cross-helicity changes sign depending on the propagation direction. In the case where the wave is propagating in the direction of the magnetic field (a), the perturbed velocity has the opposite sign in comparison with the magnetic field perturbation. The cross-helicity is negative. In contrast, when the wave changes the direction of propagation (b), the cross-helicity becomes positive.

Thus, for a system with counter propagating populations of equal intensity, H_c necessarily must vanish. Similarly, maximal cross-helicity ($|\mathbf{v}\cdot\mathbf{B}| = (|v|^2|B|^2)^{1/2}$) implies that either $N_+ = 0$ or $N_- = 0$, meaning that no Alfvén wave cascade can occur. Hereafter in this section, we take $H_c = 0$.

9.2.3 Heuristic discussion of Alfvén waves and cross-helicity

We now present a heuristic derivation of the MHD turbulence spectrum produced by the Alfvén wave cascade (Craddock and Diamond, 1990; Lazarian and Vishniac, 1999). As in the K41 theory, the critical element is the lifetime or self-correlation time of a particular mode $\mathbf{k}$. Alternatively put, we seek a time scale $\tau_{\mathbf{k}}$ such that,

$$(\mathbf{v}\cdot\nabla\mathbf{v})_{\mathbf{k}} \sim v_{\mathbf{k}}/\tau_{\mathbf{k}}. \tag{9.5}$$

This is most straightforwardly addressed by extracting the portion of the nonlinear mixing term which is *phase coherent* with the 'test mode' of interest. The method of extracting the coherent part is explained in, e.g., Section 5.3.3 or 6.1.3. Thus, we wish to determine,

$$v_{\mathbf{k}}/\tau_{\mathbf{k}} = \mathbf{k}\cdot\sum_{\mathbf{k}'} v_{-\mathbf{k}'}\, v^{(2)}_{\mathbf{k}+\mathbf{k}'}, \tag{9.6a}$$

where $v^{(2)}_{\mathbf{k}+\mathbf{k}'}$ is determined via perturbation theory for Eqs.(9.1a), (9.1b) by solving,

$$\Delta\omega_{k''}\mathbf{v}_{\mathbf{k}''}{}^{(2)} - \frac{ik''_{\|}}{4\pi\rho_0}B_0\mathbf{B}^{(2)}_{\mathbf{k}''} = \mathbf{v}^{(1)}_{\mathbf{k}'}\cdot\mathbf{k}\mathbf{v}^{(1)}_{\mathbf{k}}, \tag{9.6b}$$

$$\Delta\omega_{\mathbf{k}''}\mathbf{B}^{(2)}_{\mathbf{k}''} = B_0 i k''_{\|}\mathbf{v}^{(2)}_{k''}. \tag{9.6c}$$

Here $\mathbf{k}'' = \mathbf{k} + \mathbf{k}'$, $\Delta\omega_{\mathbf{k}''}$ is the self-correlation rate of the beat mode, and nonlinearities other than $\mathbf{v} \cdot \nabla\mathbf{v}$ are ignored. This results in no loss of generality, as all nonlinear couplings are of comparable strength in the case of nonlinear Alfvén interaction. Most important of all, we take the $\mathbf{k}''$ virtual mode to be low frequency since, as discussed above, such interactions maximize the power transfer to small scales via straining (in the perpendicular direction). Equations (9.6a, b, c) then yield,

$$1/\tau_{\mathbf{k}} = \sum_{\mathbf{k}'} |\mathbf{k} \cdot v_{\mathbf{k}'}|^2 \left[\frac{1/\Delta\omega_{\mathbf{k}''}}{1 + (k''_{\parallel} v_{\mathrm{A}}/\Delta\omega_{\mathbf{k}''})^2} \right], \tag{9.7a}$$

which, for $k_z v_A > \Delta\omega_{\mathbf{k}}$, reduces to,

$$1/\tau_{\mathbf{k}} = \sum_{\mathbf{k}'} |\mathbf{k} \cdot v_{\mathbf{k}'}|^2 \pi\delta(k''_{\parallel} v_{\mathrm{A}}). \tag{9.7b}$$

Note that Eq.(9.7b) is equivalent to the estimate $1/\tau_{\mathbf{k}} \sim \sum_{\mathbf{k}'} |\mathbf{k} \cdot v_{\mathbf{k}'}|^2 \tau_{\mathrm{ac}_{\parallel}}$, where $\tau_{\mathrm{ac}_{\parallel}} \sim 1/\left|\Delta k_{\parallel} v_A\right|$ is the auto-correlation time of the Alfvén spectrum. Here, $\Delta k_{\parallel}$ is the bandwidth of the $k_{\parallel}$ spectrum. Of course, the need for counter-propagating populations emerges naturally from the resonance condition (i.e., $k''_{\parallel} \sim 0$). Similarly, anistotropy is clearly evident, in that the coupling coefficients, (i.e. $\mathbf{k}_{\perp} \cdot \mathbf{k}'_{\perp} \times \hat{z}$), depend on $\mathbf{k}_{\perp}$, while the selection rules depend on $k_{\parallel}$. Finally, the correspondence with nonlinear Landau damping in Vlasov turbulence (discussed in Chapters 3 and 4) is also clear. For that process, nonlinear transfer is given by,

$$|E_{\mathbf{k}}|^2/\tau_k \sim \left(\sum_{\mathbf{k}'} |E_{\mathbf{k}'}|^2 F(k, k') \pi\delta(\omega_{\mathbf{k}} + \omega_{\mathbf{k}'} - (k + k')v) v_{\mathrm{T}}^2 \left. \frac{\partial \langle f \rangle}{\partial v} \right|_{v_b} \right) |E_{\mathbf{k}}|^2, \tag{9.8}$$

where $F(k, k')$ refers to a coupling function and interaction occurs at the beat phase velocity $v_b = (\omega + \omega')/(k + k)'$ (see, e.g., Section 3.3.2).

Having derived the correlation time, we can now proceed to determine the spectrum. In the interests of clarity and simplicity, we derive a scaling relation, using the expression for $\tau_{\mathbf{k}}$ given in Eq.(9.7). Despite the fact that,

(i) there are no a-priori theoretical reasons or well-documented experimental evidence that energy transfer in MHD turbulence is local in **k**,
(ii) there is no rigorously established reason whatsoever to expect that the (as yet unproven!) finite time singularity which underlies the independence of ϵ from dissipation in hydrodynamic turbulence should necessarily persist in MHD,

we intend to plunge ahead and write a cascade energy transfer balance relation.[1] Anticipating the role of anisotropy, the transfer balance relation at each scale $l_\perp$ is,

$$\epsilon = v(\ell_\perp)^2/\tau(\ell_\perp), \tag{9.9a}$$

where,

$$1/\tau(\ell_\perp) = 1/\tau_{\mathbf{k}} = \sum_{\mathbf{k}'} |k \cdot v_{\mathbf{k}'}|^2 \pi\delta(k''_{||} v_A) \cong \frac{1}{\ell_\perp^2} \frac{v(\ell_\perp)^2}{k_{||} v_A}, \tag{9.9b}$$

so,

$$\epsilon = \frac{v(\ell_\perp)^4}{\ell_\perp^2 k_{||} v_A}. \tag{9.9c}$$

Equation (9.9c) can be arrived at by the even simpler reasoning that, as is generic in weak turbulence theory, the energy transfer will have the form,

$$\begin{aligned} \epsilon \sim &\ (\textit{coupling coefficient})^2 * (\textit{interaction time}) \\ &* (\textit{scatter-er energy}) * (\textit{scatter-ee energy}). \end{aligned} \tag{9.10}$$

Taking the coupling $\sim 1/\ell_\perp$, interaction time $\sim 1/k_{||} v_A$, and scatter-er and 'scatter-ee' energy $\sim v(\ell_\perp)^2$ then yields Eq.(9.9c).

In comparison to the relation $\epsilon = v(\ell)^3/\ell$ for K41 turbulence in a neutral fluid, Eq.(9.9c) contains two new elements, namely:

(a) anisotropy – the clear distinction between perpendicular and parallel directions remains;
(b) reduction in transfer rate – notice that in comparison to its hydrodynamic counterpart, energy transfer in MHD turbulence $(v_\perp/l_\perp)$ is reduced by a factor of $v_\perp/\ell_\perp k_{||} v_A$, the ratio of a parallel Alfvén transit time to a perpendicular eddy shearing rate, which is typically much less than unity. The reduction in transfer rate in comparison to hydrodynamic turbulence is commonly referred to as the *Alfvén effect*. The Alfvén effect is a consequence of the enhanced memory of MHD turbulence, as compared to that in hydrodynamic turbulence.

9.2.4 MHD turbulence spectrum (I)

It is now possible to consider several related cases and incarnations of the MHD cascade. First, we revisit the original paradigm of Kraichnan and Iroshnikov (Kraichnan, 1965; Iroshnikov, 1964). Here, we consider a weakly magnetized system, where,

[1] The old proverb, "Fools rush in, where angels fear to tread" comes vividly to mind at this point. However, so does another ancient aphorism, "Nothing ventured, nothing gained".

$$B_{\rm rms} \gg B_0.$$

Note that in contrast to hydrodynamics, Alfvénic interaction in MHD is *not constrained* by Galilean invariance. Thus, Eq.(9.9c) applies, with $B_0 \rightarrow B_{\rm rms} = \langle \tilde{B}^2 \rangle^{1/2}$. Furthermore, as there is no *large-scale* anisotropy (B_0 is taken to be negligible!), we can dare to take $k_{||}\ell_\perp \sim 1$, so that the energy transfer balance (Eq.(9.9c)) becomes,

$$\epsilon \sim v(\ell)^4/\ell\tilde{v}_{\rm A}, \tag{9.11a}$$

where $l \sim k_{||}^{-1} \sim l_\perp$ and $\tilde{v}_{\rm A} = v_{\rm A}$ computed with $\tilde{B}_{\rm rms}$. The value of $B_{\rm rms}$ is dominated by the large eddies, and is sensitive to the forcing distribution and the geometry. In this system, the *rms* field is not straight, but does possess some large-scale local order. Thus, here the 'Alfvén waves' should be thought of as propagating along a large-scale field with some macroscopic correlation length along with a stochastic component. This in turn (via Eq.(9.11a)) immediately gives,

$$v(\ell) \sim \ell^{1/4}(\epsilon\tilde{v}_{\rm A})^{1/4}, \tag{9.11b}$$

and,

$$E(k) \sim (\epsilon\tilde{v}_{\rm A})^{1/2}k^{-3/2}, \tag{9.11c}$$

where we use the normalization $\int {\rm d}k E(k) = $ Energy. Equation (9.11c) gives the famous Kraichnan–Iroshnikov (K–I) spectrum for weakly magnetized incompressible MHD turbulence. Concomitant with the departure from $k^{-5/3}$, reconsidering the onset of dissipation when (for $P_m = 1$) $\nu/\ell_d^2 = v(\ell_{\perp d})/\ell_{\perp d}$ gives the K–I dissipation scale,

$$\ell_{\perp d} = \nu^{2/3}(\tilde{v}_{\rm A}/\epsilon)^{1/3}.$$

Notice that, this argument is not specific to three dimensions. Indeed, since the $\mathbf{J} \times \mathbf{B}$ force breaks enstrophy conservation for inviscid 2D MHD, a *forward* cascade of energy is to be expected even in 2D, *ab initio*. Thus, it is not completely surprising that the results of detailed, high resolution numerical simulations of 2D MHD turbulence are in excellent agreement with both the K–I spectrum and dissipation scale (Biskamp and Welter, 1989). The success of the K–I theory in predicting the properties of weakly magnetized 3D MHD will be discussed later in Section 9.2.6. Finally, we note that two rather subtle issues have been 'swept under the rug' in this discussion. First, the large-scale field $\tilde{B}_{\rm rms}$ is tangled, with zero *mean* direction but with a *local coherence length* set by the turbulence integral scale. Thus, while there is no *system-averaged* anisotropy, it seems likely that strong *local* anisotropy will occur in the turbulence. The theory does not account for this local anisotropy. Second, it is reasonable to expect that some minimum value of $\tilde{B}_{\rm rms}$ is necessary to arrest the inverse energy cascade, characteristic of 2D hydrodynamics, and to generate a forward cascade. (The scaling of this $\tilde{B}_{\rm rms}$ and possible dependence on the forcing scale and magnetic Prandtl number $P_{\rm m}$ are as yet unresolved.)

9.2.5 MHD turbulence spectrum (II)

We now turn to the case of strongly magnetized, anisotropic turbulence,

$$B_{\rm rms} \ll B_0.$$

In that case, Eq.(9.9c) states the energy flux balance condition, which is,

$$\epsilon \sim \frac{1}{\ell_\perp^2} \frac{v(\ell_\perp)^4}{k_{||} v_{\rm A}}. \tag{9.12}$$

Here again $v(\ell_\perp)/(\ell_\perp k_{||} v_{\rm A}) < 1$. Now using the normalization for an anisotropic spectrum where (Energy $= \int {\rm d}k_{||} \int {\rm d}k_\perp E(k_{||}, k_\perp)$), Eq.(9.12) directly suggests that,

$$E(k_{||}, k_\perp) \sim (\epsilon k_{||} v_{\rm A})^{1/2} / k_\perp^2, \tag{9.13a}$$

a steeper inertial range spectrum than that predicted by K–I for the weakly magnetized case. Note that consistency with the ordering $|\delta \mathbf{B}| < B_0$, or equivalently $v(\ell_\perp)/\ell_\perp \le k_{||} v_{\rm A}$, requires that,

$$\ell_\perp^{1/3} \epsilon^{1/3} / v_{\rm A} \le k_{||} \ell_\perp << 1, \tag{9.13b}$$

symptomatic of the *anisotropic cascade* of Goldreich and Sridhar (G–S) (1995); (1997).

It is interesting to note that Eq.(9.14) says that the *anisotropy increases as the cascade progresses toward smaller scales*, so that initially spheroidal eddies on integral scales produce progressively more prolate and extended (along B_0) eddies on smaller (cross-field) scales, which ultimately fragment into long, thin cylindrical 'rods' on the smallest inertial range scales. This anisotropic cascade process is

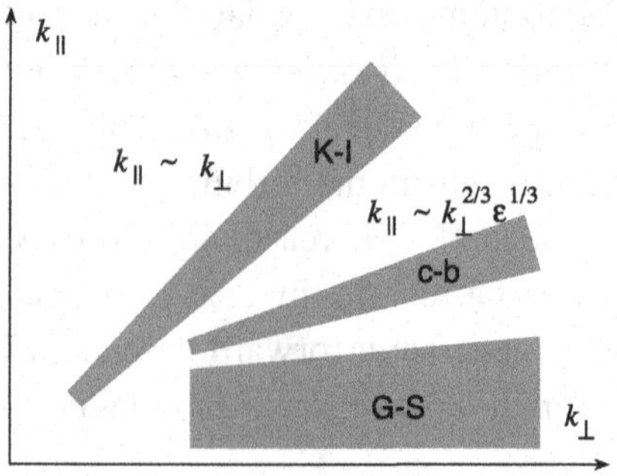

Fig. 9.4. Conceptual illustration for the regimes of the Kraichnan–Iroshnikov (K–I) spectrum and Goldreich–Sridhar (G–S) spectrum. The critically-balanced regime (c-b) is also illustrated.

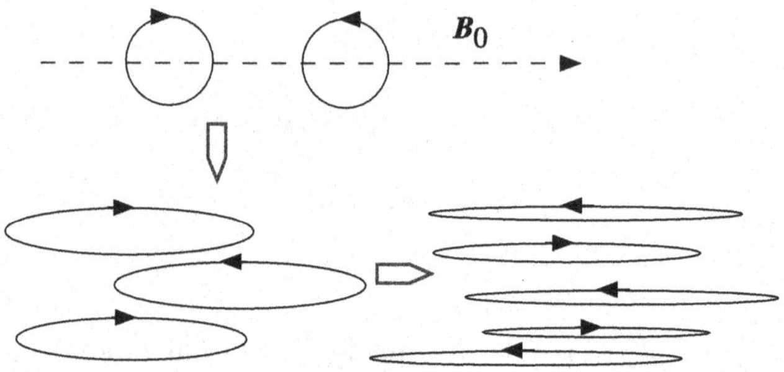

Fig. 9.5. Comparison of the isotropic Kolmogorov cascade with the anisotropic Alfvén turbulence cascade. In the latter case, anisotropy increases as the cascade progresses.

compared to the isotropic eddy fragmentation picture of Kolmgorov in Figure 9.5. Recognition of the intrinsically anisotropic character of the strongly magnetized MHD cascade was the most important contribution of the series of papers by Goldreich and Sridhar.

A particularly interesting limit of the anisotropic MHD cascade is the "critically balanced" or "marginally Alfvénic" cascade, which occurs in the limiting case where $v_{\perp}(\ell_{\perp})/\ell_{\perp} \sim k_{\parallel}v_{A}$, i.e. when the parallel Alfvén wave transit time through an (anisotropic) eddy is equal to the perpendicular straining or turn-over time of that eddy. In physical terms, "critical balance" may be thought of as a state of marginal stability to wave breaking, since critical balance implies that the characteristic nonlinear rate $v\,(l_{\perp})\,/l_{\perp}$ just balances the wave frequency $v_{A}/l_{\parallel}$, so that the exciton lives just at the boundary between a "wave" and an "eddy". Note that a state of critical balance is rather similar to the familiar mixing length limit in drift wave turbulence (discussed in Chapters 4 and 6), where $\omega_k \sim k_\perp \tilde{v}_{E\times B}$ so $e\tilde{\phi}/T \sim 1/k_\perp L_n$. In this limit, Eq.(9.12) reduces to,

$$\epsilon \sim v(\ell_\perp)^3/\ell_\perp, \tag{9.14a}$$

(i.e. back to K41!) albeit with rather different physics. Thus in the critically balanced cascade,

$$E(k_\perp) \cong \epsilon^{2/3} k_\perp^{-5/3}, \tag{9.14b}$$

and,

$$k_{||}\ell_\perp \cong \ell_\perp^{1/3}\epsilon^{1/3}/v_{\mathrm{A}},$$

so that,

$$k_{||} \sim k_\perp^{2/3}\epsilon^{1/3}/v_{\mathrm{A}},$$

which defines a trajectory or 'cone' in **k** space along which the cascade progresses. On this cone, one has the spectrum $E(k_\perp) \sim k_\perp^{-5/3}$. Note that for $v(\ell_\perp)/\ell_\perp > k_{||}v_A$, the turbulence shearing rate exceeds the Alfvén transit rate, so the dynamics are effectively 'unmagnetized' and so the spectrum *should* approach that of K–I in that limit.

Thus, in an Alfvénic cascade, eddy anisotropy increases at small scale (i.e. $k_{||}/k_\perp \sim k_\perp^{-1/3}$), so smaller eddies become more elongated along B_0 than larger eddies do. Equation (9.14b) gives the resulting spectrum for a critically balanced, anisotropic Alfvén cascade. This result was first obtained by Goldreich and Sridhar, who extended the pioneering studies of Kraichnan and Iroshnikov to the anisotropic, strongly magnetized regime. The Goldreich–Sridhar spectrum (i.e. Eq.(9.14b)) is at least semi-quantitatively consistent with the observed spectrum of ISM scintillations, better known as the "great power law in the sky" (see also Kritsuk *et al.* (2007)). The details of the Alfvén cascade remain an active area of research today (Figure 9.6).

9.2.6 An overview of the MHD turbulence spectrum

We can summarize this zoology of MHD turbulence spectra by considering a magnetized system with fixed $\nu = \eta$ and variable forcing. As the forcing strength increases, so that ϵ increases at fixed B_0, ν, η, the turbulence spectra shows

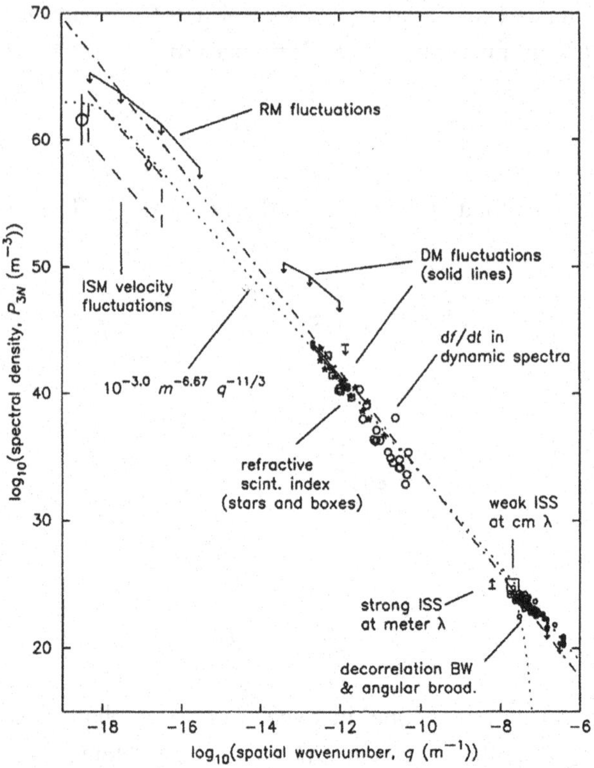

Fig. 9.6. An example of the interstellar density power spectrum in the tenuous phase of the interstellar medium. A power law with index close to the Kolmogorov value of 11/3 is shown by the dotted line (see the original article for details: quoted from Armstrong *et al.* (1995)).

a transition through three different stages. These three stages correspond, respectively, to:

(i) first, the anisotropic cascade, with $E(k_\perp) \sim (\epsilon k_\| v_A)^{1/2}/k_\perp^2$ and $k_\| \ell_\perp > \ell_\perp^{1/3}\epsilon^{1/3}/v_A$ throughout the inertial range, then;
(ii) the critically balanced anisotropic cascade, with $E(k_\perp) \sim \epsilon^{2/3} k_\perp^{-5/3}$ and $k_\| \sim k_\perp^{2/3}\epsilon^{1/3}/v_A$ throughout the inertial range, and finally;
(iii) the weakly magnetized cascade for $B_{rms} > B_0$, with $E(k_\perp) \sim (\epsilon\, \tilde{v}_A)^{1/2} k^{-3/2}$ and $\mathbf{k}$ isotropic, on average.

Note that the spectral power law index *decreases* with increasing stirring strength, at fixed B_0.

These theoretical limits are compared to numerical calculations. As discussed before, the weakly magnetized K–I cascade theory is quite successful in explaining 2D MHD turbulence at moderate R_e with $P_m = 1$. Three numerical calculations for strong B_0 in 3D have recovered results which agree with the predictions of

Goldreich and Sridhar, albeit only over intervals of scale of roughly a decade (Cho *et al.*, 2002; Maron *et al.*, 2004). Interestingly, the numerical study with the best resolution to date yields a spectrum which appears closer (for strongly magnetized 3D!) to the K–I-like prediction of $E(k_\perp) \sim k_\perp^{-3/2}$ than the G–S predictions (Muller *et al.*, 2003). The deviation from G–S scaling may be due to intermittency corrections, to alignment effects, as suggested by (Boldyrev, 2006), or to a more fundamental departure from the physical picture of G–S. In particular, it is tantalizing to speculate that the $E(k_\perp) \sim k_\perp^{-3/2}$ scaling at strong B_0 suggests that the turbulence assumes a quasi-2D structure consisting of extended columns along B_0. The viability of this speculation would be strengthened by observation of a departure from the accompanying $k_{||} \sim k_\perp^{2/3}$ scaling also predicted by G–S, although perpendicular versus parallel anisotropy clearly remains. In physical terms it seems plausible that the turbulence might form such a quasi-2D state, since:

(i) A state of extended columns aligned with the strong $\mathbf{B}_0$ is the 'Taylor–Proudman state' for the system. Such a state naturally minimizes the energy spent on magnetic field line bending, which is necessary for Alfvén wave generation.
(ii) A state of extended, field-aligned columns which are re-arranged by approximately horizontal eddy motions is also the state in which the translational symmetry along B_0, which is broken by the excitation mechanism, is restored to maximal extent.

Thus, formation of such a quasi-2D state seems consistent with considerations *of both energetics and probability*. Further detailed study of the $k_{||}$ and $k_\perp$ spectra is required to clarify the extent and causes of the apparent two dimensionalization. This issue is one of the most fundamental ones confronting researchers in MHD turbulence today.

Of course, there is a lot more to understanding MHD turbulence than simply computing spectral indexes. The nature of the dissipative structures in 3D MHD turbulence remains a mystery, and the dynamical foundations of intermittency effects are not understood. In 2D, numerical studies suggest that inertial range energy may be dissipated in current sheets, but much further study of this phenomenon is needed. In both 2D and 3D, the structure of the probability distribution function of hydrodynamic and magnetic strain (i.e. $\nabla \mathbf{v}$ and $\nabla \mathbf{B}$) at high R_m and R_e remains *terra incognita*. Dynamical alignment, namely the tendency to spontaneously form localised domains where $|\mathbf{Z}_+| \gg |\mathbf{Z}_-|$ or $|\mathbf{Z}_-| \gg |\mathbf{Z}_+|$, can surely modify the Alfvén cascade locally, thus generating intermittency. Our understanding of the physics of domain formation is still quite rudimentary. Finally, the dependence of the large-scale structure of $B_{\rm rms}$ upon stirring properties, geometry, etc. has not been addressed. Note that this structure is ultimately responsible for the breaking of local rotational symmetry and the origin and extent of domains of local anisotropy in 2D MHD turbulence.

9.3 Nonlinear Alfvén waves: compressibility, steepening and disparate-scale interaction

The nonlinear evolution of Alfvén waves in the absence of counter-propagating wave streams is an important question. This is because many physical situations and systems *do* involve nonlinear Alfvén dynamics but *do not* have counter-propagating wave streams of remotely comparable intensity. Indeed, any situation involving emission of Alfvén waves from an astrophysical body (i.e. star) falls into this category. The answer, of course, is that introduction of even modest compressibility (i.e. parallel compressibility, associated with acoustic perturbations) is sufficient to permit the *steepening* of *uni-directional* shear Alfvén wave packets (Cohen and Kulsrud, 1974). Wave steepening then generates small scales by the familiar process of shock formation. Steepening terminates in either dissipation at small scales, as in a dissipative or collisional shock, or the arrest of steepening by dispersion, as in the formation of a collisionless shock or solitary wave. Alfvén wave steepening is thus the 'mechanism of (nature's) choice' for generating small scales in uni-directional wave spectra, and naturally complements the mechanism of low-frequency beat generation (Section 9.2), which is the key to the Alfvénic wave cascade in counter-propagating wave streams. Quasi-parallel Alfvén wave steepening is especially important to the dynamics of the solar wind, since high intensity streams of outgoing Alfvén waves are emitted from solar coronal holes. These high intensity wave streams play a central role in generating and heating the 'fast solar wind'.

9.3.1 Effect of small but finite compressibility

We now present a simple, physical derivation of the theory of Alfvén wave steepening due to parallel compressibility. Alfvén wave trains steepen in response to modulations in density. The density dependence of the wave speed (here the Alfvén speed) is the focus of modulational coupling, and the method of disparate scale interaction by modulational interaction (Chapter 7) is employed. So, starting from the Alfvén wave dispersion relation,

$$\omega = k_{\|} v_A = k_{\|} B_0 / \sqrt{4\pi(\rho_0 + \tilde{\rho})}, \tag{9.15a}$$

where a localized mass density perturbation $\tilde{\rho}$ enters the wave speed, and applying the argument in Section 7.2.1 (for the subsonic limit), a straightforward expansion gives an 'envelope' equation for the slow space and time variation of the wave function of the perturbation δB, i.e.

$$\frac{\partial \delta B}{\partial t} = -\frac{v_A}{2} \frac{\partial}{\partial z} \left(\frac{\tilde{\rho}}{\rho_0} \delta B \right). \tag{9.15b}$$

We understand that, in the spirit of reductive perturbation theory, $\tilde{\rho}/\rho_0$ in Eq.(9.15b) is second order in perturbation amplitude. Here, "perturbation" refers to a modulation of the uni-directional Alfvén wave train. We assume that this modulation has parallel scale $L_{||} > 2\pi/k_{||}$. Function $\tilde{\rho}/\rho_0$ is easily determined by considering the parallel flow dynamics. In addition to the linear acoustic force, parallel forces are also induced by the gradient of the carrier Alfvén wave energy field. These terms are calculated by using expressions,

$$\mathbf{v} \cdot \nabla \mathbf{v} = \nabla \frac{|v|^2}{2} - \mathbf{v} \times \omega, \tag{9.16a}$$

$$\mathbf{J} \times \mathbf{B} = -\nabla \frac{|B|^2}{2} + \mathbf{B} \cdot \nabla \mathbf{B}. \tag{9.16b}$$

Noting the relation $\hat{z} \cdot (\mathbf{v} \times \vec{\omega}) = \hat{z} \cdot (\mathbf{B} \cdot \nabla \mathbf{B}) = 0$, to second order, we have the second-order terms on the right-hand side of Eq.(9.1a) as,

$$\frac{\partial}{\partial t} v_{||} = -c_s^2 \frac{\partial}{\partial z} \frac{\tilde{\rho}}{\rho_0} - \frac{\partial}{\partial z} \left(\frac{|\delta B|^2}{8\pi \rho_0} + \frac{|\delta v|^2}{2} \right), \tag{9.16c}$$

where δv is a modulation of the amplitude of velocity of the carrier Alfvén wave. The parallel gradient of the ponderomotive pressure of the Alfvén wave train drives the parallel flow perturbation, which then couples to the density perturbation. The feedback loop of couplings is closed by the linearized continuity equation relating $v_{||}$ to $\tilde{\rho}/\rho_0$, i.e.

$$\frac{\partial}{\partial t} \frac{\tilde{\rho}}{\rho_0} = -\frac{\partial}{\partial z} \tilde{v}_{||}. \tag{9.16d}$$

Equations (9.16c) and (9.16d) may then be combined to obtain,

$$\left(\frac{\partial^2}{\partial t^2} - c_s^2 \frac{\partial^2}{\partial z^2} \right) \frac{\tilde{\rho}}{\rho_0} = \frac{\partial^2}{\partial z^2} \left(\frac{|\delta B|^2}{4\pi \rho_0} \right), \tag{9.16e}$$

where we have used the fact that $v_\perp \sim \delta B/\sqrt{4\pi\rho_0}$ for Alfvén waves.

At this point, it is convenient to transform to a frame of reference co-moving with the Alfvén carrier wave, so that $\tilde{\rho} = \tilde{\rho}(z - v_A t)$, etc. In this frame, we can simplify Eq.(9.16e) to,

$$\tilde{\rho}/\rho_0 = \frac{1}{(1-\beta)} \left(\frac{|\delta B|^2}{B_0^2} \right), \tag{9.17a}$$

where $\beta = c_s^2/v_A^2 = 8\pi p_{th}/B_0^2$. Substituting $\tilde{\rho}/\rho_0$ into the wave equation for δB (9.15b) gives,

$$\frac{\partial}{\partial t}\delta B + \frac{\partial}{\partial z}\left[\frac{v_A}{2(1-\beta)}\left(\left|\frac{\delta B}{B_0}\right|^2 \delta B\right)\right] = 0. \tag{9.17b}$$

As mentioned above, the fast Alfvénic dependence of δB has already cancelled, so this equation almost fully describes the slow dependence of the perturbation envelope.

The physics of the steepening process encapsulated by the back-of-an-envelope (albeit a large one!) calculation presented here can also be described graphically, by a series of sketches, as shown in Figures 9.7 and 9.8. The unperturbed Alfvén wave train is shown in Figure 9.7(a), and its modulation (a parallel rarefaction) is shown in Figure 9.7(b). The modulation induces a perturbation in the ponderomotive energy field of the wave train, which in turn produces a ponderomotive force couple (i.e. dyad) along $\mathbf{B}_0$, as shown in Figure 9.8(a). Note that the resulting parallel flow is yet another example of a Reynolds stress-driven flow, though in this case, the flow is *along* $\mathbf{B}_0$ and the *diagonal* component of the Reynolds stress tensor is at work, symptomatic of the fact that the flow is compressible. The resulting

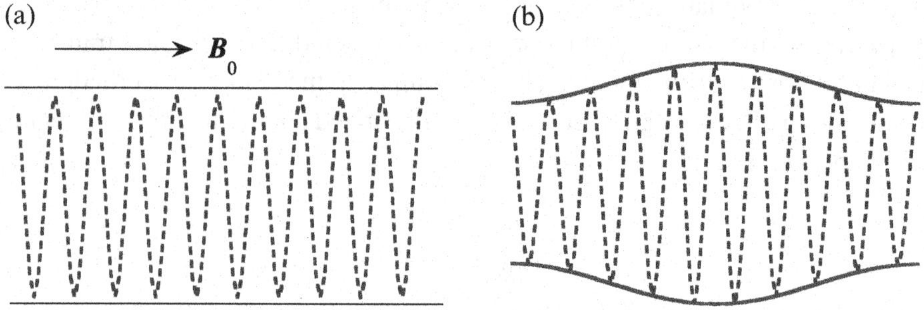

Fig. 9.7. (a) Unperturbed wave train and its envelope. (b) Localized modulational perturbation.

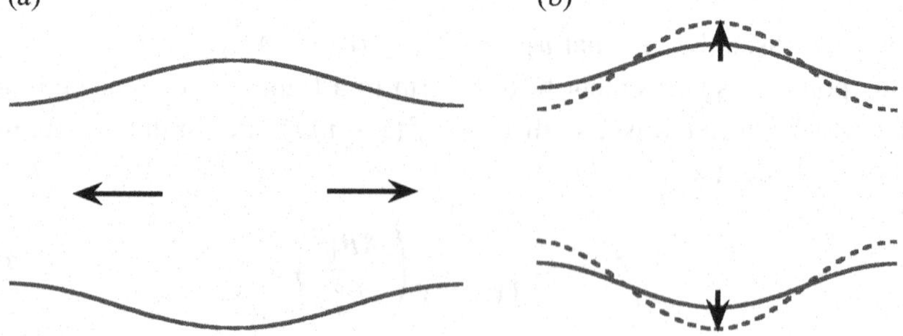

Fig. 9.8. (a) Force couple along $\mathbf{B}_0$. (b) Growth of modulation and steepening of initial perturbation.

parallel flow re-enforces δB via $\nabla \times \mathbf{v} \times \mathbf{B}$, as depicted in Figure 9.8(b), thus enhancing the initial modulation.

Equation (9.17b) describes the steepening of an Alfvén wave train. One more ingredient is necessary, however, namely a term which represents possible limitation and saturation of the steepening, once it generates sufficiently small scale. This is accomplished by adding a diffusion and/or dispersion term to Eq. (9.17b), such as $\eta \partial^2 \delta B/\partial z^2$ or $i d_i^2 \omega_{\mathrm{ci}} \partial^2 \delta B/\partial z^2$, respectively. In that case, the envelope equation for δB becomes the well-known derivative nonlinear Schrödinger (DNLS) equation,

$$\frac{\partial}{\partial t}\delta B + \frac{\partial}{\partial z}\left(\frac{v_{\mathrm{A}}}{2(1-\beta)}\left|\frac{\delta B}{B_0}\right|^2 \delta B\right) = \eta \frac{\partial^2}{\partial z^2}\delta B + i d_i^2 \omega_{\mathrm{ci}} \frac{\partial^2}{\partial z^2}\delta B. \qquad (9.18)$$

Here $d_i = c/\omega_{\mathrm{pi}}$, the ion inertial scale, and ω_{ci} is the ion cyclotron frequency (Rogister, 1971). This dispersion follows from conversion of the Alfvén wave to whistler and ion cyclotron waves (see Figure 9.9) (Stix, 1992). In most expositions and discussions, the resistive dissipation term is dropped, and ion inertial scale physics (associated with Hall currents, etc.) is invoked to saturate Alfvénic steepening by dispersion. Thus, the stationary width of a modulated Alfvén wave train is set by the balance of steepening with dispersion, and so the steepened Alfvén wave packet is often referred to as a *quasi-parallel Alfvénic collisionless shock.* In contrast to systems with counter-propagating Alfvén streams, in a uni-directional wave train modulations can generate small scales via a *coherent* process of wave envelope steepening, which is ultimately terminated via balance with small-scale dispersion.

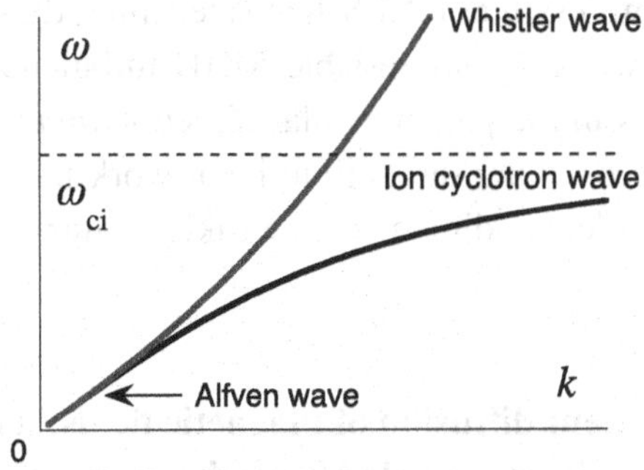

Fig. 9.9. Linear dispersion of low-frequency electromagnetic waves which are propagating in the direction of a magnetic field. When the wavelength is much longer than the ion-skin depth (or frequency is much smaller than the ion cyclotron frequency), both the right- and left-polarized waves appear as Alfvén waves. In a regime of short wavelength, the Alfvén wave continues to a whistler wave or ion cyclotron wave.

9.3.2 A short note, for perspective

Equation (9.18) stimulates the question, what happens when $\beta \to 1$?! This natural question touches on two interesting issues in the theory of Alfvénic steepening. First, it should be readily apparent that the crucial nonlinear effect in this story is the second-order parallel flow, driven by the parallel pondermotive force. Thus, any dissipation, dephasing, etc. such as parallel viscosity, Landau damping, etc., (which are surely present but not explicitly accounted for in this discussion) immediately resolves the $\beta \to 1$ singularity and can also be expected to have an impact on the steepening process for a range of β values. An extensive literature on the important topic of dissipative and kinetic modifications to the DNLS theory exists. One particularly interesting generalization of the DNLS is the kinetic nonlinear Schrodinger equation (KNLS) or the k-derivative – NLS (KDNLS) (Medvedev and Diamond, 1997; Passot and Sulem, 2003). A second point is that for $\beta = 1$, the sound and Alfvén speeds are equal, so it no longer makes sense to 'slave' the density perturbation to the Alfvén wave. Rather, the acoustic and Alfvén dynamics must be treated on an equal footing, as in the analysis in Hada (1994).

The moral of the argument in this subsection is that one should take care to avoid tunnel-vision focus on incompressible theory, alone. Indeed, in this section, we have seen that introducing weak compressibility *completely* changed the nonlinear Alfvén wave problem, by:

(i) allowing strong nonlinear interaction and wave steepening, leading to the formation of shocks, solitons and other structures;
(ii) allowing a mechanism for the nonlinear evolution of a uni-directional wave train.

One might more profitably expect that most natural Alfvénic turbulence phenomena will involve some *synergism* between the incompressible dynamics as in K–I, G–S and the compressible, DNLS-like steepening dynamics. Indeed, recent numerical studies of weakly compressible MHD turbulence have shown *both* a cascade to small scales in the perpendicular direction *and* the formation of residual DNLS-like structures along the field to be at work in nonlinear dynamics! A complete theoretical understanding of such weakly compressible MHD turbulence remains elusive.

9.4 Turbulent diffusion of magnetic fields: a first step in mean field electrodynamics

9.4.1 A short overview of issues

In this section, we discuss the mean field theory of diffusion of magnetic fields in 2D and 3D incompressible MHD. We focus primarily on the simpler 2D problem, for which the effects of diffusion are not entangled with those of field growth by dynamo action. There are many close analogies between mean flux diffusion in 2D and that of the mean field α-effect in 3D (Moffat, 1978; Krause and Radler, 1980; Yoshizawa, 1998; Roberts, 2000; Diamond *et al.*, 2005a).[2] We discuss the similarities and differences between the two problems, with the goal of eventually developing insight into the more interesting (but difficult!) 3D α-effect problem on the basis of the simpler 2D diffusion problem. Note, however, that discussion of the alpha effect and dynamo theory is deferred to Volume 2. Throughout our discussion, magnetic Prandtl number (P_m) of unity (i.e. $\nu = \eta$) and periodic boundary conditions, are assumed, unless otherwise explicitly noted. It is worth mentioning that interesting questions arise as to the resulting behaviour when either of these assumptions is relaxed.

It is appropriate to explain here why we embark on this lengthy theoretical discussion of the decidedly academic problem of flux diffusion in 2D turbulence. To see the motivation, recall that irreversibility, its origins and its role in turbulent transport have been themes which recur throughout this book. In Chapter 3, we discussed quasi-linear theory, and the important roles of dispersion, particle stochasticity, etc. in transport. In Chapter 5, we discussed three-wave resonance, its overlap and its role in nonlinear interaction models. In Chapter 6, we introduced renormalization and self-induced memory loss by nonlinear scrambling, as parametrized by the resonance broadening decorrelation factor. Here, in MHD, we encounter perhaps the simplest example of a *system with topological memory*, due to the constraints imposed by the freezing-in law. Note that the freezing-in law guarantees that for a mean $\langle \mathbf{B} \rangle$ and at high magnetic Reynolds number:

(i) displaced fluid elements will have a memory of the magnetic line they are attached to;
(ii) they cannot slip relative to that line, except via the collisional resistivity;
(iii) fluctuations will tend to Alfvénize, namely equilibrate fluid and magnetic energies.

All of (i)–(iii) suggest that cross-field transport of flux in 2D MHD is an important paradigm topic, and one worthy of extensive discussion. It is a natural complement to the examples we have previously encountered. Thus, here we pursue its exposition, with special focus on the effects of memory and the ultimate importance of the collisional resistivity, even at high Rm.

9.4.2 Flux diffusion in a two-dimensional system: model and concepts

The familiar equations of 2D MHD are deduced from Eq. (9.1) by imposing the relation $\partial/\partial z = 0$ as,

[2] The mean electro-motive force, which is in parallel to $\mathbf{B}$, is called the α-dynamo effect.

$$\frac{\partial \nabla^2 \Phi}{\partial t} + \nabla \Phi \times \hat{\mathbf{z}} \cdot \nabla \nabla^2 \Phi = \nabla A \times \hat{\mathbf{z}} \cdot \nabla \nabla^2 A + \nu \nabla^2 \nabla^2 \Phi, \tag{9.19a}$$

$$\frac{\partial A}{\partial t} + \nabla \Phi \times \hat{\mathbf{z}} \cdot \nabla A = \eta \nabla^2 A, \tag{9.19b}$$

where A is the magnetic potential ($\mathbf{B} = \nabla \times A\hat{\mathbf{z}}$), Φ is the velocity stream function ($\mathbf{v} = \nabla \times \Phi\hat{\mathbf{z}}$), η is the resistivity, ν is the viscosity, and $\hat{\mathbf{z}}$ is the unit vector orthogonal to the plane of motion. Note also that Alfvén units are employed for velocity. We shall consider the case where the mean magnetic field is in the y-direction, and is a slowly varying function of x. Eqs.(9.19a) and (9.19b) have non-dissipative quadratic invariants,

$$\text{the energy } E = \int [(\nabla A)^2 + (\nabla \Phi)^2] \mathrm{d}^2 x,$$

$$\text{mean-square magnetic potential } H_A = \int A^2 \mathrm{d}^2 x,$$

and,

$$\text{cross helicity } H_c = \int \nabla A \cdot \nabla \Phi \mathrm{d}^2 x.$$

Throughout this chapter, we take $H_c = 0$ *ab initio*, so there is no net Alfvénic alignment in the MHD turbulence. The effects of cross-helicity on MHD turbulence are discussed elsewhere in the literature (Grappin *et al.*, 1983; Yoshizawa, 1998).

The basic dynamics of 2D MHD turbulence are explained in Section 9.2. For large-scale stirring, energy is self-similarly transferred to small scales and eventually dissipated via an Alfvénized cascade, as originally suggested by Kraichnan (1965) and Iroshnikov (1964). The Kraichnan–Iroshnikov spectrum for the MHD turbulence cascade is the same in 2D as in 3D. This cascade may manifest anisotropy in the presence of a strong mean field in 3D, as predicted by Goldreich and Sridhar (1995; 1997). Mean square magnetic potential H_A, on the other hand, tends to accumulate at (or inverse-cascade toward) large scales, as is easily demonstrated by equilibrium statistical mechanics for non-dissipative 2D MHD (Fyfe and Montgomery, 1976). The explanation from Section 6 can be utilized. Here, H_A is the second conserved quadratic quantity (in addition to energy), which thus suggests a dual cascade, via a "two-temperature" statistical equilibrium spectrum.

In 2D, the mean field quantity of interest is the spatial flux of magnetic potential,

$$\Gamma_A = \langle v_x A \rangle. \tag{9.20}$$

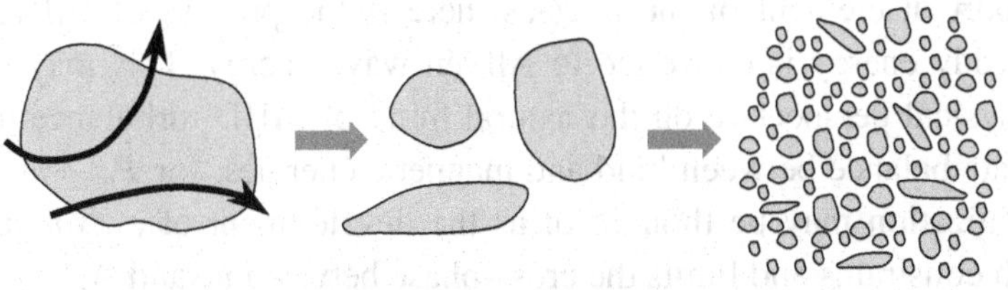

Fig. 9.10. Forward transfer: fluid eddies chop up scalar A.

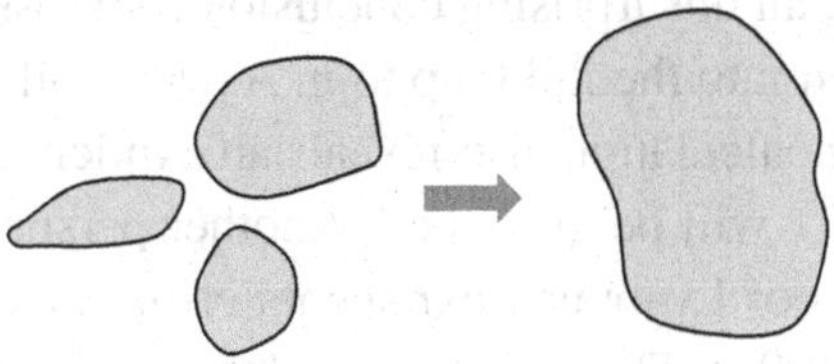

Fig. 9.11. Inverse transfer: current filaments and A-blobs attract and coagulate.

An essential element of the physics of Γ_A is the competition between advection of scalar potential by the fluid, and the tendency of the flux A to coalesce at large scales. The former is, in the absence of back-reaction, simply a manifestation of the fact that turbulence tends to strain, mix and otherwise 'chop up' a passive scalar field, thus generating small-scale structure (see Figure 9.10). The latter demonstrates the property that A is *not* a passive scalar, and that it resists mixing by the tendency to coagulate on large scales (see Figure 9.11) (Riyopoulos *et al.*, 1982). The inverse cascade of A^2, like the phenomenon of magnetic island coalescence, is ultimately rooted in the fact that like-signed current filaments attract. Not surprisingly then, the velocity field drives a positive potential diffusivity (turbulent resistivity), while the magnetic field perturbations drive a *negative* potential diffusivity. Thus, we may anticipate a relation for the turbulent resistivity of the form $\eta_T \sim \langle v^2 \rangle - \langle B^2 \rangle$, a considerable departure from expectations based upon kinematic models. (In a 'kinematic model', the response is to a prescribed fluctuating field $\tilde{v}$.) A similar competition between mixing and coalescence appears in the spectral dynamics. Note also that η_T vanishes for turbulence at Alfvénic equipartition (i.e. $\langle v^2 \rangle = \langle B^2 \rangle$). Since the presence of even a weak mean magnetic field will naturally convert some of the fluid eddies to Alfvén waves (for which $\langle v^2 \rangle = \langle B^2 \rangle$), it is thus not entirely surprising that questions arise as to the possible reduction or 'quenching' of the magnetic diffusivity relative to expectations based upon kinematics. Also, note that any such quenching is intrinsically a synergistic consequence of both:

(i) the competition between flux advection and flux coalescence intrinsic to 2D MHD;
(ii) the tendency of a mean magnetic field to 'Alfvénize' the turbulence.

An important element of the physics, here is the process of 'Alfvénization', whereby eddy energy is converted to Alfvén wave energy. This may be thought of as a physical perspective on the natural trend of MHD turbulence towards an approximate balance between fluid and magnetic energies, for $P_m \sim 1$. Note also that Alfvénization may be thought of as the development of a *dynamical memory*, which constrains and limits the cross-phase between v_x and A. This is readily apparent from the fact that $\langle v_x A \rangle$ vanishes for Alfvén waves in the absence of resistive dissipation. For Alfvén waves then, flux diffusion is directly proportional to resistive dissipation, an unsurprising conclusion for cross-field transport of flux which is, in turn, frozen into the fluid, up to η. As we shall soon see, the final outcome of the quenching calculation also reveals an explicit proportionality of η_{T} to η. For small η, then, Γ_A will be quenched. Another perspective on Alfvénization comes from the studies of Lyapunov exponents of fluid elements in MHD turbulence (Cattaneo *et al.*, 1996). This study showed that as small-scale magnetic fields are amplified and react back on the flow, Lyapunov exponents drop precipitously, so that chaos is suppressed. This observation is consistent with the notion of the development of a dynamic memory, discussed above.

9.4.3 Mean field electrodynamics for $\langle A \rangle$ in a two-dimensional system

In this section, we discuss the mean field theory of flux diffusion in 2D. In the discussion on calculation of the flux of the magnetic potential Γ_A, we do not address the relationship between the turbulent velocity field and the mechanisms by which the turbulence is excited or stirred. However, a weak large-scale field (the transport of which is the process to be studied) will be violently stretched and distorted, resulting in the rapid generation of a spectrum of magnetic turbulence. As discussed above, magnetic turbulence will likely tend to retard and impede the diffusion of large-scale magnetic fields. This, of course, is the crux of the matter, as Γ_A depends on the full spectrum arising from the external excitation and the back-reaction of the magnetic field, so the net imbalance of $\langle v^2 \rangle$ and $\langle B^2 \rangle$ determines the degree of η_{T} quenching. Leverage on $\langle B^2 \rangle$ is obtained by considering the evolution of mean-square magnetic potential density $\mathcal{H}_A = A^2$. In particular, the conservation of $H_A = \int \mathcal{H}_A \mathrm{d}^2 x$ straightforwardly yields the identity from Eq.(9.19b) as,

$$\frac{1}{2}\frac{\partial H_A}{\partial t} = -\Gamma_A \frac{\partial \langle A \rangle}{\partial x} - \eta \langle B^2 \rangle, \tag{9.21}$$

where the surface terms vanish for periodic boundaries. For stationary turbulence $\partial / \partial t = 0$, then, this gives,

$$\langle B^2 \rangle = -\frac{\Gamma_A}{\eta}\frac{\partial \langle A \rangle}{\partial x} = \frac{\eta_T}{\eta}\left(\frac{\partial \langle A \rangle}{\partial x}\right)^2, \tag{9.22}$$

which is the well-known Zeldovich (1957) theorem,

$$\langle B^2 \rangle / \langle B \rangle^2 = \eta_T / \eta,$$

for 2D MHD. The key message here is that when a weak mean magnetic field is coupled to a turbulent 2D flow, a *large mean-square fluctuation level can result*, on account of stretching iso-A or flux contours by the flow. However, while the behaviour of $\langle B^2 \rangle$ is clear, we shall see that it is really $\langle B^2 \rangle_{\mathbf{k}}$ that enters the calculation of Γ_A, via a spectral sum.

9.4.3.1 Closure approximations

To calculate Γ_A, standard closure methods in Chapter 6 (see, for example, McComb, 1990) yield,

$$\Gamma_A = \sum_{\mathbf{k}'} [v_x(-\mathbf{k}')\delta A(\mathbf{k}') - B_x(-\mathbf{k}')\delta\Phi(\mathbf{k}')] = \sum_{\mathbf{k}'} \Gamma_A(\mathbf{k}'), \tag{9.23}$$

where $\delta A(\mathbf{k})$ and $\delta\Phi(\mathbf{k})$ are, in turn, driven by the beat terms (in Eqs. (9.19a) and (9.19b)) that contain the mean field $\langle A \rangle$. The calculational approach here treats fluid and magnetic fluctuations on an equal footing, and seeks to determine Γ_A by probing an evolved state of MHD turbulence, rather than a kinematically prescribed state of velocity fluctuations alone. The calculation follows those of Pouquet (1978), and yields the result,

$$\Gamma_A = -\sum_{\mathbf{k}'} [\tau_c^{\Phi}(\mathbf{k}')\langle v^2 \rangle_{\mathbf{k}'} - \tau_c^{A}(\mathbf{k}')\langle B^2 \rangle_{\mathbf{k}'}]\frac{\partial \langle A \rangle}{\partial x} - \sum_{\mathbf{k}'} [\tau_c^{A}(\mathbf{k}')\langle A^2 \rangle_{\mathbf{k}'}]\frac{\partial}{\partial x}\langle J \rangle. \tag{9.24}$$

Here, consistent with the restriction to a weak mean field, isotropic turbulence is assumed. The quantities $\tau_c^{\Phi}(\mathbf{k})$ and $\tau_c^{A}(\mathbf{k})$ are the self-correlation times (lifetimes), at $\mathbf{k}$, of the fluid and field perturbations, respectively. These are not necessarily equivalent to the coherence time of $v_x(-\mathbf{k}')$ and $A(\mathbf{k}')$, which determines Γ_A. For a weak mean field, both $\tau_c^{\Phi}(\mathbf{k})$ and $\tau_c^{A}(\mathbf{k})$ are determined by nonlinear interaction processes, so that $1/\tau_c^{\Phi,A}(\mathbf{k}') \geq k'\langle B \rangle$, i.e. fluctuation correlation times are *short* in comparison to the Alfvén time of the mean field. (Note again that the Alfvén unit is used for velocity here. Alfvén velocity is given by the mean magnetic field $\langle B \rangle$.) In this case, the decorrelation process is controlled by the Alfvén time of the *rms* field (i.e. $[\mathbf{k}\langle B^2 \rangle^{1/2}]^{-1}$) and the fluid eddy turn-over time, as discussed by

Pouquet (1978). Consistent with the assumption of unity magnetic Prandtl number, $\tau_c^{\Phi}(\mathbf{k}) = \tau_c^{A}(\mathbf{k}) = \tau_c(\mathbf{k})$ is assumed, hereafter.

The three terms on the right-hand side of Eq.(9.24) correspond respectively (Diamond *et al.*, 1984) to:

(a) a positive turbulent resistivity (i.e. Γ_A proportional to flux gradient) due to fluid advection of flux;
(b) a negative turbulent resistivity symptomatic of the tendency of magnetic flux to accumulate on large scales;
(c) a positive turbulent hyper-resistive diffusion, which gives Γ_A proportional to *current* gradient (Schmidt and Yoshikawa, 1971; Strauss, 1986). Such diffusion of current has been proposed as the mechanism whereby a magnetofluid undergoes Taylor relaxation (Taylor, 1986; Bhattacharjee and Hameiri, 1986).

Note that terms (b) and (c) both arise from $B_x(\mathbf{k})\delta\Phi(\mathbf{k}')$, and show the trend in 2D MHD turbulence to pump large-scale H_A while damping small-scale H_A. For smooth, slowly varying mean potential profiles, the hyper-resistive term is negligible in comparison with the turbulent resistivity, (i.e. $\langle k'^2\rangle > (1/\langle A\rangle)(\partial_x^2\langle A\rangle)$), so that the mean magnetic potential flux reduces to,

$$\Gamma_A = -\eta_{\mathrm{T}}\frac{\partial\langle A\rangle}{\partial x}, \tag{9.25}$$

where,

$$\eta_{\mathrm{T}} = \sum_{\mathbf{k}'}\tau_c(\mathbf{k}')\left(\langle v^2\rangle_{\mathbf{k}'} - \langle B^2\rangle_{\mathbf{k}'}\right). \tag{9.26}$$

As stated above, the critical element in determining Γ_A is calculating $\langle B^2\rangle_{\mathbf{k}'}$ in terms of $\langle v^2\rangle_{\mathbf{k}'}$, Γ_A itself, etc. For this, mean-square magnetic potential balance is crucial! To see this, note that the flux equation may be written as,

$$\frac{\partial A}{\partial t} + \mathbf{v}\cdot\nabla A = -v_x\frac{d\langle A\rangle}{dx} + \eta\nabla^2 A, \tag{9.27}$$

so multiplying by A and summing over modes gives,

$$\frac{1}{2}\left(\frac{\partial}{\partial t}\langle A^2\rangle + \langle\nabla\cdot(\mathbf{v}A^2)\rangle\right) = -\Gamma_A\frac{\mathrm{d}\langle A\rangle}{\mathrm{d}x} - \eta\langle B^2\rangle, \tag{9.28}$$

assuming incompressibility of the flow. An equivalent, $\mathbf{k}$-space version of Eq.(9.28) is,

$$\frac{1}{2}\left(\frac{\partial}{\partial t}\langle A^2\rangle_{\mathbf{k}} + T(\mathbf{k})\right) = -\Gamma_A(\mathbf{k})\frac{\mathrm{d}\langle A\rangle}{\mathrm{d}x} - \eta\langle B^2\rangle_{\mathbf{k}}, \tag{9.29}$$

where $T(\mathbf{k})$ is the triple correlation,

$$T(\mathbf{k}) = \langle \nabla \cdot (\mathbf{v}A^2) \rangle_{\mathbf{k}}, \tag{9.30}$$

which controls the nonlinear transfer of mean-square potential, and $\Gamma_A(\mathbf{k}) = \langle v_x A \rangle_{\mathbf{k}}$ is the $\mathbf{k}$-component of the flux. Equations (9.28) and (9.29) thus allow the determination of $\langle B^2 \rangle$ and $\langle B^2 \rangle_{\mathbf{k}}$ in terms of Γ_A, $\Gamma_A(\mathbf{k})$, $T(\mathbf{k})$ and $\partial_t \langle A^2 \rangle_{\mathbf{k}}$.

9.4.3.2 Mean-field approximation for correlation time and quenching of turbulent resistivity

At the simplest, crudest level (the so-called τ-approximation), a single τ_c is assumed to characterise the response or correlation time in Eq.(9.26). In that case, we have,

$$\Gamma_A = -\left[\sum_{\mathbf{k}} \tau_c(\langle v^2 \rangle_{\mathbf{k}} - \langle B^2 \rangle_{\mathbf{k}})\right] \frac{\partial \langle A \rangle}{\partial x}. \tag{9.31}$$

For this, admittedly over-simplified case, Eq.(9.28) then allows the determination of $\langle B^2 \rangle$ in terms of Γ_A, the triplet and $\partial_t \langle A^2 \rangle$. With the additional restrictions of stationary turbulence and periodic boundary conditions (so that $\partial_t \langle A^2 \rangle = 0$ and $\langle \nabla \cdot (\mathbf{v}AA) \rangle = 0$), it follows that,

$$\langle B^2 \rangle = -\frac{\Gamma_A}{\eta} \frac{\mathrm{d}\langle A \rangle}{\mathrm{d}x}, \tag{9.32a}$$

so that magnetic fluctuation energy is directly proportional to magnetic potential flux, via H_A balance. This corresponds to a balance between local dissipation and spatial flux in the mean-square potential budget (Gruzinov and Diamond, 1994, 1996; Blackman and Field, 2000; Brandenburg and Subramanian, 2005; Subramanian and Brandenburg, 2006). Substitution of the form Γ_A in Eq.(9.25) into Eq.(9.32a) gives the relation,

$$\left\langle B^2 \right\rangle = \frac{\eta_{\mathrm{T}}}{\eta} \langle B \rangle^2, \tag{9.32b}$$

where the relation $\langle B \rangle = -\mathrm{d} \langle A \rangle / \mathrm{d}x$ is used. Inserting this into Eq.(9.26) then yields $\eta_{\mathrm{T}} = \sum_{\mathbf{k}} \tau_c \left\langle v^2 \right\rangle_{\mathbf{k}} - (\tau_c \eta_{\mathrm{T}} / \eta) \langle B \rangle^2$, i.e., the following expression for the turbulent diffusivity,

$$\eta_{\mathrm{T}} = \frac{\sum_{\mathbf{k}} \tau_c \langle v^2 \rangle_{\mathbf{k}}}{1 + \tau_c v_{\mathrm{A}0}^2 / \eta} = \frac{\eta_{\mathrm{T}}^k}{1 + R_m v_{\mathrm{A}0}^2 / \langle v^2 \rangle}, \tag{9.33}$$

where η_T^k refers to the kinematic turbulent resistivity $\tau_c\langle v^2\rangle$, v_{A0} is the Alfvén speed of the mean $\langle B\rangle$, and $R_m = \langle v^2\rangle\tau_c/\eta$. It is instructive to note that Eq.(9.33) can be rewritten as,

$$\eta_T = \frac{\eta\eta_T^k}{\eta + \tau_c v_{A0}^2}. \tag{9.34}$$

Thus, as indicated by mean-square potential balance, Γ_A *ultimately scales directly with the collisional resistivity*, a predictable outcome for Alfvénized turbulence with dynamically interesting magnetic fluctuation intensities. This result supports the intuition discussed earlier. It is also interesting to note that for $R_m v_{A0}^2/\langle v^2\rangle > 1$ and $\langle v^2\rangle \sim \langle B^2\rangle$, $\eta_T \cong \eta\langle B^2\rangle/\langle B\rangle^2$, consistent with the Zeldovich theorem prediction.

Equation (9.33) gives the well-known result for quenched flux diffusivity. There, the kinematic diffusivity η_T^k is modified by the quenching or suppression factor $[1+R_m v_{A0}^2/\langle v^2\rangle]^{-1}$, the salient dependencies of which are on R_m and $\langle B\rangle^2$. Equation (9.33) predicts a strong quenching of η_T with increasing $R_m\langle B\rangle^2$. Despite the crude approximations made in the derivation, numerical calculations indicate remarkably good agreement between the measured cross-field flux diffusivity (as determined by following marker particles tied to a flux element) and the predictions of Eq.(9.33). In particular, the scalings with both R_m and $\langle B\rangle^2$ have been verified, up to R_m values of a few hundred (Cattaneo, 1994).

9.4.3.3 Issues in the quenching of turbulent resistivity

The derivation of Eq.(9.33), as well as the conclusion of a quenched magnetic diffusivity, stimulates many questions. Special care must be taken with the treatment of the triplet term $\langle\nabla\cdot(\mathbf{v}AA)\rangle$ in Eq.(9.28). Note that $\langle\nabla\cdot(\mathbf{v}AA)\rangle$ makes no contribution to global $\mathcal{H}_A$ balance in a periodic system. However, while $\langle\nabla\cdot(\mathbf{v}AA)\rangle = 0$ in this case, $(\mathbf{v}\cdot(\nabla AA))_{\mathbf{k}}$ does not. This contribution to the $\langle A^2\rangle$ dynamics corresponds to:

(i) divergence of the flux of the mean-square potential, $\nabla\cdot\Gamma_{A^2}$, (here $\Gamma_{A^2} = \mathbf{v}AA$), when considered in a region of position space of scale $|\mathbf{k}|^{-1}$;
(ii) the spectral transport of $\langle A^2\rangle_{\mathbf{k}}$, when considered in $\mathbf{k}$-space.

In either case, a new time scale enters the mean-square magnetic potential budget Eq.(9.28), and a model with single value τ_c can be modified. This modification can, in principle, break the balance between $\Gamma_A\langle B\rangle$ and resistive dissipation, Eq.(9.22). Physically, this time scale has been associated with:

(i) the net outflow of mean-square potential at the boundaries, in the case of a non-periodic configuration. In this regard, it has been conjectured that, should the loss rate of $\langle A^2\rangle$ exceed that of $\langle A\rangle$, the quench of η_T would be weaker;

(ii) the *local* effective transport rate (on scales $\sim |\mathbf{k}|^{-1}$) of mean-square potential or, alternatively, the local spectral transport rate of $\langle A^2\rangle_{\mathbf{k}}$. Note that in this case, boundary conditions are irrelevant. Thus, local $\langle A^2\rangle$ spectral transport effects should manifest themselves in numerical calculations with periodic boundaries, such as those by Cattaneo (1994).

To address these questions, one must calculate the triplet correlations. In this regard, it is instructive to consider them from the point of view of transport in position space (i.e. $\langle \nabla \cdot (\mathbf{v}AA)\rangle$), together with the equivalent spectral transfer in **k**-space. The goal here is to assess the degree to which triplet correlations enter the relationship between resistive dissipation and magnetic flux transport, which is central to the notion of quenching.

Recall, on retaining the volume-averaged advective flux, that the equation for the mean-square potential fluctuation is,

$$\frac{1}{2}\left(\frac{\partial}{\partial t}\langle AA\rangle + \langle \mathbf{v}\cdot\nabla AA\rangle\right) = -\langle v_x A\rangle\frac{\partial\langle A\rangle}{\partial x} - \eta\langle B^2\rangle. \tag{9.35}$$

Observe that since $-\langle v_x A\rangle d\langle A\rangle/dx = \eta_{\mathrm{T}}\langle B\rangle^2$, which comes from Eq.(9.21) and (9.22), the right-hand side of Eq.(9.35) simply reduces to the Zeldovich theorem,

$$\langle B^2\rangle/\langle B\rangle^2 = \eta_{\mathrm{T}}/\eta,$$

in the absence of contributions from the triplet moment. For stationary turbulence, then, the proportionality between mean flux transport and resistive dissipation is broken by the triplet $\langle \mathbf{v}\cdot\nabla AA\rangle$, which may be rewritten as,

$$\langle \mathbf{v}\cdot\nabla AA\rangle = \nabla\cdot\langle \mathbf{v}AA\rangle = \int \Gamma_{A^2}\cdot \mathbf{ds},$$

using Gauss's law. Here $\Gamma_{A^2} = \mathbf{v}AA$ is the flux of mean-square potential and the integration $\int \mathbf{ds}$ is performed along a contour enclosing the region of averaging denoted by the bracket. This scale must, of course, be smaller than the mean field scale L for consistency of the averaging procedure. Mean-square potential evolution is thus given by,

$$\frac{1}{2}\left(\frac{\partial}{\partial t}\langle AA\rangle + \int \mathbf{ds}\cdot\Gamma_{A^2}\right) = -\langle v_x A\rangle\frac{\partial\langle A\rangle}{\partial x} - \eta\langle B^2\rangle, \tag{9.36}$$

so that the balance of mean flux transport and local dissipation is indeed broken by the *net* in/out flux of mean-square potential to the averaging region. Alternatively, the triplet correlation renders the mean-square potential balance *non-local*. The net value $\int \Gamma_{A^2}\cdot\mathbf{ds}$ is determined by the values of the turbulent velocity and potential

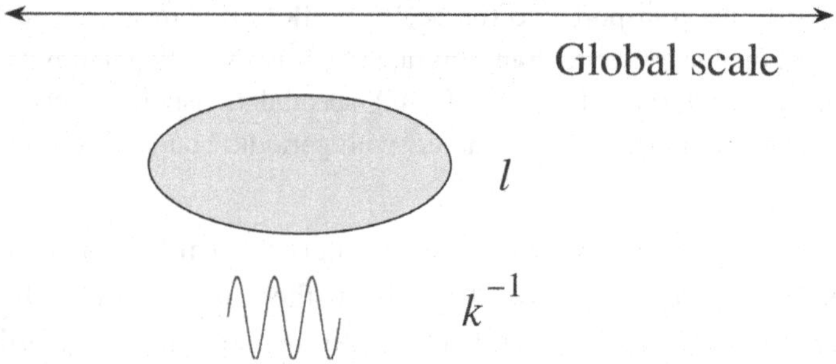

Fig. 9.12. Illustration of the scale length for the integral. It is intermediate between the global scale and the microscopic scale of fluctuations.

perturbation on the boundary of the averaging region. The non-local term in the $\mathcal{H}_A$ budget is by no means 'small' in any naive sense, either – indeed a straightforward estimate of the ratio of the second term (the A^2 flux) in Eq.(9.36) to the third term gives $(B/\langle B\rangle)(k\ell)^{-1}$, where k is a typical perturbation wave vector and ℓ is the scale of the averaging region. As $B/\langle B\rangle \sim \sqrt{R_m} \gg 1$ and $(k\ell)^{-1} \leq 1$, this ratio can certainly be large, so the triplet term is by no means a-priori negligible. However, two caveats are important. First, a *net* influx or outflux is required, these being more suggestive of an externally driven process, rather than one that is spontaneous (i.e. in 3D, of helicity injection rather than a dynamo). Second, the quantity Γ_{A^2} *may not* be calculated kinematically (i.e., by a process where $\tilde{v}$ alone is prescribed), for exactly the same reasons that the kinematic theory of Γ_A fails so miserably! This latter point is discussed at length, below.

Noting that a *net inflow or outflow* of mean-squared potential is required to break the local balance between resistive dissipation and mean potential transport (i.e. turbulent resistivity), the suggestion has appeared that a net in/out flux (Γ_{A^2}) of mean-square potential at the system boundary may weaken the quench. Implicit in this suggestion is the idea that Γ_{A^2} will exceed Γ_A, or alternatively, that the in/out flow rate of mean-square potential exceeds that of the mean potential. We shall see below that when Γ_{A^2} and Γ_A are both calculated self-consistently, this is *not* the case. While a definitive numerical test of this hypothesis has yet to be performed, the results of recent numerical calculations that relax the periodic boundary conditions used in earlier studies by prescribing A or $\partial_y A$ on the upper and lower boundaries indicate no significant departure from the predicted effective resistivity quench (Silvers, 2004; Keating and Diamond, 2007; Keating and Diamond, 2008; Keating *et al.*, 2008). Recent work indicates that flux transport *can* occur via resonant nonlinear wave interactions, and that this transport is independent of R_m, for large R_m. The key to this is that wave resonance provides the requisite irreversibility, which is independent of R_m. The resulting nonlinear transfer driven flux resembles that due to higher order quasi-linear theory, discussed in Chapter 3. However, while formally not subject to an R_m-dependent quench, the resulting flux transport is, in some sense, still "small", since wave interactions must be subcritical to wave breaking. It should be noted that while these calculations *do* suggest that the dynamics of turbulent transport are insensitive to boundary conditions under the circumstances of idealized computations, they do not actually examine the effects of external magnetic potential injection.

9.4.3.4 Evaluation of correlation time and quenching of turbulent resistivity

It is also instructive to examine the triplet correlations in **k**-space, as well as in configuration space. Indeed, it is here that the tremendous departure of Γ_{A^2} from kinematic estimates is most apparent. In **k**-space, $\mathcal{H}_A$ evolution is described by,

$$\frac{1}{2}\left(\frac{\partial}{\partial t}\langle AA\rangle_{\mathbf{k}} + T_{\mathbf{k}}\right) = -\langle v_x A\rangle_{\mathbf{k}}\frac{\partial\langle A\rangle}{\partial x} - \eta\langle B^2\rangle_{\mathbf{k}}, \tag{9.37}$$

where the triplet T_k is,

$$T_{\mathbf{k}} = \langle \mathbf{v}\cdot\nabla AA\rangle_{\mathbf{k}}. \tag{9.38}$$

In **k**-space, spectral transfer breaks the balance between resistive dissipation and turbulent transport. Thus, the key issue is the calculation of $T_{\mathbf{k}}$, which is easily accomplished by standard closure methods as discussed in Chapter 6 (Pouquet, 1978). Thus, applying EDQNM or DIA-type closures, $T_{\mathbf{k}}$ is straightforwardly approximated as,

$$\begin{aligned} T_{\mathbf{k}} = \sum_{\mathbf{k}'} \left(\mathbf{k}\cdot\mathbf{k}'\times\hat{\mathbf{z}}\right)^2 \Theta_{\substack{\mathbf{k},\mathbf{k}'\\ \mathbf{k}+\mathbf{k}'}} \left(\langle\Phi^2\rangle_{\mathbf{k}'} - \left(\frac{\mathbf{k}'^2-\mathbf{k}^2}{(\mathbf{k}+\mathbf{k}')^2}\right)\langle A^2\rangle_{\mathbf{k}'}\right)\langle A^2\rangle_{\mathbf{k}} \\ - \sum_{\substack{\mathbf{p},\mathbf{q}\\ \mathbf{p}+\mathbf{q}=\mathbf{k}}} (\mathbf{p}\cdot\mathbf{q}\times\hat{\mathbf{z}})^2\Theta_{\mathbf{k},\mathbf{p},\mathbf{q}}\langle\Phi^2\rangle_{\mathbf{p}}\langle A^2\rangle_{\mathbf{q}}, \end{aligned} \tag{9.39}$$

where $\Theta_{\mathbf{k},\mathbf{p},\mathbf{q}}$ is the triad coherence time $\Theta_{\mathbf{k},\mathbf{p},\mathbf{q}} = (1/\tau_{c\mathbf{k}} + 1/\tau_{c\mathbf{q}} + 1/\tau_{c\mathbf{p}})^{-1}$. In Eq.(9.39), the first and third terms represent advection of potential by the turbulent velocity, the first giving a turbulent resistivity, the third incoherent noise. Note that these two contributions conserve $\langle A^2\rangle$ against each other when summed over **k**, as explained in Chapter 6. The second term in Eq.(9.39) corresponds to inverse transfer of mean-square potential via flux coalescence. Note that it is negative on large scales ($k^2 < k'^2$), yielding a negative turbulent resistivity, and positive on small scales ($k^2 > k'^2$), giving a positive hyper-resistivity. Observe that the second term is manifestly antisymmetric in **k** and **k**$'$, and so conserves $\langle A^2\rangle$ individually, when summed over **k**. The second term on the right-hand side of Eq.(9.39) gives a departure from kinetic estimations.

It is clear immediately that, just as in the case of $\langle A\rangle$, $\langle A^2\rangle$ evolution is determined by the competition between advective straining and mixing of iso-A contours (the first term), together with the tendency of these flux structures to coalesce to progressively larger scales (the second term). This is hardly a surprise, since A and all its moments are frozen into the flow, up to resistive dissipation. Note also that a proper treatment of mean-square potential conservation (i.e. $\sum_{\mathbf{k}} \mathrm{T}_{\mathbf{k}} = 0$) requires that nonlinear noise due to incoherent mode coupling also be accounted for.

Equation (9.37) can be re-written in the form,

$$\frac{1}{2}\left(\frac{\partial}{\partial t}\langle AA\rangle_{\mathbf{k}} + \hat{\eta}_{\tau\mathbf{k}}\langle AA\rangle_{\mathbf{k}} - \mathrm{S}_{\mathbf{K}}\right) = -\langle v_x A\rangle_{\mathbf{k}}\frac{\partial\langle A\rangle}{\partial x} - \eta\langle B^2\rangle_{\mathbf{k}}, \tag{9.40}$$

written as a combination of the coherent term and the nonlinear noise,

$$\mathrm{T}_{\mathbf{k}} = \hat{\eta}_{\tau\mathbf{k}}\langle \mathrm{AA}\rangle_{\mathbf{k}} - \mathrm{S}_{\mathbf{k}}, \tag{9.41}$$

with,

$$\hat{\eta}_{\tau\mathbf{k}} = \sum_{\mathbf{k}'}(\mathbf{k}\cdot\mathbf{k}'\times z)^2\Theta_{\substack{\mathbf{k},\mathbf{k}'\\ \mathbf{k}+\mathbf{k}'}}\left(\langle\Phi^2\rangle_{\mathbf{k}'} - \langle A^2\rangle_{\mathbf{k}'}\right), \tag{9.42}$$

and the emission term,

$$S_{\mathbf{k}} = \sum_{\substack{\mathbf{p},\mathbf{q}\\ \mathbf{p}+\mathbf{q}=\mathbf{k}}} (\mathbf{p}\cdot\mathbf{q}\times\hat{z})^2\Theta_{\mathbf{k},\mathbf{p},\mathbf{q}}\langle\Phi^2\rangle_{\mathbf{p}}\langle A^2\rangle_{\mathbf{q}}. \tag{9.43}$$

Note that $\hat{\eta}_{\tau\mathbf{k}} \to -\frac{\partial}{\partial x}\eta_{\mathrm{T}}\frac{\partial}{\partial x}$ as $\mathbf{k}\to 0$. It is interesting to compare terms on the left and right-hand side of Eq.(9.40). Nonlinear transfer terms $\sim\langle\nabla\cdot\Gamma_{A^2}\rangle_{\mathbf{k}}$ are $O(kA^2|v|)$, while mean flux terms are $O(|vA||\langle B\rangle|)$. Thus, the ratio $|\mathrm{T}_{\mathbf{k}}|/|\langle vA\rangle_{\mathbf{k}}||\langle B\rangle| \sim O\left(\sqrt{\langle B^2\rangle}/\langle B\rangle\right)$. Here, $\sqrt{\langle B^2\rangle}/\langle B\rangle \gg 1$, as we are considering a strongly turbulent, weakly magnetised regime. Thus, to lowest order in $\left(\sqrt{\langle B^2\rangle}/\langle B\rangle\right)^{-1}$, Eq.(9.40) (at stationarity) must reduce to,

$$\mathrm{T}_{\mathbf{k}} = 0, \tag{9.44}$$

so that stationarity of nonlinear transfer determines the magnetic potential spectrum. This guarantees that the net spectral flow rate is constant in **k**, so $\mathcal{H}_A$ is conserved.

In physical terms, this means that $\langle AA\rangle_{\mathbf{k}}$ adjusts to balance nonlinear noise, which is the main source here. We formally refer to this spectrum as $\langle AA\rangle_{\mathbf{k}}^{(0)}$. Note that $\langle AA\rangle_{\mathbf{k}}^{(0)}$ is actually determined, as is usual for spectral transfer processes, by the balance between $S_{\mathbf{k}}$ (incoherent mode coupling) and $\hat{\eta}_{\tau_{\mathbf{k}}}\langle AA\rangle_{\mathbf{k}}$ (turbulent dissipation) as is illustrated in Chapter 6. Nonlinear noise is critical here (to ensure $\mathcal{H}_A$ conservation) and, in fact, constitutes the dominant source for $\langle AA\rangle_{\mathbf{k}}^{(0)}$ when $\sqrt{\langle B^2\rangle/\langle B\rangle} \gg 1$. It is interesting to observe that, as a consequence, the classical 'mean field electrodynamics' calculation of $\langle v_x A\rangle$ *cannot* be decoupled fully from the spectral transfer problem for $\langle AA\rangle_{\mathbf{k}}$. This of course follows from the constraint imposed upon the former by $\mathcal{H}_A$ conservation. To next order in $\langle B\rangle/\sqrt{\langle B^2\rangle}$ then, Eq.(9.40) gives,

$$0 = -\langle v_x A\rangle_{\mathbf{k}} \frac{\partial\langle A\rangle}{\partial x} - \eta\langle B^2\rangle_{\mathbf{k}}, \tag{9.45}$$

the solution of which yields $\langle B^2\rangle_{\mathbf{k}} = \eta^{-1}\langle v_x A\rangle_{\mathbf{k}}\langle B\rangle$. With the help of the relation,

$$\langle v_x A\rangle_{\mathbf{k}} = \tau_c(\mathbf{k})\left(\left\langle v^2\right\rangle_{\mathbf{k}} - \left\langle B^2\right\rangle_{\mathbf{k}}\right)\langle B\rangle,$$

which comes from Eqs.(9.25) and (9.26), one has,

$$\left\{1 + \tau_c(\mathbf{k})\,\eta^{-1}\langle B\rangle^2\right\}\left\langle B^2\right\rangle_{\mathbf{k}} = \tau_c(\mathbf{k})\,\eta^{-1}\langle B\rangle^2\left\langle v^2\right\rangle_{\mathbf{k}}.$$

Ultimately, this yields a quenched turbulent resistivity of the form,

$$\eta_{\mathrm{T}} = \sum_{\mathbf{k}} \frac{\tau_{c\mathbf{k}}\langle v^2\rangle_{\mathbf{k}}}{1 + \tau_{c\mathbf{k}} v_{A0}^2/\eta}. \tag{9.46}$$

Note that this is basically equivalent to the result in Eq.(9.33), with, however, the quench factor varying with $\mathbf{k}$.

Several comments are in order here. First, it cannot be over-emphasised that a self-consistent calculation of $\langle\nabla\cdot\Gamma_{A^2}\rangle_{\mathbf{k}}$ is crucial to this conclusion. Such a calculation necessarily must include both coherent nonlinear response and nonlinear noise. A kinematic calculation would leave $\hat{\eta}_{\tau\mathbf{k}} > -\eta_{\mathrm{T}}\partial^2/\partial x^2$, which is incorrect. Likewise, neglecting noise would violate $\mathcal{H}_A$ conservation. It is also amusing to note that the question of the relation between $\langle v_x A\rangle_{\mathbf{k}}\mathrm{d}\langle A\rangle/\mathrm{d}x$ and $\eta\langle B^2\rangle_{\mathbf{k}}$ does *not* hinge upon boundary conditions or inflow/outflow, at all. Hence, the available numerical experiments, already published, constitute a successful initial test of the theory of flux diffusivity quenching in 2D, at least for modest values of R_m and for smooth $\langle A\rangle$ profiles.

It is instructive to return to configuration space now, in order to compare the rates of transport of $\langle A\rangle$ and $\langle AA\rangle$. The analysis given above may be summarised by writing the equations of evolution for $\langle A\rangle$, i.e.,

$$\frac{\partial}{\partial t}\langle A\rangle = \frac{\partial}{\partial x}\eta_{\mathrm{T}}\frac{\partial\langle A\rangle}{\partial x}, \tag{9.47}$$

where,

$$\eta_{\mathrm{T}} = \sum_{\mathbf{k}'}\tau_{c\mathbf{k}'}(\langle v^2\rangle_{\mathbf{k}'} - \langle B^2\rangle_{\mathbf{k}'}); \tag{9.48}$$

and for $\langle AA\rangle_{\mathbf{k}}$, i.e.,

$$\frac{1}{2}\left(\frac{\partial}{\partial t}\langle AA\rangle_{\mathbf{k}} + \hat{\eta}_{\tau\mathbf{k}}\langle AA\rangle_{\mathbf{k}}\right) = \frac{1}{2}S_{\mathbf{k}} - \langle v_x A\rangle_{\mathbf{k}}\frac{\mathrm{d}\langle A\rangle}{\mathrm{d}x} - \eta\langle B^2\rangle_{\mathbf{k}}, \tag{9.49}$$

where,

$$\hat{\eta}_{\tau\mathbf{k}} = \sum_{\mathbf{k}'}(\mathbf{k}\cdot\mathbf{k}'\times\hat{\mathbf{z}})^2\Theta_{\substack{\mathbf{k},\mathbf{k}'\\ \mathbf{k}+\mathbf{k}'}}(\langle\Phi^2\rangle_{\mathbf{k}'} - \langle A^2\rangle_{\mathbf{k}'}). \tag{9.50}$$

Not surprisingly, $\hat{\eta}_{\tau\mathbf{k}} \to -\eta_{\mathrm{T}}\nabla^2$ as $\mathbf{k} \to 0$. This is, of course, a straightforward consequence of the fact that the same physics governs the dynamics of both $\langle A\rangle$ and $\langle A^2\rangle$, since A is conserved along fluid trajectories, up to resistive dissipation. Hence, the total diffusive loss rates for $\langle A\rangle$ and $\langle A^2\rangle$ are simply $1/\tau_A = \eta_{\mathrm{T}}/L_A^2$ and $1/\tau_{A^2} = \eta_{\mathrm{T}}/L_{A^2}^2$, where L_A and L_{A^2} are the gradient scale lengths for $\langle A\rangle$ and $\langle AA\rangle$, respectively. Here L_{A^2} is set either by the profile of forcing or injection, or by the profile of $\langle A\rangle$. For the latter, $\tau_A = \tau_{A^2}$, so that preferential loss of $\langle AA\rangle$ is impossible. For the former, inflow of flux at the boundary, say by plasmoid injection, could however decouple L_{A^2} from L_A. In this case, however, the magnetic dynamics are not spontaneous but, rather, strongly driven by external means.

9.4.4 Turbulent diffusion of flux and field in a three-dimensional system

In this section, we discuss flux and field diffusion in three dimensions. In 3D, $\mathbf{A}$ is *not* conserved along fluid element trajectories, so the flux diffusion problem becomes significantly more difficult. With this in mind, we divide the discussion of 3D diffusion into two subsections; one on turbulent diffusion in 3D reduced MHD (RMHD) (Strauss, 1976), the other on weakly magnetised, full MHD. This progression facilitates understanding, as 3D RMHD is quite similar in structure to 2D MHD, allowing us to draw on the experience and insight gained in the study of that problem.

9.4.4.1 Flux diffusion in three-dimensional reduced MHD

The reduced MHD equations in 3D are,

$$\frac{\partial}{\partial t}\nabla^2\Phi + \nabla\Phi \times \hat{\mathbf{z}} \cdot \nabla\nabla^2\Phi = \nu\nabla^2\nabla^2\Phi + \nabla A \times \hat{\mathbf{z}} \cdot \nabla\nabla^2 A + B_0\partial_z\nabla^2 A, \quad (9.51)$$

and,

$$\frac{\partial A}{\partial t} + \nabla\Phi \times \hat{\mathbf{z}} \cdot \nabla A = B_0\partial_z\Phi + \eta\nabla^2 A. \quad (9.52)$$

These equations describe incompressible MHD in the presence of a strong field $\mathbf{B}_0 = B_0\hat{\mathbf{z}}$, which is externally prescribed and fixed. The 'test field' undergoing turbulent diffusion is $\langle\mathbf{B}\rangle = \langle B(x)\rangle\hat{\mathbf{y}}$, where $\langle B(x)\rangle = -\partial\langle A\rangle/\partial x$. Obviously, $\langle B\rangle \ll B_0$ here.

The presence of a strong $\mathbf{B}_0$ renders 3D RMHD dynamics quite similar (but *not* identical!) to those in 2D. In particular, note that one can define a mean-square magnetic potential in 3D RMHD, i.e.,

$$H_A = \int \mathrm{d}^2x \int A^2\mathrm{d}z, \quad (9.53)$$

and that H_A *is conserved up to resistive dissipation and Alfvénic coupling*, so that the fluctuation H_A balance becomes,

$$\frac{1}{2}\frac{\partial H_A}{\partial t} = -\langle v_x A\rangle\frac{\partial\langle A\rangle}{\partial x} - \eta\langle B^2\rangle + B_0\langle A\partial_z\Phi\rangle, \quad (9.54)$$

where the third term on the right-hand side remains in 3D. In contrast to its 2D counterpart (i.e. Eq.(9.21)), H_A balance is achieved by a competition between cross-field transport and resistive dissipation *together with Alfvénic propagation along* B_0 (i.e. observe that the last term in Eq.(9.54) is explicitly proportional to B_0). It is interesting to note, however, that total H_A conservation is broken only by local dissipation (as in 2D) and by a *linear* effect, which corresponds to wave propagation along B_0. Thus, although H_A is not conserved (even as $\eta \to 0$), the potential equation nonlinearity (i.e. the nonlinearity in Eq.(9.54)) is *still* annihilated in 3D RMHD, as it is in 2D (i.e. $\langle\mathbf{v}\cdot\nabla AA\rangle \to 0$, up to boundary flux terms)! Hence, the mean-square potential budget is still a powerful constraint on flux diffusion in 3D.

9.4.4.2 Closure approximations

For simplicity and brevity, the discussion of flux diffusion in 3D is limited to the case of constant τ_c. Proceeding as in the previous section (Eq.(9.31)) straightforwardly yields,

$$\Gamma_A = -\tau_c(\langle v^2\rangle - \langle B^2\rangle)\frac{\partial\langle A\rangle}{\partial x}. \tag{9.55}$$

Here, the current diffusivity has been dropped, as for 2D. To relate $\langle B^2\rangle$ to $\langle v^2\rangle$ etc., mean-square potential balance (Eq.(9.54)) and stationarity give,

$$\langle B^2\rangle = \frac{-\langle v_x A\rangle}{\eta}\frac{\mathrm{d}\langle A\rangle}{\mathrm{d}x} + \frac{B_0}{\eta}\langle A\partial_z\Phi\rangle. \tag{9.56}$$

Thus, the new element in 3D is the appearance of Alfvénic coupling (i.e. the last term on the right-hand side) in the H_A balance. This coupling is non-zero only if there is a net directivity in the radiated Alfvénic spectrum, or equivalently, an imbalance in the two Elsasser populations, which account for the intensity of wave populations propagating in the $\pm\hat{\mathbf{z}}$-directions.

This contribution may be evaluated as before, i.e.,

$$\langle A\partial_z\Phi\rangle = \langle A\partial_z\delta\Phi\rangle + \langle\delta A\partial_z\Phi\rangle, \tag{9.57}$$

where $\delta\Phi$ and δA are obtained via closure of Eqs.(9.51), (9.52). A calculation along the line of Chapter 6 to extract coherent contribution gives,

$$\langle A\partial_z\Phi\rangle = \tau_c B_0\Big(\varepsilon_v\langle v^2\rangle - \varepsilon_B\langle B^2\rangle\Big), \tag{9.58}$$

where,

$$\varepsilon_v = \frac{\int k_z^2\langle\Phi^2\rangle_{\mathbf{k}}\mathrm{d}^3k}{\int(k_x^2+k_y^2)\langle\Phi^2\rangle_{\mathbf{k}}\mathrm{d}^3k}, \tag{9.59}$$

and ε_B similarly, with $\langle A^2\rangle_{\mathbf{k}}$. Note that this approximation to $\langle A\partial_z\Phi\rangle$ clearly vanishes for equal Elsasser populations with identical spectral structure. This, of course, simply states that, in such a situation, there is no net imbalance or directivity in the Alfvénically radiated energy, and thus no effect on the H_A budget. Taking $\varepsilon_v = \varepsilon_B$ and substituting Eqs.(9.55) and (9.58) into the right-hand side of Eq.(9.56) gives,

$$\left\langle B^2\right\rangle = \frac{\tau_c}{\eta}\left(\langle B\rangle^2 + \varepsilon_B B_0^2\right)\left(\left\langle v^2\right\rangle - \left\langle B^2\right\rangle\right).$$

The intensity on magnetic perturbation is given as,

$$\left\langle B^2\right\rangle = \frac{\tau_c\eta^{-1}\left(\langle B\rangle^2 + \varepsilon_B B_0^2\right)}{1+\tau_c\eta^{-1}\left(\langle B\rangle^2 + \varepsilon_B B_0^2\right)}\left\langle v^2\right\rangle,$$

Upon substitution of this relation into Eq.(9.55), one has,

$$\Gamma_A = \frac{-\eta_T^k \partial \langle A \rangle / \partial x}{1 + \frac{\tau_c}{\eta}\left(\varepsilon_B B_0^2 + \langle B \rangle^2\right)}, \tag{9.60}$$

where,

$$\eta_T^k = \tau_c \langle v^2 \rangle. \tag{9.61}$$

In 3D, τ_c is also a function of B_0^2, i.e. $\tau_c = \tau_{NL}/(k_z^2 v_A^2 \tau_{NL}^2 + 1)$, where τ_{NL} is the amplitude-dependent correlation time.

The message of Eqs.(9.60) and (9.61) is that in 3D RMHD, the strong guide field $\mathbf{B_0}$ contributes to the quenching of η_T. The presence of the factor ε_B implies that this effect is sensitive to the parallel–perpendicular anisotropy of the turbulence, a result which is eminently reasonable. Thus, the degree of quenching in 3D RMHD is stronger than in 2D, as $B_0 \gg \langle B \rangle$. Finally, note however that the upshot of the quench is still that η_T scales with η, indicative of the effects of the freezing of magnetic potential into the fluid.

9.4.4.3 A short note on implications

Given the attention paid to turbulence energy flux through the system boundary, it is worthwhile to comment here that the Alfvénic radiation contribution to the H_A budget ($\langle A \partial_z \Phi \rangle$) could be significantly different if there were a net imbalance in the two Elsasser populations. For example, this might occur in the solar corona, where Alfvén waves propagate away from the Sun, along 'open' field-lines. In this case, a local balance between such Alfvénic leakage and cross-field transport could be established in the H_A budget. Such a balance would, of course, greatly change the scalings of η_T from those given here.

Three-D effects have various influences on the magnetic reconnection process as pointed out, e.g., in (Drake *et al.*, 1994). We note here an interesting application of mean field electrodynamics within RMHD which is to the problem of fast, turbulent reconnection in 3D (Lazarian and Vishniac, 1999; Kim and Diamond, 2001). As, in essence by definition, reconnection rates are measured globally (i.e. over some macroscopic region), they are necessarily constrained by conservation laws, such as that of H_A conservation. It is not surprising, then, that one upshot of the quenching of η_T (i.e. Eq.(9.34)) is that the associated magnetic reconnection velocity $V \leq \left(\langle v^2 \rangle / \langle v_A \rangle^2\right)^{1/2} v_{SP}$, where v_{SP} is the familiar Sweet–Parker velocity $v_{SP} = \langle v_A \rangle / \sqrt{R_m}$, where $R_m = \langle v_A \rangle L/\eta$ (Kulsrud, 2005). Note that this result states that the reconnection rate is enhanced beyond the prediction of collisional theory, but still exhibits the Sweet–Parker-type scaling with resistivity, indicative of the effects of flux freezing.

9.4.4.4 *Limit of the weakly magnetized case*

Moving now to consider the case of weakly magnetised, incompressible, 3D MHD, magnetic potential is no longer conserved, even approximately. Detailed calculations (Gruzinov and Diamond, 1994) predict that,

$$\eta_T \cong \eta_T^k, \tag{9.62}$$

or, equivalently, that the kinematic turbulent resistivity is unchanged and unquenched, to leading order. The obvious question then naturally arises not as to why α is quenched (see (Diamond *et al.*, 2005a)), but rather why η_T (or, equivalently, β) is not quenched! Here, we note that β being a scalar, and not a pseudo-scalar like α, plays no role in magnetic helicity balance. As magnetic helicity balance, which forces a balance between α and resistive dissipation of magnetic helicity ($\sim \eta\langle \mathbf{B} \cdot \mathbf{J}\rangle$), together with stationarity, is the origin of α-quenching, it is thus not at all surprising that η_T is not quenched in 3D, for weak fields. The weak-field result stated here must necessarily pass to the strong field RMHD case discussed earlier, as a strong guiding field is added. The analytical representation of β that smoothly connects these two limiting cases has yet to be derived, and remains an open question in the theory.

Computational studies have not yet really confronted the physics of magnetic flux diffusion in 3D. Although results indicate some tendency toward reduction of β as $\langle B\rangle^2$ increases, it is unclear whether or not the onset of this occurs in the 'weak' or 'strong' field limit. Further work is clearly needed.

9.4.5 *Discussion and conclusion for turbulent diffusion of a magnetic field*

In Section 9.4, the theory of turbulent transport of magnetic flux and field in 2D and 3D MHD is explained. The 2D flux diffusion problem has been given special attention, both for its intrinsic interest and relative simplicity, and for its many similarities to the problem of the α-effect in 3D. Several issues have been addressed in detail. These include: boundary in-flow and out-flow effects on the mean-square potential budget, the role of nonlinear spectral transfer in the mean-square potential budget, and the dynamics of magnetic flux in 3D reduced MHD. Several topics for further study have been identified, including, but not limited to:

(i) the derivation of an expression for diffusion in 3D that unifies the weak and strong field regimes;
(ii) a numerical study of transport in 2D that allows a net flux of turbulence through the system boundary;
(iii) both a theoretical and numerical study of flux diffusion in 3D with balanced and unbalanced Elsasser populations, for various along-field boundary conditions;
(iv) a study of η_T quenching for $P_m \neq 1$ and consideration of non-stationary states.

Appendix 1

Charney–Hasegawa–Mima equation

Drift waves and their nonlinear interactions are one of the most fundamental elementary processes in magnetized inhomogeneous plasmas. The simplest model equation that includes a fundamental nonlinear process is known as the Charney–Hasegara–Mima equation. The analysis of this model equation appears repeatedly in the text, in various contexts of plasma turbulence. The key feature of the equation is illustrated here.

Among various nonlinear interaction mechanisms, the advective nonlinearity (Lagrange nonlinearity) associated with $E \times B$ motion plays a fundamental role in drift wave dynamics. This nonlinearity appears in the fluid description as well as in the Vlasov description of plasmas. In the latter formalism, a large number of degrees of freedom is kept (as a velocity distribution), while the wave nonlinearity is possibly studied without considering this degree of freedom. An elementary nonlinearity associated with drift waves can be studied by use of fluid models.

A1.1 Model

The simplest model equation is constructed for the inhomogenous slab plasma, which is magnetized by a strong magnetic field in the z-direction, and the density has a gradient in the x-direction, the scale length of which is given by L_n (Figure A1). Plasma temperature is constant, and temperature perturbation is not considered. Ion temperature is assumed to be much smaller than that of electrons. The perturbation is mainly propagating in the (x, y) plane, and has a small wave number in the direction of the magnetic field $k_z \ll k_\perp$. The small but finite k_z is essential, so that the drift wave turbulence is a quasi-two-dimensional turbulence. The electrostatic perturbation $\tilde{\phi}$ is considered. Under these specifications, the dynamical equation of plasmas is investigated and the nonlinear equation is deduced.

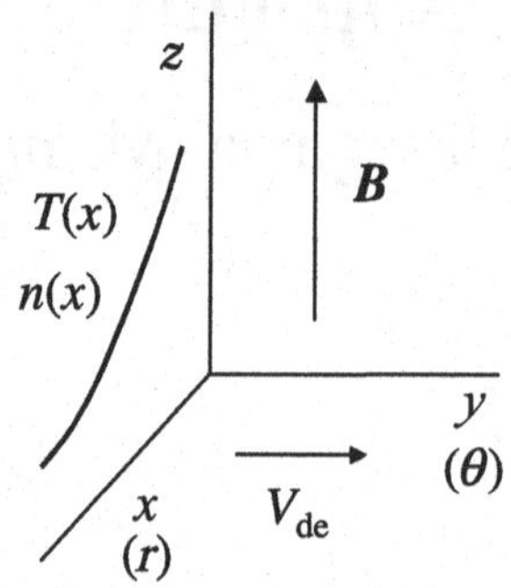

Fig. A1.1. Geometry of inhomogeneous plasma and magnetic field.

First, the electron response is considered. The thermal velocity is taken to be much faster than the phase velocity of waves, $v_{\mathrm{Te}} \gg \omega / |k_z|$, so that the pressure balance of electrons along the magnetic field line provides the Boltzmann response of electrons as,

$$\frac{\tilde{n}_e}{n_0} = \frac{e\tilde{\phi}}{T_e}, \tag{A.1}$$

where n_0 is the unperturbed density and T_e is the electron temperature. The ion dynamics is studied by employing the continuity equation,

$$\frac{\partial}{\partial t} n_i + \nabla \cdot (n_i \boldsymbol{v}_\perp) = 0, \tag{A.2}$$

and the equation of motion,

$$m_i \frac{\mathrm{d}}{\mathrm{d}t} \boldsymbol{v}_\perp = e\left(-\nabla\phi + \boldsymbol{v}_\perp \times \boldsymbol{B}\right). \tag{A.3}$$

Ions are immobile in the direction of the magnetic field line. Time scales are assumed to be much longer than the period of the ion cyclotron motion, and the equation of motion is solved by expansion with respect to $\omega_{\mathrm{ci}}^{-1}\mathrm{d}/\mathrm{d}t$, where $\omega_{\mathrm{ci}} = eB/m_{\mathrm{i}}$ is an ion cyclotron frequency, as,

$$\boldsymbol{v}_\perp = -\frac{1}{B}\nabla\phi \times \hat{z} - \frac{1}{\omega_{\mathrm{ci}} B}\frac{\mathrm{d}}{\mathrm{d}t}\nabla\phi + \cdots, \tag{A.4a}$$

with,

$$\frac{\mathrm{d}}{\mathrm{d}t} = \frac{\partial}{\partial t} - \frac{1}{B}\left(\nabla\phi \times \hat{z}\right) \cdot \nabla + \cdots, \tag{A.4b}$$

where ... remains a higher-order correction. The first and second terms in Eq.(A.4a) are the $E \times B$ and polarization drift motions, respectively.

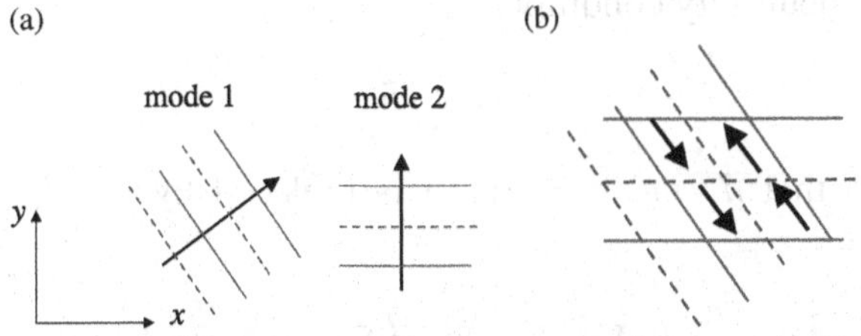

Fig. A1.2. Two waves are propagating on the (x, y) plane (left). Lines indicate the wave front (equi-potential surface) (a). The $E \times B$ drift owing to mode 1 is shown by thick arrows in (b), which causes an additional polarization drift by mode 2.

A1.2 Hasegawa–Mima equation

In addition to the smallness parameter $\omega_{\mathrm{ci}}^{-1}\mathrm{d}/\mathrm{d}t$, the normalized perturbation amplitude $e\tilde{\phi}/T_e$ and the density gradient (normalized to the wavelength) $1/L_n k$ are also taken as smallness parameters. The ordering here is to assume that they are in the same order of magnitude, i.e.,

$$\frac{1}{\omega_{ci}}\left|\frac{\mathrm{d}}{\mathrm{d}t}\right| \sim \frac{e\left|\tilde{\phi}\right|}{T_e} \sim O\left(k^{-1}L_n^{-1}\right). \qquad \text{(A.5)}$$

This assumption, $\tilde{n}_e/n_0 \sim O\left(k^{-1}L_n^{-1}\right)$, means that we consider that the turbulence amplitude is of the order of the mixing length estimate. With this ordering, Eq.(A.4) takes a form as,

$$\boldsymbol{v}_\perp = -\frac{1}{B}\nabla\phi \times \hat{z} - \frac{1}{\omega_{ci}B}\left(\frac{\partial}{\partial t} - \frac{1}{B}\left(\nabla\phi \times \hat{z}\right)\cdot\nabla\right)\nabla\phi, \qquad \text{(A.6)}$$

where ∇ indicates the derivative with respect to (x, y). The second term in parentheses on the right-hand side of Eq.(A.6) indicates the nonlinear interaction term. This mechanism is illustrated in Figure A1.2. Consider there are two modes obliquely propagating. Owing to the electric perturbation of mode 1, ions are subject to the $E \times B$ motion so that they drift along the equi-potential surface of mode 1. This $E \times B$ motion induces a Doppler shift, so that the electric field of mode 2 has additional temporal oscillation. This temporal variation causes the polarization drift due to mode 2. Thus, this polarization drift turns to be a beat of two modes. Putting $n_i = n_0 + \tilde{n}_i$ into Eq.(A.2) gives,

$$\frac{\partial}{\partial t}\frac{\tilde{n}_i}{n_0} + \boldsymbol{v}_\perp \cdot \left(\frac{\nabla n_0}{n_0} + \frac{\nabla\tilde{n}_i}{n_0}\right) + \left(1 + \frac{\tilde{n}_i}{n_0}\right)\nabla\cdot\boldsymbol{v}_\perp = 0. \qquad \text{(A.7)}$$

Now, the charge neutrality condition,

$$\tilde{n}_i = \tilde{n}_e \tag{A.8}$$

is employed, so that $\tilde{n}_i/n_0$ in Eq.(A.7) is replaced by $e\tilde{\phi}/T_e$. Thus, Eq.(A.7) becomes,

$$\begin{aligned}&\frac{\partial}{\partial t}\frac{e\tilde{\phi}}{T_e} - \rho_s c_s\left(\frac{e\nabla\tilde{\phi}}{T_e}\times\hat{z}\right)\cdot\left(\frac{\nabla n_0}{n_0}+\frac{e\nabla\tilde{\phi}}{T_e}\right)\\ &\quad+\left(1+\frac{e\tilde{\phi}}{T_e}\right)\nabla\cdot\left\{-\rho_s c_s\left(\frac{e\nabla\tilde{\phi}}{T_e}\times\hat{z}\right)\right.\\ &\quad\left.-\rho_s^2\left(\frac{\partial}{\partial t}-\rho_s c_s\left(\frac{e\nabla\tilde{\phi}}{T_e}\times\hat{z}\right)\cdot\nabla\right)\frac{e\nabla\tilde{\phi}}{T_e}\right\}=0,\end{aligned} \tag{A.9}$$

where c_s is the ion sound speed and ρ_s is the ion cyclotron radius at sound speed. There are several nonlinear terms. The second-order term in the second term on the left-hand side of Eq.(A.9) vanishes, because of the relation $\nabla\tilde{\phi}\times\hat{z}\cdot\nabla\tilde{\phi}=0$. The $E\times B$ motion is perpendicular to the gradient of the density perturbation (when Eq.(A.1) holds), so that the term $V_\perp\cdot\nabla\tilde{n}_i$ vanishes in the ordering of Eq.(A.5). The nonlinear terms which come from $\left(eT_e^{-1}\tilde{\phi}\right)\nabla\,\cdot$ in the third term on the left-hand side of Eq.(A.9) either vanish or are higher-order smallness terms. Thus, Eq.(A.9) turns out to be,

$$\begin{aligned}&\frac{\partial}{\partial t}\frac{e\tilde{\phi}}{T_e}+\rho_s c_s\left(\frac{\nabla n_0}{n_0}\times\hat{z}\right)\cdot\frac{e\nabla\tilde{\phi}}{T_e}-\rho_s^2\frac{\partial}{\partial t}\frac{e\nabla^2\tilde{\phi}}{T_e}\\ &\quad+\rho_s^3 c_s\nabla\cdot\left(\left(\frac{e\nabla\tilde{\phi}}{T_e}\times\hat{z}\right)\cdot\nabla\right)\frac{e\nabla\tilde{\phi}}{T_e}=0.\end{aligned} \tag{A.10}$$

The second term is a linear term with the derivative in the direction of electron diamagnetic drift, and yields the drift frequency. The third term denotes the effect of the polarization drift of ions. The fourth term is the nonlinear coupling term, which is originated by the combination of the $E\times B$ drift by one mode and the polarization drift by another. With the normalization,

$$\frac{\rho_s}{L_n}\omega_{ci}t\to t,\quad \left(\frac{x}{\rho_s},\frac{y}{\rho_s}\right)\to(x,y),\quad \frac{L_n}{\rho_s}\frac{e\tilde{\phi}}{T_e}\to\varphi, \tag{A.11}$$

by which φ takes the value of the order of unity, Eq.(A.10) is rewritten as,

$$\frac{\partial}{\partial t}\left(\varphi-\nabla_\perp^2\varphi\right)+\nabla_y\varphi-\nabla_\perp\varphi\times\hat{z}\cdot\nabla_\perp\nabla_\perp^2\varphi=0, \tag{A.12}$$

where the derivative on the (x, y) plane is explicitly written. This equation is also written as,

$$\frac{\partial}{\partial t}\left(\varphi - \nabla_{\perp}^{2}\varphi\right) + \nabla_{y}\varphi - \left[\varphi, \nabla_{\perp}^{2}\varphi\right] = 0, \tag{A.13}$$

where $\left[g, h\right] = \frac{\partial g}{\partial x}\frac{\partial h}{\partial y} - \frac{\partial h}{\partial x}\frac{\partial g}{\partial y}$. Equation(A.12) or Eq.(A.13) is called the Charney–Hasegawa–Mima equation (or Hasegawa–Mima equation).

A1.3 Fourier decomposition

The fluctuation is decomposed into Fourier components in space,

$$\varphi\left(\boldsymbol{x}, t\right) = \sum_{\boldsymbol{k}} \varphi_{\boldsymbol{k}}\left(t\right) \exp\left(i\boldsymbol{k}\cdot\boldsymbol{x}\right), \tag{A.14}$$

where the variable $\boldsymbol{x}$ is in two-dimensional space, so that $\boldsymbol{k}$ covers the two-dimensional Fourier space. (The suffix $\perp$, which denotes the direction perpendicular to the main magnetic field, is suppressed for simplicity of notation.) By this transformation, Eq.(A.12) turns to be,

$$\frac{\partial}{\partial t}\varphi_{\boldsymbol{k}}\left(t\right) + i\omega_{\boldsymbol{k}}\varphi_{\boldsymbol{k}}\left(t\right) + \frac{1}{2}\sum_{\boldsymbol{k}=\boldsymbol{k}'+\boldsymbol{k}''} N_{\boldsymbol{k},\boldsymbol{k}',\boldsymbol{k}''}\varphi_{\boldsymbol{k}'}\left(t\right)\varphi_{\boldsymbol{k}''}\left(t\right) = 0, \tag{A.15}$$

where,

$$\omega_{\boldsymbol{k}} = \frac{k_y}{1+k^2}, \tag{A.16a}$$

is the linear wave dispersion relation and the nonlinear coupling coefficient is given as,

$$N_{\boldsymbol{k},\boldsymbol{k}',\boldsymbol{k}''} = \frac{-1}{1+k^2}\left(\boldsymbol{k}' \times \boldsymbol{k}'' \cdot \hat{z}\right)\left(k''^2 - k'^2\right). \tag{A.16b}$$

In Eq.(A.16a), the term $1+k^2$ on the right-hand side denotes the 'effective mass' (i.e., the first parenthesis) on the left-hand side of Eq.(A.12), the second term $\boldsymbol{k}' \times \boldsymbol{k}'' \cdot \hat{z}$ comes from the operator $\nabla_{\perp}\varphi \times \hat{z} \cdot \nabla_{\perp}$, and the coefficient $k''^2 - k'^2$ represents the symmetrization. Equation (A.16b) shows that the nonlinear coupling vanishes if $\boldsymbol{k}'$ and $\boldsymbol{k}''$ are parallel (or if $k''^2 = k'^2$ holds).

A1.4 Conservation relation

It is known that the H–M equation has the following conservation quantities,

$$E = \iint \mathrm{d}^2x\left(\varphi^2 + \left(\nabla\varphi\right)^2\right), \tag{A.17a}$$

$$\Omega = \iint d^2x \left(\varphi - \nabla^2\varphi\right)^2, \tag{A.17b}$$

where $\mathcal{E}$ and Ω are interpreted as the energy and enstropy, respectively. It is noted that the energy and enstropy are written as,

$$E_f = \iint d^2x (\nabla\varphi)^2 \quad \text{and} \quad \Omega_f = \iint d^2x \left(\nabla^2\varphi\right)^2, \tag{A.18}$$

respectively, in the fluid dynamics. The difference in Eq.(A.17) from Eq.(A.18) is that the density perturbation is associated in the plasma dynamics, $\tilde{n}/n \simeq e\tilde{\phi}/T_e$. The density perturbation (i.e., the pressure perturbation) appears as the first term in the parenthesis of Eq.(A.17a).

The conservation relation is understood by calculating the weighted integral of the nonlinear term $\nabla_\perp\varphi \times \hat{z} \cdot \nabla_\perp\nabla_\perp^2\varphi$ in Eq.(A.12). Multiplying by φ and integrating over space, and performing a partial integral, one obtains,

$$\iint d^2x \left[\varphi\left(\nabla_\perp\varphi \times \hat{z} \cdot \nabla_\perp\nabla_\perp^2\varphi\right)\right] = -\iint d^2x \left[\nabla_\perp\varphi \cdot \nabla_\perp\varphi \times \hat{z}\nabla_\perp^2\varphi\right] = 0. \tag{A.19}$$

The linear wave term $\nabla_y\varphi$ is an odd function in y so that the integral of $\varphi\nabla_y\varphi$ over a space vanishes. Thus, multiplying φ to Eq.(A.12) and integrating it over space, one finally obtains,

$$\iint d^2x \left[\varphi\frac{\partial}{\partial t}\left(\varphi - \nabla_\perp^2\varphi\right)\right] = 0, \tag{A.20a}$$

i.e.,

$$\frac{\partial}{\partial t}E = 0. \tag{A.20b}$$

In a similar manner, one can show that multiplication of $\nabla_\perp^2\varphi$ to Eq.(A.12) leads to the identity,

$$\iint d^2x \left[\nabla_\perp^2\varphi\frac{\partial}{\partial t}\left(\varphi - \nabla_\perp^2\varphi\right)\right] = 0. \tag{A.21a}$$

With the help of Eq.(A.20), one obtains from Eq.(A.12) the relation,

$$\frac{\partial}{\partial t}\Omega = 0. \tag{A.21b}$$

In a Fourier representation, one has two conservation quantities,

$$E = \sum_k \left(1 + k^2\right) I_k, \tag{A.22a}$$

$$\Omega = \sum_k \left(1 + k^2\right)^2 I_k, \tag{A.22b}$$

where $I_k = \left\langle \varphi_k(t)\, \varphi_k^*(t) \right\rangle$.

A1.5 Order of magnitude

The dynamical equation (A.13) indicates that in stationary turbulence, if it exists, scaling properties hold such that (apart from numerical factors),

$$\frac{e\tilde{\phi}}{T_e} \sim \frac{\rho_s}{L_n},$$

and,

$$\text{diffusivity} \sim \frac{c_s \rho_s^2}{L_n}, \tag{A.23}$$

because the latter has the dimension of $[length]^2[time]^{-1}$. Note that Eq.(A.23) gives only a possible *scaling* of the diffusivity. Rigorously speaking, the particle diffusivity vanishes for the H–M system, on account of Boltzmann electrons. These relations show that the fluctuation level and turbulence-driven diffusivity are induced by plasma inhomogeneity, which is characterized by the normalized gradient scale length ρ_s/L_n. The diffusivity in Eq.(A.23) is often called 'gyro-reduced Bohm diffusion'.

A1.6 Propagating solitary structure

A solution of interest is a solitary structure, which moves with a constant velocity,

$$\varphi(x, y, t) = \varphi\left(x, \hat{y}\right), \quad \text{and} \quad \hat{y} = y - v_{**}t. \tag{A.24}$$

(x- and z-coordinates are taken in the direction of decreasing density and in the direction of magnetic field, respectively.) The velocity v_{**} stands for the propagating velocity in the direction of diamagnetic drift. An example is,

$$\begin{aligned} \varphi(x, \hat{y}) &= v_{**} a_m K_1^{-1}(b)\, K_1\left(b r a_m^{-1}\right) \cos\theta && \text{for} \quad r > a_m, \\ &= v_{**}\left[r + b^2\gamma^{-2}\left(r - a_m J_1^{-1}(\gamma)\, J_1\left(\gamma r a_m^{-1}\right)\right)\right] \cos\theta && \text{for} \quad r < a_m, \end{aligned} \tag{A.25}$$

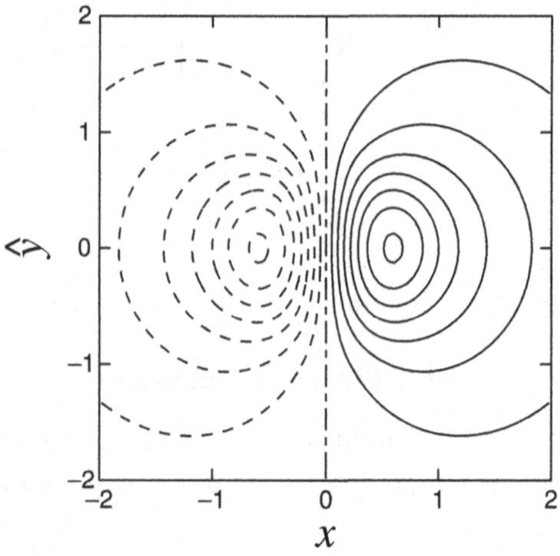

Fig. A1.3. Electrostatic perturbation for the drift wave modon. It propagates in the y-direction. (Parameters: $A_m = 1$, $v_{**} = (3/2)V_{\text{de}}$)

where $r^2 = x^2 + \hat{y}^2$, $x = r\cos\theta$, $b = a_m\sqrt{1 - V_{\text{de}}/v_{**}}$, γ is a solution of the equation $\gamma J_1(\gamma) J_2^{-1}(\gamma) = -bK_1(b) K_2^{-1}(b)$, and functions J and K are the Bessel function and modified Bessel function, respectively. This propagating solitary structure is a pair of oppositely-rotating vorteces. It is often called drift wave *modon*. It is anti-symmetric in the x-direction (in the direction of the density gradient) and symmetric in the $\hat{y}$-direction. A equi-contour plot of the drift wave modon is illustrated in Figure A1.3. Away from the vortex, the perturbation decays with the asymptotic form,

$$\left|\varphi\left(x, \hat{y}\right)\right| \to \exp\left(-r\sqrt{1 - V_{\text{de}}v_{**}^{-1}}\right). \tag{A.26}$$

Thus this structure is localized in the region of $r \sim a_m/b$.

This solitary solution is characterized by the velocity v_{**} and the size a_m. The magnitude of the potential perturbation, the velocity and the size are coupled to each other. For the same size a_m, the stronger vortex (with larger $|\phi|$) has the larger velocity of v_{**}. The vortex is propagating either in the direction of the ion diamagnetic drift $v_{**} < 0$ (ion modon) or of the electron diamagnetic drift $v_{**} > 0$ (electron modon). The velocity of the drift wave vortex (electron modon), is faster than the drift velocity V_{de}. The dynamics and lifetime have been discussed (Makino *et al.*, 1981; Meiss and Horton, 1983). Figure A1.3 illustrates a contour of electrostatic perturbation of a modon.

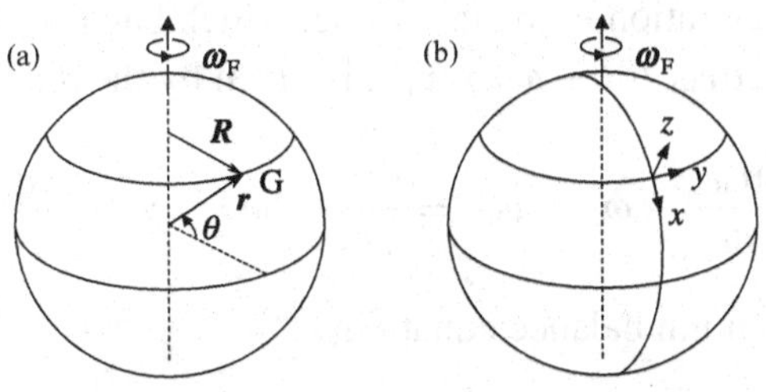

Fig. A1.4. Rotating sphere (a). Coordinates on a rotating sphere (b). The x-axis in the direction of latitude (from pole to equator), y-axis in the direction of longitude, and z-axis in the vertical direction.

A1.7 Waves in a rotating atmosphere

We next explain Rossby waves in a rotating atmosphere of planets.

A1.7.1 Introduction for rotating coordinates

The observers of geophysical phenomena are rotating in space, with the Earth. To understand observations of geophysical, planetary and astrophysical objects, which are rotating on their own axes, description using coordinates rotating with the objects is convenient. Thus, we begin with a brief introduction of the rotating coordinates for those not familiar with the rotating sphere.

We consider the rotating sphere, as shown in Fig.A1.4(a). The radius is given by r, latitude is given by θ, and the distance from the rotation axis is denoted by R. The point G on the ground has a velocity,

$$\boldsymbol{v}_{\mathrm{G}} = \omega_{\mathrm{F}} \times \boldsymbol{r} = \omega_{\mathrm{F}} \times \boldsymbol{R}, \tag{A.27}$$

where ω_{F} is the angular frequency of rotation. The acceleration of the point G is given as,

$$\frac{\mathrm{d}\boldsymbol{v}_{\mathrm{G}}}{\mathrm{d}t} = \frac{\mathrm{d}}{\mathrm{d}t}(\boldsymbol{\omega}_{\mathrm{F}} \times \boldsymbol{R}) = \boldsymbol{\omega}_{\mathrm{F}} \times \boldsymbol{v}_{\mathrm{G}}. \tag{A.28}$$

When the observed velocity relative to G (i.e., the velocity in the rotating frame) on the surface is $\boldsymbol{v}$, the total velocity $\boldsymbol{v}_{\mathrm{tot}}$ is given by $\boldsymbol{v}_{\mathrm{tot}} = \boldsymbol{v} + \boldsymbol{\omega}_{\mathrm{F}} \times \boldsymbol{r}$. The acceleration observed in the rotating frame $\mathrm{d}\boldsymbol{v}/\mathrm{d}t$ is then written as,

$$\frac{\mathrm{d}\boldsymbol{v}_{tot}}{\mathrm{d}t} = \frac{\mathrm{d}\boldsymbol{v}}{\mathrm{d}t} + \boldsymbol{\omega}_{\mathrm{F}} \times \boldsymbol{v}. \tag{A.29a}$$

The motion $\boldsymbol{v}_{\mathrm{tot}} = \boldsymbol{v} + \boldsymbol{\omega}_{\mathrm{F}} \times \boldsymbol{r}$ has centripetal acceleration $\boldsymbol{\omega}_{\mathrm{F}} \times \boldsymbol{v}_{\mathrm{tot}}$, i.e.,

$$\boldsymbol{\omega}_{\mathrm{F}} \times \boldsymbol{v}_{\mathrm{tot}} = \boldsymbol{\omega}_{\mathrm{F}} \times \boldsymbol{v} + \frac{\mathrm{d}\boldsymbol{v}_{\mathrm{G}}}{\mathrm{d}t}. \tag{A.29b}$$

The total vector acceleration in the lab frame, which balances against the external force per mass F/ρ, $\mathrm{d}\boldsymbol{v}_{\mathrm{tot}}/\mathrm{d}t + \boldsymbol{\omega}_{\mathrm{F}} \times \boldsymbol{v}_{\mathrm{tot}}$, is given by the relation,

$$\frac{\mathrm{d}\boldsymbol{v}_{\mathrm{tot}}}{\mathrm{d}t} + \boldsymbol{\omega}_{\mathrm{F}} \times \boldsymbol{v}_{\mathrm{tot}} = \frac{\mathrm{d}\boldsymbol{v}}{\mathrm{d}t} + 2\boldsymbol{\omega}_{\mathrm{F}} \times \boldsymbol{v} + \frac{\mathrm{d}\boldsymbol{v}_{\mathrm{G}}}{\mathrm{d}t}. \tag{A.30}$$

This gives the momentum balance equation,

$$\frac{\mathrm{d}\boldsymbol{v}}{\mathrm{d}t} + 2\boldsymbol{\omega}_{\mathrm{F}} \times \boldsymbol{v} = \frac{1}{\rho}\vec{F} - \frac{\mathrm{d}\boldsymbol{v}_{\mathrm{G}}}{\mathrm{d}t}. \tag{A.31}$$

Usually the right-hand side denotes the force which includes the centrifugal force, which is additive with pressure. The term,

$$\boldsymbol{f} = -2\boldsymbol{\omega}_{\mathrm{F}} \times \boldsymbol{v}, \tag{A.32}$$

is called the Coriolis force. It is straightforward to see that the Coriolis force for horizontal motion is given as,

$$\boldsymbol{f}_{\perp} = -2\omega_{\mathrm{F},z}\hat{z} \times \boldsymbol{v}, \tag{A.33}$$

where the $\hat{z}$-axis is taken in the vertical direction. Note the structural similarity of the Coriolis force to the Lorentz force.

The equation of motion on a rotating sphere takes the form,

$$\underset{(1)}{\rho\frac{\partial \boldsymbol{v}}{\partial t}} + \underset{(2)}{\rho\boldsymbol{v}\cdot\nabla\boldsymbol{v}} - \underset{(3)}{\boldsymbol{J}\times\boldsymbol{B}} + \underset{(4)}{\nabla p} - \underset{(5)}{\rho\boldsymbol{g}} + \underset{(6)}{2\rho\boldsymbol{\omega}_{\mathrm{F}}\times\boldsymbol{v}_{\mathrm{tot}}} - \underset{(7)}{\mu\nabla^2\boldsymbol{v}} = 0, \tag{A.34}$$

where $\boldsymbol{g}$ is the gravitational acceleration and p is the pressure. Combining some of the terms (1)–(7) in Eq.(A.34), various phenomena in fluid dynamics can be addressed. Characteristic examples are listed in Table A1.1.

A1.7.2 Rossby wave

The Rossby wave has special importance in geophysical fluid dynamics. This wave has a strong similarity to drift waves in magnetized plasmas and provides a bridge from the study of plasmas to many other applications in nature (Pedlosky, 1987; Vallis, 2006).

Local Cartesian coordinates are chosen on a rotating sphere. In this monograph, we choose the x-axis in the direction of latitude (from pole to equator), the y-axis in the direction of longitude (from west to east) and z-axis in the vertical direction (Fig.A1.4(b).) This distinction is for the convenience of plasma physicists who are

Table A1.1. *Terms in a dynamical equation (A.34) and related equations, flow and waves (Gossard and Hooke, 1975)*

(1)	(2)	(3)	(4)	(5)	(6)	(7)	
			◇	◇			Hydrostatic equilibrium
	◇		◇	◇			Bernoulli equation
◇	◇		◇	◇		◇	Navier–Stokes equation
			◇	◇	◇		Geostrophic and thermal wind eq.
	◇		◇		◇		Gradient-driven flow
	◇				◇		Inertial flow
			◇			◇	Stokes flow
					◇	◇	Eckman spiral
◇	◇					◇	Fick's law
◇	◇		◇				Sound wave
◇	◇				◇		Rossby wave
◇	◇	◇					Alfvén wave

accustomed to using the y-axis for the ignorable coordinate and the x-axis along the direction of inhomogeneity. (The convention in fluid dynamics is to choose the x-axis in the direction of longitude and the y-axis in the direction of latitude.)

Neglecting the gravitational force and viscosity in Eq.(A.34), the dynamical equation for the Rossby wave is derived. Rossby waves occur in rapidly rotating systems, where the Coriolis frequency ω_{F} is the fastest in the system. In such a case, the motion is primarily *geostrophic*, so the dominant balance in the momentum equation is between the pressure gradient and the Coriolis force ((4) and (6) in Eq.(A.34)). Such dynamics are classified to as geostrophic, and occur in regimes where $R_o < 1$. Here R_o is the Rossby number, which is the ratio of the vorticity of the motion to the rotation frequency ω_{F}. One may assume incompressible motion on the horizontal plane (V_x, V_y). One can then relate the fluid velocity to a stream function,

$$V_x = -\frac{\partial}{\partial y}\phi, \quad V_y = \frac{\partial}{\partial x}\phi. \tag{A.35}$$

Taking a curl of Eq.(A.34) eliminates the pressure term, and assuming independence of the z-direction, one has,

$$\frac{d}{dt}\nabla_\perp^2\phi - 2\frac{\partial \omega_{\mathrm{F},z}}{\partial x}\frac{\partial \phi}{\partial y} = 0, \tag{A.36}$$

(here, (3), (5) and (7) are neglected). Noting that the Coriolis force is stronger near the pole and weaker near the equatorial plane, (i.e., $\omega_{\mathrm{F},z}$ is a decreasing function of x), the coefficient $\partial\omega_{\mathrm{F},z}/\partial x$ is negative. If the perturbation is not constant in the

z-direction, Eq.(A.36) has a form,

$$\frac{d}{dt}\left(\nabla_{\perp}^{2}\phi - \frac{{\omega_{\mathrm{F},z}}^{2}}{gH_m}\phi\right) - 2\frac{\partial\omega_{\mathrm{F},z}}{\partial x}\frac{\partial\phi}{\partial y} = 0, \tag{A.37}$$

where H_m is the eigenvalue in the vertical waveform, being of the order of the vertical thickness. This introduces a spatial scale, the Rossby radius,

$$\rho_{\mathrm{R}} = \frac{\sqrt{gH_m}}{\omega_{\mathrm{F},z}}. \tag{A.38}$$

Equation (A.37) is equivalent to the Hasegawa–Mima equation for drift wave turbulence. In the context of geophysical fluid dynamics, the equation is commonly referred to as the 'quasi-geostrophic equation', and was first derived by J. Charney (Charney, 1948). The gradient of the Coriolis force in the direction of latitude plays the role of the density gradient in magnetized plasmas. The Rossby radius ρ_{R} is the analogue of the gyro-radius for drift wave systems. Taking a perturbation to have the form,

$$\phi \propto \cos(k_x x)\exp\left(ik_y y - i\omega t\right), \tag{A.39}$$

the dispersion relation of the linear perturbation is given as,

$$\omega = \frac{2k_y\rho_{\mathrm{R}}^2}{1+\rho_{\mathrm{R}}^2 k_{\perp}^2}\frac{\partial\omega_{F,z}}{\partial x}. \tag{A.40}$$

where the quantity $\left|2\rho_{\mathrm{R}}^2(\partial\omega_{\mathrm{F},z}/\partial x)\right|$ plays the role of the diamagnetic drift velocity in confined plasmas. The wave is propagating in the y-direction (westward), because $\partial\omega_{\mathrm{F},z}/\partial x$ is negative. The propagation of the Rossby wave is illustrated in Fig.A1.5.

It is useful to evaluate various scale lengths for the Earth's atmosphere:

- vertical height: $H_m \simeq 10^4$m, angular frequency: $\omega_{\mathrm{F},z} \simeq 1.5 \times 10^{-4}$sec;
- gradient of frequency: $(\partial\omega_{\mathrm{F},z}/\partial x) \sim 10^{-11}\mathrm{m}^{-1}\mathrm{sec}^{-1}$;
- Rossby radius: $\rho_{\mathrm{R}} \sim 2 \times 10^6$m;
- phase velocity: $\left|2\rho_{\mathrm{R}}^2(\partial\omega_{\mathrm{F},z}/\partial x)\right| \sim 10^2\mathrm{msec}^{-1}$. (A.41)

The Rossby radius is about 10% of the arc length of the equator.

With introduction of the normalization,

$$\left|2\rho_{\mathrm{R}}\frac{\partial\omega_{\mathrm{F},z}}{\partial x}\right|^{-1} t \to t,\ \frac{x}{\rho_{\mathrm{R}}} \to x,\ \frac{y}{\rho_{\mathrm{R}}} \to y,\ \text{and}\ \left(2\left|\frac{\partial\omega_{\mathrm{F},z}}{\partial x}\right|\right)^{-1}\rho_{\mathrm{R}}^{-3}\phi \to \varphi, \tag{A.42}$$

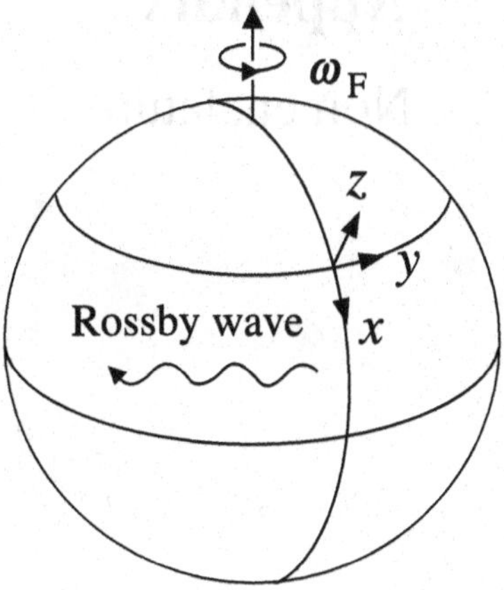

Fig. A1.5. Propagation of Rossby wave in the westward direction.

Eq.(A.37) takes the form,

$$\frac{\partial}{\partial t}\left(\nabla_\perp^2\varphi - \varphi\right) + \left[\varphi, \nabla_\perp^2\varphi\right] + \frac{\partial \varphi}{\partial y} = 0, \tag{A.43}$$

where $\left[f, g\right] = (\nabla f \times \nabla g) \cdot \hat{z}$. Eq.(A.43) has a close similarity to Eq.(A.13) (Rhines, 1975; Hasegawa *et al.*, 1979; Horton and Hasegawa, 1994).

Appendix 2

Nomenclature

$\boldsymbol{A}$	vector potential
A	amplitude of vector potential in 2D MHD turbulence, Eq.(9.19)
$\boldsymbol{B}_0$	mean magnetic field (§9.2)
$B_{\alpha,\beta}$	cross-section of collision (§2.2.3.6)
$C(\boldsymbol{k}, \omega)$	velocity-integrated discreteness correlation function, Eq.(2.14)
C_k^{P}	turbulent collision operator through wave–particle interaction
C_k^{W}	turbulent collision operator through wave–wave interaction
c_{s}	ion sound speed
$\boldsymbol{D}$	diffusion coefficient of quasi-particle in disparate-scale interaction, Eq.(7.35)
d_{i}	ion collisionless skin depth, c/ω_{pi}
$\boldsymbol{D}_k$	k-space diffusivity, Eq.(5.68)
$\boldsymbol{D}_{k,\omega}$	electric displacement vector, $\boldsymbol{D}_{k,\omega} = \epsilon(k, \omega)\boldsymbol{E}_{k,\omega}$
$d_{k,\omega}$	rate of test particle scattering in magnetized plasma, Eq.(4.43)
D_{QL}	quasi-linear diffusion coefficient
$D(v)$	fluctuation-induced diffusion (§2.2.3)
$D_{\mathrm{NR}}(v)$	quasi-linear diffusion coefficient for non-resonant particles, Eq.(3.22)
D_{vv}	velocity diffusion of magnetized plasma, Eq.(3.68)
$D_{v,\mathrm{col}}$	collisional diffusion coefficient in velocity space (§8.1.2.3)
$D_{\mathrm{rel}}(x_-)$	relative diffusion function, Eq.(8.56)
D_{jj}, $j = 1, 2$	quasi-linear diffusion coefficient at phase space 1 or 2, Eq.(8.47)
D_{12}, D_{21}	resonant correlated diffusion coefficient, Eq.(8.47)

e	unit charge
$E, \boldsymbol{E}$	electric field
E	total energy of turbulence (§2.3.2.1)
E	total energy of turbulence for H–M equation, (§6.4.5.1)
E	total energy of MHD turbulence (§9.4.2)
E_{k}	total electric field energy per mode, Eq.(2.23)
E^{f}	electric field energy density
E_{kin}	particle kinetic energy density
$E_{\mathrm{kin,k}}$	particle kinetic energy density per mode (§2.2.2)
$E_{\mathrm{kin}}^{\mathrm{nr}}$	non-resonant particle kinetic energy density
$E_{\mathrm{kin}}^{\mathrm{res}}$	resonant particle kinetic energy density
E_{p}	ponderomotive energy density, Eq.(3.23)
f	distribution function in phase space
$\langle f \rangle$	mean distribution function
δf	deviation of f from mean
f^{c}	coherent Vlasov response, Eq.(2.1)
$\tilde{f}$	noise due to particle discreteness, deviation owing to granulations
f^{gc}	guiding centre distribution function
$\left\langle \tilde{f}(1)\, \tilde{f}(2) \right\rangle$	discreteness correlation function
$\left\langle \delta f^{2} \right\rangle$	phasestrophy; mean-square phase space density (§8.1.2.3)
$\langle \delta f(1)\, \delta f(2) \rangle$	two-point phase space density correlation function, Eq.(8.16)
$F(v)$	normalized average distribution function, $v_{\mathrm{T}} n^{-1} \langle f(v) \rangle$ (§2)
$F(x_{-}, v_{-})$	probability density function of relative dispersion, Eq.(8.57)
$F_{r}(v)$	dynamical friction term (§2.2.3)
F_{k}	fluctuating force by nonlinear interaction on a test mode (§6.4.2.2)
$\hat{F}_{k}$	fluctuating force (contribution of one triad suppressed) (§6.4.2.2)
F_{k}^{ext}	external forcing on a Fourier component (§6.1.1)
$\mathbf{F}_{m,v}^{\mathrm{ext}}$	magnetic/kinetic forcing in MHD turbulence, Eq.(9.1)
$G(t)$	response function for random oscillator, Eq.(6.3)
H_{A}	mean-square magnetic potential (§9.4.2)
$\mathcal{H}_{A}$	mean-square magnetic potential density, A^{2} (§9.4.3)

H_c	cross-helicity (§9.4.2)
I_k	one-time spectral function, Eq.(6.128)
$J(v)$	flux of the course-grained phase space density (§2.2.3)
$K(x-x', t-t')$	convolution (nonlocal interaction) kernel, Eq.(6.29)
$\mathcal{K}$	Kubo number
$\mathcal{K}_\theta$	phase space Kubo number (§6.1.6)
K	kinetic energy density for neutral fluid turbulence (§7.4.4)
$\tilde{\mathbf{k}}$	modulation of wave vector for microscopic modes by large-scale waves, Eq.(7.40)
k_B	Boltzmann constant. Unit of $k_\mathrm{B}=1$ for temperature is employed unless specified.
k_cap	wave number of transition for capillary wave (for surface gravity wave)
k_w	wave number of the wind wave (for surface gravity wave)
l_d	dissipation scale, Eq.(2.56)
l_d	microscale for enstrophy cascade in fluid turbulence
l_E	energy input scale in neutral fluid turbulence (§7.4.4)
$l_{\perp d}$	Kraichnan–Iroshnikov dissipation scale of MHD turbulence (via perpendicular cascade)
L	characteristic scale length of mean distribution
L_A	gradient scale length for $\langle A\rangle$
L_{A^2}	gradient scale length for$\langle AA\rangle$
L_k	linear operator, Eq.(3.67)
L_n	density-gradient scale length
L_s	magnetic shear length, Eq.(4.11)
L_T	temperature-gradient scale length
$L_\perp$	characteristic scale length (perpendicular to main magnetic field) of mean distribution
$L_\parallel$	parallel scale of the modulation of Alfvén wave envelope, (Section 9.3.1)
m	particle mass
n	plasma density
n	refractive index, $n=c/v_\mathrm{ph}$

$N(\mathbf{k}, \mathbf{x}, \mathbf{t})$	wave population density;
$\langle N \rangle$	mean wave population density
$\tilde{N}$	deviation of $N(\mathbf{k}, \mathbf{x}, \mathbf{t})$ $from$ $\langle N \rangle$
N_{BV}	BV buoyancy frequency, Eq.(5.86b)
N_i	number of wave quanta at i-th mode,
$N_{k,\omega}$	nolinearlity in three-wave coupling, Eq.(4.19)
p	plasma (thermal) pressure
p_f	net momentum associated with the phase space density fluctuation δf, Eq.(8.13)
$p_{f,e}$, $p_{f,i}$	net momentum associated with the electron/ion phase space density fluctuation, Eq.(8.13)
p_{pw}	radiation pressure of plasma waves (§7.2.1)
$\mathcal{P}$	projection operator, Eq.(6.59), Eq.(6.72)
$P(1, 2)$	production term for $\langle \delta f(1)\, \delta f(2) \rangle$,Eq.(8.21)
$P_{\mathrm{c}}(1, 2)$	coherent production term for $\langle \delta f(1)\, \delta f(2) \rangle$, Eq.(8.27)
$P_{\mathrm{G}}(1, 2)$	incoherent, granulation-induced production term for $\langle \delta f(1)\, \delta f(2) \rangle$, Eq.(8.33)
$P_i(1, 2)$	production term for ion phasestrophy $\langle \delta f(1)\, \delta f(2) \rangle$, Eq.(8.37)
$P_{i,i}(1, 2)$	production term for ion phasestrophy $\langle \delta f(1)\, \delta f(2) \rangle$ by ions, Eq.(8.39)
$P_{i,e}(1, 2)$	production term for ion phasestrophy $\langle \delta f(1)\, \delta f(2) \rangle$ by electrons, Eq.(8.39)
P_m	magnetic Prandtl number (ratio of viscosity to resistivity) (§9.2)
P_{W}	wave momentum density
q	particle charge
$\boldsymbol{q}$	wave number of (large-scale) scattering field (in disparate scale interaction)
$q(r)$	safety factor of toroidal magnetic field
$\mathcal{Q}$	projection operator, Eq.(6.61)
Q	dissipation $\langle E \cdot J \rangle$, Eq.(3.37)
$Q(x, t)$	potential vorticity in quasi-geostrophic system, Eq.(8.2c)
$Q_{e,i}$	fluctuation-induced energy flux of (electron or ion), Eq.(3.62)
r	minor radius of torus

R	major radius of torus
R	radius of sphere
R_e	Reynolds number
R_m	magnetic Reynolds number (§9.2)
$R_{k,\omega}$	linear response function, Eq.(2.1)
$\tilde{R}(t)$	fluctuating force, Eq.(6.78)
$\mathcal{S}$	Chirikov parameter, Eq.(7.46)
S	entropy for wave population density, Eq.(5.54)
S	entropy functional for H–M system, Eq.(6.147)
S_r	radial wave energy density (§3.5.1)
T	temperature
T_k	triad correlator for k-mode, Eq.(5.58), nonlinear transfer of energy, Eq.(6.35)
T_k	triplet interaction operator in MHD turbulence, Eq.(9.38)
T_{1k}, $T_k^{(I)}$	triad correlator which gives nonlinear noise (incoherent emission), Eq.(5.65), Eq.(6.38)
T_{2k}, $T_k^{(C)}$	triad correlator which gives nonlinear relaxation, Eq.(5.64), Eq.(6.38)
$T(\Delta k, \Delta t)$	transition probability for a step, Eq.(5.68)
$T(\Delta v_1, v_1; \Delta v_2, v_2; \Delta t)$	two-particle transition probability, Eq.(8.49)
$T_{1,2}(x_-, v_-)$	relative evolution/dispersion operator, Eq.(8.9), Eq.(8.19)
$v(l)$	fluctuation velocity in fluid turbulence at the scale l (§2.3.1)
v_{A}	Alfvén velocity
V_{de}	diamagnetic drift velocity of electrons
v_g	group velocity of a wave
$\mathbf{V}_k$	mean flow velocity in k-space, Eq.(5.68)
$V_{\mathbf{k},\mathbf{k}',\mathbf{k}''}$, $V_{\mathbf{k},\mathbf{k}'}$	coupling function for three mode nonlinearity
v_{ph}	phase velocity of a wave
v_{SP}	Sweet–Parker velocity, $v_{\mathrm{A}}/\sqrt{R_m}$ (§9.4.4.3)
v_{T}	thermal velocity
$v_{\mathrm{T}e,i}$	electron (ion) thermal velocity

V_*	friction velocity for pipe-flow turbulence, (§2.3.3.2)
Δv	separatrix width of phase space island, $(q\phi/m)^{1/2}$
Δv_T	width of velocity of the broadened resonance, Eq.(4.8)
W	wave energy density, Eq.(3.24)
W_k	total wave energy per mode (§2.2.2)
x_+, v_+	centroid coordinates, Eq.(8.18)
x_-, v_-	relative coordinates, Eq.(8.18)
y_d	viscous sub-layer width for pipe-flow turbulence, Eq.(2.74)
Z_k	enstrophy density, Eq.(6.144)
$Z_\pm$	Elsasser variable, Eq.(9.4)
α	constant phasestrophy cascade flux (analogous to enstrophy cascade flux), Eq.(8.11)
$\alpha_{k,\omega}$	phase factor of fluctuation at $k, \omega, \alpha_{k,\omega} = \exp i\theta_{k,\omega}$
α_T	coefficient for electron thermal force, Eq.(3.76)
β	gradient of mean potential vorticity in quasi-geostrophic system, Eq.(8.2)
β	ratio of kinetic pressure to magnetic pressure (§9.3.1)
$\chi_\parallel$	parallel thermal conductivity
$\chi_{e,i}$	electron and ion susceptibilities, Eq.(2.31)
χ_e	linear susceptibility of electrons, Eq.(8.30)
ϵ	mean rate of dissipation per unit mass in fluid turbulence
$\epsilon\,(k, \omega)$	dielectric function, Eq.(2.6)
ϕ	electrostatic potential
ϕ	electrostatic perturbation in plasma turbulence
ϕ	stream function in 2D fluid turbulence
Φ	velocity stream function in 2D MHD turbulence, Eq.(9.19)
$\langle\phi^2\rangle_{k,\omega}$	potential fluctuation correlation function
γ_k	linear growth rate of k-Fourier mode
γ_k^L	linear growth rate (damping rate) of k-Fourier mode
γ_k^NL	non-linear growth rate (damping rate) of k-Fourier mode
γ_k^eff	effective growth rate of granulations in the stationary state, Eq.(8.70)
γ_decay	growth rate of decay instability, Eq.(5.21)
γ_E	energy exchange rate of three-wave coupling, Eq.(5.18)

$\gamma_{\mathrm{E}}^{\mathrm{coh}}$	energy exchange rate via coherent interaction
$\gamma_{\mathrm{E}}^{\mathrm{stoch}}$	energy exchange rate via stochastic interaction, Eq.(5.39), Eq.(5.40)
$\gamma_{g,k}$	growth rate of granulations, Eq.(8.73)
γ_{PS}	growth rate of parametric subharmonic instability
$\gamma_{\mathrm{PS}}^{\mathrm{coh}}$	parametric subharmonic instability via coherent interaction
$\gamma_{\mathrm{PS}}^{\mathrm{stoch}}$	parametric subharmonic instability via stochastic interaction
$\hat{\gamma}_k$	eddy damping rate of a test mode (§6.4.2.2)
$\hat{\hat{\gamma}}_k$	eddy damping rate of a test mode (contribution of one triad suppressed) (§6.4.2.2)
γ_T	specific heat ratio of electrons, Eq.(7.2)
Γ	circulation in quasi-geostrophic system, Eq.(8.4)
Γ	circulation in Vlasov system, Vlasov eddy, Eq.(8.6)
$\Gamma(s)$	memory function Eq.(6.65), Eq.(6.83)
Γ_k, $\Gamma_{k,\omega}$	effective damping rate of a test mode by nonlinearity, Eq.(6.19)
$\hat{\Gamma}$	decay rate of quasi-particle (by linear or like-scale-nonlinear interactions), Eq.(7.27)
Γ_A	spatial flux of magnetic potential, Eq.(9.20)
Γ_{A^2}	spatial flux of mean-square potential, Eq.(9.36)
η	enstrophy cascade rate, Eq.(2.68)
η	electric resistivity
$\eta_{e,i}$	temperature gradient parameter, $\eta_{e,i} = d\ln T_{e,i}/d\ln n_e$
η_{T}	turbulent resistivity
η_{T}^k	kinematic turbulent resistivity, Eq.(9.33), Eq.(9.61)
$\hat{\eta}_{\tau,k}$	diffusion rate by turbulent resistivity in MHD turbulence, Eq.(9.42)
λ_{De}	electron Debye length
λ_{Di}	ion Debye length
ν	(molecular) viscosity
ν	model of collisional damping rate for $\langle\delta f(1)\,\delta f(2)\rangle$, Eq.(8.17)
ν_{ii}	ion–ion collision frequency, Eq.(6.160)
ν_{T}	turbulent (eddy) viscosity
θ	poloidal angle
$\theta_{k,\omega}$	phase of fluctuation at k, ω
$\Theta_{k,k',k''}$	triad coherence time, Eq.(5.63), Eq.(6.44)

$\rho\,(a, b; t)$	probability density function of variables (a, b), (§6.2.1.1)
ρ_0	mass density of plasma, Eq.(9.1)
$\tilde{\rho}$	mass density perturbation by the modulation of Alfvén wave envelope, Eq.(9.15)
ρ_i	ion gyroradius
ρ_s	ion gyroradius at electron temperature, Eq.(6.107)
ρ_w	density of water (for surface gravity wave)
τ_{ac}	spectral autocorrelation time (Table 3.1), Eq.(3.20). for drift waves, Eq.(3.71)
$\tau_{\mathrm{ac}\parallel}$	auto-correlation time of Alfvén spectrum, Eq.(9.7)
τ_b	period of particle bounce motion (table 3.1)
τ_c	correlation time of interaction or fluctuations (in general)
τ_{ck}	correlation time of k-Fourier mode (in general)
τ_c	characteristic decorrelation rate for granulations, Eq.(8.59)
$\tau_c^{\Phi}\,(\mathbf{k})$	self-correlation time for fluid perturbation in MHD turbulence, Eq.(9.24)
$\tau_c^{A}\,(\mathbf{k})$	self-correlation time for field perturbation in MHD turbulence, Eq.(9.24)
τ_{cascade}	time for cascade
τ_{coll}	collision time, $1/\nu$ (§8.2.3.4)
τ_{D}	diffusion time in the velocity space, $\tau_{\mathrm{D}} \sim \Delta v^2/D_v$
τ_{E}	energy transfer time, γ_{E}^{-1}
τ_{et}	eddy-turn-over time
τ_k	eddy lifetime in MHD turbulence (§9.2)
$\tau_{\mathbf{k},\mathbf{q}}$	coherence time of scattering field with scattered ray $\boldsymbol{k}$
τ_{L}	characteristic time for the wave pattern to change (§3.2.2)
τ_{relax}	relaxation time of average distribution function
$\tau_{\mathrm{R}k}$	relaxation time for k-mode, (§5.3.3), Eq.(5.67)
τ_{Tc}	triad coherence time
$\tau\,(\Delta x, \Delta v)$	lifetime of phase space element, Eq.(8.25)
$\tau_{\mathrm{sep}}\,(\Delta x, \Delta v)$	lifetime of phase space element, determined by dispersion of particles (§8.2.3)
$\tau_{\mathrm{sep}}\,(x_-, v_-)$	scale-dependent separation time, Eq.(8.59)

ω	vorticity
ω_b	bounce frequency of quasi-particle in disparate-scale nonlinear interactions, Eq.(7.42)
ω_c	cyclotron frequency
ω_{ce}, ω_{ci}	electron and ion cyclotron frequency
ω_k	linear eigenfrequency that satisfies $\epsilon\ (k, \omega) = 0$
ω_{MM}	mismatch frequency, Eq.(5.37)
$\omega_p, \omega_{p,e}$	electron plasma frequency
ω^*	drift frequency
ω_e^*	electron diamagnetic drift frequency
$\Delta\omega_k$	nonlinear scrambling rate (due to convective nonlinearity) (§6.1.1)
$\Delta\omega_k$	line width at fixed k driven by granulations, Eq.(8.70)
$\Delta\omega_k$	self-decorrelation rate of the mode in MHD turbulence, Eq.(9.6)
$\Delta\omega_{\mathrm{phase}}$	rate of frequency wondering by straining filed, $v_g\ \lvert q \rvert \sqrt{D_\phi / \gamma_{\mathrm{E}}}$, p.198
$\Delta\omega_{\mathrm{T}}$	rate of dispersion of ω_{MM}, Eq.(5.38)
Ω	total enstrophy (§2.3.2.1)
Ω	potential enstrophy of H–M equation, (§6.4.5.1)
Ω	frequency of large-scale scattering field (in disparate scale interaction)

References

Abe, H. (2004). *Int. J. Heat Fluid Flow* **25**, 404.
Adam, J. C., Laval, G., and Pesme, D. (1979). *Phys. Rev. Lett.* **43**, 1671.
Angioni, C. and Peeters, A. G. (2006). *Phys. Rev. Lett.* **96**, 095003.
Arimitsu, T. and Arimitsu, A. (2001). *Prog. Theor. Phys.* **105**, 355.
Armstrong, J. A., Bloembergen, N., Ducuing, J., and Pershan, P. S. (1962). *Phys. Rev.* **127**, 1918.
Armstrong, J. W. *et al.* (1995). *Astrophys. J.* **443**, 209.
Bak, P., Tang, C., and Wiesenfeld, K. (1987). *Phys. Rev. Lett.* **59**, 381.
Balescu, R. (1963). *Statistical Mechanics of Charged Particles*. Inter Science, New York.
Balescu, R. (2005). *Aspects of Anomalous Transport in Plasmas*. Inst. of Phys. Publ., Bristol.
Balescu, R. and Misguish, J. H. (1984). In *Statistical Physics and Chaos in Fusion Plasmas*, ed. C. W. Horton and L. Reichl. Wiley, New York.
Barenblatt, G. I. (1979). *Similarity, Self-similarity, and Intermediate Asymptotics*. Consultants Bureau, New York. (translated by N. Stein; translation editor, M. van Dyke).
Berk, H. L. *et al.* (1999). *Phys. Plasmas* **6**, 3102.
Bernstein, I. B., Greene, J. M., and Kruskal, M. (1957). *Phys. Rev.* **108**, 546.
Bespalov, V. and Talanov, V. (1966). *JETP* **3**, 471.
Bhattacharjee, A. and Hameiri, E. (1986). *Phys. Rev. Lett.* **57**, 206.
Biglari, H., Diamond, P. H., and Terry, P. W. (1990). *Phys. Fluids B* **2**, 1.
Biskamp, D. (1993). *Nonlinear Magnetohydrodynamics*. Cambridge University Press, Cambridge.
Biskamp, D. and Welter, H. (1989). *Phys. Fluids B* **1**, 1964.
Blackman, E. G. and Field, G. F. (2000). *Astrophys. J.* **534**, 984.
Boldyrev, S. (2006). *Phys. Rev. Lett.* **96**, 115002.
Borue, V. (1994). *Phys. Rev. Lett.* **71**, 1475.
Boutros-Ghali, T. and Dupree, T. H. (1981). *Phys. Fluids* **24**, 1839.
Bowman, C., Krommes, J. A., and Ottaviani, M. (1993). *Phys. Fluids B* **5**, 3558.
Braginskii, S. I. (1965). *Review of Plasma Physics*, volume 1. Consultants Bureau, New York.
Brandenburg, A. and Subramanian, K. (2005). *Phys. Reports* **417**, 951.
Callen, H. B. and Welton, T. A. (1951). *Phys. Rev.* **83**, 34.
Candy, J. and Waltz, R. http://fusion.gat.com/comp/parallel/gyro_gallery.html.
Carnevale, G. F. (1982). *J. Fluid Mech.* **122**, 143.

Carreras, B. A. *et al.* (1998). *Phys. Plasmas* **5**, 3632.

Cattaneo, F. (1994). *Astrophys. J.* **434**, 200.

Cattaneo, F., Hughes, D. W., and Kim, E. (1996). *Phys. Rev. Lett.* **76**, 2057.

Champeaux, S. and Diamond, P. H. (2001). *Phys. Lett. A* **288**, 214.

Chang, Z. and Callen, J. D. (1992). *Phys. Fluids B* **4**, 1167.

Chapman, S. and Cowling, T. G. (1952). *The Mathematical Theory of Non-uniform Gases*, 2nd ed. Cambridge Univ. Press, Cambridge.

Charney, J. G. (1948). *Geophys. Publ. Oslo* **17**, 1.

Chekhlov, A. and Yakhot, V. (1995). *Phys. Rev. E* **51**, R2739.

Chirikov, B. V. (1960). *J. Nucl. Energy* **1**, 253.

Chirikov, B. V. (1979). *Physics Reports* **52**, 265.

Cho, J., Lazarian, A., and Vishniac, E. T. (2002). *Astrophys. J.* **564**, 291.

Cohen, R. H. and Kulsrud, R. H. (1974). *Phys. Fluids* **17**, 2215.

Collins, J. (1984). *Renormalization*. Cambridge University Press, Cambridge.

Coppi, B., Rosenbluth, M. N., and Sagdeev, R. Z. (1967). *Phys. Fluids* **10**, 582.

Coppi, B. and Spight, C. (1978). *Phys. Rev. Lett.* **41**, 551.

Craddock, G. and Diamond, P. H. (1990). *Plasma Phys. Control Fusion* **13**, 287.

Craik, A. D. D. (1985). *Wave Interactions and Fluid Flows*. Cambridge University Press, Cambridge.

Davidson, P. A. (2004). *Turbulence*. Oxford University Press, Oxford.

Davidson, R. C. (1972). *Methods in Nonlinear Plasma Theory*. Academic Press, New York.

del Castillo-Negrete, D., Carreras, B. A., and Lynch, V. E. (2004). *Phys. Plasmas* **11**, 3854.

Dendy, R. O. and Helander, P. (1997). *Plasma Phys. Contr. Fusion* **39**, 1947.

Diamond, P. H. *et al.* (1982). In *Plasma Phys. and Controlled Nuclear Fusion Research*, IAEA, Vienna **1**, 259.

Diamond, P. H., Hazeltine, R. D., An, Z. G., Carreras, B. A., and Hicks, H. R. (1984). *Phys. Fluids* **27**, 1449.

Diamond, P. H., Hughes, D. W., and Kim, E. (2005a). In *Fluid Dynamics and Dynamos in Astrophysics and Geophysics*, A. Soward, *et al.* ed., volume 145. CRC Press.

Diamond, P. H., Itoh, S.-I., Itoh, K., and Hahm, T. S. (2005b). *Plasma Phys. Contr. Fusion* **47**, R35.

Diamond, P. H. and Malkov, M. A. (2007). *Astrophys. J.* **654**, 252.

Diamond, P. H., McDevitt, C. J., Gurcan, O. D., Hahm, T. S., and Naulin, V. (2008). *Phys. Plasmas* **15**, 012303.

Diamond, P. H., Rosenbluth, M. N., Hinton, F. L., Malkov, M., Fleisher, J., and Smolyakov, A. (1998). In *Plasma Phys. and Controlled Nuclear Fusion Research*. IAEA, Vienna. IAEA-CN-69/TH3/1.

Drake, J. F., Kleva, R. G., and Mandt, M. E. (1994). *Phys. Rev. Lett.* **73**, 1251.

Drummond, W. E. and Pines, D. (1962). *Nucl. Fusion Supplement* **2**, 1049.

Dubois, D. F., Rose, H. A., and Russell, D. (1988). *Phys. Rev. Lett.* **61**, 2209.

Dupree, T. H. (1961). *Phys. Fluids* **4**, 696.

Dupree, T. H. (1966). *Phys. Fluids* **9**, 1773.

Dupree, T. H. (1970). *Phys. Rev. Lett.* **25**, 789.

Dupree, T. H. (1972). *Phys. Fluids* **15**, 334.

Dyachenko, A. I., Nazarenko, S. V., and Zakharov, V. E. (1992). *Phys. Lett. A* **165**, 330.

Elskens, Y. and Escande, D. (2003). *Microscopic Dynamics of Plasmas and Chaos*. Inst. of Phys. Publ., Bristol.

Escande, D. F. and Sattin, F. (2007). *Phys. Rev. Lett.* **99**, 185005.

Estrada-Mila, C., Candy, J., and Waltz, R. E. (2005). *Phys. Plasmas* **12**, 022305.
Falkovich, G., Gawedski, K., and Vergassola, M. (2001). *Rev. Mod. Phys.* **73**, 913.
Freidberg, J. (1989). *Ideal MHD*. Plenum Press, New York.
Frisch, U. (1995). *Turbulence*. Cambridge University Press, Cambridge.
Frisch, U., Sulem, P.-L., and Nelkin, M. (1978). *J. Fluid Mech.* **87**, 719.
Fukuyama, A., Momota, H., Itatani, R., and Takizuka, T. (1977). *Phys. Rev. Lett.* **38**, 701.
Fyfe, D. and Montgomery, D. (1976). *J. Plasma Phys.* **16**, 181.
Galeev, A. A. and Sagdeev, R. Z. (1965). Nonlinear Plasma Theory, in *Reviews of Plasma Physics, Vol. 7, page 1*. ed. M. A. Leontovich, Consultants Bureau, New York.
Gang, F. Y., Diamond, P. H., Crotinger, J. A., and Koniges, K. E. (1991). *Phys. Fluids B* **3**, 955.
Garbet, X. (2003). *Phys. Rev. Lett.* **91**, 035001.
Garbet, X., Dubuit, N., Asp, E., Sarazin, Y., Bourdelle, C., Ghendrih, P., and Hoang, G. T. (2005). *Phys. Plasmas* **12**, 082511.
Garret, C. and Munk, W. (1975). *J. Geophys. Res.* **80**, 291.
Goldenfeld, N. (1992). *Lectures on Phase Transitions and the Renormalization Group*. Addison-Wesley, Reading Mass.
Goldman, M. V. (1984). *Rev. Mod. Phys.* **56**, 709.
Goldreich, P. and Sridhar, S. (1995). *Ap. J.* **458**, 763.
Goldreich, P. and Sridhar, S. (1997). *Ap. J.* **485**, 680.
Goldston, R. J. and Rutherford, P. H. (1995). *Introduction to Plasma Physics*. Inst. of Phys. Publ., Bristol.
Gossard, E. E. and Hooke, W. H. (1975). *Waves in the Atmosphere*. Elsevier Scientific Pub., Amsterdam.
Gotoh, T. (2006). A Langevin model for turbulence. *20th Symposium of numerical fluid dynamics, JSFM*, pages A7–3.
Gotoh, T. *et al.* (2002). *Phys. Fluids* **14**, 1065.
Grappin, R., Pouquet, A., and Leorat, J. (1983). *Astron. Astrophys.* **126**, 51.
Gruzinov, A. V. and Diamond, P. H. (1994). *Phys. Rev. Lett.* **72**, 1651.
Gruzinov, A. V. and Diamond, P. H. (1996). *Phys. Plasmas* **3**, 1853.
Gurcan, O. D. and Diamond, P. H. (2004). *Phys. Plasmas* **11**, 4973.
Hada, T. (1994). *Geophys. Res. Lett.* **21**, 2275.
Hahm, T. S. and Burrell, K. H. (1995). *Phys. Plasmas* **2**, 1648.
Hahm, T. S., Diamond, P. H., Gurcan, O. D., and Rewoldt, G. (2007). *Phys. Plasmas* **14**, 072302.
Hamaguchi, S. and Horton, C. W. (1990). *Phys. Fluids B* **2**, 1833.
Hammett, G. W. and Perkins, F. W. (1990). *Phys. Rev. Lett.* **64**, 3019.
Hasegawa, A. (1985). *Adv. Phys.* **34**, 1.
Hasegawa, A., Maclennan, C., and Kodama, Y. (1979). *Phys. Fluids* **22**, 2122.
Hasegawa, A. and Mima, K. (1978). *Phys. Fluids* **21**, 87.
Hasselmann, K. (1962). *J. Fluid Mech.* **12**, 481.
Hasselmann, K. (1968). *Rev. Geophys.* **4**, 1.
Hazeltine, R. D. and Meiss, J. D. (1992). *Plasma Confinement*. Addison Wesley, Redwood City.
Hinton, F. L. (1984). In *Handbook of Plasma Physics*, volume 1. ed. M. N. Rosenbluth and R. Z. Sagdeev, North Holland, Amsterdam, New York.
Hinton, F. L. and Rosenbluth, M. N. (1999). *Plasma Phys. Control. Fusion* **41**, A653.
Hirshman, S. P. and Molvig, K. (1979). *Phys. Rev. Lett.* **42**, 648.

Horton, C. W. (1984). In *Handbook of Plasma Physics*, volume 2. ed. M. N. Rosenbluth and R. Z. Sagdeev, North Holland, Amsterdam, New York.

Horton, C. W. (1999). *Rev. Mod. Phys.* **71**, 735.

Horton, C. W. and Hasegawa, A. (1994). *Chaos* **4**, 227.

Ichimaru, S. (1973). *Basic Principles of Plasma Physics, A Statistical Approach, Frontiers in Physics*. Benjamin, Reading, Mass.

Ichimaru, S. and Rosenbluth, M. N. (1970). *Phys. Fluids* **13**, 2778.

Iroshnikov, T. S. (1964). *Sov. Astron.* **7**, 566.

Ishichenko, M. B., Gruzinov, A. V., and Diamond, P. H. (1996). *Phys. Rev. Lett.* **74**, 4436.

Isichenko, M. B. (1992). *Rev. Mod. Phys.* **64**, 961.

Itoh, K. and Itoh, S.-I. (1996). *Plasma Phys. Control. Fusion* **38**, 1.

Itoh, K., Itoh, S.-I., and Fukuyama, A. (1999). *Transport and Structural Formation in Plasmas*. Inst. of Phys. Publ., Bristol.

Itoh, K., Itoh, S.-I., Diamond, P. H., Hahm, T. S., Fujisawa, A., Tynan, G. R., Yagi, M., and Nagashima, Y. (2006). *Phys. Plasmas* **13**, 055502.

Itoh, S.-I. and Itoh, K. (1988). *Phys. Rev. Lett.* **60**, 2276.

Itoh, S.-I. and Itoh, K. (1990). *J. Phys. Soc. Jpn.* **59**, 3815.

Itoh, S.-I. and Itoh, K. (2000). *J. Phys. Soc. Jpn.* **69**, 427.

Itoh, S.-I. and Itoh, K. (2001). *Plasma Phys. Contr. Fusion* **43**, 1055.

Jenko, F. (2005). *J. Plasma Fusion Res. SERIES* **6**, 11.

Jenko, F. *et al.* (2001). *Phys. Plasmas* **8**, 4096.

Kadomtsev, B. B. (1965). *Plasma Turbulence*. Academic Press, New York.

Kadomtsev, B. B. (1992). *Tokamak Plasma: A Complex Physical System*. Inst. of Phys. Publ., Bristol.

Kadomtsev, B. B. and Pogutse, O. P. (1970). *Phys. Rev. Lett.* **25**, 1155.

Kaneda, Y. (1981). *J. Fluid Mech.* **107**, 131.

Kartashova, E. (2007). *Phys. Rev. Lett.* **98**, 214502.

Kaw, P., Singh, R., and Diamond, P. H. (2002). *Plasma Phys. Contr. Fusion* **44**, 51.

Kaw, P. K. *et al.* (1975). *Phys. Rev. Lett.* **34**, 803.

Keating, S. R. and Diamond, P. H. (2007). *Phys. Rev. Lett.* **99**, 224502.

Keating, S. R. and Diamond, P. H. (2008). *J. Fluid Mech.* **595**, 173.

Keating, S. R., Silvers, L. J., and Diamond, P. H. (2008). *Astrophys. J. Lett.* **678**, L137.

Kim, E. and Diamond, P. H. (2001). *Astrophys. J.* **556**, 1052.

Kim, E. and Dubrulle, B. (2001). *Phys. Plasmas* **8**, 813.

Klimontovich, Y. L. (1967). *Statistical Theory of Nonequilibrium Process in Plasmas.* Pergamon Press, Oxford.

Kolmogorov, A. N. (1941). *C. R. Acad. Sci. USSR* **30**, 301. English translations; Kolmogorov, A. N. (1991). *Proc. Roy. Soc. A* **434**, 9, 15.

Kolmogorov, A. N. (1958). *Doklady Akad. Nauk SSSR* **119**, 861.

Kraichnan, R. H. (1961). *J. Math. Phys.* **2**, 124.

Kraichnan, R. H. (1965). *Phys. Fluids* **8**, 1385.

Kraichnan, R. H. (1967). *Phys. Fluids* **10**, 1417.

Kraichnan, R. H. (1970). *J. Fluid Mech.* **41**, 189.

Kraichnan, R. H. (1977). *J. Fluid Mech.* **83**, 349.

Krall, N. A. and Trivelpiece, A. W. (1973). *Principles of Plasma Physics*. McGraw-Hill, New York.

Krause, F. and Radler, K.-H. (1980). *Mean-Field Magnetohydrodynamics and Dynamo Theory*. Pergamon, Oxford.

Kritsuk, A. *et al.* (2007). *Astrophys. J.* **665**, 416.

Krommes, J. A. (1984). In *Handbook of Plasma Physics, Vol. 2, 183*, ed. M. N. Rosenbluth and R. Z. Sagdeev, North Holland, Amsterdam, New York.
Krommes, J. A. (1996). *Phys. Rev. E* **53**, 4865.
Krommes, J. A. (1999). *Plasma Phys. Contr. Fusion* **41**, A641.
Kubo, R. (1957). *J. Phys. Soc. Jpn.* **12**, 570.
Kulsrud, R. M. (2005). *Plasma Physics for Astrophysics*. Princeton Univ. Press, Princeton.
Kuramoto, Y. (1984). *Chemical Oscillations, Waves and Turbulence*. Springer, New York.
Landau, L. D. and Lifshitz, E. M. (1976). *Mechanics*. Elsevier, Amsterdam.
Landau, L. D. and Lifshitz, E. M. (1980). *Statistical Physics, Part 1*. Pergamon Press, Oxford.
Laval, G. and Pesme, D. (1983). *Phys. Fluids* **26**, 52.
Laval, J.-P., Dubrulle, B., and Nazarenko, S. (2001). *Phys. Fluids* **13**, 1995.
Lazarian, A. and Vishniac, E. (1999). *Astrophys. J.* **517**, 700.
Lee, G. S. and Diamond, P. H. (1986). *Phys. Fluids* **29**, 3291.
Lee, W. W. and Tang, W. M. (1988). *Phys. Fluids* **31**, 612.
Lenard, A. (1960). *Ann. Physics (NY)* **3**, 390.
Lesieur, M. (1997). *Turbulence in Fluids (3rd rev. and enl. ed)*. Kluwer Academic Publishers.
Li, J. and Kishimoto, Y. (2002). *Phys. Rev. Lett.* **89**, 115002.
Lichtenberg, A. J. and Lieberman, M. A. (1983). *Regular and Stochastic Motion*. Springer, New York.
Lifshitz, E. M. and Pitaevskii, L. P. (1981). *Physical Kinetics*, translated by J. B. Sykes and R. N. Franklin. Pergamon Press, Oxford.
Lighthill, J. (1978). *Waves in Fluids*. Cambridge Univ. Press, Cambridge.
Lin, Z., Holod, I., Chen, L., Diamond, P. H., Hahm, T. S., and Ethier, S. (2007). *Phys. Rev. Lett.* **99**, 265003.
Longuet-Higgins, M. S., Gill, A. E., and Kenyon, K. (1967). *Proc. Roy. Soc. A London* **299**, 120.
Lynden-Bell, D. (1967). *Mon. Notic. Roy. Astron. Soc.* **136**, 101.
Ma, S. K. (1976). *The Modern Theory of Critical Phenomena*. Benjamin, Reading.
Makino, M., Kamimura, T., and Taniuti, T. (1981). *J. Phys. Soc. Jpn.* **50**, 980.
Manfredi, G., Roach, C. M., and Dendy, R. O. (2001). *Plasma Phys, Control. Fusion* **43**, 825.
Manheimer, W. M. and Dupree, T. H. (1968). *Phys. Fluids* **11**, 2709.
Maron, J. *et al.* (2004). *Ap. J.* **603**, 569.
Mattor, N. and Parker, S. E. (1997). *Phys. Rev. Lett.* **79**, 3419.
McComas, C. H. and Bretherton, F. P. (1977). *J. Geophys. Res.* **82**, 1397.
McComb, W. D. (1990). *The Physics of Fluid Turbulence*. Oxford University Press, Oxford.
McWilliams, J. C. (1984). *J. Fluid Mechanics* **146**, 21.
Medvedev, M. V. and Diamond, P. H. (1997). *Phys. Rev. Lett.* **78**, 4934.
Meiss, J. D. and Horton, C. W. (1983). *Phys. Fluids* **26**, 990.
Mikailovski, A. B. (1992). *Electromagnetic Instabilities in an Inhomogeneous Plasma*. Inst. of Phys. Publ., Bristol.
Mima, K. and Nishikawa, K. (1984). In *Handbook of Plasma Physics, Vol. 2, 451*. ed. M. N. Rosenbluth and R. Z. Sagdeev, North Holland, Amsterdam, New York.
Miyamoto, K. (1976). *Plasma Physics for Nuclear Fusion*. MIT Press, Cambridge.
Miyamoto, K. (2007). *Controlled Fusion and Plasma Physics*. Inst. of Phys. Publ., Bristol.
Moffat, H. K. (1978). *Magnetic Fields Generation in Electrically Conducting Fluids*. Cambridge University Press, Cambridge.

Moiseev, S. S., Pungin, V. G., and Oraevsky, V. N. (2000). *Nonlinear Instabilities in Plasmas and Hydrodynamics*. Inst. of Phys. Publ., Bristol.

Montgomery, D., Turner, L., and Vahala, G. (1979). *J. Plasma Phys.* **21**, 239.

Mori, H. (1965). *Prog. Theor. Phys.* **33**, 423.

Mori, H. (2008). Private communication.

Mori, H. and Fujisaka, H. (2001). *Phys. Rev. E* **63**, 026302.

Mori, H., Kurosaki, S., Tominaga, H., Ishizaki, R., and Mori, N. (2003). *Prog. Theor. Phys.* **109**, 333.

Mori, H. and Okamura, M. (2007). *Phys. Rev. E* **76**, 061104.

Muller, W. G., Biskamp, D., and Grappin, R. (2003). *Phys. Rev. E* **67**, 066302.

Naulin, V. *et al.* (1998). *Phys. Rev. Lett.* **81**, 4148.

Nevins, W. M. *et al.* (2005). *Phys. Plasmas* **12**, 122305.

Newell, A. C. (1985). *Solitons in Mathematics and Physics*. Society for Industrial and Applied Mathematics, Philadelphia.

Newell, A. C. and Zakharov, V. E. (2008). *Phys. Lett. A* **372**, 4230.

Okubo, A. (1970). *Deep Sea Research and Oceanographic Abstracts* **17**, 445.

Orszag, S. A. (1970). *J. Fluid Mech.* **41**, 363.

Ott, E. (1993). *Chaos in Dynamical Systems*. Cambridge University Press, Cambridge.

Ottaviani, M. *et al.* (1991). *Phys. Fluids B* **3**, 2186.

Parker, E. N. (1979). *Cosmic Magnetic Fields*. Clarendon, Oxford.

Parker, S. E., Dorland, W., Santoro, R. A., Beer, M. A., Liu, Q. P., Lee, W. W., and Hammett, G. W. (1994). *Phys. Plasmas* **1**, 1461.

Passot, T. and Sulem, P. L. (2003). *Phys. Plasmas* **10**, 3906.

Pedlosky, J. (1987). *Geophysical Fluid Dynamics*, 2nd ed. Springer, New York.

Phillips, O. M. (1966). *The Dynamics of the Upper Ocean*. Cambridge Univ. Press, Cambridge.

Podlubny, I. (1998). *Fractional Differential Equations*. Academic Press, New York.

Pope, S. B. (2000). *Turbulent Flows*. Cambridge University Press, Cambridge.

Pouquet, A. (1978). *J. Fluid Mech.* **88**, 1.

Prandtl, L. (1932). *Ergebn. Aerodyn. Versuchsanst.* **4**, 18.

Rhines, P. B. (1975). *J. Fluid Mech.* **69**, 417.

Richardson, L. F. (1926). *Proc. Roy. Soc. London, A* **110**, 709.

Riyopoulos, S., Bondeson, A., and Montgomery, D. (1982). *Phys. Fluids* **25**, 107.

Roberts, P. H. (2000). *Rev. Mod. Phys.* **72**, 1081.

Robinson, P. A. (1997). *Rev. Mod. Phys.* **69**, 507.

Rogister, A. (1971). *Phys. Fluids* **14**, 2733.

Rosenbluth, M. N., MacDonald, W. M., and Judd, D. (1957). *Phys. Rev.* **102**, 1.

Rostoker, N. and Rosenbluth, M. N. (1960). *Phys. Fluids* **3**, 1.

Rudakov, L. I. and Sagdeev, R. Z. (1960). *Sov. Phys.-JETP* **10**, 952. [ZhETF **37**, 1337 (1959)].

Sagdeev, R. Z. and Galeev, A. A. (1969). *Nonlinear Plasma Theory*, ed. T. M. O'Neil and D. L. Book. Benjamin, New York.

Sagdeev, R. Z., Shapiro, V. D., and Shevchenko, V. I. (1978). *Sov. J. Plasma Phys.* **4**, 306. [Fiz. Plasmy **4**, 551 (1978)].

Sanchez, R. (2005). *Phys. Plasmas* **12**, 056105.

Schmidt, J. and Yoshikawa, S. (1971). *Phys. Rev. Lett.* **26**, 753.

Shaing, K. C., Crume, Jr. E. C., and Houlberg, W. A. (1990). *Phys. Fluids B* **2**, 1492.

She, Z. S. and Leveque, E. (1994). *Phys. Rev. Lett.* **72**, 336.

Silvers, L. (2004). Ph. D. Thesis, University of Leeds.

Sinai, Y. G. (1959). *Doklady Akad. Nauk SSSR* **124**, 768.

Sinai, Y. G. (1994). *Topics in Ergodic Theory*. Princeton Univ. Press, Princeton.
Smith, G. R. and Kaufman, A. N. (1975). *Phys. Rev. Lett.* **34**, 1613.
Smolyakov, A. *et al.* (1999). *Phys. Plasmas* **6**, 4410.
Stix, T. H. (1992). *Waves in Plasmas*. Springer, New York.
Strauss, H. R. (1976). *Phys. Fluids* **19**, 134.
Strauss, H. R. (1986). *Phys. Fluids* **29**, 3668.
Sturrock, P. A. (1994). *Plasma Physics: an introduction to the theory of astrophysical, geophysical, and laboratory plasmas*. Cambridge University Press, Cambridge.
Subramanian, K. and Brandenburg, A. (2006). *Astrophys. J.* **648**, L71.
Sugama, H., Watanabe, T.-H., and Horton, C. W. (2001). *Phys. Plasmas* **6**, 2617.
Sulem, C. and Sulem, P.-L. (1999). *The Nonlinear Schrödinger Equation: self-focusing and wave collapse*. Springer, New York.
Suzuki, M. (1984). In *Statistical Physics and Chaos in Fusion Plasmas*, ed. C. W. Horton and L. Reichl. Wiley, New York.
Tajima, T. and Shibata, K. (2002). *Plasma Astrophysics*. Addison-Wesley, Reading.
Taylor, G. I. (1915). *Eddy motion in the atmosphere in* G. I. Taylor Scientific Papers (ed. G. Batchelor) Vol II, pages 1-23. Cambridge University Press, 1960 Cambridge.
Taylor, J. B. (1986). *Rev. Mod. Phys.* **58**, 741.
Terry, P. W. (1989). *Phys. Fluids B* **1**, 1932.
Terry, P. W. (2000). *Rev. Mod. Phys.* **72**, 109.
Terry, P. W. and Diamond, P. H. (1984). In *Statistical Physics and Chaos in Fusion Plasmas*, ed. C. W. Horton and L. Reichl. Wiley, New York.
Trullinger, S. E., Zakharov, V. E., and Pokrovsky, V. L. (ed.) (1986). *Solitons*. North-Holland, Amsterdam.
Vallis, G. K. (2006). *Atmospheric Motion and Ocean Fluid Dynamics*. Cambridge University Press, Cambridge.
Vedenov, A. A., Velikov, E. P., and Sadgeev, R. Z. (1961). *Nucl. Fusion* **1**, 82.
Vedenov, A. A., Velikov, E. P., and Sadgeev, R. Z. (1962). *Nucl. Fusion Supplement* **2**, 465.
Vlad, M. *et al.* (2004). *Plasma Phys. Contr. Fusion* **46**, 1051.
Wagner, F. *et al.* (1982). *Phys. Rev. Lett.* **49**, 1408.
Waltz, R. (1988). *Phys. Fluids* **31**, 1963.
Watanabe, T.-H. *et al.* (2002). *Phys. Plasmas* **9**, 3659.
Watanabe, T.-H., Sugama, H., and Sato, T. (2000). *Phys. Plasmas* **7**, 964.
Wei, T. and Willmarth, W. W. (1989). *J. Fluid Mechanics* **204**, 57.
Weiland, J. (2000). *Collective Modes in Inhomogeneous Plasmas*. Inst. of Phys. Publ., Bristol.
Weiss, J. (1991). *Physica D* **48**, 273.
Wesson, J. (1997). *Tokamaks*. Clarendon, Oxford.
White, R. (1989). *Theory of Tokamak Plasmas*. North Holland, New York.
Yagi, M. *et al.* (2005). *Nucl. Fusion* **45**, 900.
Yamada, T., Itoh, S.-I., Maruta, T., Kasuya, N., Nagashima, Y., Shinohara, S., Terasaka, K., Yagi, M., Inagaki, S., Kawai, Y., Fujisawa, A., and Itoh, K. (2008). *Nature Phys.* **4**, 721.
Yoshizawa, A. (1998). *Hydrodynamic and Magnetohydrodynamic Turbulent Flows*. Kluwer, Dordrecht.
Yoshizawa, A. (1999). Private communications.
Yoshizawa, A. (2005). *Phys. Fluids* **17**, 075113.
Yoshizawa, A., Itoh, S.-I., and Itoh, K. (2003). *Plasma and Fluid Turbulence*. Inst. of Phys. Publ., Bristol.

Yoshizawa, A., Itoh, S.-I., Itoh, K., and Yokoi, N. (2004). *Plasma Phys. Contr. Fusion* **46**, R25.

Zakharov, V. E. (1984). In *Handbook of Plasma Physics, Vol. 2, 8.* ed. M. N. Rosenbluth and R. Z. Sagdeev, North Holland, Amsterdam, New York.

Zakharov, V. E. (1985). *Phs. Reports* **129**, 285.

Zakharov, V. E. and Filonenko, N. N. (1967). *Sov. Phys. - DOKLADY* **11**, 881. [Doklady Akademii Nauk SSSR **170**, 1292 (1966)].

Zakharov, V. E., L'vov, V. S., and Falkovich, G. (1992). *Wave turbulence*. Springer-Verlag, Berlin.

Zaslavsky (2005). *Hamiltonian Chaos and Fractional Dynamics*. Oxford University Press, Oxford.

Zaslavsky, G. M. and Filonenko, N. N. (1968). *Sov. Phys. JETP* **27**, 851. [Zh. Eksp. Teor. Fiz. **54**, 1590 (1968)].

Zeldovich, Y. B. (1957). *Sov. Phys. JETP* **4**, 460.

Zhang, Y. Z. and Mahajan, S. M. (1992). *Phys. Fluids B* **4**, 1385.

Zwanzig, R. (2001). *Nonequilibrium Statistical Physics*. Oxford University Press, Oxford.

Index

Printed in the United States
By Bookmasters